Conference Proceedings of the Society for Experimental Mechanics Series

For other titles published in this series, go to
www.springer.com/series/8922

Tom Proulx
Editor

Structural Dynamics and Renewable Energy, Volume 1

Proceedings of the 28th IMAC, A Conference on Structural Dynamics, 2010

Springer

Editor
Tom Proulx
Society for Experimental Mechanics, Inc.
7 School Street
Bethel, CT 06801-1405
USA
tom@seml.com

ISSN 2191-5644 e-ISSN 2191-5652
ISBN 978-1-4419-9715-9 e-ISBN 978-1-4419-9716-6
DOI 10.1007/978-1-4419-9716-6
Springer New York Dordrecht Heidelberg London

Library of Congress Control Number: 2011929945

Printed on acid-free paper

Springer is part of Springer Science+Business Media (www.springer.com)

PREFACE

Structural Dynamics and Renewable Energy represents one of five clusters of technical papers presented at the 28th IMAC, A Conference and Exposition on Structural Dynamics, 2010 organized by the Society for Experimental Mechanics, and held at Jacksonville, Florida, February 1-4, 2010. The full proceedings also include volumes on Nonlinear Modeling and Applications, Dynamics of Bridges, Dynamics of Civil Structures and, Structural Dynamics.

Each collection presents early findings from experimental and computational investigations on an important area within Structural Dynamics. The current volume on *Structural Dynamics and Renewable Energy* includes studies on Wind Turbine Blades, Energy Harvesting, Wind Turbine Dynamics, Electromagnetic and Magnetostrictive Energy Harvesting, Piezoelectric Energy Harvesting, and Operational Modal Analysis Applied to Wind turbines

In recent years, renewable energy has become a major research area. This volume brings together researchers and engineers interested in the structural dynamics aspects of energy systems and materials, and provides a forum to facilitate technical interaction and exchange.

The organizers would like to thank the authors, presenters, session organizers and session chairs for their participation in this track.

Bethel, Connecticut

Dr. Thomas Proulx
Society for Experimental Mechanics, Inc

CONTENTS

Proceedings of the IMAC-XXVIII
February 1–4, 2010, Jacksonville, Florida USA

Experimental Modal Analysis of 9-meter Research-sized Wind Turbine Blades

D. Todd Griffith
Thomas G. Carne

Sandia National Laboratories*
Albuquerque, NM 87185-0557

The dominant and persistent trend with wind turbine technology, particularly in the past three decades, has been growth in the length of the blades. In order to investigate design choices which reduce blade weight, Sandia Labs initiated a study, which is near completion, to evaluate innovative concepts for large blades. The innovations include strategic use of carbon fiber in the spar caps, bend-twist coupling in the composite layup, and thick, flatback airfoils. Several large blades were designed and then built at a down-scaled 9-meter length. Each blade design has undergone a full series of structural tests including modal tests, static tests, and fatigue tests. The modal tests performed for evaluation of these blades is the focus of this paper. Major findings from these tests are summarized, and they include: (1) techniques for experimental quantification of uncertainty in the modal parameters, (2) insight into model calibration using both static load-deflection data and the modal parameters, (3) novel test techniques for reducing the uncertainty in the root boundary condition, and (4) the development of validated structural models. This paper will provide a summary of blade modal testing and structural model validation, and will emphasize recent validation tests using a seismic-mass-on-airbags boundary condition.

Introduction

The trend in wind energy technology continues to be larger machines with larger blades as the technology has continued to improve and larger machines have become cost effective. Also, the capital investment for each individual wind power plant is increasing as machines become larger (and as a result of market factors as well). To address the technology needs in development of larger blades, Sandia Wind Energy Technology Department has focused on design innovations to improve structural efficiency. In particular, the focus is on reduction of blade weight to mitigate gravitational loads while maintaining acceptable tip deflections. While designing to aerodynamic loads has been the predominant design driver, future blade designs will need to consider gravitational loads to a larger extent. To address blade reliability, a focus has been on validation of blade models. Wind turbines are the largest rotating structures in the world. It is of paramount importance that new designs be developed without major systematic flaws because new designs will be fielded in large numbers in remote locations and the capital investment per mega-watt has been in the range of $1 to $2 million (US) in recent years. The development of accurate predictive analysis tools, with support from the testing program, is crucial for reliability through improved modeling and simulation in design of large blades.

A blade development program has been underway at Sandia Labs for several years to evaluate innovative structural mechanics concepts for wind turbine blades. These 9-meter research-sized blades have been evaluated with static, fatigue, and free boundary condition modal tests. See Figure 1 for a sketch of the planform for each of the blades developed in this effort. The CX-100 design incorporates carbon fiber in the spar cap, as indicated in the sketch. The TX-100 blade incorporates off-axis fiber in the skins to produce bend-twist coupling. The CX-100 and TX-100 blades have identical external geometries. The BSDS (Blade System Design Study) blade was designed with a new planform, new

* Sandia is a multiprogram laboratory operated by Sandia Corporation, a Lockheed Martin Company for the United States Department of Energy's National Nuclear Security Administration under contract DE-AC04-94AL85000.

T. Proulx (ed.), *Structural Dynamics and Renewable Energy, Volume 1*, Conference Proceedings of the Society for Experimental Mechanics Series 10, DOI 10.1007/978-1-4419-9716-6_1, © The Society for Experimental Mechanics, Inc. 2011

selection of airfoils, and a larger root diameter. Figure 2 shows the modal test setups for free boundary condition modal tests conducted at Sandia to evaluate each of the three research blades. Each modal test was designed with particular focus on the unique innovation used in each design. Testing and modeling of the BSDS blade is the focus of this paper. The BSDS blade is nominally 8.325 meters (27.3 feet) and 127 kilograms (290 lbs). A key feature in the design of the BSDS blade are the flatback airfoils, which are noted in Figure 2(c) (and also in Figure 5). The flatback airfoil provides a structural advantage as a flap-wise stiffener; however, the flatback airfoil has increased drag and is more sensitive to soiling. This airfoil can provide a reasonable trade-off because the loss in aerodynamic performance is less in the inboard section of the blade while the flatback airfoil provides the structural benefit where it is most needed inboard. Modal testing of the CX-100 and TX-100 blades was reported in References 1 and 2.

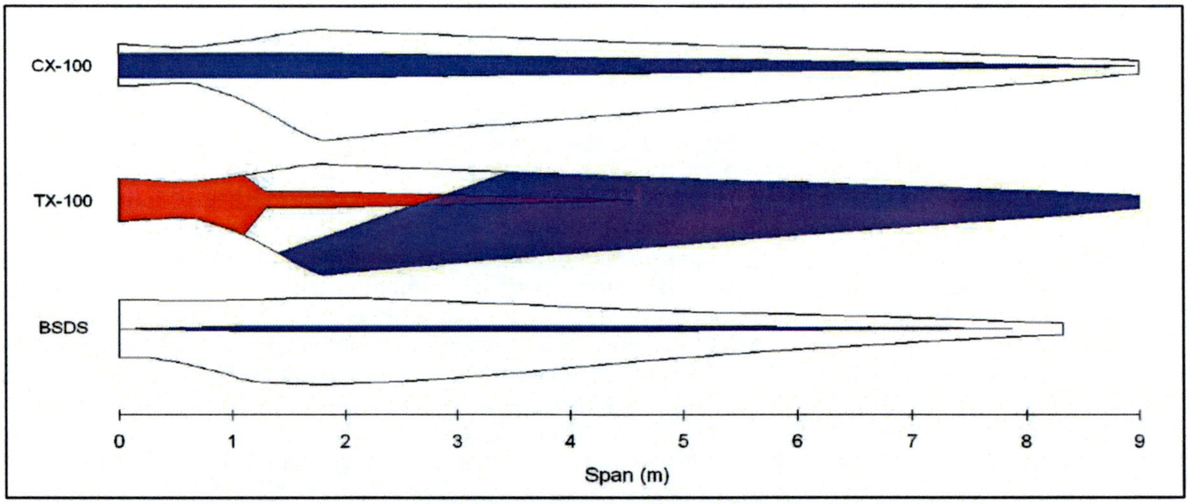

Figure 1. Research-sized blades as part of Blade Innovation Study

Figure 2. Photos of Modal Tests Performed on Sandia Research-sized Blades (a) CX-100 (left), (b) TX-100 (center), (c) BSDS (right)

This paper summarizes previously published work in a number of areas of blade testing and structural model validation. The outline of the paper is as follows. We first present an overview of a validation methodology developed for wind turbine blades, which was applied to the BSDS blade. Then, we discuss the modal tests and the experimental efforts to quantify uncertainty in the measured modal parameters. Next, we present a hybrid calibration approach which is used to determine the blade span-wise stiffness properties using both static and modal test data. Finally, we describe and present results from validation experiments and model development using a novel blade root boundary condition

A Methodology for Blade Structural Model Validation

Model validation is a comprehensive undertaking which requires carefully designing and executing experiments, proposing appropriate physics-based models, and applying correlation techniques to improve these models based on the test data. A methodology for rigorous blade structural model validation was presented in Reference 3, and each of the three components of model validation is reviewed in the following sections.

A. Experiment Design

A primary concern with any test for model validation is correspondence between the conditions of the test and the conditions of the analysis. For example, it is important that loads and boundary conditions be well-characterized in a test for inclusion in the analysis. Further, coordinate systems must be carefully considered. Bias errors must be well-characterized or eliminated. Typically, modal tests are performed to validate structural models although static testing has also been incorporated in this work.

An important issue with modal testing is assessment of uncertainty in the modal parameters as test observations and analysis predictions must be compared. The decisions made in the design of the test setup are critical to validation of blade structural models. For example, the design of the instrumentation layout, the type of support conditions (boundary conditions), and choice of excitation type are important considerations in the design of a modal test. It is important to quantify the bias errors resulting from the test setup (e.g. boundary conditions and instrumentation effects) when validating models because the bias errors can hinder suitable comparison with model predictions.

B. Model Development

With analysis, one is concerned with the chosen form of the model and level of detail in the model, its correspondence with the test article and the conditions of the test, and the parameters that comprise the model. Wind turbine blade modeling capabilities exist for levels of detail ranging from low fidelity beam models to geometrically accurate high fidelity finite element models. The decision depends on what type of analysis is needed and the availability of resources. If desired, the precise geometry of the blade airfoils, placement of materials, and internal structural geometry can be represented in a high fidelity finite element model as shown in Figure 3. These types of models predict a wide range of phenomena including detailed stress contours and local buckling. A highly automated program developed at Sandia for blade modeling can be used to speed up the modeling effort [4]. However, lower dimension finite element models, such as the beam model shown in Figure 4, are suitable for other purposes. This type of model represents the equivalent, averaged properties of the blade cross sections and is useful for calculating, for example, deflections and natural frequencies, but does not capture the detailed local behavior of the high fidelity model. As a comparison, the high fidelity model shown in Figure 3 contains 35 material regions, 47,426 elements, and 141,454 nodes, whereas the low fidelity model contains 20 elements and 21 nodes, with each element representing a unique material region representative of the averaged properties of that section. The time required to develop the high fidelity model is several days while the low fidelity model can be generated in a matter of hours.

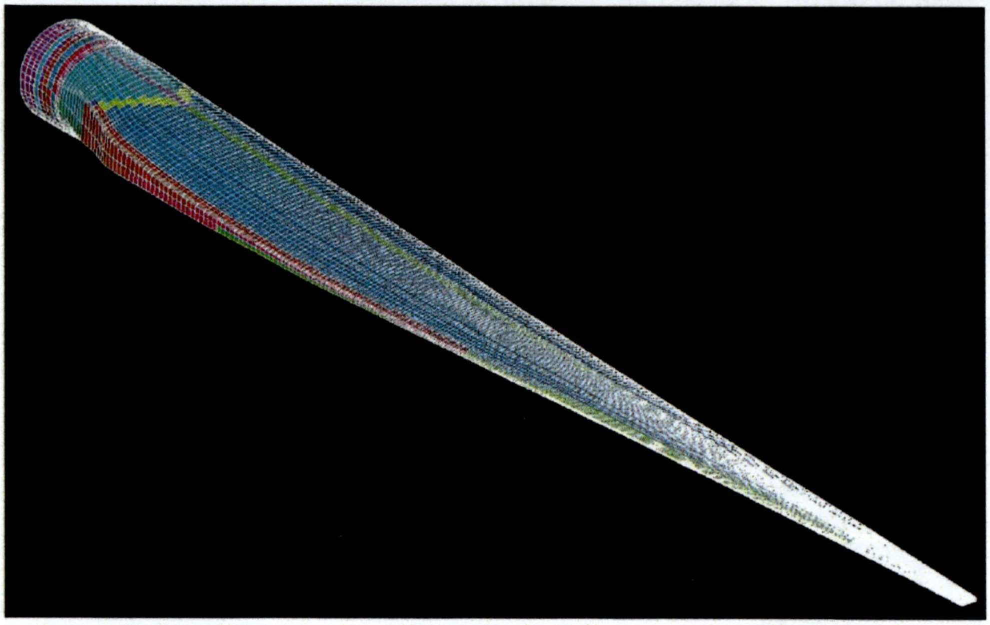

Figure 3. High Fidelity Finite Element Model of a Wind Turbine Blade

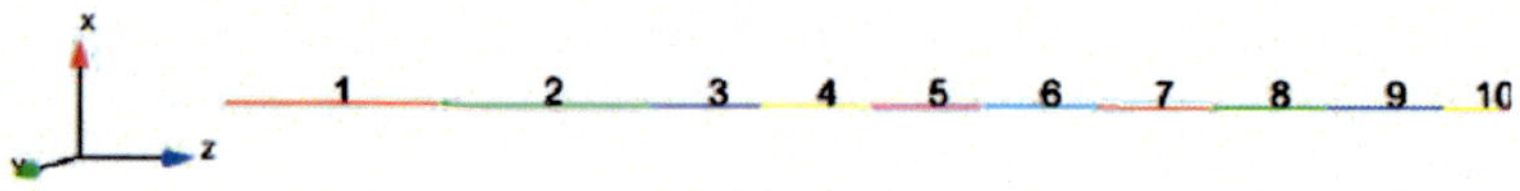

Figure 4. Low Fidelity Finite Element Model of a Wind Turbine Blade

C. Test-Analysis Correlation

Once tests have been conducted, models can be analyzed using the conditions (e.g. boundary conditions) of the test to make predictions and assess the credibility of the model. Typically a model does not adequately predict all aspects of the test to the predetermined adequacy criterion required for validation. Thus one is then concerned with improvements by choosing a model form and/or model parameters (material properties and geometric properties) which best represent reality. Updating model parameters enables one to improve the model such that it agrees with the test data. However, it must be done in a physically meaningful manner. Parameters with well known values are typically held constant; examples include the total mass of the blade because it can be accurately measured. On the other hand, material and geometric properties typically have some uncertainty. These are the parameters which one would consider varying in order to calibrate the model. Model calibration for a wind turbine blade was performed in a previous study. Uncertain material parameters were estimated using modal test data for a pultruded blade section with uniform cross-section [5]. More recent work for the BSDS blade study was conducted to calibrate a beam structural model [3]. Later in this paper, we present results from this study of different approaches for calibrating models using modal test data, static test, and a combination of modal and static test data. We note that the traditional means of updating structural dynamics models uses only modal test data. Also, recent work was performed to assess model validity with respect to predetermined adequacy criteria [6, 7]. Both the calibration and validation studies for the BSDS blade are summarized in later sections of this paper.

Blade Testing and Experimental Uncertainty Quantification

In this section, we describe the free boundary condition modal tests and static tests which were performed to provide calibration data for the BSDS blade structural model development. A modal test of the BSDS blade was conducted using a free boundary condition, which is shown in Figure 5. The support conditions were designed to minimize their effect on the modal parameters by optimal placement and low stiffness at the two support locations using bungee rope. Experimental quantification of the uncertainty in the measured modal parameters is discussed later in this section.

Figure 5. Free Boundary Condition Modal Test of BSDS Blade

The static test setup for the BSDS blade is shown in Figure 6. This test was conducted at the National Wind Technology Center (NWTC) in Golden, CO. The whiffle-tree apparatus is visible above the blade which provides the upward vertical load at three locations, while the blade is constrained at the root. From this test, deflection data was obtained as a function of the measured load input, which provides a means to calibrate the stiffness properties of the blade model. Note, however, that this test provided no information regarding the properties of the blade section outboard of the outer loading position because this portion of the blade is not stressed in this loading arrangement. This is important to consider when calibrating structural models based on static tests. The uncertainty in the root boundary and the measured loads/deflections was not quantified in these tests.

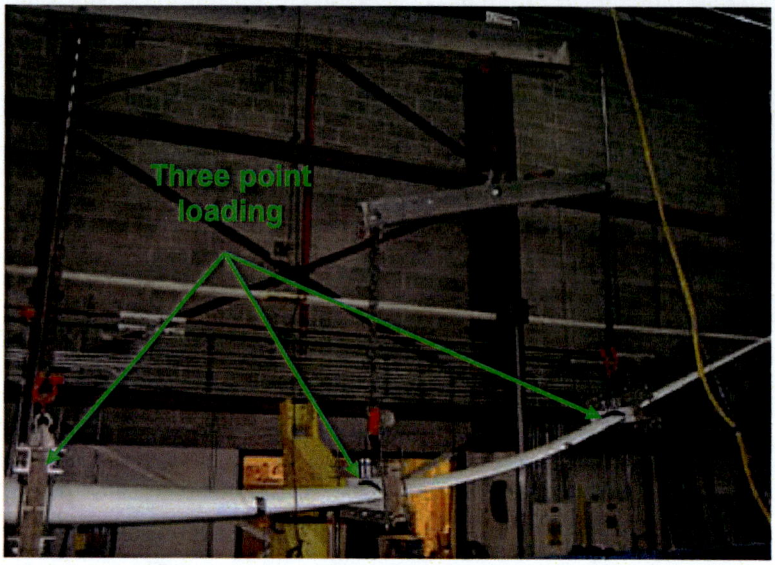

Figure 6. Static Test of the BSDS Wind Turbine Blade

We now turn our attention to characterization of uncertainty in experimental modal tests. Proper pre-test design and test technique are critical for the validation of blade models. In Reference 8, we presented an experimental study for quantifying the uncertainty in the modal parameters for the BSDS blade. In that study, we considered test-setup uncertainty, measurement uncertainty, and data analysis uncertainty. Bias errors in the test setup were found to be the largest sources of uncertainty. The principal sources of bias error were due to the support conditions (boundary conditions) and instrumentation (mass-loading and cable damping).

Support conditions for free boundary condition modal tests can introduce large bias errors if not designed properly. The traditional rule of thumb is that the highest frequency mode of the system due to the supports has frequency less than one-tenth the frequency of the lowest elastic mode. This rule of thumb only considers the support stiffness; however, one should consider the location (and direction) of the supports as well as their stiffness. For our tests, we evaluated different choices for stiffness and location of the two supports (as shown in Figure 5). In summary, we found that poor choices in the support design could lead to a 2% shift in frequency and 35% increase in damping for the principally affected mode. For the optimal support configuration which minimized the bias errors, we chose soft supports placed on the nodes of the principally affected mode (first edge-wise bending) in-line with the bungee supports. In addition to the experimental study, analytical formulas for assessing support condition effects in a modal test (including support stiffness and location) were developed and verified [9].

The instrumentation was also found to be a significant potential source of bias error [8]. Instrumentation mass tends to lower the measured modal frequencies while instrumentation cables tend to increase damping. Tests were conducted to identify the frequency and damping bias of instrumentation cables independent of the effect of the accelerometer mass-loading as the experiment was concluding. This was accomplished by performing a baseline test with the fully instrumented sensor set. Then, a large fraction of the cables were removed while removing no accelerometers to evaluate the cable effect alone. Twelve accelerometers remained in the reduced sensor set. Lastly, the set of accelerometers for which the cables had been removed in the previous step were removed to evaluate the combined effect of the accelerometers and cables. It was found that the instrumentation cables added more than 30% damping to some modes, and accounted for about 25% of the total mass-loading. Lastly, we mention that traditional rules of thumb for calculating the mass-loading effect cannot always be trusted. For the modal test of the 127 kg blade with approximately 0.9 kg of mass-loading due to the full set of accelerometers and associated cables, we estimate only 0.4% frequency shift. In reality, this was an underestimation by a factor of 3 or more for the bending modes, and a factor of 10 for torsional modes based on experiments. Although pre-test analysis of the test setup using traditional rules of thumb can be useful, one should conduct additional tests in order to experimentally measure the bias errors in the test setup.

Hybrid Calibration Method and Results

Within the framework of the validation methodology presented earlier, we now focus on test-analysis correlation. In this section, we provide details on a hybrid approach for calibration of a blade structural model [3].

A beam finite element model (FEM) was chosen to investigate calibration of a blade structural model. The blade span-wise mass distribution of the model was determined from measurements of a sectioned blade which had been tested to failure. With mass properties known, the objective of the calibration was to determine the blade span-wise stiffness properties, Young's Modulus multiplied by second moment of inertia ($E*I$). Each section moment of inertia was estimated using a thin-walled ellipse with dimensions equivalent to the corresponding blade section shape, which varies from a circular shape to an airfoil shape. Given simplified measurements of the blade cross sectional geometric properties and the blade actual mass properties, the calibration process was simplified to the determination of the equivalent Young's Modulus for each beam element. The calibrated model could then be utilized to make predictions under different conditions; for example, for a different root boundary condition for validation assessment.

The results of different calibration approaches in this calibration study are given in Table 1, which were described in more detail in Reference 3. We consider three approaches for calibration of the beam model: 1) using only load-deflection data from static tests, 2) using only natural frequencies from free boundary condition modal tests, and 3) a hybrid approach using both load-deflection and natural frequencies. When only static test data was used, the post calibration static deflection residuals were, of course, small; however, prediction of the first flap-wise mode for a free boundary condition was under predicted by 12.3%. Likewise, when only natural frequency data from free boundary condition tests was used for calibration, the natural frequency was in close agreement, but the static deflection prediction errors were large. This suggested that static test data and modal data may be used together to create a better model. The result of the third calibration approach indicates that the residual errors can be reduced for natural frequency and static deflection predictions. Static test data and modal data from these tests are complimentary for calibration. Static data provides key calibration data for root end property estimates; however, in these static test the blade was not strained outboard of the third (outermost) load location. Free boundary condition modal tests do not result in large strains at the root; however, they do provide important data in the outboard section of the blade. Results of the third, hybrid approach demonstrate the benefit of combining static test data and free boundary condition modal test data during calibration. Also, the boundary condition has a significant effect on the modal frequencies. The first flap-wise frequency for a free boundary condition is 47% higher than the predicted first flap-wise frequency for a cantilever root boundary condition. The boundary condition needs to be well-characterized in a test for inclusion in a model.

Table 1. Predictions of Calibrated Models and Prediction Residuals

Calibration Type	Calibration Error for First Flap-wise Mode with Free Boundary Conditions (%)	Norm of Static Deflection Error (meters)
1) Statics Updating	-12.3%	0.01
2) Modal Updating	1.1%	0.31
3) Modal and Statics Updating	-1.1%	0.03

Description of Validation Tests Using Seismic-Mass-on-Airbags Boundary Condition

In this section, we review the key aspects in the test design of independent validation tests performed on the BSDS blade. These tests were designed with the objective of evaluating the calibrated BSDS blade structural model. A focus of the test design was to provide a boundary condition which exercised the root while minimizing boundary condition uncertainty. The test design includes the selection of the boundary condition, test fixturing, the lift procedure, pre-test model prediction, test instrumentation layout, and test execution, topics that are each discussed in Reference 6.

Again, it was essential that the boundary condition for the validation experiment be very well-characterized for inclusion in the structural model. This requirement was accomplished by attaching the blade root to a seismic mass on airbags. The seismic mass is composed of steel and has a mass of 21,740 lbs (9858 kg) with dimensions of 66 inches by 72 inches (1.67 by 1.83 meters) and 24 inches (0.61 meters) thick at the thickest point. Four airbags were placed near the corners of the mass; the center-to-center distances between the airbags were 51.5 inches (1.31 meters) and 59.75 inches (1.51 meters). When pressurized, the seismic mass is lifted from the floor, providing a flexible boundary condition. The natural frequencies of the six rigid body modes associated with the seismic mass on airbags are completely dependent on the mass properties of the seismic mass, and the stiffness properties and placement of the airbags. The schematic in Figure 7 demonstrates the approach of the test, showing the blade mounted on the seismic mass/airbag system.

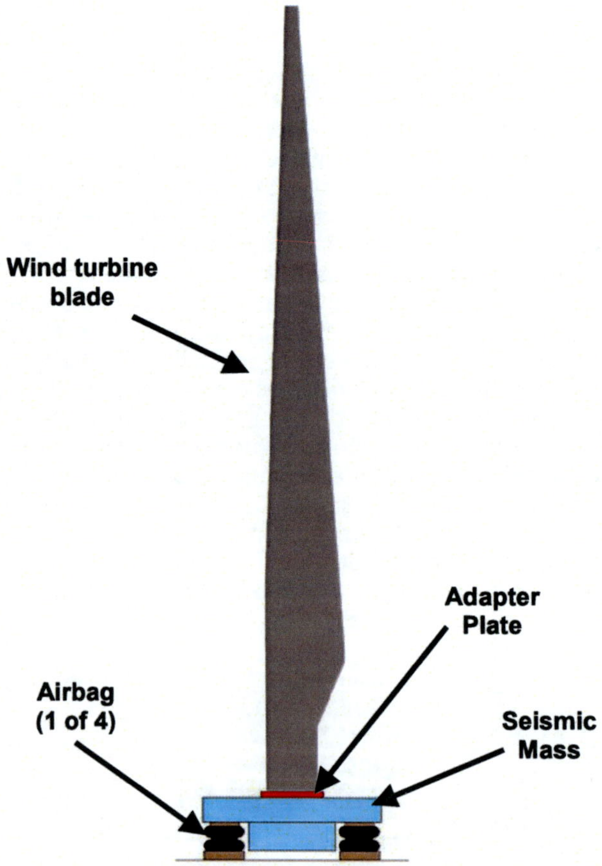

Figure 7. Schematic showing Seismic-Mass-on-Airbags Boundary Condition

The first test design issue considered was the test fixture. A very stiff connection between the blade root and the seismic mass was desired, and the adapter plate was designed accordingly. The correctness of the assumption of a fixed connection between the blade and seismic mass was verified by redundant accelerometer measurements – one on the blade root and a co-located measurement on the adapter plate.

It was desired to place the blade in a vertical position on the seismic mass during the modal test (as shown in Figure 7); therefore, another key aspect of the test was the design of the lift. A simple lifting device and safe procedure were desired for full control of the blade during the lift. The procedure was simplified by placing a hinge between the adapter plate and the seismic mass; this limited the motion of the blade to motion about the hinge axis.

A pre-test analysis was performed to assist in the instrumentation layout. A 20 element FEM of the blade calibrated using static test load-deflection data and natural frequencies from the free boundary condition modal tests as described in the previous section was used for the analysis. This calibrated blade model was then combined with a preliminary model of the seismic mass boundary condition to make predictions for the validation test. Based on the mode shapes of this model, the span-wise layout of accelerometers was optimized for measurement of the bending modes (as shown in Figure 8). In addition to sensors placed on the blade, high sensitivity tri-axial accelerometers were placed at each of the four corners of the seismic mass to measure the rigid body motions of the blade/mass system as a check of the boundary condition. Finally, as mentioned before, instrumentation was also placed on the root end of the blade and additional sensors were co-located with these on the adaptor plate as a check of the rigidity of the blade connection to the seismic mass. Figure 8 shows a plot of the sensor layout. The blade was instrumented while in the horizontal position before lifting into the test configuration.

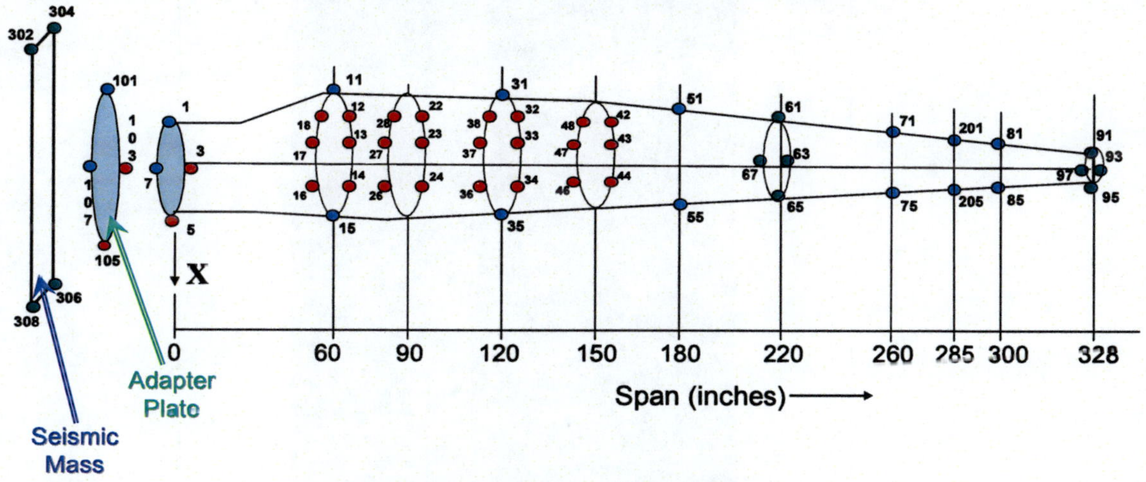

Figure 8. Instrumentation Layout

The execution of the test will now be discussed. Before the blade was placed on the seismic mass, tests were performed to characterize the modes of the seismic mass/airbags system for the airbag pressurization level and airbag placement chosen for the full system test with the blade. This test only characterized the boundary condition. It is known from previous tests that the seismic mass on airbags can be suitably characterized by a simple model as its rigid body modes depend completely on the mass properties of the seismic mass and the stiffness/damping properties and placement of the airbags. These modes include a vertical bounce mode, two lateral motion modes, a twist mode about the vertical axis, and two pitching modes. Modes related to the elastic deformation of the seismic mass are present; however, these modes are well above the frequency band of interest for this test and were not included in the seismic mass boundary condition model. Impact tests of the seismic mass on airbags provided natural frequencies and damping ratios for the six rigid body modes. The mass and mass moments of the inertia of the seismic mass have been measured previously, thus the unknown parameters in development of the boundary condition model were the stiffness and damping of the airbags. It was determined that a simple mass-damper-spring model with six degrees of freedom was appropriate for the boundary condition model, thus the six stiffness constants and six damping constants were determined using the known mass properties and the measured modal parameters. This calibrated model of the boundary condition was then combined with the calibrated blade FEM to make a final set of pre-test predictions of the validation experiments.

With final predictions complete, a number of impact modal tests were performed on the blade with the seismic-mass-on-airbags boundary condition. A photo of the test is shown in Figure 9. More than 20 modes were measured below 160 Hz in these tests, which included rigid body modes, flap-wise and edge-wise bending modes, torsional modes, and localized panel modes. The results for these tests, focused on the flap-wise bending modes, are summarized in the following section.

Figure 9. Photo of BSDS Blade Being Modal Tested with Seismic-Mass-on-Airbags Boundary Condition

Results of Validation Tests with Seismic-Mass-on-Airbags Boundary Condition

The primary objective of the validation test was to evaluate the seismic-mass-on-airbags boundary condition for validation of a blade structural model. Mass properties measurements were used to develop an accurate mass representation for the blade FEM [3]. Then, static tests and free boundary condition modal tests were conducted, and the data from these tests was then used to calibrate the stiffness properties of the blade FEM. This calibration resulted in a pre-test calibrated blade FEM to be evaluated in this validation effort. Then, a model of the seismic mass boundary condition, calibrated with test data, was combined with pre-test calibrated blade FEM to make predictions for the combined assembly for the new test. This is a validation experiment – if the pre-test calibrated blade model is valid, then our predictions with the new boundary condition should be in agreement with the measurements according to our predetermined adequacy criterion of less than 5% error in the first 4 flap-wise bending modal frequencies. Furthermore, the results of this study will demonstrate the feasibility of this test approach for blade model validation in addition to validation of the BSDS model. The results, as reported in References 6 and 7, are summarized in this section.

A large set of modes were measured for comparison with the model predictions. It should be made clear that these modes represent the combined assembly of the blade and seismic-mass-on-airbags boundary condition. We are principally focused on comparing the six rigid body modes and the first 5 flap-wise bending modes of the blade system. The rigid body mode natural frequencies are examined first. Table 2 lists the predictions of rigid body modes for the system using the pre-test calibrated model for the blade with the seismic-mass-on-airbags boundary condition; compared with the measurements from the validation test. This serves as a check of the boundary condition model and the mass properties of the blade model. The first three listed modes changed less than 1.5% by the presence of the blade on the seismic mass, that is, these modes are at nearly the same frequency with or without the blade on the mass. These modes depend on the total mass of the combined system, which is nearly that of the seismic mass. Conversely, the two pitch modes are significantly affected by the presence of the blade on the mass because these modes depend on the mass moments of the system. The blade has a significant contribution to the system mass moment of inertia, through its mass and CG offset from the seismic mass, which reduce the pitch rigid body mode frequencies by 20% with the addition of the blade to the mass. Note that even though these modes change in frequency by 20% with the inclusion of the blade, the model accurately predicts the pitch mode frequencies with errors of 1.1% and 1.9%. The twisting mode is in error by 3.6%, which indicates a possible inaccuracy in the mass moment of inertia of the blade model about the span-wise axis. However, this mode has an insignificant effect on the flap-wise bending modes. Again, the pitch mode in the flap-wise direction is in error by only 1.1% which provides certainty in this boundary condition model as this mode most strongly affects the flap-wise bending modes.

Table 2. System Rigid Body Modes Natural Frequency: Prediction versus Measured

Mode	Prediction	Measured	Percent Difference
Lateral motion flap-wise	1.08	1.08	0.0%
Lateral motion edge-wise	1.07	1.09	1.5%
Axial motion in vertical direction	1.75	1.77	1.1%
Twisting about vertical	1.80	1.86	3.6%
Pitch in flap-wise direction	1.79	1.81	1.1%
Pitch in edge-wise direction	2.26	2.22	-1.9%

An important consideration of the test was the orientation of the blade on the mass. The frequency separation between the low frequency bending modes of the blade and the highest frequency rigid body modes of the boundary condition should be as large as possible. Therefore, the blade was mounted such that the softer flap-wise direction was aligned with the motion of the lower frequency pitch mode (1.81 Hz) and the stiffer edge-wise direction was aligned with the higher frequency pitch mode (2.22 Hz).

Given that the rigid body modes of the combined system were accurately predicted, we conclude that the boundary condition model is accurate and continue by comparing the natural frequencies of the flap-wise bending modes. This provides an evaluation of the calibrated blade model mass and stiffness properties. Table 3 lists the predictions of the flap-wise bending modes of the pre-test calibrated blade model with the seismic-mass-on-airbags boundary condition and compares them with the measurements from the validation test measurements.

Table 3. Pre-test Flap-wise Bending Modes Natural Frequency: Prediction versus Measured

Mode	Prediction	Measured	Percent Difference
1st Flap-wise	3.95	4.20	5.9%
2nd Flap-wise	9.56	9.57	0.0%
3rd Flap-wise	18.25	18.29	0.2%
4th Flap-wise	29.38	29.77	1.3%
5th Flap-wise	44.84	43.66	-2.7%

The predictions were quite good for the flap-wise bending modes overall as the largest error was found in the 1st flap-wise mode with an error of 5.9%. However, this error is slightly larger than our predetermined adequacy criterion of 5% error in the first 4 flap-wise bending mode frequencies. Possible contributors to this discrepancy include the size of the elements in the root section of the FEM and the correctness of the assumption of the adapter plate rigidity. FRFs of the co-located root instrumentation showed that there was no significant difference in these sensors. In the initial model, the first two elements starting at the root were chosen to be 1.0 meter in length to match the measured mass properties of the sectioned blade because it was cut at the 1-meter and 2-meter span-wise locations. Therefore, it was decided that improvements in the model could be made by considering re-calibration of a blade model with refined geometry. To test this idea, the two elements at the root end of the model were split in two, which created four elements. This provided better resolution for the root end elements, and a better ability to capture the large gradients in the root end mass and stiffness properties. The outboard elements of the model were already well defined with 0.25 meter length, so only the root element lengths were modified for another set of predictions to be validated.

A slightly different approach was taken for the re-calibration process [6]. In the initial calibration, a hybrid calibration was used as the static test and modal test data were used simultaneously to calibrate the model. In the re-calibration, the new 22 element FEM was first calibrated with the static test load-deflection data. The stiffness parameters obtained from this calibration were held constant for the portion of the blade from the root to the third loading point of the static test (approximately at the 6 meter (236 inches) span location), while the parameters for the remaining outboard elements were calibrated using the free boundary condition natural frequencies. The re-calibrated 22 element model was then used to make predictions for the blade combined with the seismic-mass-on-airbags boundary condition. These predictions are listed in Table 4.

Table 4. Re-calibrated Flap-wise Bending Modes Natural Frequency: Prediction versus Measured

Mode	Prediction	Measured	Percent Difference
1st Flap-wise	4.05	4.20	3.5%
2nd Flap-wise	9.73	9.57	-1.7%
3rd Flap-wise	17.88	18.29	2.2%
4th Flap-wise	29.11	29.77	2.2%
5th Flap-wise	43.27	43.66	0.9%

Improvements were found for the 1st flap-wise mode, but at some expense to the accuracy of the other modes. One does not expect to predict all natural frequencies precisely; however, from the standpoint of reduction of the maximum prediction error the re-calibrated model is an improvement. Also, all errors are below 3.5% which indicates that this model is valid for our chosen adequacy criterion. Additionally, the fifth flap-wise bending mode also meets our adequacy criterion. It is worth noting that for a free boundary condition the 1st flap-wise frequency was measured at 5.25 Hz in contrast to the fixed root boundary condition whose frequency was predicted to be 3.57 Hz. For the free boundary condition, the natural frequency is 47% higher. This demonstrates the significance of the boundary condition on the natural frequencies.

Discussion and Conclusions

As wind turbine blades grow longer and more costly, it is crucial that adequate models be developed for use in the design phase. This paper summarized work in modal testing of research-sized wind turbine blades designed and tested at Sandia Labs. Modal tests were performed to evaluate new blade designs which incorporated innovations for blade weight reduction. In this paper, a methodology for validation of blade structural models was reviewed. Then, studies related to various aspects of a model validation effort were summarized. These modal tests have resulted in (1) techniques for experimental quantification of uncertainty in the modal parameters, (2) insight into model calibration using both static load-deflection data and the modal parameters, (3) novel test techniques for reducing the uncertainty in the root boundary condition, and (4) the development of a validated structural model. A summary of the modal testing results was provided with emphasize on recent tests using a seismic-mass-on-airbags boundary condition.

Quantification of uncertainty in the modal parameters was performed by conducting a small number of additional experiments. The largest bias errors were due to the support conditions for free boundary condition modal tests and instrumentation mass-loading and damping. Traditional rules of thumb for evaluating these bias errors were evaluated, and it was found that in some cases these traditional formulas were not highly accurate. New formulas for assessing the effects of both support stiffness and location were developed [9].

A hybrid calibration approach using both modal and static test data was developed and evaluated for updating of blade structural models. Traditionally, only the modal parameters are used for calibration; however, we found that static load-deflection data can improve the calibration process. The hybrid calibration approach was compared to calibration with only static data or modal data. The resulting calibrated model was then evaluated by performing an independent validation modal test.

The boundary condition is one of the important considerations when performing structural dynamics analysis; any test designed to validate a structural dynamics model should provide information that well characterizes the boundary condition for inclusion in the model. A seismic-mass-on-airbags boundary condition was introduced for a blade validation modal test. The results of this work demonstrate that this boundary condition can be accurately characterized from test observations and simply modeled with properties derived from the modal parameters of the rigid body modes. When the blade was placed on the seismic mass, the rigid body modes of the system and the bending modes of the blade were measured and then compared with analytical predictions of a calibrated model. This provided an evaluation of the boundary condition model as well as the blade model. This capability could be considered a new test technique for modal testing of wind turbine blades because it shows promise as an alternative to boundary conditions traditionally selected for wind turbine blade modal testing, which are free and fixed boundary conditions. Free boundary condition tests do not exercise the root end of the blade as is present in service. Cantilever boundary conditions cannot be realized in practice because of fixture compliance, although it provides service type strain at the root. The seismic mass boundary condition offers the advantage of the fixed boundary conditions by straining the root as in service, and it also reduces uncertainty in the boundary condition model.

References

[1] Casias, M., Smith, G., Griffith, D.T., and Simmermacher, T.W., "Modal Testing of the CX-100 Wind Turbine Blade," Sandia National Laboratories Technical Report.

[2] Griffith, D.T., Smith, G., Casias, M., Reese, S., and Simmermacher, T.W., "Modal Testing of the TX-100 Wind Turbine Blade," Sandia National Laboratories Technical Report, Report # SAND2005-6454.

[3] Griffith, D.T., Paquette, J.A., and Carne, T.G., "Development of Validated Blade Structural Models," *46th AIAA Aerospace Sciences Meeting and Exhibit*, 7-10 January 2008, Reno, NV, USA, AIAA 2008-1297.

[4] Laird, D. and Ashwill, T., "Introduction to NuMAD: A Numerical Manufacturing and Design Tool," *ASME/AIAA Wind Energy Symposium*, Reno, NV, 1998, pp. 354-360.

[5] Veers, P.S., Laird, D.L., Carne, T.G., and Sagartz, M.J., "Estimation of Uncertain Material Parameters Using Modal Test Data," *ASME/AIAA Wind Energy Symposium*, Reno, NV, 1998, AIAA-98-0049.
[6] Griffith, D.T., Hunter, P.S., Kelton, D.W., Carne, T.G., and Paquette, J.A., "Boundary Condition Considerations for Validation of Wind Turbine Blade Structural Models," Society for Experimental Mechanics Annual Conference, June 2009, Albuquerque, NM, USA.
[7] Griffith, D.T., "Structural Dynamics Analysis and Model Validation of Wind Turbine Structures," 50th AIAA Structures, Structural Dynamics, and Materials Conference, May 4-7, 2009, Palm Springs, CA, AIAA-2009-2408.
[8] Griffith, D. T. and Carne, T.G, "Experimental Uncertainty Quantification of Modal Test Data," *25th International Modal Analysis Conference*, February 2007, Orlando, FL, USA.
[9] Carne, T.G., Griffith, D.T. and, Casias, M.E., "Support Conditions for Free Boundary-Condition Modal Testing," *25th International Modal Analysis Conference*, February 2007, Orlando, FL, USA.

Proceedings of the IMAC-XXVIII
February 1–4, 2010, Jacksonville, Florida USA
©2010 Society for Experimental Mechanics Inc.

Modal Analysis of CX-100 Rotor Blade and Micon 65/13 Wind Turbine

J. R. White[1], D. E. Adams[2], and M. A. Rumsey[1]

[1]Wind and Water Power Technologies, Sandia National Laboratories, Albuquerque, New Mexico 87123
[2]Department of Mechanical Engineering, Purdue University, West Lafayette, IN 47907-2031

NOMENCLATURE

HP	=	High pressure side of the rotor blade, predominantly upwind during operation
LP	=	Low pressure side of the rotor blade, predominantly downwind during operation
Flap-wise	=	Direction of rotor blade oriented from high-pressure to low-pressure side of the airfoil
Lead-lag	=	Direction of rotor blade oriented from lead-edge to trailing-edge along the chord of the airfoil
Span	=	Direction of rotor blade oriented from root to tip along the length of the rotor blade
SNL	=	Sandia National Laboratories Wind and Water Power Technologies Department

ABSTRACT

At the end of 2008 the United States became the largest producer of wind energy with 25,369 MW of electricity. This accounts for 1.25% of all U.S. electricity generated and enough to power 7 million homes. As wind energy becomes a key player in power generation and in the economy, so does the performance and reliability of wind turbines. To improve both performance and reliability, smart rotor blades are being developed that collocate reference measurements, aerodynamic actuation, and control on the rotor blade. Towards the development of a smart blade, SNL has fabricated a sensored rotor blade with embedded distributed accelerometer measurements to be used with operational loading methods to estimate the rotor blade deflection and dynamic excitation. These estimates would serve as observers for future smart rotor blade control systems. An accurate model of the rotor blade was needed for the development of the operational monitoring methods. An experimental modal analysis of the SNL sensored rotor blade (a modified CX-100 rotor blade) with embedded DC accelerometers was performed when hung with free boundary conditions and when mounted to a Micon 65/13 wind turbine. The modal analysis results and results from a static pull test were used to update an existing distributed parameter CX-100 rotor analytical blade model. This model was updated using percentage error estimates from cost functions of the weighted residuals. The model distributed stiffness parameters were simultaneously updated using the static and dynamic experimental results. The model updating methods decreased all of the chosen error metrics and will be used in future work to update the edge-wise model of the rotor blade and the full turbine model.

1 INTRODUCTION

The following work was performed to develop the approaches for producing an experimentally updated wind turbine model for the development of operational monitoring methods. The following work was focused on developing the model updating approach for the flap-wise stiffness of the CX-100 rotor blade.

1.1 Background & Motivation

Wind energy accounts for a significant portion of U.S. power generation capacity, currently 1.25%, and is predicted to grow to 20% by 2030 [1]. Consequently, wind turbine performance and reliability will become increasingly important. Estimates of wind forces applied to a wind turbine and the corresponding reaction forces/moments in the rotor blades, drivetrain bearings, and gearbox while in operation will be needed to identify and resolve damage and failure modes in the wind turbine. There are many general approaches to wind force estimation, such as: wind speed and direction measurements with inference models to estimate aerodynamic loads, aerodynamic sensors located on the rotor blade for direct estimation of aerodynamic loads, structural based sensor systems for loading estimation through structural models, and loading inference with typical wind turbine sensors (drivetrain rpm, generator torque, nacelle inertial measurement unit, etc.) used with turbine global state-space models for force estimation. In the following work a structural model is developed for an accelerometer based structural monitoring system for operational wind turbine rotor blades. The structural model will be used in future work to validate the expected sensor signals and then to evaluate the corresponding feature extraction methods (deflection and loading estimates). This system is envisioned to provide wind turbine operators and designers direct information on the dynamic and quasi-static rotor deflection and loading for monitoring of turbine performance, fatigue, and to facilitate potential design improvements. Additionally, the sensor system could serve as an operational observer in future advanced smart rotor blades that will actively adapt based on the wind force being applied to mitigate damaging loads.

1.2 Previous Work

Relevant prior works are those which acquire wind turbine modal data, those which use the modal data to validate a model, and those which improve on modal testing of wind turbines. Molenaar [2] performed an experimental modal analysis of a wind turbine with accelerometers distributed over the rotor blades and tower. The natural frequencies of the test were used for comparison with a state-space model of the same turbine. In addition, the damping from the experimental results was used to compare the transient response of the modeled wind turbine and actual wind turbine to a step input. The natural frequencies were used to validate the model parameters of the wind turbine; however, active updating of the model parameters was not discussed and the mode shapes were not used for any of the validation. Griffith et al. [3,4] presented results on updating a BSDS wind turbine rotor blade model using modal parameters and static deflection data with weighted residual cost functions. Furthermore, Griffith [5] continued that work by mounting the rotor blade in a vertical span-wise direction and mounting it to a known boundary mass on airbags to validate the accuracy of the model. The approach was used to validate the model using static and modal features and then estimate the effect of adding a well defined mass and spring onto the root. The hypothesis was that the model would be validated if the predicted change caused by the mass addition was the same as the actual change. Lastly, Griffith et al. [6] discussed the influence of a boundary condition that is, assumed to be free with soft elastic support straps, and the effects of mass loading from sensors on the natural frequencies and damping ratios. These works were used when planning the free and mounted modal testing that is presented in the following sections.

The model validated in this paper will be used for developing operational monitoring approaches. White et al. [7,8] discussed accelerometer based methods for the estimation of static and dynamic deflections and loads. The methods extract operational information from the quasi-static and dynamic accelerations. The coordinate transformations from the ground reference to a sensor mounted to the rotating blade were also defined.

1.3 Sensored Rotor Blade

SNL has fabricated a sensored rotor blade to investigate load monitoring and damage detection capabilities of accelerometers and Fiber Bragg sensors. The instrumented rotor blade is a 9 m CX-100 design with integrated carbon composite along the shear web. The rotor blade was fabricated by TPI Composites, RI with one array of fiber optic sensors installed during layup and a second fiber optic array and an accelerometer array surface mounted on the interior surface. The surface mount sensors were installed prior to joining the HP and LP sides of the rotor blade. The surface mounted sensor arrays were fully functional after blade manufacture and shipment to the USDA Conservation and Production Research Laboratory in Bushland, TX. The Fiber Bragg sensor array that was installed during layup suffered micro-bending failure of the fiber optic cable that limited the number of functioning sensors. The accelerometer array also suffered sensor failures once the rotor blade was mounted on the wind turbine. Although the failure mode has not been fully defined, it is believed that electric static discharge from the wind passing over the blades produced a charge buildup on the rotor blade that was dissipated by discharging through the sensors, causing the sensor failures. Future work will be focused on mitigating the susceptibility of accelerometers to ESD.

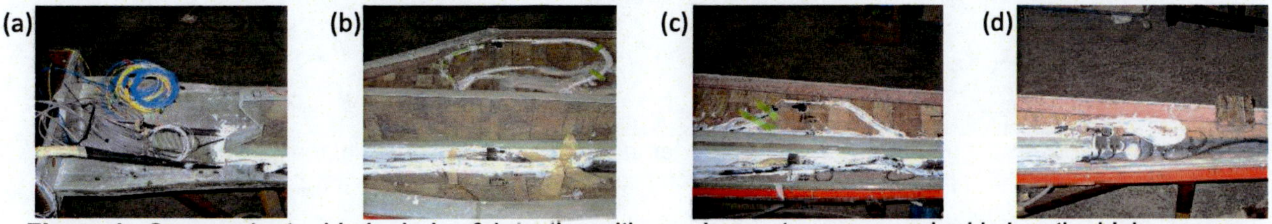

Figure 1: Sensored rotor blade during fabrication with accelerometer arrays embedded on the high-pressure sides at span-wise distances of: (a) 0 m, (b) 1.74 m, (c) 6.5 m, and (d) 8 m.

2 EXPERIMENTAL TESTING

Experimental modal tests of the full wind turbine and rotor blade subcomponent were needed to validate a wind turbine model. In the following sections, a modal test of the CX-100 sensored rotor blade was performed. The blade was hung by lifting straps with elastic bungee cord to replicate a free boundary condition. This setup was useful in validating the stiffness and mass properties of the blade with minimal influence by the boundary conditions. A modal test of the full wind turbine was also performed to account for the turbine tower and drivetrain effects on the rotor dynamics. The analysis showed that the rotor plane modes dominate the dynamics of the wind turbine through coupling of the hub and drivetrain.

2.1 Free-free Modal Testing

A modal test of the sensored rotor blade was performed to estimate the natural frequencies, damping ratios, and mode shapes of the rotor blade. To estimate these characteristics independent of mounting conditions, the rotor blade was hung by elastic cables with the rotor blade oriented such that the desired excitation and response directions were perpendicular to gravity, as shown in Figure 2 and Figure 3a. By aligning gravity perpendicular to the direction of excitation, the effect of the elastic boundary condition was minimized as was the influence of the internal body forces produced by weight loading in the flap-wise direction.

Figure 2: Experimental setup for modal test of sensored rotor blade with free boundary conditions.

In addition to the seven triaxial and three uniaxial accelerometers embedded in the rotor blade, eight ICP® triaxial sensors (PCB 356A14 and 356A65) were temporarily mounted to the exterior surface, as shown in Figure 3b. The sensor additions provided additional reference for modal parameter estimation. The rotor blade was excited by a modal impact hammer (PCB 086C03) with a soft rubber impact tip for low frequency excitation and a rear mounted mass for increased impact energy transfer as shown in Figure 3c. A VXI Technologies data acquisition system was used to measure the excitation and response signals. X-Modal2© software from the University of Cincinnati Structural Dynamics Research Laboratory was used to curve fit the frequency response data and estimate natural frequencies, damping ratios, and mode shapes. The results of the test are shown in Table 1.

(a)  (b)  (c) 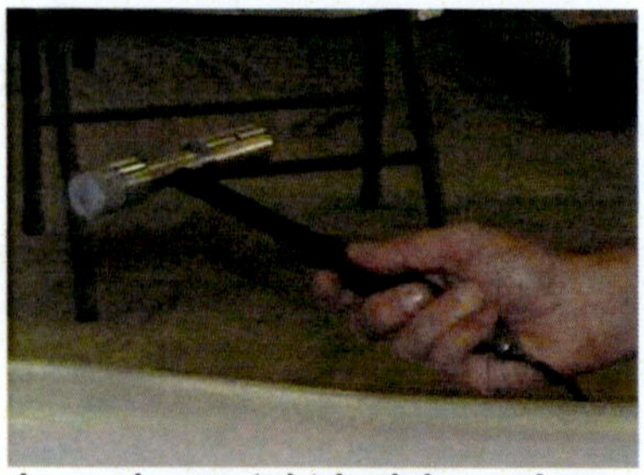

Figure 3: (a) Elastic support strap for test. (b) Externally mounted supplemental triaxial accelerometer. (c) Modal impact hammer with soft compliant tip and additional hammer mass.

The first mode of vibration of the rotor blade was a flap-wise bending mode, which was similar to the first beam-bending mode shape, at 8.2 Hz. The modal results showed that flap-wise and lead-lag bending mode shapes accounted for the first six modal features of the rotor blade. The seventh mode was a torsional oscillation about the span-wise direction. The flap-wise modes occurred at lower frequencies than the lead-lag modes because the second moment of area was smaller about the lead-lag axis than about the flap-wise axis for all airfoil cross sections (cross-sections outboard of 10% span-wise length). The damping of the first mode was also smaller than the other modes, which meant that the first flap bending mode was likely to respond for the longest period of time after excited.

As mentioned previously, the rotor blade was tested with the flap-wise direction perpendicular to gravity and the lead-lag direction parallel to gravity. This orientation could have led to errors in the lead-lag direction as support straps were directly actuated when the rotor blade was excited in the lead-lag direction. The rigid body mode shape in the lead-lag direction was at a frequency less than one-tenth of the first lead-lag flexible mode of 16.8 Hz. A general rule of thumb for free boundary conditions is that they separate the rigid and flexible body modes by a factor of at least ten. This characteristic of free supports is used to minimize the influence of the boundary condition on the natural frequencies. Additionally, the support straps were located at the nodes of the first flap-wise bending mode to further minimize the influence of the boundary conditions. By hanging the rotor blade vertically with the tip down, the most ideal orientation for modal testing is realized because the flap and lead-lag directions are perpendicular to gravity in this orientation thereby minimizing gravitational and boundary effects.

Table 1: Modal results for CX-100 rotor blade with free boundary conditions.

Mode	Frequency (Hz)	Damping (% critical)	Description
1	8.2	0.28	1st Flap Bending
2	16.8	0.62	1st Lag Bending
3	20.3	0.49	2nd Flap Bending
4	33.8	0.51	3rd Flap Bending
5	42.2	1.17	2nd Lag Bending
6	52.2	0.58	4th Flap Bending
7	60.6	1.16	1st Torsion
8	69.9	0.74	3rd Lag Bending

The estimated natural frequencies and mode shapes were assumed to be sufficiently accurate to update the distributed parameter model available for the CX-100 rotor blade. However, in order to investigate the full wind turbine distributed parameter model generated by FAST in MSC.ADAMS, a full wind turbine modal analysis was performed with the sensored rotor blade and two unsensored rotor blades mounted to the wind turbine.

2.2 Full Turbine Testing

A modal test of the sensored rotor blade mounted to the Micon 65/13 wind turbine located in Bushland, TX was performed. Figure 4a shows the sensored rotor blade mounted to the wind turbine and oriented vertically downward. Two unsensored CX-100 rotor blades were also mounted to the wind turbine for the modal and operational testing. In addition to the sensors embedded in the sensored blade, a triaxial accelerometer was temporarily mounted to the tip of each rotor blade, four biaxial sensors were mounted to the drivetrain, the inertial measurement unit mounted in the nacelle was also used as a reference accelerometer, and two triaxial sensors were mounted across the mounting flange of the sensored rotor blade for a total of forty-seven simultaneously sampled acceleration measurements. Thirty-four locations on the sensored rotor blade, fourteen locations on each of the unsensored rotor blades, four locations on the drivetrain, and sixteen locations on the tower for a total of eighty-two locations were excited with an impact hammer to estimate the modal parameters. Five averages were acquired at each impact location and the data was recorded with a VXI Technologies HPE 1432a data acquisition system.

Figure 4: (a) Micon 65/13 Wind Turbine with the Sensored Rotor Blade mounted. (b) Modal impact testing of the wind turbine rotor blades. (c) Modal impact testing of the wind turbine tower.

Data acquired during the modal test was used to estimate frequency response functions, which were imported into X-Modal2© along with the spatial location of each impact and reference sensor location and orientation. An example of the mode shapes produced, the first umbrella mode at 4.5 Hz, is shown in Figure 5. Modal analysis results showed that the tower of the turbine was very stiff and generally behaved as a rigid boundary condition, excluding the first two tower modes. Due to the rigid tower, the excitation and reference data from measurement degrees of freedom located on the rotor plane were selected as the input data for the experimental modal analysis.

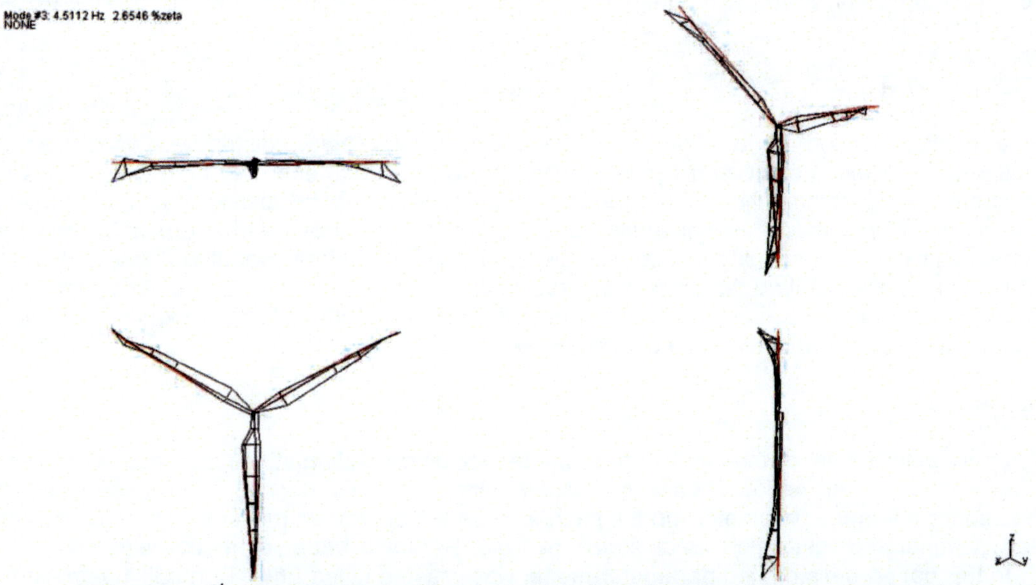

Figure 5: 1st Umbrella Mode Shape of rotor plane from modal test experiment.

The results of the modal analysis are shown in Table 2. The table indicates that the first two modes were side-to-side and fore-aft tower modes, which were dominated by the stiffness of the tower and foundation and the mass of the tower and nacelle. Beyond the first two modes, the tower did not play a significant role in the motion of any other modes. The coupled system of the three rotor blades attached to the hub and connected to the low speed shaft dominated the modal features. Individual rotor blade modes were not evident, but instead the system of rotor blades produced the modal frequencies and shapes. Experimental results showed that the stiffness of either the hub or low speed shaft was small enough to allow coupling of the rotor blades. If the hub and low speed shaft were truly rigid, with an infinitely large stiffness, then each blade would have had individual uncoupled modes. These modes would likely be very close to the same natural frequency and would be very similar to the frequencies and shapes observed in a cantilevered rotor blade modal test.

Table 2: Modal results for CX-100 rotor blades mounted to Micon 65/13 wind turbine.

Mode	Frequency (Hz)	Damping with Window (% critical)	Damping without Window (% critical)	Description
1	3.19	2.90	1.96	Side-Side Tower Mode
2	3.40	3.21	2.33	For-Aft Tower Mode
3	4.51	2.12	1.46	1st Umbrella Mode
4	5.51	1.05	0.51	1st Spoke Mode
5	6.36	3.30	2.84	1st Vertical Antisymmetric
6	7.16	1.66	1.25	1st Horizontal Antisymmetric
7	9.96	2.50	2.20	2nd Horizontal Antisymmetric
8	10.30	2.08	1.79	2nd Vertical Antisymmetric
9	11.49	1.05	0.79	2nd Umbrella Mode
10	15.41	1.31	1.12	1st Rotary Torsion

The results of this modal test provided sufficient information to extract the natural frequencies, damping ratios, and shapes for model verification and updating. In future work, excitation methods other than a modal impact hammer will be evaluated. An impact excitation produces a finite length time response. The length of the response time dictates the amount of low frequency information available in the estimated frequency response function. A higher energy impact would produce a higher signal to noise ratio time response; however, this type of impact would also require a lower sensitivity accelerometer for the initial spike in acceleration. The lower sensitivity accelerometer will increase the noise floor for both the modal testing and the operational monitoring that followed. In future work, alternate modal excitation methods will be attempted to improve the results.

3 MODEL UPDATE

The purpose of this section is to show how the CX-100 rotor blade distributed parameter model was simultaneously updated using static and dynamic local and global features. The emphasis of this section is a proposed approach using commonly installed sensors and typical "pre-flight" testing of a wind turbine rotor blade to update a computational model. Additionally, cost functions and error metrics are described that produce physically meaningful error percentages that can be compared to update model stiffness properties. Lastly, weighting functions are described to emphasize residuals when testing errors are less significant and de-emphasizing residuals when testing errors are more significant. Some of the weighting functions are also adjusted during model updating to expedite convergence.

3.1 Initial Model

A distributed parameter model of the CX-100 rotor blade including bending stiffness, mass, airfoil geometry, wall thickness, and span-wise section locations was available prior to model updating. The properties described could have been used for a lumped parameter model, such as a series of lumped masses connected by springs, or for a distributed parameter model, such as a series of Euler-Bernoulli beam elements with stiffness and mass properties. In this paper, a distributed parameter model was created using Euler-Bernoulli beam elements, which maintained smooth displacement at the nodes; that is, the displacement and slope at the nodes were continuous [8]. For example, the deflection of the rotor blade model caused by a 4.45 N (1 lbf, used for comparison with experimental data acquired in units of lbf) force at the 7.25 m span-wise location is shown in Figure 6. This plot shows the slope and curvature along the rotor blade. It also illustrates the typical cantilevered beam bending shape for a discrete location force input, where there was zero deflection and slope at the root and constant slope and zero curvature at the tip. The curvature plot was discontinuous with each element showing a decrease in curvature in the outboard span-wise direction. For an Euler-Bernoulli beam element with linear-elastic material properties, the curvature plot was related to the moment, stress, and strain through the following relationship:

$$EI\frac{\partial^2 u}{\partial z^2}=\frac{EI\varepsilon}{c}=\frac{I\sigma}{c} \tag{1}$$

where E was the modulus of elasticity, I was the second area moment, u was the displacement in the flap-wise direction, z was the span-wise location, ε was the elastic strain, c was the distance from the neutral bending axis, and σ was the elastic stress. In a simple form, the curvature and strain were related by:

$$\frac{\partial^2 u}{\partial z^2}=\frac{\varepsilon}{c}\,. \tag{2}$$

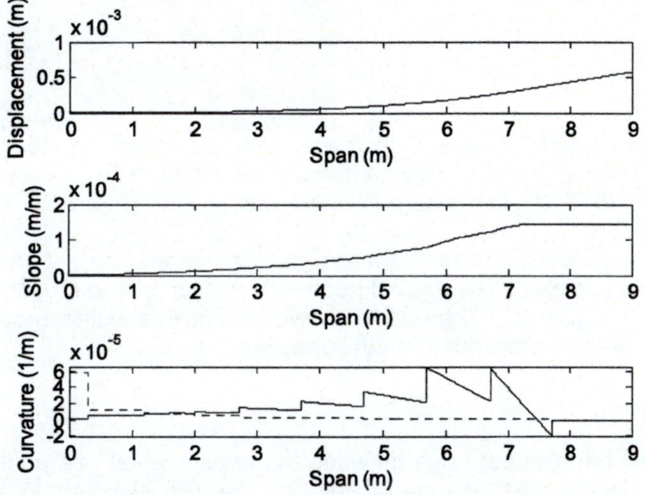

Figure 6: Displacement, slope, and curvature of distributed parameter finite element CX-100 rotor blade. The distribution of bending stiffness prior to model updating is shown in the curvature plot (dashed).

The moment from Equation (1) produced by a 4.45 N force applied to the rotor blade is linearly increased from a value of zero at a span-wise location of 7.25 m to a maximum value of 32 N-m at the rotor cantilevered connection as shown in Figure 7a. In this figure, the theoretical moment distribution along the rotor blade (dashed) was compared with the moment estimates from the HP and LP distributed strain gages that use the relationships in Equation (1). Experimental strain gage measurements were acquired during a static pull test at the 7.25 m span-wise location in the flap direction as shown in Figure 7b. Several deflection distances and force levels were tested to establish the strain-force (1/lbf) calibration of the rotor blade with strain gages. The slope of this calibration was the measured strain for a 1 lbf (4.45 N) applied force. The sensor estimates of moment from this data depended on the accuracy of the model bending stiffness, EI, and of the distance from the neutral axis, c. Strain gages located symmetrically across the neutral axis separated by a total distance h were used to improve the accuracy of the neutral axis distance through the relationship:

$$\begin{aligned} h &= c_1 - c_2 \\ c_1 &= \frac{\varepsilon_1 h}{\varepsilon_1 - \varepsilon_2} \end{aligned} \tag{3}$$

assuming that the line connecting the strain gages was perpendicular to the neutral axis and parallel to the cross sectional plane. The moment estimates from the strain gages also assumed that a distributed parameter Euler-Bernoulli approach was valid, which meant that during deflection the cross-sectional planes rotated an amount equal to the slope of the deflection. Future work may evaluate whether this approximation was valid in wind turbine rotor blades, but for the scope of this work the model was assumed to be valid.

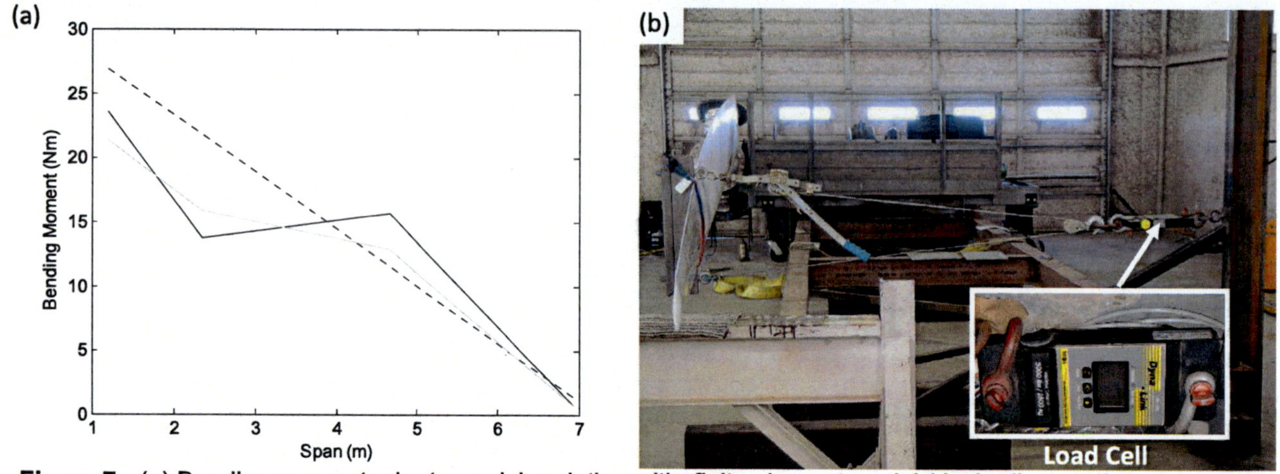

Figure 7: (a) Bending moment prior to model updating with: finite element model (dashed), experimental results from high pressure side strain gages (black), and experimental results from low pressure side strain gages (gray). (b) Static flap-wise pull fixture with in-line load cell.

The initial distributed parameter model stiffness and mass matrices were also used to estimate the free boundary condition modal frequencies and shapes. The natural frequency results prior to model updating are listed in Table 3 and the shapes are shown in Figure 10. The results showed a significant difference between the experimental and model modal frequencies and shapes prior to model updating.

3.2 Model Updating

The previous sections showed the discrepancies between the experimental and analytically modeled static pull and modal analysis results. Having quantified the differences, the next step was to determine quantifiable error indicators from several sources of experimental and analytical errors that could be equally compared to determine how to best update the distributed property parameters. The following work was mostly focused on updating the eleven distributed bending stiffness properties. The distributed mass property could be directly measured from section cuts of the rotor blade when the testing was completed. However, the distributed mass applied to each element length and then summed was 20% lower than the actual measured mass. To account for this deviation, each distributed mass property was increased by 25%.

To update the stiffness properties, a percentage error metric was created from the difference in critical features of the experimental and model results. For each of the static and modal results, one feature dominated by the overall stiffness and one feature dominated by the local stiffness was estimated, for a total of four types of features. The balance of local and global stiffness updates was assumed to improve the model updating by minimizing the constraints on the individual stiffness parameter changes.

The cost function J was used to calculate the weighted square of the residuals:

$$J = e^T W e \tag{4}$$

where e was the residual and W was the weighting function. An example of a residual that was used is the comparison of the experimental and model mode shapes:

$$e = \begin{Bmatrix} U_{\text{model } 1,r} - U_{\text{experiment } 1,r} \\ \vdots \\ U_{\text{model } p,r} - U_{\text{experiment p},r} \end{Bmatrix} \tag{5}$$

for a mode *r* where *U* was the mode shape displacement coefficient at a location *p*. The weighting function was a diagonal matrix with each diagonal term corresponding to the weighting of an individual residual:

$$W = diag\{w_1 \quad \cdots \quad w_p\}. \tag{6}$$

For equally weighted residuals, this matrix was the identity matrix. For the case of the mode shapes, the weighting was based on the mode shape itself because locations with small modal deflection corresponded to node locations that were sensitive to small errors in testing location accuracy, whereas locations with large modal deflection corresponded to anti-node locations and were not nearly as sensitive to testing inaccuracy. Weighting by the modal deflection therefore emphasized the model updating on the residuals that were most likely attributed to model errors and de-emphasized the residuals that corresponded to possible testing limitations, i.e. issues with the dynamic range of experimental data. To produce weighting that did not bias residuals in comparison to other modes and other error metrics, the absolute value of the mode shape displacement coefficient for a mode *r* was divided by the sum of the absolute value of all displacement coefficients for that mode shape:

$$W = diag\left\{ \frac{\left|U_{\text{model } 1,r}\right| \cdot n}{\sum_{i=1}^{n}\left|U_{\text{model i},r}\right|} \quad \cdots \quad \frac{\left|U_{\text{model n},r}\right| \cdot n}{\sum_{i=1}^{n}\left|U_{\text{model i},r}\right|} \right\} \tag{7}$$

and then multiplied by the total number of displacement coefficients of that mode, *n*. The division of each modal displacement coefficient by the total produced an absolute percentage of the total model displacement distributed at each location. If all *n* locations were weighted equally, then the total weight would be *n*. In this case the percentage distribution of *n* was established. For example a location of significant modal response had a weighting function greater than one, but a location with small modal response had a weighting function less than one. The total weighting however was the same as for an equally weighted case. The other advantage of this weighting approach was that the model mode shape, not the experimental shape, itself was used to establish the weighting function. In this case, the weighting function changed its distribution as the model was updated.

The cost function, *J*, that resulted from this process was the sum of the weighted square of the residuals for a given mode shape. To compare this residual with other error metrics, it was converted into an estimate of the percentage error of the modeled mode shape. To do this the square root of the cost function was divided by the absolute sum of the modal displacement coefficients for the experimental mode shape. The percentage error of all three mode shapes was compared with the percentage error between the experimental and model natural frequencies. As mentioned previously, local and global stiffness error metrics were used to minimize the constraints placed on the parameter updating. For the modal results, the mode shape was used as the local error metric as it was more sensitive to localized changes in the stiffness parameters and the natural frequencies were used as the global error metric as they were less sensitive to localized changes to the stiffness parameters.

A similar approach was used for the static moment comparison. For the global error metric a first order polynomial was fit through the theoretical and experimentally estimated moment distributions. The percentage difference between the slopes of the two polynomials, using the theoretical estimate as a reference, was used for comparison. For the local error metric, the cost function was calculated for the residuals between the individual moment estimates (HP and LP estimates at 1.2 m, 2.35 m, 4.65 m, and 6.95 m) and the theoretical moment at those span-wise locations and weighted by the normalized theoretical moment distribution. This weighting function emphasized measurements near the root area where the moment magnitude was a maximum and de-emphasized measurements near the applied force location where the moment magnitude was a minimum.

The preceding calculation produced local and global percentage error estimates, *E*, for the first three mode shapes and natural frequencies, for the slope of the moment distribution estimated from the strain gages, and for the individual moment estimates. The global error metric that was optimized was:

$$E_{total} = \sum E_{natural\ frequency} + \sum E_{mode\ shape} + 3\sum E_{moment\ slope} + 3\sum E_{moment\ residual} \tag{8}$$

where the static errors were weighted by a factor of three so that the static error was equally weighted with the total dynamic error produced from the summation of the error of the first three mode shapes. To minimize the total error, the **fminsearch** function in MATLAB© was used. This function implemented the Nelder-Mead simplex multiple degree of freedom nonlinear optimization method [10].

3.3 Updated Model

The simultaneous model update was performed on a standard desktop PC with dual-core 2.0 Ghz processor and 2 GB of RAM with a run time of approximately one minute. Figure 8a illustrates the original distribution of blade stiffness (dashed) and the updated distribution of stiffness (solid) on a logarithmic plot. This plot shows that the stiffness was increased outboard of the 4.5 m location and was decreased significantly in the first 1 m of the blade length. The moment estimated by the strain gages is shown in Figure 9, which showed closer agreement between the estimated moment and theoretical moment than for the original stiffness properties that were used in Figure 7a. The natural frequency estimates were also significantly closer to the experimental results after model updating as listed in Table 3. The updated mode shapes (solid) also showed closer agreement with the experimental mode shapes (circles) as shown in Figure 10.

The updated model showed closer agreement to all of the static and dynamic error metrics calculated. The significant decrease in the stiffness of the inboard region of the rotor blade was concerning as that much of a reduction in stiffness was not realistic. However in that region there were minimal measurements because all three mode shapes de-emphasized the weighting in that region and there were only two strain measurements located inboard of the 1 m station at 0.35 m. Additionally, the distributed mass was not updated, which could have affected the changes to this region where the mass distribution was most significant as shown by Figure 8b. In the future, constraints could be defined so that individual stiffness could not be changed without affecting the adjacent stiffnesses or additional sensor locations and measurements could be acquired specifically focused on the inboard spans of concern.

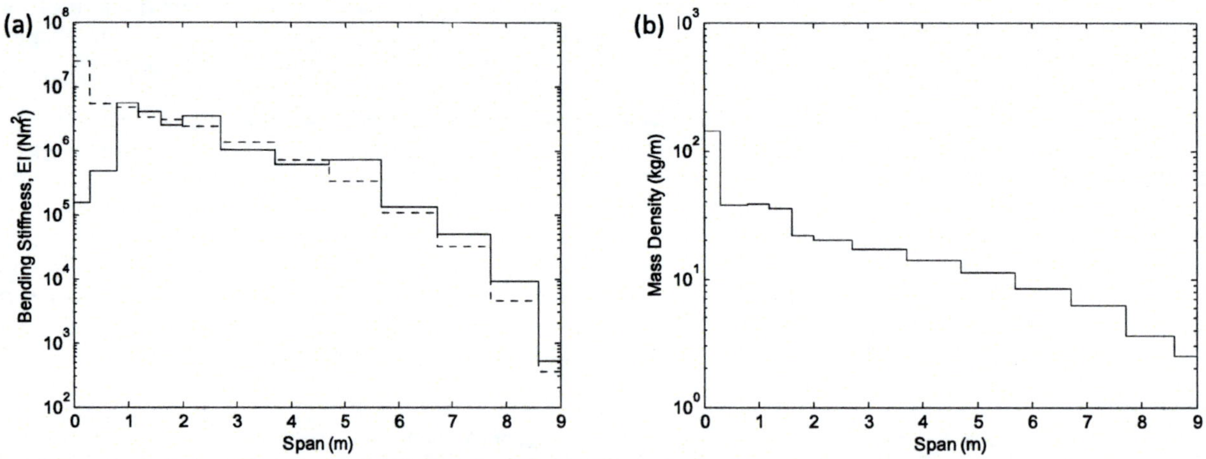

Figure 8: (a) Comparison of distributed bending stiffness properties before (dashed) and after (solid) model updating. (b) Distributed mass density (kg/m) of CX-100 rotor blade.

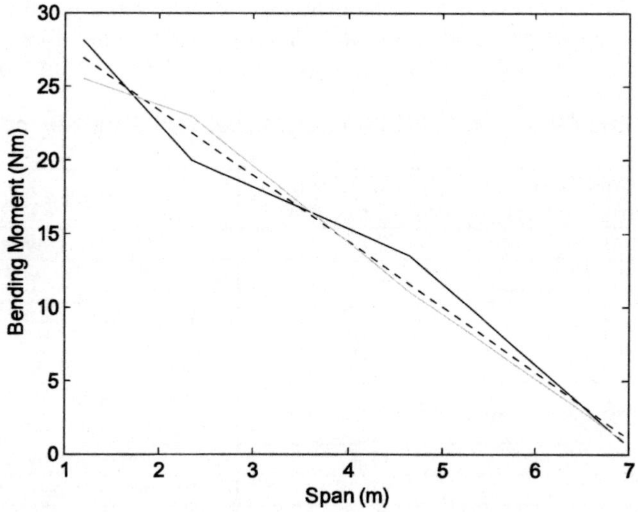

Figure 9: Bending moment after model updating with: finite element model (dashed), experimental results from high pressure side strain gages (black), and experimental results from low pressure side strain gages (gray).

Table 3: Comparison of experimental and finite element modal frequencies before and after model updating.

Mode	Experimental Frequency (Hz)	Model Frequency Prior to Updating		Model Frequency After Updating		Description
		Hz	% Difference	Hz	% Difference	
1	8.2	6.94	-16%	8.13	-1%	1st Flap Bending
2	20.3	15.90	-22%	18.28	-10%	2nd Flap Bending
3	33.8	29.53	-13%	33.84	0%	3rd Flap Bending

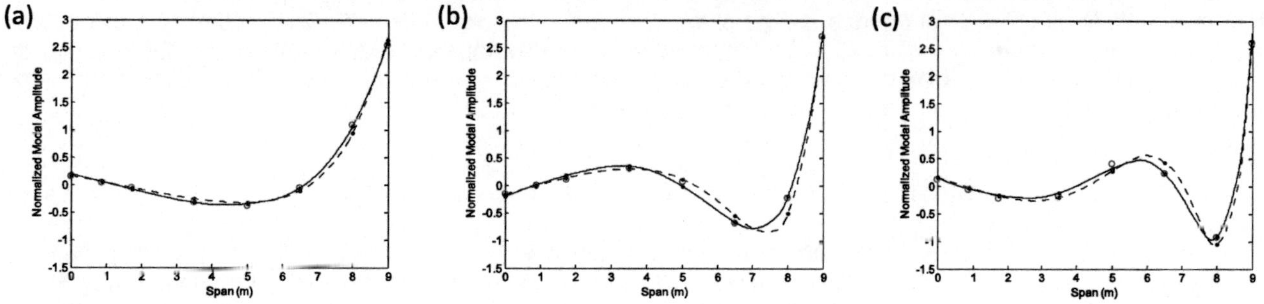

Figure 10: Comparison of the first three flap-wise bending modes at (a) 8.2 Hz, (b) 20.3 Hz, and (c) 33.8 Hz where the experimental mode shape coefficients are (point circle), mode shape before model updating is (dashed), and mode shape after model updating is (solid).

4 FUTURE WORK

The preceding work demonstrated the use of a combination of static and dynamic test data for the updating of a CX-100 rotor blade flap-wise stiffness properties. In future efforts, these approaches will be expanded to update the flap-wise and edge-wise stiffness and mass properties of the CX-100 rotor blade. Additionally, the full turbine experimental modal analysis results will be used in conjunction with a previous wind turbine tower modal analysis to update the stiffness and mass properties of the combined wind turbine system. The individually updated rotor blade and tower models will be used to determine the degrees of freedom, stiffness, and mass properties of the connections that are needed to construct a full turbine model. Table 4 lists the experimental natural frequencies of the turbine modal analysis in comparison to the natural frequencies of the current CX-100 Micon 65/13 wind turbine model. Both the experimental and computational models contain the same modes, however the frequencies have significant errors that will require boundary condition / attachment model updating.

Table 4: Comparison of experimental and MSC.ADAMS frequencies for full turbine model.

Mode	Modal Test Frequency (Hz)	MSC.ADAMS Frequency (Hz)	Description
1	3.19	2.62	Side-Side Tower Mode
2	3.40	2.62	For-Aft Tower Mode
3	4.51	4.69	1st Umbrella Mode
4	5.51	7.28	1st Spoke Mode
5	6.36	4.58	1st Vertical Antisymmetric
6	7.16	4.62	1st Horizontal Antisymmetric
7	9.96	11.95	2nd Horizontal Antisymmetric
8	10.30	11.77	2nd Vertical Antisymmetric
9	11.49	12.08	2nd Umbrella Mode
10	15.41	7.19	1st Rotary Torsion

5 CONCLUSIONS

To aid in future operational monitoring system development, an experimentally updated wind turbine model was needed that accurately reflected the operational dynamics of the CX-100 Micon 65/13 wind turbine. Experiments were performed to acquire the modal frequencies, damping ratios, and mode shapes of the CX-100 sensored rotor blade in free boundary conditions and of the same rotor blade mounted to the wind turbine with two other unsensored rotor blades. Static pull tests were also performed to calibrate the strain-force relationships of the embedded strain gages. The experimental results were used to update the distributed stiffness properties using percentage errors generated from cost functions of weighted residuals. The results of the model update made improvements to all of the error metrics. Some of the changes to the distributed stiffness properties may not have been physically accurate and will need to be improved in future development of the methods. In future work, the methods presented will be expanded to include edge-wise motion of the rotor blade and a full turbine model including the wind turbine rotor plane, tower, and connections.

6 ACKNOWLEDGEMENTS

The results presented in this work were made possible by the US Department of Energy Sandia National Laboratories System Modeling Tools & Analysis Sensors Task performed by the Wind and Water Power Technologies Department. The experimental results were made possible with help from B. Neal and A. Holman of the US Department of Agriculture Conservation and Production Research Laboratory. The strain gage experimental data analysis was performed by J. Berg of Sandia National Laboratories.

REFERENCES

1. American Wind Energy Association. "American Wind Energy Association Annual Wind Industry Report Year Ending 2008." US Energy Information Administration. AWEA. 2009. http://www.awea.org/publications/reports/AWEA-Annual-Wind-Report-2009.pdf (accessed September, 2009)
2. Molenaar, D. P., "Experimental Modal Analysis of a 750 kW Wind Turbine for Structural Model Validation," 41st AIAA Aerospace Sciences Meeting and Exhibit, 2003, Reno, NV.
3. Griffith, D. T., Paquette, J. A., and Carne, T. G., "Development of Validated Blade Structural Models," 46th AIAA Aerospace Sciences Meeting and Exhibit, 2008, Reno, NV.

4. Griffith, D. T. and Carne, T.G, "Experimental Uncertainty Quantification of Modal Test Data," 25th International Modal Analysis Conference, February 2007, Orlando, FL, USA.
5. Griffith, D. T., "Structural Dynamics Analysis and Model Validation of Wind Turbine Structures," 50th AIAA/ASME/ASCE/AHS/ASC Structures, Structural Dynamics, and Materials Conference, 2009, Palm Springs, FL.
6. Griffith, D. T., Carne, T. G., and Paquette, J. A., "Modal Testing for Validation of Blade Models," Society for Experimental Mechanics Annual Conference, 2009, Albuquerque, NM.
7. White, J. R., Adams, D. E., and Rumsey, M. A., "Operational Load Estimation of a Smart Wind Turbine Rotor Blade," SPIE Smart Structures and Materials & Nondestructive Evaluation and Health Monitoring: Health Monitoring of Structural and Biological Systems, 2009, San Diego, CA.
8. White, J. R., Adams, D. E., and Rumsey, M. A., "Sensor Acceleration Potential Field Identification of Wind Turbine Rotor Blades," 7th International Workshop on Structural Health Monitoring, 2009, Stanford, CA.
9. Cook, R. D., Malkus, D. S., Plesha, M. E., and Witt, R. J., Concepts and Applications of Finite Element Analysis Fourth Edition, John Wiley & Sons, Inc., New York, NY.
10. Lagarias, J. C., Reeds, J. A., Wright, M. H., and Wright, P. E., "Convergence Properties of the Nelder-Mead Simplex Method in Low Dimension," SIAM Journal of Optimization, Vol. 9 Number 1, pp. 112-147, 1998.

Proceedings of the IMAC-XXVIII
February 1–4, 2010, Jacksonville, Florida USA
©2010 Society for Experimental Mechanics Inc.

Modeling, Estimation, and Monitoring of Force Transmission in Wind Turbines

Colin Haynes, Dept. of Structural Eng, University of California – San Diego
Nick Konchuba, Dept. of Mechanical Eng, Virginia Tech
Gyuhae Park, The Engineering Institute, Los Alamos National Laboratory
Kevin M. Farinholt, The Engineering Institute, Los Alamos National Laboratory

ABSTRACT

Recent studies by the Department of Energy suggest that wind energy could make up as much as 20% of total U.S. power generation by 2030 [1]. By better understanding the load transmission in wind turbines, subsequent improvements in design could result in more efficient turbines and reduced maintenance costs. The objective of this study is twofold. First, a low-cost monitoring system to assess the condition of the bolted joint connections used to attach the blades to the hub is devised in order to characterize the force transmission into the hub and to detect when bolt loosening for in-service wind turbines. A section of the CX100 blade is used for the joint monitoring study. Second, a method to estimate the tip deflection of the blade is developed to better understand the dynamic loads acting on the rotor hub. A one-meter-long blade is used to estimate the tip deflection under laboratory conditions. Using the results from these tests, this study demonstrates how a better understanding of load transmission from the blade to the hub may be achieved.

Introduction

In the coming decades, the number of online wind turbines is expected to increase dramatically as the nation seeks to expand its renewable energy generation. The trend in wind energy is also toward larger, longer, and heavier blades in order to generate more power per unit. With demand increasing and designs growing larger, ensuring the reliability of wind turbine design is of critical importance to achieving the goal of 20% wind power. Although failure in the gearboxes or rotor blades is not the most common type of damage that occurs in wind turbines, it is among the most difficult, time-consuming, and expensive damage to repair [2]. Designers therefore need a better understanding of the forces acting on these devices to develop improved systems. Furthermore, a rotational imbalance due to blade damage can cause serious secondary damage to the turbine if not corrected promptly, emphasizing the need for monitoring to protect against bolt loosening and other damage [3].

In order to monitor the state of a structure and to predict failure before it occurs, modern structures may implement health monitoring. The Structural Health Monitoring (SHM) process would bring significant advantages to the wind turbine application by moving the maintenance paradigm to condition-based maintenance rather than time-based maintenance. The current time-based maintenance might prescribe a set time for replacement of blades or other components, whether or not these components are in need of replacement. However, wind turbines are typically located in remote and windy sites which make them expensive and dangerous to repair. Current inspection techniques cost about two percent of the initial wind turbine cost annually [5]. A SHM system would identify the state of damage of turbine components and reduce the uncertainty associated with time-based maintenance, allowing repairs to be made when a part is actually damaged. Such a system would also mitigate the high cost of unexpected catastrophic failure caused by undetected damage.

Many health monitoring systems rely on the use of "smart" materials called piezoelectrics (PZT) patches. These materials strain when a voltage is applied; conversely an applied strain causes the piezoelectric (in this case, a PZT patch) to produce a charge. In this way, PZT tranducers bonded to a host structure can serve the dual purposes of actuation and sensing and consequentially are called "smart" or "active" materials. PZT wafers are used widely in structural dynamics applications to measure high-frequency dynamic responses because they are lightweight, inexpensive, and robust. Park et al. reviewed PZT sensors and their application to the impedance method in great detail [6].

To better understand the causes of rotating imbalance and premature gearbox damage, designers are particularly interested in the blade tip deflection and the forces transmitted through the hub to the gearbox. Therefore, methods are proposed here to measure both quantities.

Mascarenas et al. were able to identify loose bolt damage by implementing the impedance method using frequency shifts as a feature for damage identification [4]. A PZT patch bonded to an aluminum washer ('smart washer') was used as a damage detection sensor. A smart washer is advantageous in turbine applications because it is cheap to manufacture compared to other alternatives such as load cells. The simplicity and repeatability of the washer result in the need for only one baseline measurement of the impedance and not a baseline measurement for each washer. The unique aspect of this study is modifying the approach of Mascarenas et al. by implementing the impedance method with a smart washer using damping as a feature for both damage identification and quantifying the static forces transmitted through the hub. A mathematical relationship correlating the damping with the bolt torque is determined so that a simple red-yellow-green light SHM system could be implemented to provide turbine operators a condition-based warning system for loose bolt damage.

In order to monitor the tip deflection of the blades, an accelerometer is applied to the tip of the blade. The acceleration data and a state-space model of the blade structure are the input to a Kalman filter, which will be used to extract the desired deflection [7]. The Kalman filter has several advantages over other possible numerical methods. First, the algorithm does not require knowledge of the initial conditions as do integration routines, which prevents it from diverging over time. Also, the Kalman filter requires no knowledge of the forcing function applied to the structure—quite convenient considering that it is impractical to measure the wind pressure on an actual wind turbine blade. Finally, there is significant robustness with respect to the parameters that make up the model, meaning that not all parameters need to be known exactly to obtain a solution. In this study, the algorithm is subjected to two verification tests: an analytical 2-DOF spring-mass-damper system and an aluminum cantilever beam test. Subjecting the algorithm to a test on an actual wind turbine is the next step in verification.

Experimental Procedures for Loose Bolt Damage Identification

The smart washer shown in Figure 1 first implemented by Mascarenas et al. was used as a bolt damage detection sensor. Initially, an experiment was performed to validate the capabilities of the washer for loose bolt damage identification. Using a CX-100 turbine hub shown in Figure 2, 6 bolts spaced evenly around the hub were tightened to 25 ft-lbs to simulate an undamaged (tight) condition. Five baseline measurements were taken by recording the mechanical impedance of the smart washers under these loads at the washers' first resonant frequency. Bolt/washer assembly labeled 4 was loosened to 0 ft-lbs of torque to simulate a damaged (loose) condition. Then bolts 4 and 9 were loosened to 0 ft-lbs of torque to simulate two bolts in the damaged condition. The impedance of the damaged bolts showed significantly larger resonant amplitudes than the undamaged bolts, indicating that by measuring the impedance of a smart washer, a damage index could be developed to identify which bolts were in the damaged state and undamaged state.

Figure 1 - A smart washer used for damage identification. A PZT patch is bonded on the right, and a strain gauge is bonded on the left.

Secondly, the impedance response of the smart washer was measured under varying bolt torque conditions at the first and second modes. Using the ABAQUS finite element code the mode shapes and frequencies of the first and second mode were estimated. The frequencies estimated were on the order of 20 kHz and 50 kHz.

For the experiment, an Agilent 4294 impedance analyzer was used. The analyzer was connected to the PZT patch on the smart washer. A MATLAB program controlled the operation of the impedance analyzer which actuated the PZT over a frequency range and then sensed the response using the same patch.

Figure 2 - A CX-100 turbine hub, approximately 15" OD, used for testing.

Procedure:

1. Input a frequency range for the first and second modes into MATLAB® program controlling the Agilent® impedance analyzer
2. Measure impedance response with analyzer at first and second modes at each torque level
   (Torque Levels: 0, hand-tight (~1), 5, 10, 15, 20 [ft-lbs])
3. Repeat 1-2 for seven Smart Washers and bolt assemblies

With increasing torque level, a slight shift in natural frequency to higher frequencies and substantial amplitude attenuation were observed in the measurements. Additionally, the 20 kHz mode was lost in noise above 10 ft-lbs and the second mode shows coupled resonance peaks above 10 ft-lbs. The experimental setup was modified to include a piece of rubber isolating the washer and the bolt and the washer and the test plate to isolate structural coupling with the washer dynamics. The experiment was then repeated with the new setup. The first mode was again lost in noise after 10 ft-lbs. The second mode has one resonance peak with greater amplitude attenuation as torque increases. The damping of the second mode was chosen as a good indicator of torque level over the torque test range of interest because the amplitude attenuation is significant between torque levels; however, the rubber isolators resulted in somewhat inconsistent torque measurements. Therefore, a strain gauge was applied to the smart washer and the impedance response of the washer under varying loading conditions was measured after the strain had reached a steady state. The strain was recorded for each impedance measurement. The strain was calibrated with the actual torque in the bolt by recording the strain while the torque in the bolt was maintained with a torque wrench.

Experimental Procedures for Kalman Filter

Tip deflection estimation was accomplished by the application of an algorithm known as the Kalman filter. The Kalman filter allows particular states (e.g. displacement, temperature, strain) to be estimated when it is impossible to obtain direct measurements. A dynamic model of the system must first be produced, which is used by the algorithm together with the measured data to produce the estimates of the desired states. The algorithm itself uses an iterative method that seeks to minimize the variance of the estimation error to predict and update the values of the state variables in each time step of the simulation [7]. There are several significant advantages of using a Kalman filter in the wind turbine application. First, the filter does not require knowledge of the forcing function, which is highly variable and dependent on wind conditions in a turbine. Second, the filter can estimate tip deflection—a quantity that is infeasible to measure directly in a large-scale wind turbine—using only tip acceleration measurements. Furthermore, convergence of the solution does not depend on knowledge of the initial conditions and is robust with respect to variation in parameters (geometrical, material, etc.) whose values are not well known.

The process of tip deflection estimation began with the implementation of the Kalman filter in the Python language. The code was then verified by running simple test problems. The first test problem was an analytical, two DOF spring-mass-damper system with a sinusoidal input applied. The input was assumed to be known, the "measured" quantity is the displacement of the first mass. The solution from this problem was compared to a known reference solution and the performance of the algorithm was confirmed. Next, the algorithm was modified to use a model constructed from beam finite elements. After assembling the stiffness and mass matrices, the model is converted into a state-space representation and passed to the Kalman filter. Code verification for this revised Kalman filter was accomplished using a rectangular aluminum beam test structure, shown in Figure 3. The beam was fixed on one end and the cantilevered end was 15 cm long. A shaker was attached 5 cm from the fixed end, and an accelerometer and laser vibrometer were placed at the free end of the beam. Three inputs were applied to the beam: a chirp, an impulse chain, and a sine wave. The tip acceleration, tip displacement, and input force data were recorded imported to Python for the data analysis.

Figure 3 - Aluminum cantilever beam test set-up.

Smart Washer Analytical Modeling

In order to determine the modes and frequencies of the smart washer that is measured by the impedance analysis, an analytical model was constructed. ABAQUS finite element software was used to create the geometry and mesh it with about 5 elements through the thickness at the thinnest point. The meshed model is shown in Figure 4. Then a linear modal analysis was run to determine the frequencies and mode shapes for various boundary conditions.

For initial frequency estimates used to determine a reasonable range for the experiment, the washer was simulated in the free-free condition. Later, attempts were made to replicate the experimental phenomenon of damping and frequency shifts with changing torque in the model. To make a comparison of damping, a frequency response function was obtained by selecting two nodes in the model as well as a frequency range over which the dynamics would be obtained. The output data were then processed to extract the FRF.

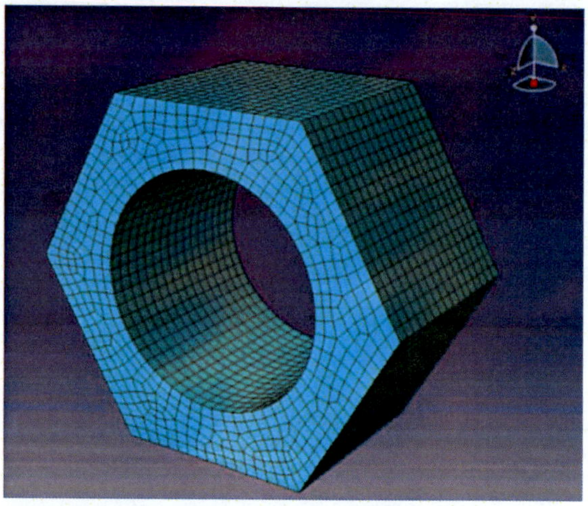

Figure 4 - Finite element model of the washer.

However, it was observed in all cases that the frequency and damping did not change significantly and that whatever change was exhibited was highly dependent on the boundary conditions applied. The washer was first modeled with an applied compressive pressure to simulate the bolt torque, but this boundary condition did not change the frequency response. The washer was then modeled using 3D spring-to-ground elements at each node of the contact faces. The spring stiffnesses were varied, resulting in noticeable changes in frequency. It is hoped that by understanding the boundary conditions better, the model can be calibrated to show better agreement with experimental observations.

Data Analysis - Bolt Joint Monitoring

In the bolt monitoring experiments, the impedance of a smart washer was measured under two loading conditions—loose and tight—to investigate the capacity of a smart washer for identifying loose bolt damage. The smart washer impedance was measured under multiple loading conditions at the first and second modes. In the analysis, the damping of the impedance response is used as the feature to correlate with torque load. The

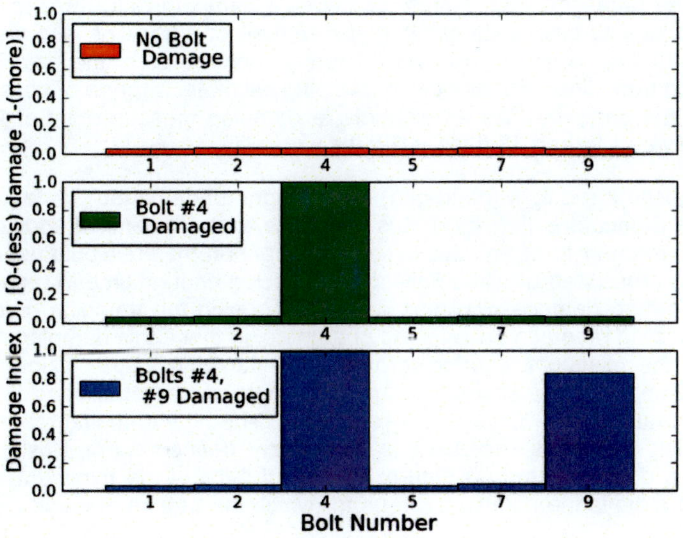

Figure 5 - Damage detection in turbine hub bolted joints.

frequencies from the modal analysis of the washer were used to determine the experimental frequency range. However, as stated in the previous section, the resonance peaks of the first and second mode of the washer become coupled with the test structure and the bolt at higher torque levels; as a result, damping estimation becomes difficult. The washer was isolated from the test structure and the bolt with two pieces of rubber material. Damping was extracted from the impedance response, but results were not repeatable. Finally, a strain gauge was attached to the washer and damping was correlated to the axial strain in the washer.

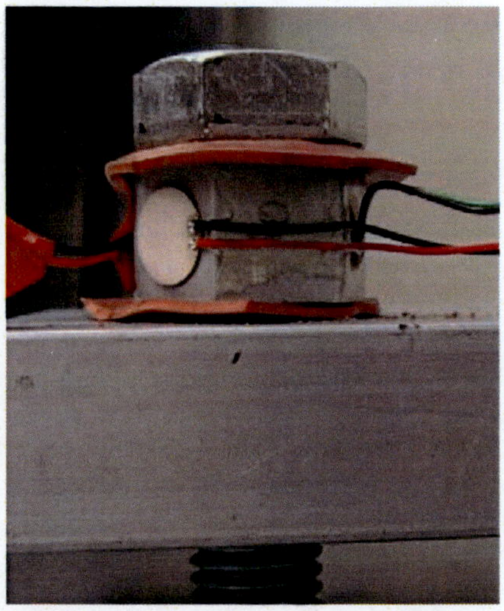

Figure 6 - Test assembly showing the rubber isolators (dark orange) between the bolt and washer and the washer and aluminum test plate.

A simple damage index was developed to demonstrate the effectiveness of a smart washer at detecting damage. The magnitude of the impedance for all data was normalized by the peak amplitude of the loose bolt impedance response; in this way, a damage index of 1 indicated complete loose bolt damage and a 0 indicates infinite bolt torque load. Figure 5 shows the undamaged baseline data (all bolts torqued to 25 ft-lbs) near a damage index of 0. When bolts 4 and 9 are loosened to 0 ft-lbs, the damage index clearly indicates loose bolt damage. There is some variability in the damage index value of bolts 4 and 9; however, the values are relatively close validating the claim that smart washers provide repeatable data.

The next test, which measured the impedance of the washer at varying torques, demonstrated coupling in the impedance response between the structure, bolt, and washer. The half power method was used to calculate the damping of the impedance response at each torque level. A script written in the Python language implemented the half-power method for the experimental data. Another Python script was used to estimate the variability from using the half-power damping method. In this script, the damping ratio of a 1-DOF mass-spring-damper system was varied and the half-power damping script was implemented to estimate the response. The half-power method estimate of damping was determined not to diverge significantly from the true damping value until approximately 10% damping. The damping in the test range of torques was between 0 and 2.5 %; therefore, the half-power method is valid for our test. In addition, the half-power method resulted in only a small amount of error in the torque estimates. The most significant error in torque estimation came from the torque wrench and the rubber isolators. Using a 350 ohm strain gauge in a quarter bridge setup, the torque was correlated with the strain in the bolt. A maximum variability of 2 ft-lbs was determined from our strain-torque calibration. Finding a more suitable damping material than the rubber that would both isolate the washer but demonstrate fewer viscoelastic characteristics might be necessary to implement this measurement method on an actual turbine hub. Curve fitting techniques might also be tried on the impedance response; this might reduce the need for a correlation to strain and the use of the isolators all together.

Figure 7 presents a correlation between damping ratio and bolt torque load. The data in red represents damping measurements when the bolt was damaged (0-1 ft-lbs). The data in blue represents damping measurements when the bolt undamaged (>1 ft-lbs). All of the damaged data points fell below a damping ratio threshold of 1.25%. In implementing this system on an actual turbine a damping measurement of 1.25% or less would indicate loose bolt damage is present. The undamaged data points were linearly curve fit; all of the data falls between ±2.5 standard deviations from the mean trend line. For any given damping estimate, a range of possible torque values can be identified with close to 95% certainty. As an example, a damping measurement of 2 % would indicate that the torque in the bolt was between 14 and 18 ft-lbs, ± 2 ft-lbs variability.

Previous bolt damage indication systems have used a red light (damaged) - green light (undamaged) system [4]. In order to better service a wind turbine before it becomes damaged, it is advisable to implement an additional 'yellow' light that a bolt is soon to be damaged. In order to design this system several parameters would need to be identified. A minimum torque value would be prescribed based on giving a technician enough time to respond to the warning and fix the loose bolt damage before damage becomes critical. By using the damping ratio to torque relationship established previously, a damping ratio can be selected guaranteeing that if this damping ratio is measured the torque will not be less than the minimum warning torque. Any damping values below this threshold would have to indicate the bolt is damaged. The second parameter that must be established is the minimum torque level which can be considered undamaged. Once this level is established the damping ratios for which the red, yellow, and green light warning indicators represent is completely defined. For illustrative purposes, the red, yellow, green light system is demonstrated on Figure 7 where 6 ft-lbs is the minimum safe torque for potential damage to be fixed, and 14 ft-lbs is the minimum torque for which the turbine blade is safely tightened to the hubs.

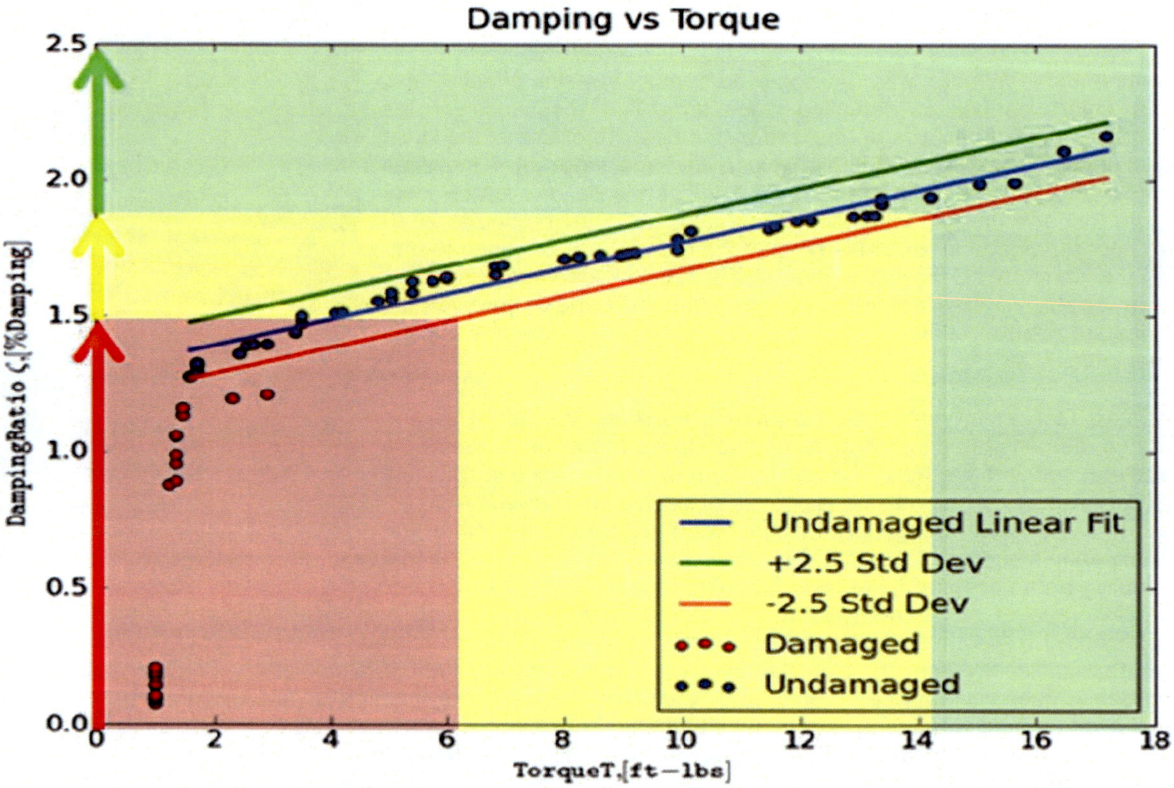

Figure 7 - Correlation between damping ratio and torque. The red, yellow, and green sections show how a potential warning system could be developed for loose bolt damage.

Future work would include improving the variance of the torque measurements by implementing an improved smart washer vibration isolator or curve fitting to eliminate the need for the isolators all together. In addition the following should be investigated: the repeatability and variability of damping-torque curve experiments on the CX-

100 hub, validation of the red-yellow-green light warning system when bolts/washers are on a turbine hub, and the capacity of a smart washer for determining dynamic loads through the hub.

Data Analysis - Kalman Filter

After making measurements on a credible test structure (in this case, the aluminum beam), the data were analyzed. For the Kalman filter, the input force is assumed to be unknown; therefore, the input does not play a role in the algorithm. The measured quantity used for comparison by the filter is the acceleration at the tip of the beam. Finally, the displacement data is not used directly, instead being used as a comparison for the estimated deflection.

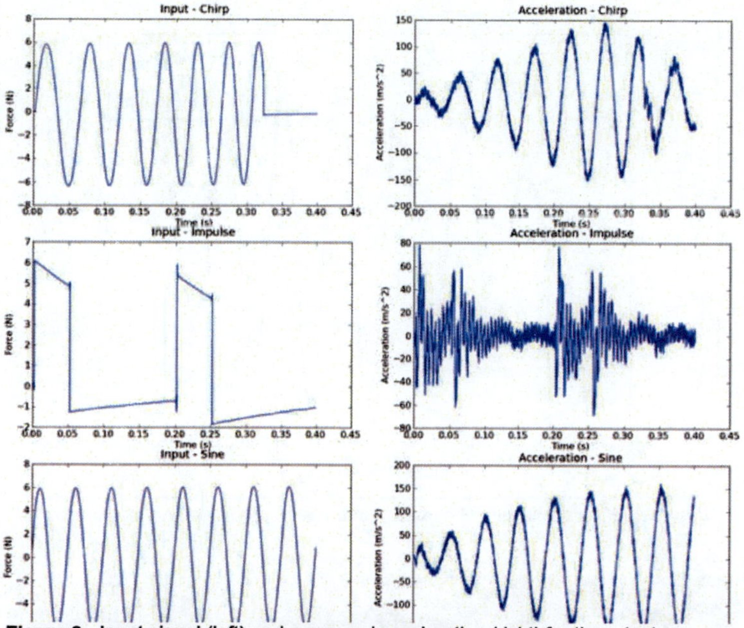

Figure 8 - Input signal (left) and measured acceleration (right) for three tests conducted on the aluminum cantilever beam.

The most notable feature of the input data is the noisiness of the accelerometer data, as shown in Figure 8. This figure shows measurements of force for all three input types in the left column compared to the measured tip acceleration in the right column. Speculation suggested that such high levels of noise might lead to a poor convergence of the Kalman filter solution, but in fact the solution still converged quite well for the chirp and sine inputs. However, the impulse chain input did not converge well, probably because the sudden changes in the (unknown) input combined with noisy accelerometer data did not allow the Kalman filter to adjust properly in the time window being studied. For the remainder of the discussion of the aluminum beam test, the chirp input will be the example chosen.

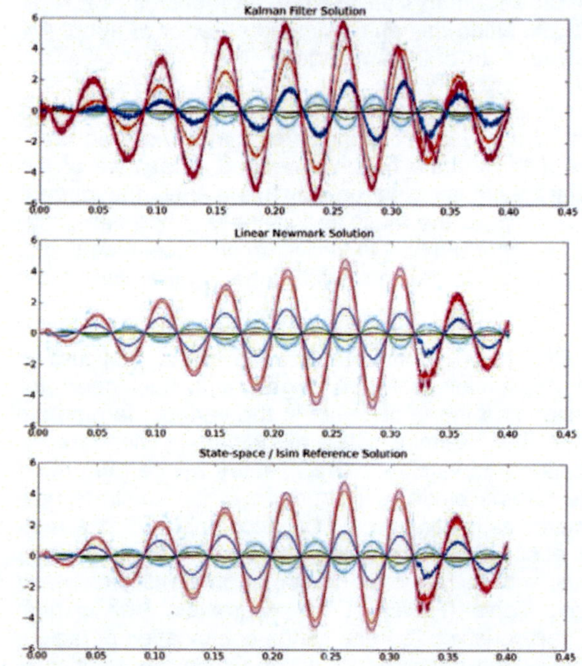

Figure 9 - Comparison of state vector predicted by Kalman filter and by two integration routines.

To further verify the performance of the Kalman filter, the estimated state vector was compared with two separate integrated reference solutions, as shown in Figure 9. It is important to note that integrated solutions will be more accurate in this test application, but are infeasible for use in wind turbines for two reasons. First, unlike the Kalman filter, the integrated solutions require a measurement of the input force. Second, the integrated solutions will begin to diverge over time due to the uncertainty in the initial conditions. The solutions would be sure to diverge in a wind turbine because the initial conditions are not known as they are in the simple beam test and because of the much longer time duration. However, these solutions are useful for comparative purposes, and it can be seen that they agree well.

A more quantitative assessment is obtained by comparing the tip deflection prediction of the Kalman filter to the values measured by the laser vibrometer during the test. This comparison is shown in Figure 10. The error between the measured and predicted values is shown in yellow. The peak value of the deflection is predicted with only a

2.0% error, an encouraging number considering the peak deflection is the quantity of most interest in determining the loads transferred to the wind turbine hub.

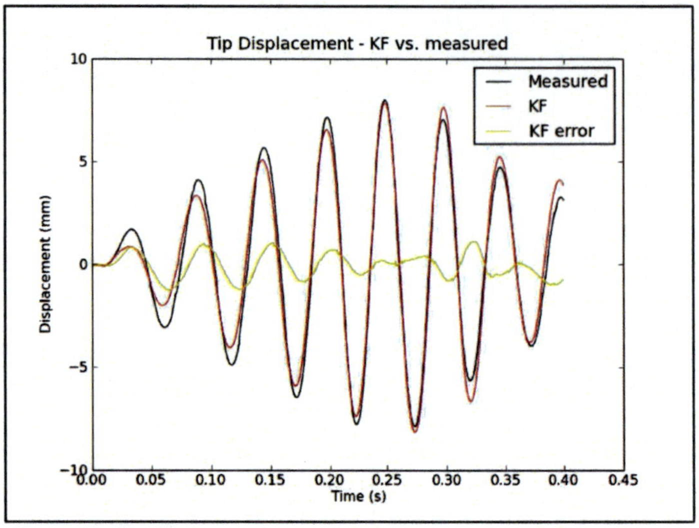

Figure 10 - Tip displacement comparison between predicted and measured values.

Given that the results of the Kalman filter simulation correlate quite well with the various references, the algorithm can be considered verified for the simple cantilever beam problem. Next, the Kalman filter must be applied to the one-meter-long wind turbine blade structure. The test set-up, shown in Figure 11, was very similar to the aluminum cantilever beam test. A laser vibrometer was placed at the end of the blade. An accelerometer and a shaker were attached 21 cm from the tip of the blade. Tests were run and acceleration data taken for a variety of sine and chirp inputs.

The next challenge was to construct the dynamic model of the blade. Geometrical properties were estimated for a five-element discretization. Material properties were unknown, so the modulus of elasticity had to be estimated. To do so, a simple finite element beam model was constructed to compare natural frequencies to the modal properties of the blade. The blade properties were found experimentally by conducting a roving impact hammer modal test. The value of the modulus of elasticity in the analytical model was then calibrated to best match the first natural frequency (which is the mode excited the most in the Kalman filter testing). A modulus of E = 17 GPa was settled on, which yielded an error of about 1% in the frequency of the first mode.

Figure 11 - Wind turbine test set-up for Kalman filter validation.

However, the first round of simulations did not give good results. Such a difficulty indicates that a more accurate model of the turbine blade is needed. Because of the irregular geometry and unknown material, the current model estimates the mass and stiffness properties of the blade rather crudely. In order to produce meaningful results, a more detailed model of the turbine blade must be developed.

This difficulty will undoubtedly arise in the real turbine blade application as well, highlighting the need for understanding the uncertainty in the various parameters used in the Kalman filter algorithm. Therefore, a verification and validation (V&V) study will be performed on the turbine blade test structure to demonstrate how parameter estimation would be accomplished in a real-world application. First, the important parameters are selected with the aid of a Phenomenon Identification and Ranking Table (PIRT). The parameters with a high "UxS" factor, which represents the combination of factors that have high uncertainty and on which the solution is sensitive, will be assessed to see how they may be improved. Performing a sensitivity analysis will be particularly important to establish how the solution error changes as parameters are changed.

Conclusions

A low-cost smart washer has been demonstrated as an effective impedance sensor for detecting damage in the hub of a CX-100 wind turbine. The repeatability of the smart washers eliminates the need for a baseline measurement when implementing multiple washers, making smart washers ideal for wind turbine application. Because of this minimal baseline requirement, a simple statistical model for estimating bolt torque was developed by extracting the damping ratio of impedance measurements. After establishing a model, an advance-warning system was developed by which the parameters were defined so that turbine blade operators would successfully be able to identify undamaged bolted connections, damaged bolted connections, and bolted connections that could potentially become damaged soon if not attended to.

Sources of error in the model were identified as coming from the torque measurements. Finding a suitable vibration isolator for the washer which would also allow a consistent bolt load to be maintained would help improve model variance and result in more accurate torque estimation. Future work would include improving the torque measurements and validating the statistical model and the warning system in an actual turbine hub.

The Kalman filter technique for measuring tip deflection has made substantial progress toward implementation on wind turbine blades. The algorithm was selected to make dynamic estimations of tip deflection using only acceleration data and a model of the structure. It is particularly well suited to the wind turbine application because it requires no knowledge of the input force or of the initial conditions and can tolerate substantial error in the input parameters. For a wind turbine, with its complex loading conditions, geometry, and material properties, these advantages are critical for obtaining a good estimate.

The algorithm was first implemented in Python code. The code was tested on the analytical 2-DOF spring-mass-damper system and the results matched the known solution. The aluminum cantilever beam was used to generate actual test data for the Kalman filter. Again, the algorithm performed well, predicting the peak deflection to within 2.0%. Finally, a model was constructed for the one-meter-long turbine blade. It was determined after data was taken that the system model requires significant refinement to better comprehend the complex geometrical and material properties. A sensitivity analysis will be performed to quantify the effect that uncertainty in various parameters has on the final solution. After the final two steps—improvement of the model and sensitivity analysis—the technique will be ready for application on full-scale, in-situ wind turbine blades. In short, dynamic tip displacement can be obtained by only measuring acceleration with proper application of a Kalman filter to an appropriate model of the system.

Acknowledgments

The authors would like to thank Dr. Charles Farrar and the Los Alamos Dynamics Summer School for the opportunity to conduct this research. The following companies generously provided various software packages to aid in modeling and data analysis: Vibrant Technologies, SIMULIA, and Enthought. The authors would also like to acknowledge, Abraham Light-Marquez (New Mexico Tech) and Alexandra Sobin (Daniel Webster College) for their assistance. Finally, the authors would like to thank by Mark Rumsey from Sandia National Laboratory for providing the test structure and guidance of this research.

References

1. Energy, U.S. Department of. "20% Wind Energy by 2030: Increasing Wind Energy's Contribution to U.S. Electricity Supply." July 2008. Wind & Hydropower Technologies Program. <http://www1.eere.energy.gov/windandhydro/pdfs/41869.pdf>.

2. Hahn, Berthold, Michael Durstewitz, Kurt Rohrig. "Reliability of Wind Turbines." Joachim Peinke, Peter Schaumann and Stephan Barth. Wind Energy. Springer Berlin Heidelberg, 2007.

3. Ciang, Chia Chen, Jung-Ryul Lee, and Hyung-Joon Bang. "Structural health monitoring for a wind turbine system: a review of damage detection methods." Measurement Science and Technology (2008).

4. Mascarenas, D.L., Park, G., Farinholt, K.M., Todd, M.D., Farrar, C.R., 2009, "A Low-Power Wireless Sensing Device for Remote Inspection of Bolted Joints," *Journal of Aerospace Engineering, Part G of the Proceedings of the Institution of Mechanical Engineers*, Vol.233, No. 5, pp. 565-575.

5. Operation and Maintenance Costs for Wind Turbines. 12 May 2003. <http://www.windpower.org/EN/tour/econ/oandm.htm>.

6. Park, Gyuhae, Hoon Sohn, Charles R. Farrar, and Daniel J. Inman. "Overview of Piezoelectric Impedance-Based Health Monitoring and Path Forward." The Shock and Vibration Digest (2003): 451-463.

7. Welch, Greg and Gary Bishop. "An Introduction to the Kalman Filter." 24 July 2006. The Kalman Filter. <http://www.cs.unc.edu/~welch/media/pdf/kalman_intro.pdf>.

Proceedings of the IMAC-XXVIII
February 1–4, 2010, Jacksonville, Florida USA

Optical Non-contacting Vibration Measurement of Rotating Turbine Blades II

Chris Warren, Chris Niezrecki, Peter Avitabile
Structural Dynamics and Acoustic Systems Laboratory
University of Massachusetts Lowell
One University Avenue
Lowell, Massachusetts 01854

ABSTRACT

Identifying the structural dynamics of rotating components can be difficult. Often times, structural dynamic measurements are obtained while the structure is in a static configuration. There are differences that exist in the structural behavior when comparing these statically performed tests and the dynamic characteristics when in operation. In order to evaluate the actual system while in operation, slip-rings are used during testing with measurements made at only a very few selected points. But this slip-ring configuration can be problematic, suffer from measurement noise and the attached sensors can obscure the true dynamic response due to mass loading and aerodynamic effects.

3D digital image correlation (DIC) has been used to capture the out-of-plane motion on the surface of a small scale rotating fan blade. This work extends prior efforts, by quantifying the performance of the optical measurement on a 46 in (1.17m) diameter, rotating wind turbine. The optical measurements are made using DIC (10,000+ measurement points) and dynamic photogrammetry (providing dozens of effective measurement locations). The motion of the turbine as measured using DIC, photogrammetry and accelerometers is compared at several discrete points. The proposed measuring approaches via DIC and dynamic photogrammetry enable full-field dynamic measurement and monitoring of rotating structures in operation.

INTRODUCTION

Although stereophotogrammetry has been used for many years in the field of solid mechanics to measure displacement and strain, only very recently has the technique been exploited for dynamic applications to measure vibration. The DIC technique is able to measure the non-contacting static and dynamic 3D motion of virtually any surface. Prior to testing, the photogrammetric principles of triangulation and bundle adjustment [1] are used in the determination of a camera pair's relative position and to account for the internal distortion parameters of each lens. The calibration process is essentially a ray-tracing process to find unique intersection points, similar to how a GPS system triangulates coordinates. A speckle pattern is applied to a test object prior to imaging for DIC while dynamic photogrammetry tracks high-contrast, circular targets. Examples of each can be seen in Figure 1.

After calibration occurs, a series of image pairs are taken during the course of an experiment. The first pair of images is set as the reference stage to which all subsequent stages are compared. The images are divided into overlapping facets (or subsets), typically 10-20 pixels square. For correlation to work, the corners of each facet must be matched within the surrounding "fingerprint" of light intensity values. This is why a speckle pattern is applied to the object prior to imaging. Software is able to recognize and track a specific point on a series of images through a correlation process which is well documented for two-dimensional [2, 3] and three-dimensional [4, 5] measurements. By tracking discrete points in images taken by a stereo pair of cameras and applying photogrammetric principles, shape and strain can be measured. Subsequently, displacement, velocity, and acceleration can be computed. For cameras with 1280 x 1024 pixels, the overall accuracy of the system can be conservatively stated as 1/30,000 the field of view [6]. Multiple image sets provide a progressive measurement of structural deformation and strain.

The method is extremely robust and has wide dynamic range that is not affected by rigid body motions, ambient vibrations, etc. As long as non-blurred images can be captured, 3D coordinates, displacements, and strains can be measured on virtually any surface. Because the technique is non-contacting, it is possible to measure the full-field vibration of a rotating object without significant mass loading or the use of slip rings.

EXPERIMENTAL RESULTS

The test article used was a 46 inch (1.17m) diameter Southwest Windpower Airbreeze wind turbine, mounted to the shaft of a commercial electromechanical fan motor. During non-rotational testing, traditional modal tests and forced normal mode testing were performed using accelerometers and a laser Doppler vibrometer as references for the performance of the optical measurement techniques. Rotating tests were also performed using accelerometers mounted at the tips of the three blades and monitored via a slip ring.

Figure 1. Example of prepared measurement surfaces: speckling for DIC and circular retroreflective targets used for dynamic photogrammetry.

Stationary Testing

Initially, a three-shaker multiple input, multiple output (MIMO) modal test was performed. A laser Doppler vibrometer was used to measure the response at roughly 18 points per blade. Three shakers were used to more evenly distribute the input energy to all three blades of the turbine. Also, using one shaker on each blade allowed for a more appropriate set up where the shaker-stinger interaction with the structure was minimized [7]. The shaker mounting points are indicated by the three large, circular red dots shown in Figure 2; the measurement points are specified by small green squares. Once frequency response functions (FRFs) were calculated, a frequency domain polynomial curvefitter was used to extract modal parameters and mode shapes.

Subsequently, forced normal mode testing (FNMT) was performed to excite the structure such that it exhibited single degree of freedom behavior at the first two modes [8]. The full-field mode shapes for each frequency were obtained by using the data extracted via DIC measurements. The displacement and phase were then used as feedback to appropriate the input forces more accurately and drive the turbine in a forced normal mode. Figure 3 displays an overlay of the extracted MIMO mode shapes (wire-frame) and the shapes measured using the DIC system (ARAMIS™) [9]. In these tests, the motion of approximately 6,000 data points per blade was tracked, providing a high-resolution depiction of how the rotor was moving. After several iterations of force appropriation adjustment, the normal mode shapes were produced. For modes 1 (16.8 Hz) and 2 (17.2 Hz), MAC values of 99.3% between the laser vibrometer measured mode shape and the DIC measured mode shape were obtained using forced normal mode testing.

Figure 2. Full view of the wind turbine with measurement locations.

Rotational Testing

Past work has shown that dynamic photogrammetry can be used to measure the motion of large rotating wind turbines while in operation using relatively high speed cameras (measured at 20~30 times the rotational frequency) [10]. For tests herein, a camera pair operating at 500 frames per second (FPS) was used to monitor the rotor as it was being driven by the electric motor. A separate data acquisition system was synchronized with the optical system and measured the accelerometer outputs via a slip ring at 10 kHz (20 times the camera frame rate). The accelerometers were placed at the tips of the 3 blades, indicated by red arrows in Figure 2. The frequency of rotation was set arbitrarily to approximately 10 Hz.

Figure 4 depicts a typical output from the dynamic photogrammetry software (PONTOS™) used [10]. 3D motion of any visible target can be tracked from stage to stage; the plot shows the position in the x-direction as a function of time. Each oscillation is 120° out of phase with the other two, and the amplitudes are all nominally equal, as expected. Additionally, motion in the y- and z-directions can be tracked. The z-axis is defined by the vector about which the rotor turns. Figure 5 displays the time response at one of the blade tips in the z-direction (out-of-plane) measured by the optical system. The synchronized accelerometer measurement is displayed in Figure 6.

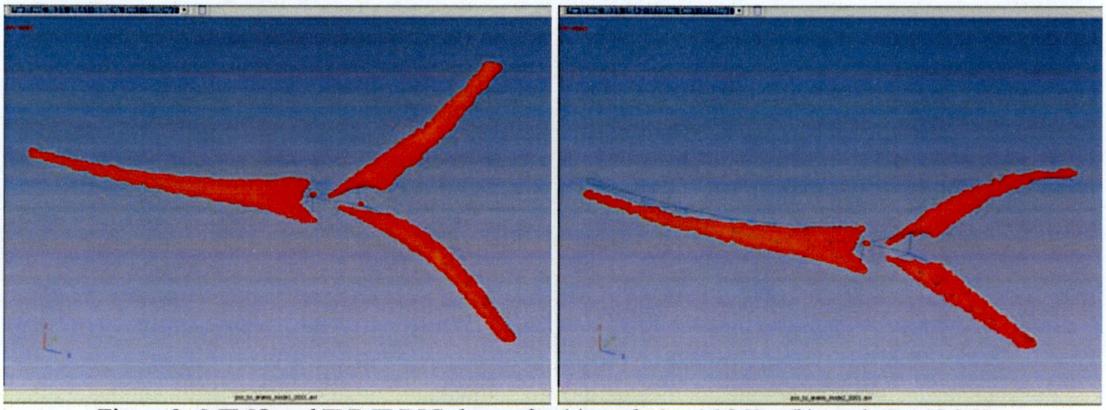

Figure 3. MIMO and FNMT DIC shapes for (a) mode 1 – 16.8 Hz; (b) mode 2 – 17.2 Hz.

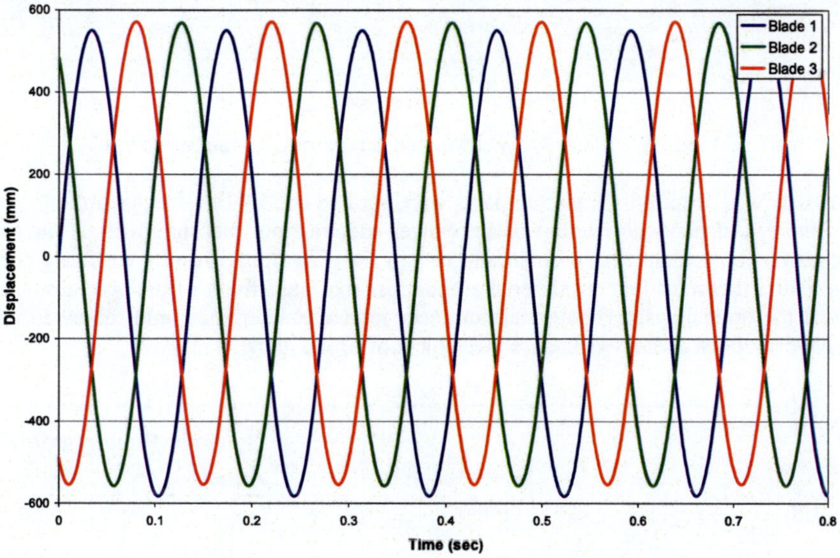

Figure 4. Typical dynamic photogrammetry results, x-displacement.

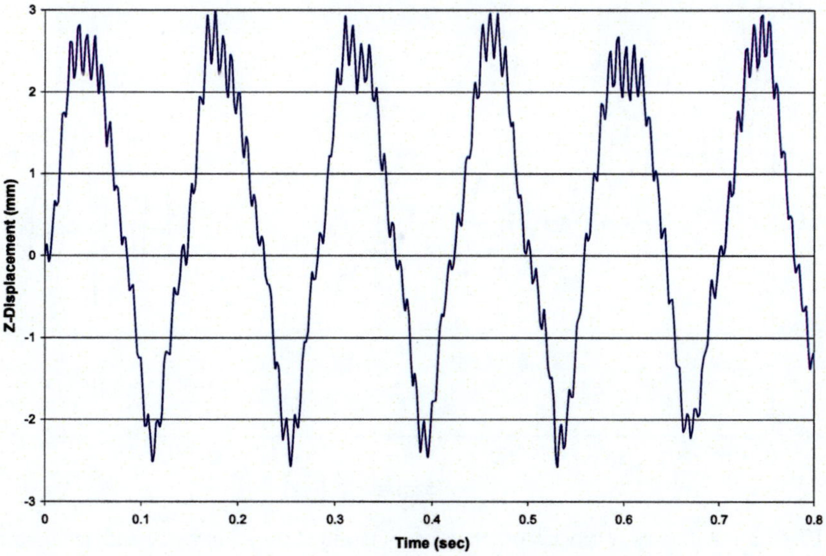

Figure 5. Blade 3 Results, z-displacement.

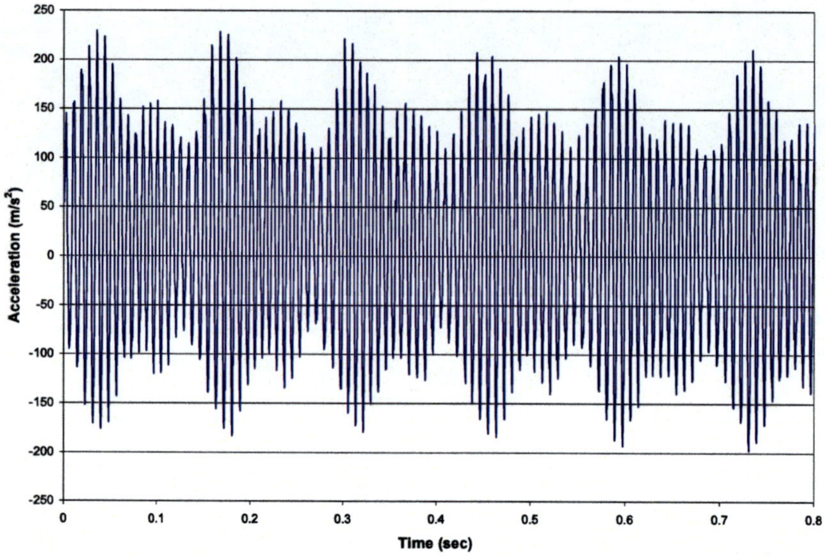

Figure 6. Blade 3, synchronized accelerometer measurement.

Once the time domain data was acquired, flattop windows were applied and FFTs were performed to compare the data in the frequency domain. Figure 7 displays an overlay of autopower spectra from measurements at the tip of one of the blades using dynamic photogrammetry and an accelerometer up to 150 Hz. The initial results are fairly consistent but differences are apparent. Known issues that must be considered are: the cross-axis sensitivity of the accelerometers; noise from the slip ring; the noise floor of the optical system; potential nonlinear joint characteristics; and the useful ranges of both sensors. These discrepancies have not been addressed and are the subject of future work.

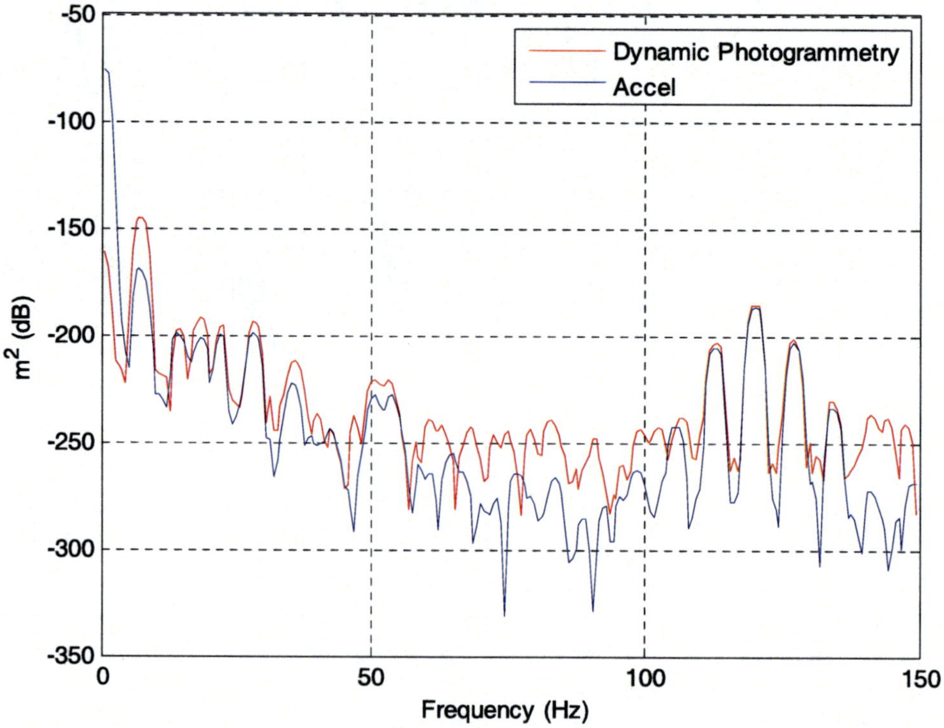

Figure 7. Blade 3 – Autopower spectra overlay for dynamic photogrammetry and accelerometer results.

All the measurements above are shown, in general, for only one particular point on the structure. A significant advantage of PONTOS is that all of the measured points are collected simultaneously which allows for 3D visualization of the motion. This enables a very clear understanding of the structure deformation while rotating. Figures 8 and 9 display one step of the time-varying displacement vector fields for the xy – and z – components, respectively, superimposed on an image of the wind turbine studied. Shown are vectors of specific length scaled to the motion as well as color coded to allow for easy visualization of the data. In Figure 9, one can clearly see that at this point in time, the top and right hand blades are roughly 180° out of phase while the left blade is deformed very little.

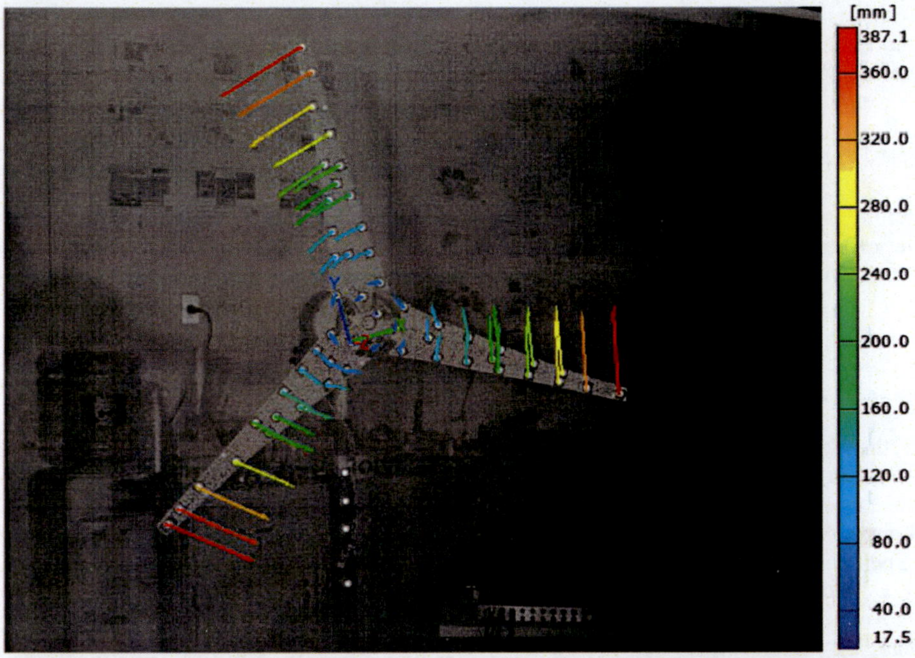

Figure 8. XY-component motion of the rotor shown in color to represent vector length.

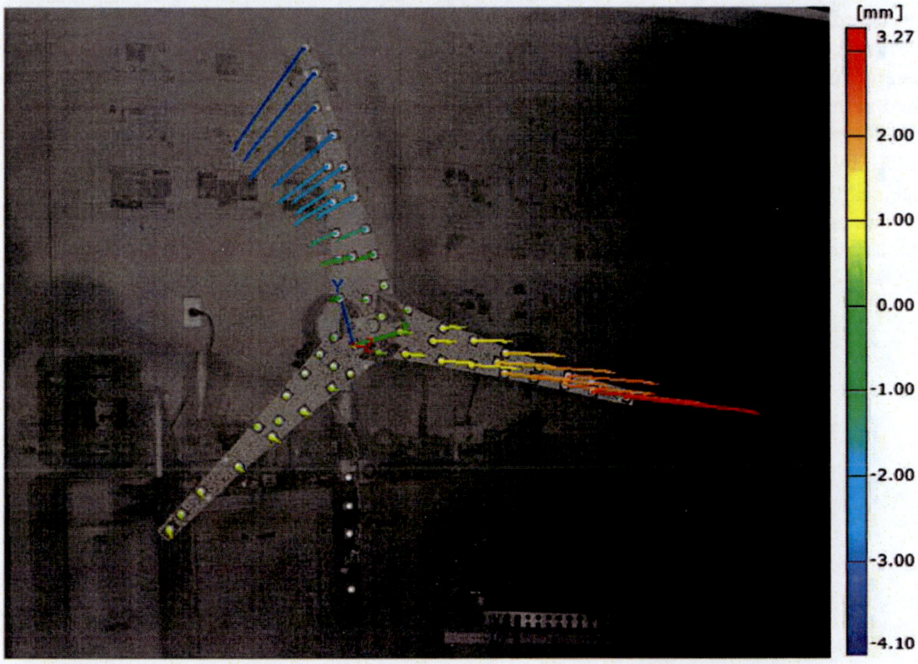

Figure 9. Z-displacement vector field shown in color to represent vector length.

CLOSING REMARKS

Optical based measurements have been shown to be consistent with more traditional measurement techniques such as accelerometers and laser Doppler vibrometers while measuring both rotating and non-rotating structures. When used in conjunction with forced normal mode testing, DIC can provide a high-fidelity measurement of mode shapes at tens of thousands of points. Likewise, dynamic photogrammetry allows for the measurement of the full-field motion of the rotating blade at numerous points, without contact. MAC values greater than 99% were obtained for the first two modes of the wind turbine studied. Dynamic photogrammetry was used to measure the rotor while spinning and yielded results consistent with collocated accelerometers. Slight discrepancies are evident and will be addressed in future work. While these initial results are encouraging, more studies must be completed while working towards understanding the strengths, weaknesses, and useful ranges of these systems. Additional measurements need to be made to validate and confirm these preliminary findings.

ACKNOWLEDGEMENTS

The authors gratefully appreciate the financial support for this work provided by the U.S. Army Research Office Nanomanufacturing of Multifunctional Sensors Ref. Award Number: W911NF-07-2-0081. The authors would also like to thank Tim Schmidt of Trilion Quality Systems for providing insight and providing the necessary equipment to conduct the measurements.

REFERENCES

1. Mikhail, E., Bethel, J., and McGlone, J., Introduction to Modern Photogrammetry, John Wiley and Sons, 2001.
2. Peters, W.H., Ranson, W.F., Sutton, M.A., Chu, T.C., Anderson, J., Application of Digital Image Correlation Methods to Rigid Body Mechanics, Opt. Eng., Vol. 22, No 6, 1983, pp. 738–42.
3. Chu, T.C., Ranson, W.F., and Sutton, M.A., 'Applications of Digital-image-correlation Techniques to Experimental Mechanics, Experimental Mechanics, Vol. 25, No 3, 1985, pp. 232–244.
4. Kahn-Jetter, Z.L., and Chu, T.C., 'Three-dimensional Displacement Measurements Using Digital Image Correlation and Photogrammic Analysis,' Experimental Mechanics, Vol. 30, No 1, 1990, pp. 10–16.
5. Luo, P.F., Chao, Y.J., Sutton M.A., and Peters, W.H., Accurate Measurement of Three-Dimensional Deformations in Deformable and Rigid Bodies using Computer Vision, Experimental Mechanics, Vol. 33, No 2, June 1993.
6. Schmidt, T., Tyson, J., Revilock, D. M., Padula S. II, Pereira, J. M., Melis, M., and Lyle, K., "Performance Verification of 3D Image Correlation using Digital High-Speed Cameras," Proceedings of the SEM Annual Conference and Exposition, Portland, OR, June 7-9, 2005.
7. Warren, Chris and Peter Avitabile. "Effects of Shaker Test Set Up on Measured Natural Frequencies and Mode Shapes." To be published in the Proceedings of IMAC XVIII. February 1-4, 2010. Jacksonville, Florida, USA.
8. LMS Test.Lab – Leuven Measurement Systems, Leuven, Belgium
9. ARAMIS – GOM mbH, Mittelweg 7-8, 38106 Braunschweig, Germany
10. Paulsen, U. S., Erne, O., Moeller, T., Sanow, G., Schmidt, T., "Wind Turbine Operational and Emergency Stop Measurements Using Point Tracking Videogrammetry," *Proceedings of the 2009 SEM Annual Conference and Exposition*, Albuquerque, NM, June 4, 2009.
11. PONTOS – GOM mbH, Mittelweg 7-8, 38106 Braunschweig, Germany

Proceedings of the IMAC-XXVIII
February 1–4, 2010, Jacksonville, Florida USA
©2010 Society for Experimental Mechanics Inc.

Application of a Wireless Sensor Node to Health Monitoring of Operational Wind Turbine Blades

S.G. Taylor[1,2], K.M. Farinholt[1], G. Park[1], C.R. Farrar[1], M.D. Todd[2]

[1] The Engineering Institute, Los Alamos National Laboratory, NM 87545
[2] Dept. of Structural Engineering, University of California, San Diego, CA 92093

NOMENCLATURE

$x(n)$	*discrete time signal*	$\bar{x}$	*feature vector*
a_i	i^{th} *AR coefficient*	$\hat{\mu}$	*sample mean*
ϵ_n	*error signal*	$\{S\}$	*sample covariance matrix*
d	*Mahalanobis distance*		

ABSTRACT

Structural health monitoring (SHM) is a developing field of research with a variety of applications including civil structures, industrial equipment, and energy infrastructure. An SHM system requires an integrated process of sensing, data interrogation and statistical assessment. The first and most important stage of any SHM system is the sensing system, which is traditionally composed of transducers and data acquisition hardware. However, such hardware is often heavy, bulky, and difficult to install *in situ*. Furthermore, physical access to the structure being monitored may be limited or restricted, as is the case for rotating wind turbine blades or unmanned aerial vehicles, requiring wireless transmission of sensor readings. This study applies a previously developed compact wireless sensor node to structural health monitoring of rotating small-scale wind turbine blades. The compact sensor node collects low-frequency structural vibration measurements to estimate natural frequencies and operational deflection shapes. The sensor node also has the capability to perform high-frequency impedance measurements to detect changes in local material properties or other physical characteristics. Operational measurements were collected using the wireless sensing system for both healthy and damaged blade conditions. Damage sensitive features were extracted from the collected data, and those features were used to classify the structural condition as healthy or damaged.

INTRODUCTION

Structural health monitoring (SHM) is the process of detecting damage in structures. The goal of SHM is to improve the safety and reliability of aerospace, civil, and mechanical infrastructure by detecting damage before it reaches a critical state. To achieve this goal, technology is being developed to replace qualitative visual inspection and time-based maintenance procedures with more quantifiable and automated damage assessment processes. These processes are implemented using both hardware and software with the intent of achieving more cost-effective condition-based maintenance. A more detailed general discussion of SHM can be found in (1).

Wind turbines represent a significant investment in the energy production infrastructure of the United States and countries around the world. In the U.S., a recent Department of Energy report examined the issues surrounding the production of 20% of the nation's electricity from wind energy by the year 2030(2). In order to maintain the sheer number of operating wind turbines required to meet that goal, automated and continuous structural health monitoring systems must become a reality. A typical horizontal-axis wind turbine sits atop a tower and has three rotor blades connected to a hub, which drives a low-speed shaft connected to a gearbox. The output of the gearbox drives a generator, which produces the desired electricity. Gearboxes in installed wind turbines seem to have unreasonably high failure rates (3) and are particularly expensive to repair, both in terms of incidental cost and lost productivity.

T. Proulx (ed.), *Structural Dynamics and Renewable Energy, Volume 1*, Conference Proceedings of the Society for Experimental Mechanics Series 10, DOI 10.1007/978-1-4419-9716-6_5, © The Society for Experimental Mechanics, Inc. 2011

Although blade replacement is less common and less expensive than gearbox repair (3), it must be the case that the loads causing gearbox failures are transmitted from the wind first through the blades. It may be the case that sophisticated monitoring of the blades' loading and structural condition could prevent all manner of woes for wind turbine maintenance. This study applies a compact wireless data acquisition system to health monitoring of wind turbine blades, with the ultimate goal of improving the overall ability to assess wind turbine health. Although large-scale wind turbines are actually equipped with slip rings, providing power and communications from the tower into the hub, wireless communication systems provide notable advantages. These advantages include fewer additional complications to the blade manufacturing process, no additional maintenance issues caused by failed wires and cabling, and preserving the insulation of the turbine hub from lightning strikes at the blade tip. This study takes a significant step toward understanding overall wind turbine health by monitoring the start of the load path and vehicle for introduction of debilitating damage: the blade.

METHODOLOGY

Any damage detection or classification scheme requires an integrated process of sensing, data interrogation and statistical assessment. Inherent in the statistical assessment process is some comparison to a baseline condition; in order to determine that a given structure or system is damaged, there must be some knowledge concerning its behavior when operating in a healthy condition. This study presents experimental data from wind turbine blades tested in both healthy and damaged conditions. Features were extracted from these collected data, and those features were classified to determine both the condition of the turbine blade.

Two methods of feature extraction were implemented in this study. The first was based on conventional spectral estimation, whereby the power spectral density (PSD) of the measured acceleration data was estimated, and the location of the first resonance was identified using a simple peak-picking algorithm. The state of the wind turbine blade was classified using scalar thresholding on the location of the first resonance. The threshold value was determined, depending on the test, using either visual inspection or a decision rule based on separating two normally distributed random variables. A more detailed explanation of separating random variables can be found in (4).

The second method was based on fitting an auto-regressive (AR) model to the measured acceleration data. An AR model determines the coefficients a_i such that, for a signal $x(n)$,

$$x(n+1) = \sum_{i=n-p}^{n} a_i x(i) + \epsilon_n, \tag{1}$$

where ϵ_n is a Gaussian distributed random variable for a well-fit model. The vector of AR parameters was then used to compute the square Mahalanobis distance (5) from the mean of the set of healthy training data. The square Mahalanobis distance is given by

$$d^2 = [\bar{x} - \hat{\mu}]^T \{S\}^{-1} [\bar{x} - \hat{\mu}], \tag{2}$$

where $\bar{x}$ is the feature vector of interest, $\hat{\mu}$ is the sample mean of the training set, and $\{S\}$ is the sample covariance matrix of the training set. Using the Mahalanobis distance requires an assumption that the underlying data are Gaussian, which may be reasonable if the data are collected without deterministic changes in the experimental apparatus.

The state of the wind turbine blade was then classified by using a one-dimensional clustering algorithm using the square Mahalanobis distance as a scalar feature. The clustering method attempted to minimize the sum of the variances of each cluster group. Given n cluster groups, and denoting as C_j the set of points belonging to the jth cluster, the objective function to be minimized is

$$f(C) = \sum_{j=1}^{n} \sum_{i \in C_j} \left\| \bar{x}_i - \hat{\mu}_j \right\|^2, \tag{3}$$

where C defines all cluster sets, x_i is a cluster member, and $\hat{\mu}_j$ is a cluster mean. A more detailed treatment of cluster analysis can be found in (6).

EXPERIMENTAL SETUP

Acquisition Hardware

Time domain data were collected from the rotating blade using the WiDAQ, or wireless data acquisition system, which was developed previously by the authors (7). The WiDAQ was designed as an expansion to the wireless impedance device (WID3), intended to extend its capabilities beyond high-frequency impedance measurements to collect low-frequency data, such as structural vibration data from accelerometers. The WiDAQ and its major components are shown in Figure 1. The WiDAQ is controlled by an ATmega1281v μCu, but it lacks its own radio. The WiDAQ can function as a stand-alone device using wired communication, but it is primarily intended to be used in combination with the WID3, which provides the wireless communication capability. The WID3 and the WiDAQ have the ability to share resources, such as processing power, data storage, and access to peripheral devices.

Figure 1. Major components of the WiDAQ

The primary functions of the WiDAQ are data acquisition and signal generation, the key components for which are the Analog Devices AD7924 analog-to-digital (A/D) and AD5621 digital-to-analog (D/A) converters. By excluding sensor-specific conditioning circuitry from the WiDAQ, sensor data can be acquired from any transducer that provides a voltage output; sensor-specific conditioning, such as that required for Integrated Circuit Piezoelectric (ICP) accelerometers, must be included on a third module. The four-channel AD7924 has a 12-bit resolution over a range from 0 to 2.5 V, and it would consume a maximum of 6 mW while sampling at one million times per second. However, the combined system with the ATmega1281v has a maximum useful sampling frequency of 40 kHz. After each measurement sequence, the recorded data can be stored in on-board flash memory for later retrieval. With these capabilities, a network of sensor nodes could wake on schedule, acquire data, and return to sleep. At a later time, a data mule, such as an unmanned aerial vehicle (8), could approach the sensor network and request it to transmit the recorded data.

The sensors used in this study were PCB Piezotronics model 352A24 ICP accelerometers with a nominal sensitivity of 100 mV/g. A separate module, the WiDAQ ICP, provided the excitation power required for the ICP accelerometers, low-pass filtered the signals to prevent aliasing during A/D conversion, and adjusted the output voltage to lie within the 0 to 2.5 V input range of the AD7924. In order to maintain low power consumption, no additional amplification was implemented prior to A/D conversion; however, this lack of amplification resulted in underuse of the A/D converter's dynamic range for low-excitation test conditions. The WiDAQ ICP and the three modules assembled together are shown in Figure 2.

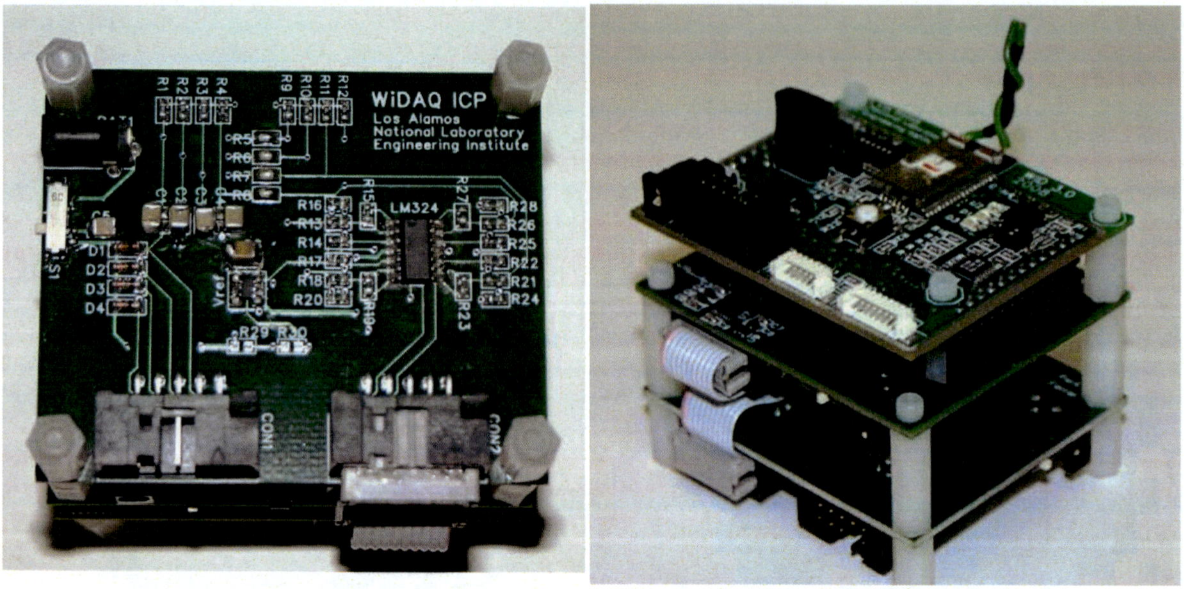

Figure 2. WiDAQ ICP (left) and assembled modules for vibration data acquisition (right)

Test Article and Instrumentation

The test article for this paper was the Whisper 100, manufactured by Southwest Windpower (9), which has a tip-to-tip diameter of 2.1 m and a rated power output of 900 W under 12.5 m/s (28 mph) sustained winds. Four wind turbine blades were instrumented in turn with three accelerometers each, located at 69, 38, and 1 cm from the blade tip. The test setup is shown in Figure 3. Three blades were tested in an undamaged condition, and the fourth was subjected to two successive levels of damage. For each test condition, five replicates were collected using a sampling rate of 504 Hz, and each record contained 2048 points. Each of the three undamaged blades was tested at both 20 and 26 °C, for a total of 30 replicates in the baseline condition.

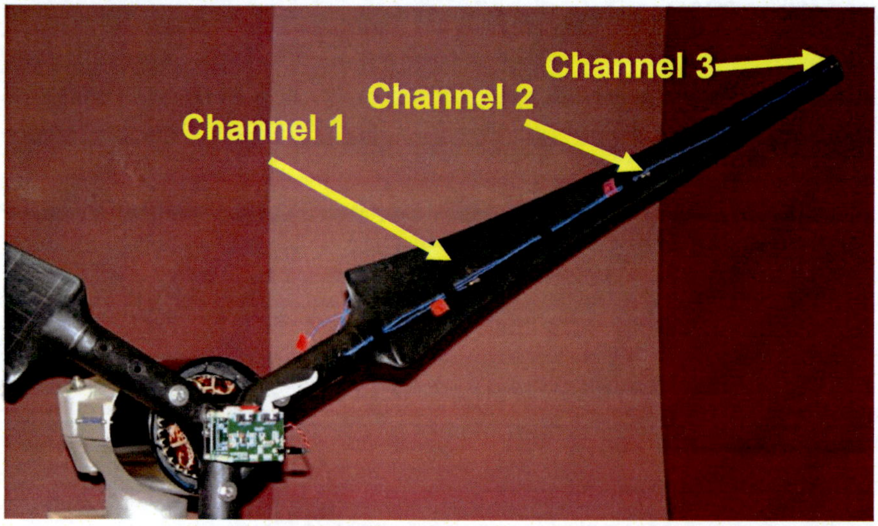

Figure 3. Whisper 100 instrumentation and test setup

Damage

Damage was inflicted by progressively cutting the wind turbine blade perpendicular to its longitudinal direction at a location 70 cm from the blade tip. Two damage cases were tested: in the first damage case, the cut extended 40% of the width of the blade; in the second case, the cut extended 60% of the width of the blade. A photograph of the blade in the second damage condition is shown in Figure 4.

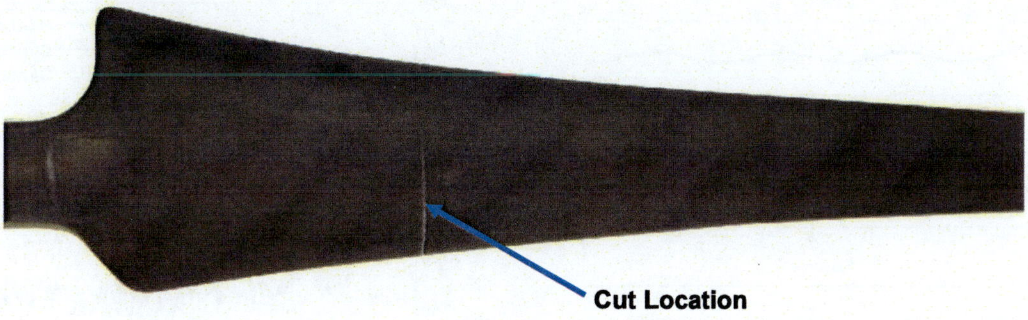

Figure 4. Blade damage case 2

DAMAGE DETECTION METHODOLOGY

Resonant Peaks

All tests in this study were conducted in an output only sense; there was no measurement of input force or initial conditions. Two types of tests were conducted: a free vibration, or "pluck" test, and an ambient excitation, or "free-spin" test. Of the two, only the ambient excitation test would have useful applicability to an operating wind turbine, but the results of the pluck test are informative. To perform the pluck test, the blade was given an initial displacement and subsequently released. This type of test provided a much higher signal-to-noise ratio (SNR) than the ambient excitation test. Results of a typical pluck test are shown in Figure 5. The time histories are shown for each sensor in the left plot of Figure 5, and the PSD estimates are shown on the right. Note that the SNR for sensor 1 is quite low; its response for the first resonance is about 50 dB lower than that for sensor 3, and very little of the A/D converter's 0 to 2.5 V dynamic range was able to be utilized.

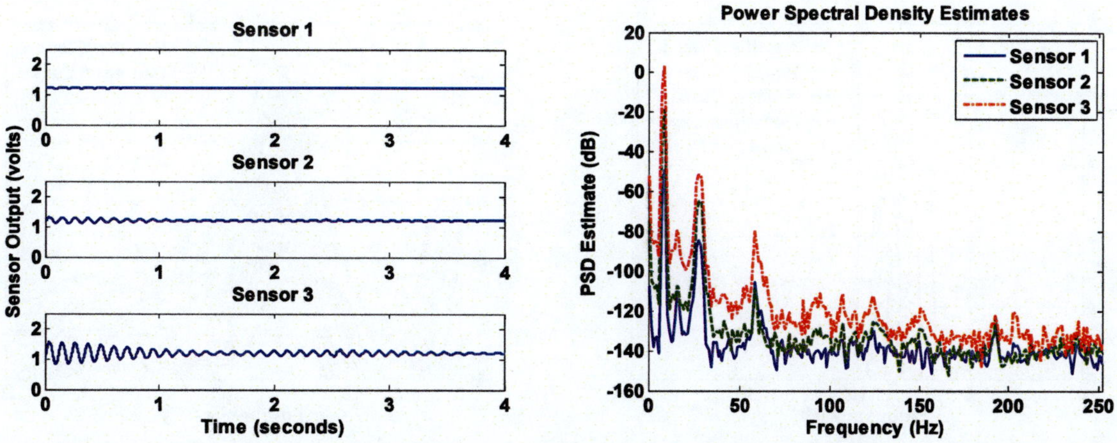

Figure 5. Time histories (left) and power spectral density estimates (right) for a pluck test

To perform the ambient excitation test, the wind turbine was spun by hand and allowed to free-spin while the data were collected. This excitation is similar to the type of excitation the wind turbine would undergo during normal operating conditions. Results of a typical ambient excitation test are shown in Figure 6. The time histories are shown for each sensor in the left plot of Figure 6, and the PSD estimates are shown on the right. Note that only a small portion of the 0 to 2.5 V dynamic range could be utilized by any sensor; the resulting power of the first resonance, visible in Figure 6 (right) is 40 dB lower than that for the corresponding sensor in the pluck test, shown in Figure 5 (right). The low frequency oscillation visible in Figure 6 (left) is likely a low-frequency structural mode related to the turbine's rotation.

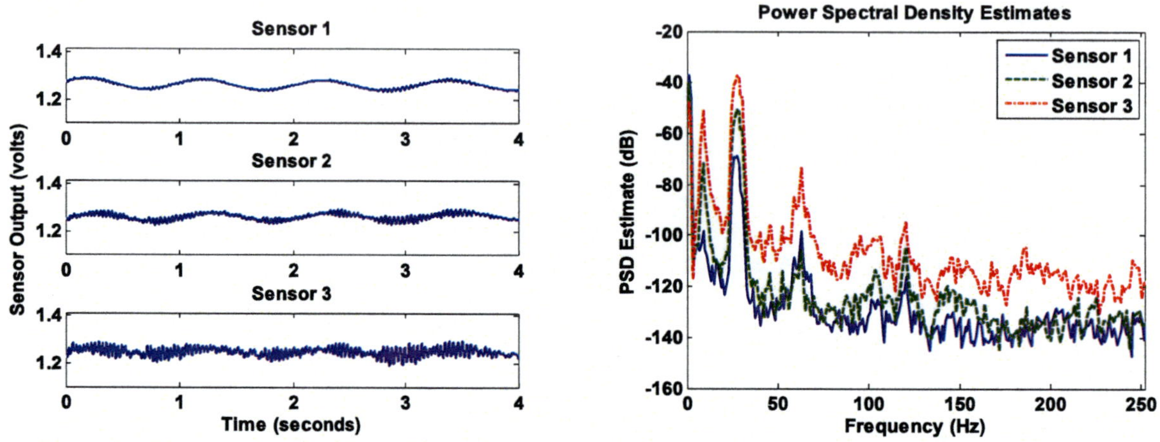

Figure 6. Time histories (left) and power spectral density estimates (right) for a free-spin test

Feature Extraction

The most basic method of structural health monitoring might be observing shifts in structural resonances. Using the pluck test data, shifts in the first resonant peak can be visibly identified for each damage case. A plot of the first resonant peak for the three healthy blades and the two damage cases is shown in Figure 7 (left). As expected for a decrease in structural stiffness, the first structural resonance decreases for each successive damage case. Under theses excitation conditions, the damage can be detected quite easily. However, for the test data obtained under ambient excitation conditions, the same damage is much more difficult to detect. For 30 healthy cases and 10 damaged cases, the first resonant peaks are shown in Figure 7 (right). Although there is a generally visible trend for the damaged cases to show a decrease in the frequency of the first resonance, the low SNR for the free-spin test introduces much greater variability than was apparent in the pluck test.

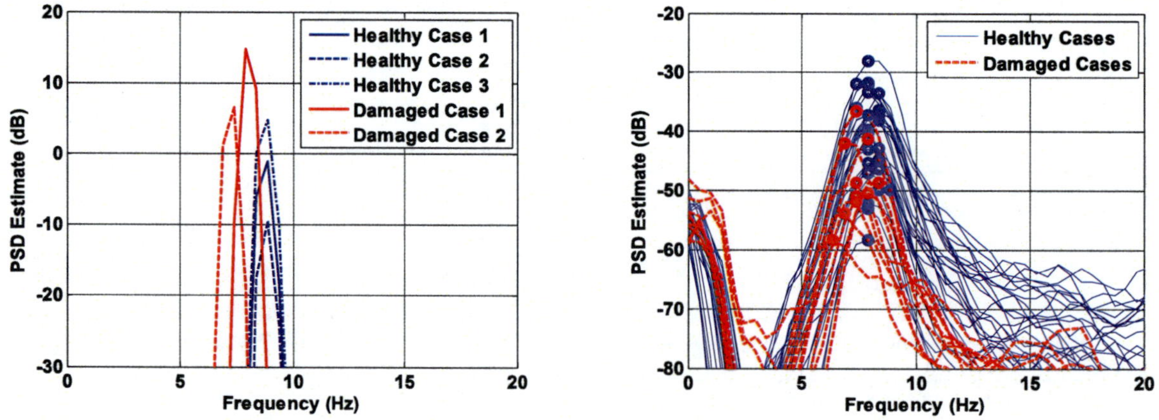

Figure 7. First resonant peaks for the healthy and damaged cases: pluck test (left) and run test (right)

Damage Classification

In the case of the pluck test, the peaks in Figure 7 (left) are coincident for the three healthy cases, and they decrease for the two successive levels of damage. However, for the free-spin test, the peaks shown in Figure 7 (right) are not in easily separable clusters. If one were to assume that the data for both the healthy and damaged conditions were normally distributed, and that training data in both the healthy and the damaged configuration were available, the structural condition could be classified by setting a threshold at the intersection of the probability density functions describing the damaged and healthy data, respectively (4). This method carries the significant disadvantage that training data from the damaged structure must be available in order to classify future measurements. This method also requires that the expected damage be of a predictable and consistent form, rather than simply being a measureable deviation from healthy behavior. Figure 8 shows the results of such a classification scheme. Using this method, two healthy cases would be incorrectly identified as damaged, while three damaged cases would have been missed, having been misclassified as healthy.

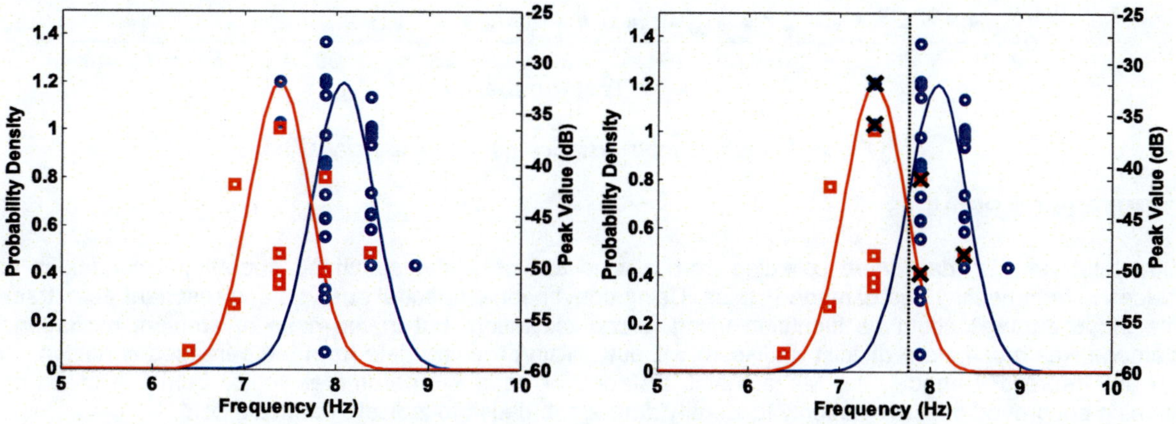

Figure 8. Classification of structural condition using known damaged test cases as training data

Autoregressive Model

Feature Extraction

As an alternative to conventional spectral analysis of structural vibration data, a linear auto-regressive (AR) model of order 14 was determined for each of the 30 healthy test cases and the 10 damaged test cases. Of the 30 healthy test cases, 25 were chosen to train the damage detection algorithm, with the first of each set of five replicate tests was in a control group. Supposing that each of the 14-element vectors in the healthy training set were sampled from the same multidimensional Gaussian distribution, the sample covariance matrix was estimated for the 25 training samples. Using that covariance matrix, the Mahalanobis distance from the mean of the training set was computed for each of the 30 healthy cases and the 10 damaged cases. The Mahalanobis distance was chosen as a damage indicating feature, or damage index, and it is shown plotted versus test number in Figure 9. Visually, the control cases appear to be correctly identified as belonging to the healthy test group, with the exception of test 16. Furthermore, the damaged test cases appear easily distinguishable from the healthy ones, and the two distinct levels of damage are distinct from one another.

Damage Classification

The state of the wind turbine blade was classified using a simple clustering algorithm, which minimized the sum of the variances of each cluster. The algorithm does not require training data taken from the structure in a damaged condition; it only requires that the features extracted from each structural condition be separable. Partitioning the damage indices into three groups, the clustering algorithm separated the two levels of damage from one another as well as from the baseline data. The clusters separating the two damage levels are indicated in Figure 9; the expected false indication of damage occurred at test number 16.

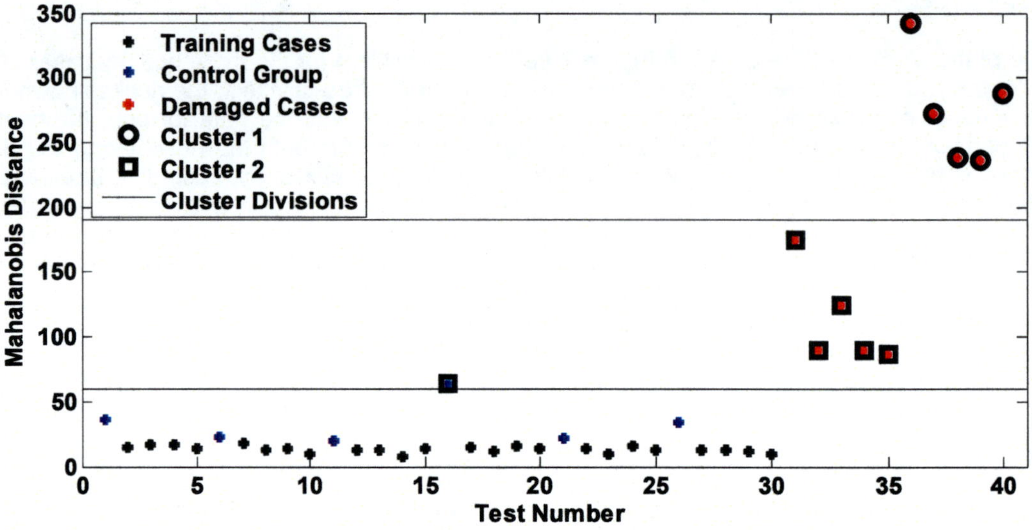

Figure 9. Mahalanobis distance as a damage classifier

SUMMARY OF RESULTS

Structural vibration data were collected from a wind turbine under simulated operational conditions using blades in both healthy and damaged states. Using conventional spectral analysis and resonant peak tracking, the blade damage could be identified using forced excitation, but in the case of ambient excitation, the damage was much more difficult to discern without utilizing training data from the damaged condition. Using an auto-regressive model, the Mahalanobis distance of each AR parameter vector from the mean of the training set proved effective not only to identify damage, but also to indicate relative extent.

FUTURE WORK

This study will be combined with other work utilizing active sensing methods, whereby the wind turbine blade will be actively excited and interrogated using a wireless low-power platform similar to that used for the current study. Using this two-pronged sensing approach, the application will be extended to larger wind turbine blades with composite construction. In addition the ambient temperature, other operational and environmental variabilities will be introduced. After establishing the ability to detect damage in such cases, more realistic damage conditions such as fatigue cracking or delamination will be considered.

REFERENCES

1. *An Overview of Intelligent Fault Detection in Systems and Structures.* **Worden, K. and Dulieu-Barton, J.M.** 1, 2004, International Journal of Structural Health Monitoring, Vol. 3, pp. 85-98.

2. *20% Wind Energy by 2030: Increasing Wind Energy's Contribution to U.S. Electricity Supply.* Department of Energy, United States of America. 2008.

3. *Wind Energy's New Role in Supplying the World's Energy: What Role will Structural Health Monitoring Play?* **Butterfield, S., Sheng, S. and Oyague, F.** Stanford, CA : s.n., 2009. 7th International Workshop on Structural Health Monitoring.

4. **Kay, Steven M.** *Fundamentals of Statistical Signal Processing.* Englewood Cliffs, N.J. : Prentice-Hall, 1993. Vol. 2.

5. *Damage Detection using Outlier Analysis.* **Worden, K. and Manson, G.** 3, 2000, Journal of Sound and Vibration, Vol. 229, pp. 647-667.

6. **Spath, H.** *Cluster Dissection and Analysis.* [trans.] J. Goldschmidt. New York : Halsted Press, 1985.

7. *Impedance-based Wireless Sensor Node for SHM, Sensor Diagnostics, and Low-frequency Vibration Data Acquisition.* **Taylor, Stuart G., et al.** Stanford, CA : s.n., 2009. 7th International Workshop on Structural Health Monitoring.

8. *A Different Approach to Sensor Networking for SHM: Remote Powering and Interrogation with Unmanned Arial Vehicles.* **Todd, M.D., et al.** Stanford, CA : s.n., 2007. 6th International Workshop on Structural Health Monitoring.

9. Southwest Windpower. [Online] http://www.windenergy.com.

Proceedings of the IMAC-XXVIII
February 1–4, 2010, Jacksonville, Florida USA
©2010 Society for Experimental Mechanics Inc.

Structural Damage Identification in Wind Turbine Blades Using Piezoelectric Active Sensing

Abraham Light-Marquez[1], Alexandra Sobin[2], Gyuhae Park[3], Kevin Farinholt[3]

[1] Dept. of Mechanical Engineering, New Mexico Institute of Mining and Technology
[2] Dept. of Aeronautic Engineering, Daniel Webster College
[3] The Engineering Institute, Los Alamos National Laboratory, Los Alamos, NM 87545

Abstract

This paper presents a variety of structural health monitoring (SHM) techniques, based on the use of piezoelectric active-sensors, used to determine the structural integrity of wind turbine blades. Lamb wave propagation, frequency response functions, and a time series based method are utilized to analyze the wind turbine blade. For these experiments, a 1m section of a 9m CX100 blade is used. Different types of simulated damage are introduced into this structure and a performance matrix is created to compare the validity and functionality of each technique. Overall, these three methods yielded sufficient damage detection to warrant further investigation into field deployment. Time series analysis shows the most effective and reliable to detect damage while lamb wave testing could locate, as well as detect the onset of damage. This paper summarizes considerations needed to design such SHM systems, experimental procedures and results, and additional issues that can be used as guidelines for future investigations.

Introduction

In past years, wind turbine technology has become a topic of extensive investigation. The DOE projects that 20% of the US electrical supply could be produced via wind power by 2030 [1]. In order to reach this goal, an increase in the efficiency of wind power generation will be required. Therefore, a nondestructive, structural health monitoring (SHM) technique that can be used in real-time during operation could be very useful in the wind turbine industry [2]. These structures are not only immensely difficult to install due to their large size but are normally placed in remote locations. It would be cost-effective to design and implement an SHM system which would enable the operator to monitor these structures from an offsite location, while safely maintaining the system. An important component of the wind turbine system to monitor is the blade because it contributes 15 – 20% to the total cost, is one of the most expensive components to repair, and can create secondary damage to other components due to rotational imbalance [3].

The goal of this study is to assess the advantages and disadvantages of using high-frequency SHM techniques, including lamb wave, frequency response function, and time series based measurements as a way to nondestructively monitor the health of a wind turbine blade with piezoelectric sensors [4-6]. An array of piezoelectric sensors on a 1m section of a 9m CX100 blade are used for the purposes of this research. The assessment will delve into how to provide damage detection effectively with each method. Optimization of issues which arise with items such as testing procedures, data collection and data processing are attempted. Lamb wave propagation uses waves traveling through the thickness of the structure. Similarly, frequency response functions measure the frequency response of the structure given an excitation. On the other hand, time series analysis uses time domain responses from the structure to determine a data-driven model and use that model to determine if the system is damaged. These three techniques measure real-time data at high-frequency ranges (>5 kHz). When a system response is compared against baseline, undamaged, measurements a noticeable variation can imply structural damage.

A performance matrix is then created to look into the three SHM techniques and compares a defined set of damage detection parameters. Damage detection has been defined with three major parameters which will be detailed in the later section: the detection area of an individual sensor, the ability to locate the onset of damage, and the minimum damage size for damage verification. Industry is optimally looking for a technique that can detect the location of minor damage with a small number of sensors within a large area. The wind turbine blade itself can easily be damaged with no signs due to the delamination occurring within the composite material. Moreover, the turbine blade could catastrophically fail, cause an imbalance, and render the system inoperable.

In addition to these parameters, commercialization and implementation concerns (weight, cost, installation, etc.) are addressed after analyzing the three techniques. For instance, it is crucial for complex SHM systems to self-diagnose and interpret errors within sensors. Sensor fracture is the most common type of failure, which can be attributed to the brittle nature of many piezoelectric devices. Another issue that arises is the bonding condition of the PZT sensors to the structure. During use, the adhesive can easily fail which would result in inaccurate data. If there is no self diagnostic tool within the system, then sensor damage would be interpreted as structural damage and the health monitoring system would be ineffective. These concerns will also be part of the performance matrix in order to accurately assess the commercial feasibility of each technique.

2 Experimental Procedure

The basic experimental setup for the comparison of structural health monitoring systems consisted of a 1m section of a CX100 wind turbine blade instrumented with piezoelectric patches, as shown in Figure 1 below. The patches were connected to a National Instruments PXI data acquisition system, shown in Figure 2 below.

2.1 Lamb Wave Propagation

The lamb wave propagation experiment was performed, which utilizes one of the piezoelectric patches as an actuator and another as a sensor. The user interface for this program can be seen in Figure 2 below.

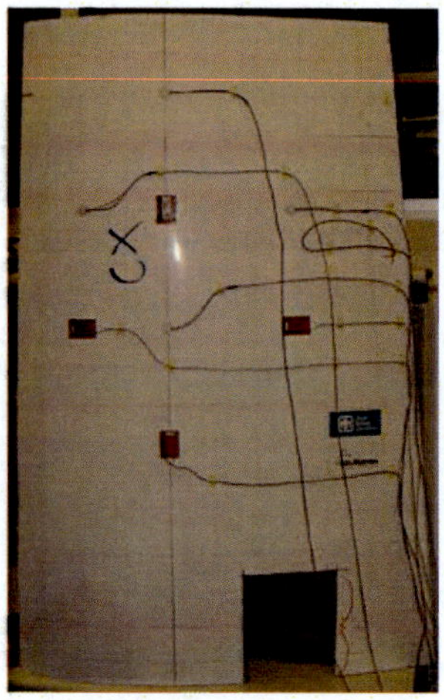

Figure 1: Turbine Blade Section

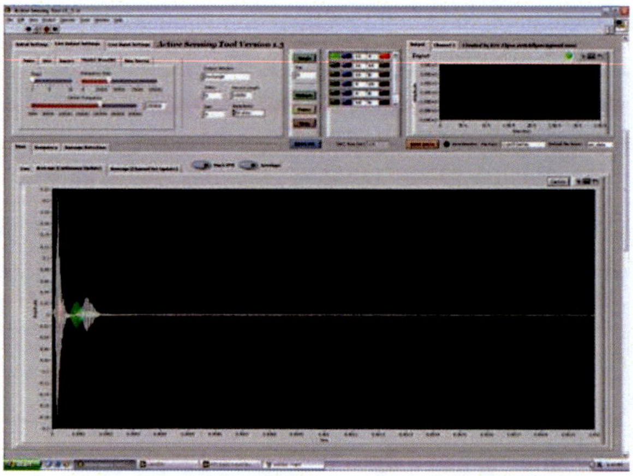

Figure 2: Lamb Wave Testing Interface

The interface allows for the modification of the excitation waveform, both shape and magnitude. For this set of experiments a Morlet Wavelet was selected and the interface provides control over the waveform's center frequency. The tests used four piezoelectric patches which results in six wave propagation paths, as seen in Figure 3 below. The vertical path is located directly on the center support spar of the blade, as seen in the cross section picture in Figure 4.

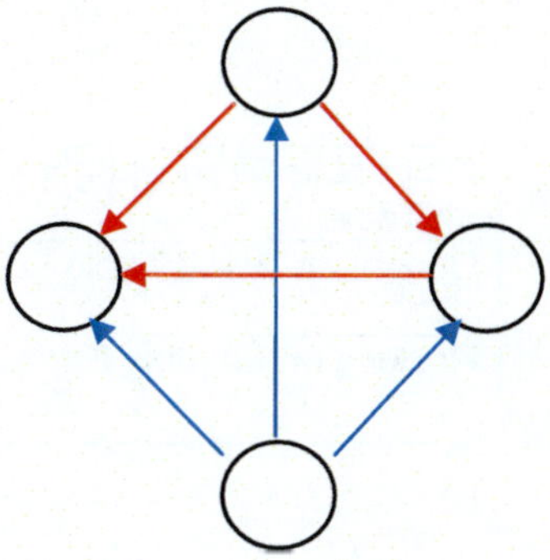

Figure 3: Propagation Directions

Figure 4: Cross Section View of Turbine Blade

2.1.1 Input Waveform Frequency

To maximize the effectiveness of the lamb wave technique the excitation frequency needs to provide a response that is capable of indicating damage. In order for damage detection to be possible the magnitude of the response must be of a sufficient magnitude, and the response must be separated by a sufficient amount from the electromagnetic interference to allow for proper identification of the arrival waveform. Due to the complexity of the blade section, traditional methods of predicting the ideal excitation frequency for homogeneous material are not applicable and the ideal frequency must be determined by experimentation.

The aforementioned setup was used to modify the center frequency of the input waveform and the response was recorded along each of the propagation paths. The response was recorded for a range of frequencies from 15 kHz to 250 kHz. The frequencies which displayed the best response were selected for further testing to identify damage detection capability.

To determine the damage detection capability of each frequency the natural response of the system was used as the baseline. Damage was then simulated by applying a piece of industrial putty to the surface of the turbine blade in the path of propagation of the lamb waves. The putty simulates changes in the damping of the structure in a localized area, similar to the effects of delamination formation. For this set of tests the putty section was approximately a 3.5cm square with 0.5cm thickness. The putty was placed in the path of the propagating waves; this was done for all of the paths. Figure 5 illustrates an example of the putty acting as damage on the structure.

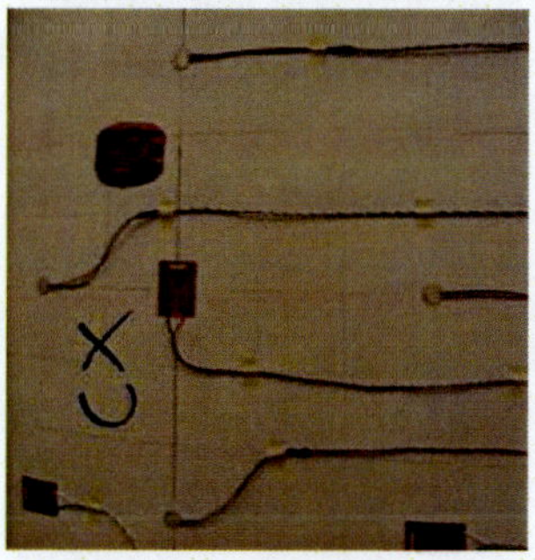

Figure 5: Putty attached to the blade to simulate Damage

2.1.2 Baseline Measurements and Environmental Variations

After the ideal excitation frequency has been determined, multiple baselines for the undamaged structure were recorded. Thus these baselines contain some variations in the boundary conditions to attempt to simulate potential real world variability. Several damaged measurements were also taken during this set of testing to determine the damage detection capabilities of each technique. Table 1 contains a list of environmental variations and damage locations that were tested.

Table 1: Environmental Variations

Upright on Table	On Pink Foam Upright on Table	On White Foam Upright on Table	On Bubble Wrap Upright on Table
Upside-down on Wood Table	Laying on Back on Table	Upside-down on Table	Upright on Wood Table
Leading Edge on Table	Hammering + Upside-down	Rocking Upright	Rocking on Leading Edge

The tests were conducted along all six paths for the frequency(s) of interest.

2.2 Frequency Response Method

The frequency response method was investigated using a National Instrument data acquisition system. The user interface for this program can be seen in Figure 6 below.

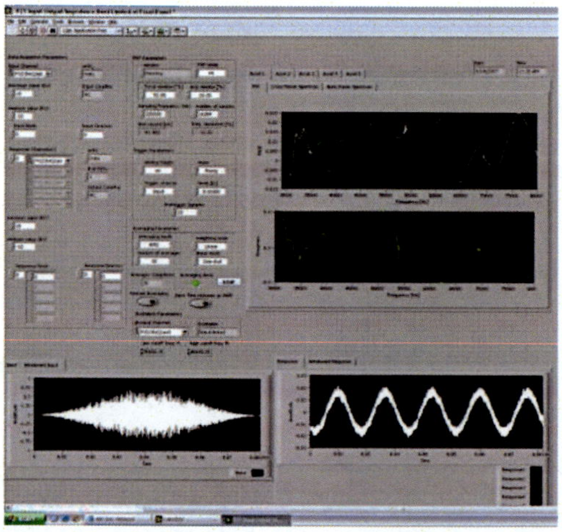

Figure 6: FRF User Interface

The interface uses one of the piezoelectric patches as an actuator and a different one as a sensor. The actuator is connected through an amplifier to the DAQ function generator. The waveform sent to the actuator by the amplifier is recorded by connecting the amp to one of the DAQ's inputs. The sensor is connected directly to one of the DAQ's input channels. The program allows for the manipulation of the excitation waveform, specifically the excitation bandwidth. The program creates the frequency response function (FRF) based on the recorded excitation waveform and sensor output.

Four paths, shown in Figure 7, were experimentally tested and as in the lamb wave propagation method, a variety of boundary conditions was used to determine variations in the baseline data for each path (See Table 1). These data sets were then used to compare with the damaged data, simulated by applying a square piece of industrial putty to the surface of the turbine blade.

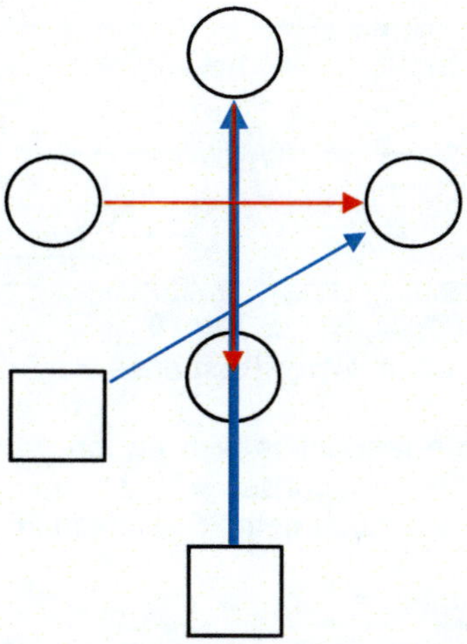

Figure 7: FRF Propagation Directions

2.3 Time Series Analysis

The system used to capture the time domain response of the structure was a Dactron LabVIEW program. The structure was excited with a chirp signal from 5 – 20 kHz over a voltage range of 30 volts. The sampling frequency was 48 kHz, a total of 8192 points data points were recorded. Two paths were analyzed, as seen in Figure 8.

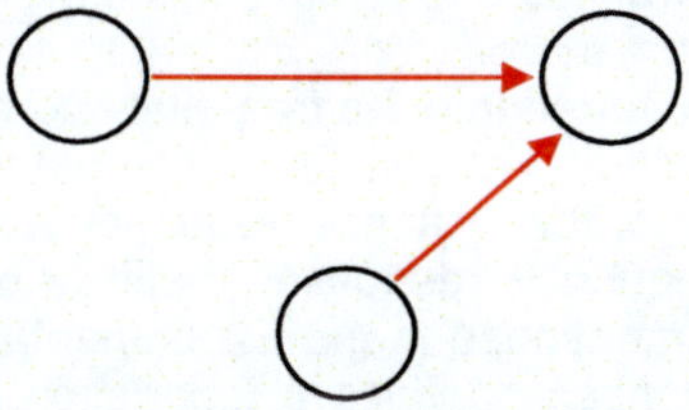

Figure 8: Time Series Propagation Paths

3 Results

The data analysis for this project was conducted using EPDLab, an interactive Python environment.

3.1 Lamb Wave Propagation

Lamb wave testing consisted of two phases. The first was determining the ideal excitation frequency. As previously stated, the ideal excitation frequency could not be determined with traditional methods and had to be determined experimentally with an anisotropic material. The second phase was the collection of multiple baseline cases as well as damage identification testing. The purpose of this peculiarity in the testing was to see if the high frequency source could excite the structure so that even small damage could be detected.

3.1.1 Preliminary Testing Analysis

The preliminary testing consisted of a qualitative analysis on the systems response. The response at a range of frequencies (15-250 kHz) was visually analyzed to determine the frequency which gave the best response for each path (See Figure 3: Propagation Directions). The parameters monitored were the time offset, noise disturbance, and magnitude of the systems response. See Figure 9 for a graphical description of some of these parameters.

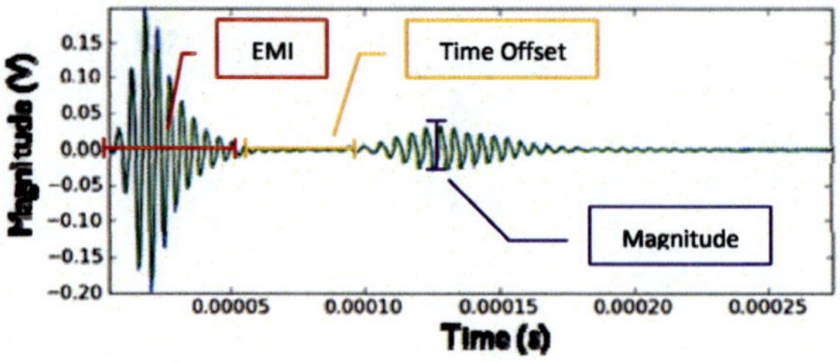

Figure 9: Lamb Wave Response Parameters

Overall, most of the paths showed desirable responses with an input frequency of 25 kHz. Two paths did not follow this trend and gave more advantageous responses with an input frequency of 200 kHz. Therefore, the second phase of lamb wave testing was conducted at both 25 kHz and 200 kHz.

3.1.2 Baseline and Damage Testing Analysis

Once a reasonable frequency for each path was determined from the preliminary qualitative analysis, additional testing was performed on the structure. The testing consisted of collecting baseline and damaged cases. The purpose of this study was to detect damage within the structure. In order to accomplish this task, the data were reduced by transforming time-domain data into the frequency domain, and performed a cross-correlation analysis between a "true" baseline measurement and newly acquired data.

The first step in the process was to reduce the collected data sample. This step was required because the experimental set up yielded an output signal with a large electromagnetic interference (EMI) (as seen in the first portion of Figure 9). Thus the data were truncated to remove this unwanted portion of the data.

Next, the truncated output signal was converted into the frequency domain with a discrete Fourier transform. The system acts like a 'band pass filter', allowing some frequencies to pass more readily than others. The frequency domain allows a graphical representation of the system's frequency response to a given input. To determine an acceptable variation in the signal's frequency response, multiple baseline measurements were taken with changing environmental parameters, as mentioned in the Experimental Procedure section, to produce a bank of responses. This bank of baseline responses provides a foundation to compare with the damaged cases. If damaged is induced, the Fourier transform can potentially depict a distinct change in either the magnitude or the frequency content of the response signal versus the baseline measurements. See Figure 10 to visually describe an example of how the data were processed.

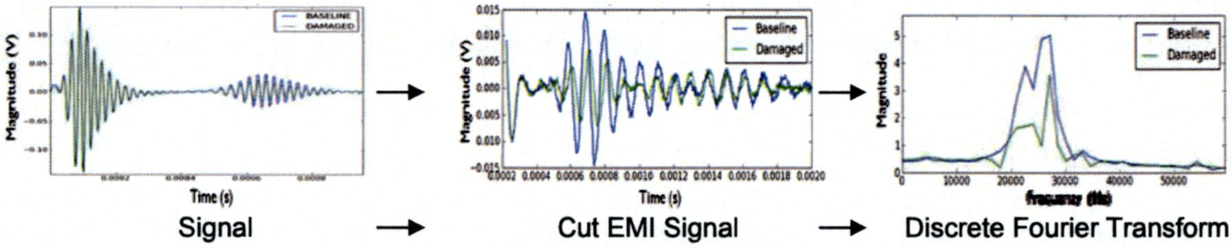

Figure 10: Lamb Wave Data Reduction

Once the data were transformed into the frequency domain, a cross correlation analysis was made between the baseline and damaged frequency domain data. Cross correlation is a measure of linear relationship between two waveforms, while a time-lag function is applied to one of them. One can potentially see a drop in correlation magnitude between the data sets as well as a lag or lead in the position of the largest magnitude correlation. Figure 11 illustrate an example. Each baseline and damaged case is compared to a single 'true' baseline case to determine if a significant difference in correlation is evident. The red dots represent the damaged case while the other nine dots of varying colors represent the baseline cases. One can see a lower correlation of the damaged case with respect to the baseline cases.

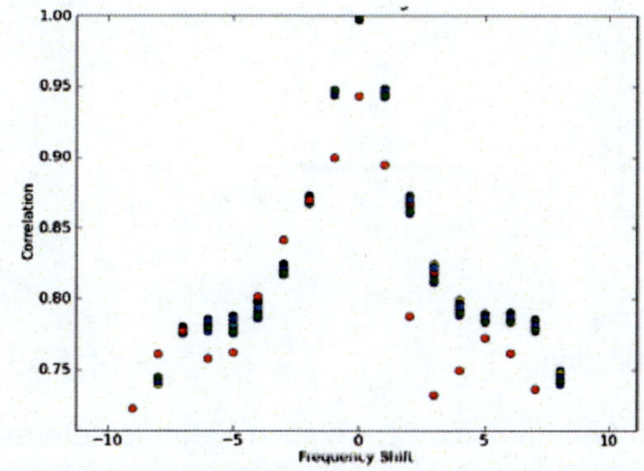

Figure 11: Lamb Wave Cross Correlation Graph

A damage index was created, which uses the maximum value of each cross correlation and subtracts the value from one. In Figure 12, one can clearly observe a large difference in the damage index between the damaged measurement and the other baseline comparisons. The last measurement was taken when the putty was attached in the direct path between the sensor and actuator. The value of the damage index allows for a decision to be made on the structural integrity of the system and if damaged has occurred.

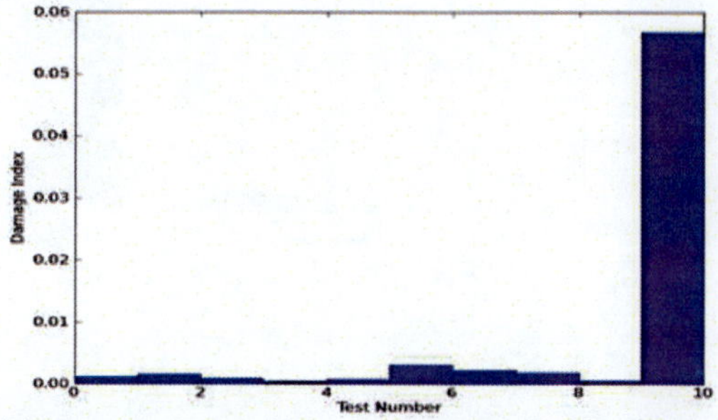

Figure 12: Lamb Wave Damage Index Graph

3.2 Frequency Response Testing Analysis

Initially, the testing was conducted using an input frequency bandwidth of 30-80 kHz on each path, as shown in Figure 7. The data were captured in the time (to perform an auto-regressive time series analysis found in the next section) and frequency domain. For the purposes of frequency responses, only the frequency domain data were used.

3.2.1 Frequency Response Functions (FRF) Analysis

Much like the lamb wave propagation data, a cross correlation was performed to compare a bank of baseline responses (all with different environmental procedures) with damaged responses. See Figure 13 for an example. One can see on Figure 13 lines 1-10 are baseline cases while line D is a damaged case. The same applies for Figure 13 where the red dots show a low cross correlation with respect to the nine baseline cases, which correlate very well.

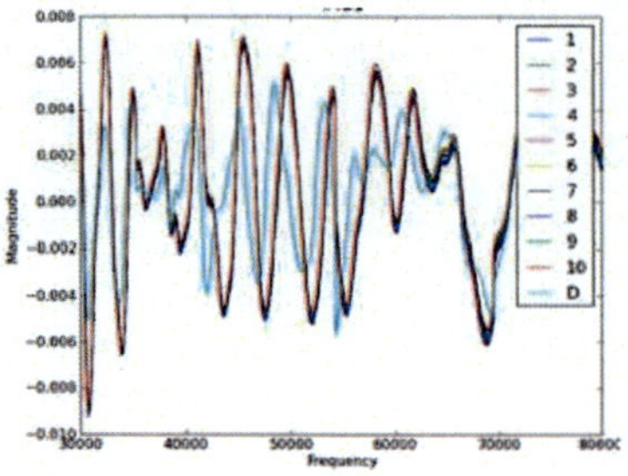

Figure 13: Frequency Response Function

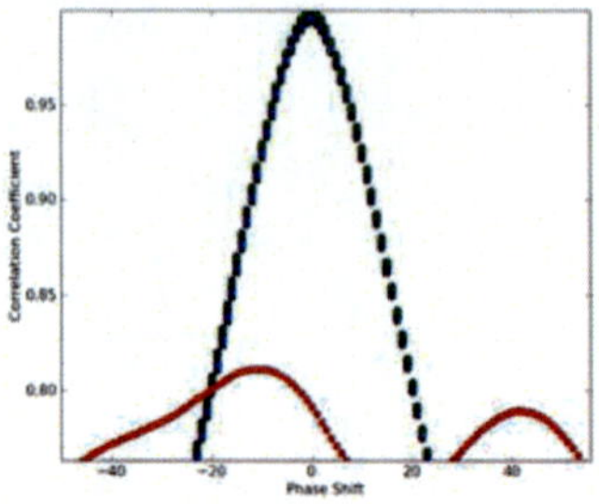

Figure 14: FRF Cross Correlation Graph

Next, a damage index was created from the cross correlation graph to determine a qualitative value between the damaged and undamaged cases as in Figure 15. This once again shows likelihood for damage in the tenth response.

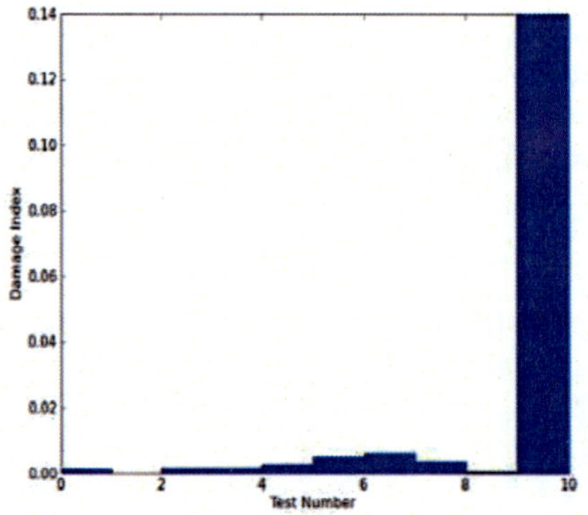

Figure 15: FRF Damage Index Graph

3.3 Time Series Analysis

An auto-regressive with exogenous input (ARX) time series analysis uses the data obtained during the frequency response testing (in the time domain) to calculate a predictive model and apply it to potentially determine damage in the structure. An ARX model is used to predict the output response of the system using data gathered in the time domain from both the output and input sensors. This predicted output response from the ARX model can then be compared to the actual response initially captured during testing to determine if significant error is present. If the model is incapable of predicting the newly measured response, then the system has changed in some way that damage may be present.

3.3.1 Normalizing Data

The time domain data was first cut to include just 60% of the data (from 20% - 80 % of the data set) in order to decrease computational time and ensure the noise seen at the extreme excitation frequencies did not significantly affect the model. This reduced data set was then normalized with respect to the standard deviation so that each set could be compared during the statistical portion to determine damage. See Equation1 for the normalization procedure.

$$X_{normalized} = (X_{original} - X_{mean})/\sigma \tag{1}$$

Lastly, a high and low pass filter of 5 and 20 kHz, respectively was applied to the resultant data set.

3.3.2 Creating an ARX Model

Using time domain data from a random excitation input signal from one sensor to another, an ARX model can be formulated. Theoretically, the equation can be written as:

$$x(t) = \sum_{j=1}^{p} A_j * x(t-j) + \sum_{i=1}^{p} B_i * y(t-i) + e(t)$$

Initially to create the model, baseline data needs to be taken. The baseline data from the input and output sensors can be placed on both the right and left hand side of the equation in order to solve for the unknown coefficients (A_j and B_i). These coefficients model the system's baseline output response. Applying new data to the ARX model, a predicted output response can be created. If one were to apply a different baseline case to the newly created ARX model, the predicted response would be expected to highly correlate to the actual response of the system. Figure 16 illustrates the predicted and actual baseline response.

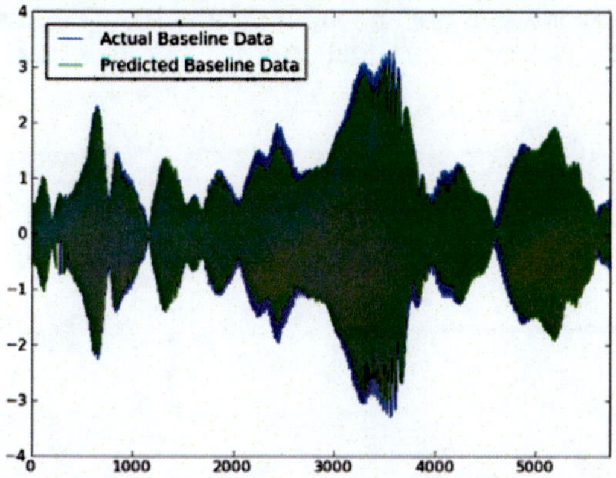

Figure 16: Predicted and Actual Baseline Responses

3.3.3 Determining Order

To perform this time series analysis, one must first determine the optimal order of the model. For the purposes of this project, The Akaike's Information Criterion (AIC) was used with built-in Matlab functions. An order of 218 for the input and 177 for the output was found. However, an order of 218 will be used for both the input and output for this project, since it provides better results.

A normal probability plot of the residual error between the actual and predicted data should show a normal distribution if the model is working accurately. As can be seen in Figure 17, the model order chosen works accurately and provides a normal probability of error.

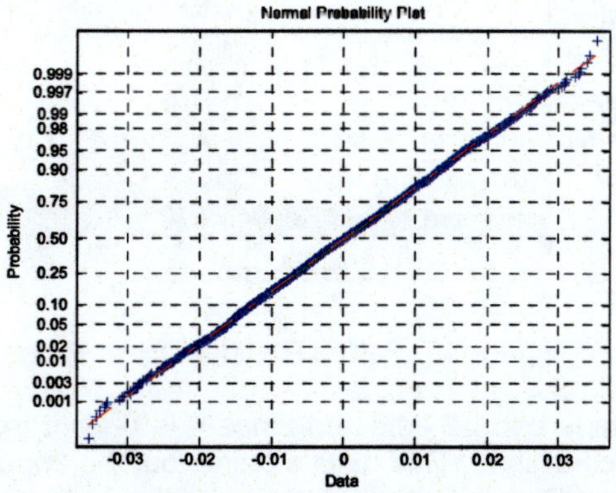

Figure 17: Residual Error Normal Probability Plot

3.3.4 Correlation Technique: Residual Error

The ARX model can be used to make a predicted response as stated before from an actual response of the system. The predicted response can be correlated to the actual response using the residual error between the actual and predicted responses of the system. In practice, the actual and predicted responses from a baseline case should correlate well and have a relatively small residual error value. Using this method, a set of correlated baseline cases can be used to determine a threshold residual error value, much like the lamb wave and frequency response methods. Damaged cases and their corresponding residual error values can be compared to the bank of baseline residual error values to find any significant difference to determine if damage is present. Potentially, the actual and predicted damaged cases should yield a relatively large residual error value.

This trend can successfully be seen in the data, shown in Figure 18. The baseline residual error is significantly smaller on average compared to the damaged residual error for a representative example baseline and damaged case.

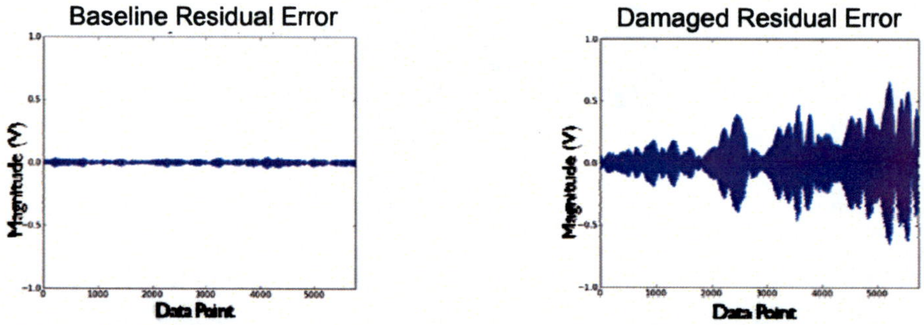

Figure 18: Baseline and Damaged Residual Error Value Comparison

Comparing all the cases, both undamaged and damaged, damaged cases close to the path could be detected, while damaged placed farther away or with little surface area could not. Figure 19 shows the RMSE (root mean square error) values of the undamaged cases in green and damaged cases in red. The blue line was chosen by visual inspection of the chart and determining the threshold residual error value so that no false positives would occur. A value of 0.038 was used for damage detection in this study.

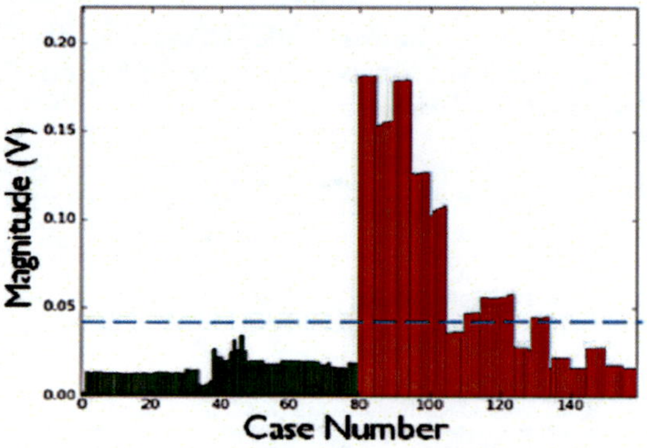

Figure 19: Path 2 Residual Error Plot

As seen in Figure 19, not all damage can be detected since it is distant from the path, although some causes larger variations than undamaged conditions. This results points out the importance of identifying better threshold limits for damage identification. In addition, a further study is needed in order to quantify the sensing range of this method.

4 Conclusions

It can be shown that all three techniques (lamb wave, frequency response, and time series analysis) can detect damage, have localized sensing capability, and are less sensitive to operational variations. Lamb wave testing was capable of detecting damage only when it was close the path of the propagating wave. For this reason lamb waves could be used to determine the location of the damage. The frequency response method showed an intriguing ability to detect damage when it was located anywhere along the spar of the blade section. This is significant because the majority of the delamination in turbine blades occurs when the skin detaches from the spar. The time series analysis is the simplest of the three techniques and thus the memory and power usage of the system is minimal. This is ideal for a SHM system that needs to be self powered when in operation on a real structure. Table 2 depicts the advantages and disadvantages of each technique to give a comparison between the techniques and illustrate which technique is best for a given scenario.

Table 2: Performance Matrix

	Pros		Cons Memory	Power	Damage Detection	Other
Lamb Wave	Locate Damage		High Memory Usage	High Power Consumption	Damage Very Close to Path	Multiple Frequencies
Frequency Response	Sensitive to Spar Damage		Moderate Memory Usage	Moderate Power Consumption		
Time Series	Memory Usage	Power Consumption			Damage Close to Path	EMI Effect

In order to determine a viable damage detection package for commercial use, actual system testing within operating wind turbine blades needs to be performed. Also, an investigation into wireless sensing and power harvesting for the sensing network should be looked into. Additionally, a more rigorous statistical analysis to create a threshold value should be determined.

4 References

1. Energy, U.S. Department of. "20% Wind Energy by 2030: Increasing Wind Energy's Contribution to U.S. Electricity Supply." July 2008. Wind & Hydropower Technologies Program. <http://www1.eere.energy.gov/windandhydro/pdfs/41869.pdf>.

2. Rumsey, M.A., Paquette, J.A., "Structural Health Monitoring of Wind Turbine Blades," *Proceedings of SPIE*, V.6933, 2008.

3. Ciang, Chia Chen, Jung-Ryul Lee, and Hyung-Joon Bang. "Structural health monitoring for a wind turbine system: a review of damage detection methods." Measurement Science and Technology (2008).

4. Park, G., Rutherford, C.A., Wait, J.R., Nadler, B.R., Farrar, C.R., 2005, "The Use of High Frequency Response Functions for Composite Plate Monitoring with Ultrasonic Validation," *AIAA Journal*, Vol. 43, No. 11, pp. 2431-2437.

5. Raghavan, A., Cesnik, C.E., 2007, "Review of Guided-wave Structural Health Monitoring," *The Shock and Vibration Digest*, Vol. 39, No. 2, pp. 91-114.

6. Park, G., Sohn, H., Farrar, C.R., Inman, D.J., 2003, "Overview of Piezoelectric Impedance-based Health Monitoring and Path Forward," *The Shock and Vibration Digest*, Vol. 35, No. 6, pp. 451-463.

Proceedings of the IMAC-XXVIII
February 1–4, 2010, Jacksonville, Florida USA

ENERGY RECOVERING FROM VIBRATIONS IN ROAD VEHICLE SUSPENSIONS

Ferdinando Mapelli*, Edoardo Sabbioni*, Davide Tarsitano*

*Department of Mechanical Engineering
Politecnico di Milano, Campus Bovisa, Via La Masa 1, 20156 Milano, Italy
e-mail: ferdinando.mapelli@polimi.it, edoardo.sabbioni@polimi.it, davide.tarsitano@mecc.polimi.it

ABSTRACT

Over the last years, energy saving has become a very important issue. One possibility for limiting energy consumption is recovering it from systems where it is dissipated (energy harvesting). One of the most effective methods for implementing energy harvesting is to convert kinetic energy produced by mechanical vibrations into useful electric energy which can be stored in accumulators and then used to power sensors and/or active systems and/or on board auxiliary electrical loads.

The application of an energy harvesting device to a road vehicle suspension system is presented in this paper. The device consists in a mass damper excited in resonance, a linear permanent magnet alternator and a power factor controlled rectifier (electromagnetic vibration driven generator).

In a first stage of the research, the capability of the device of recovering energy from road induced vibrations of the suspension unsprung mass has been explored. The parameters of the device have then been tuned in order to optimize energy harvesting, taking into account the physical constraints concerned with the application (suspension geometry, electromagnetic vibration driven generator mass, etc.).

INTRODUCTION

In the last years, the interest for harvesting energy from mechanical systems has grown, due to the increasing attention in limiting energy consumption. Different energy harvesting devices have been proposed, analyzed and designed in the literature ([1],[2],[3],[4]). One of the most common approaches is to drive an electromechanical converter from ambient motion or vibration. This allows to convert kinetic energy produced by mechanical vibrations into useful electric energy which can be stored in accumulators and then used to power electric loads.

Several transduction methods (eventually affecting the amount of electrical power which can be generated) suitable for implementing vibration-driven generators have been proposed:

- piezoelectric: a piezoelectric material is used to convert strain in the spring into electricity ([1],[2],[7]);
- electromagnetic: a magnet is attached to the moving mass, thus inducing a voltage in a coil as it moves ([2],[3],[5]);
- electrostatic: an electric arrangement with a permanent charge embedded in the moving mass induces a voltage in the plates of a capacitor as it moves ([8]).

Most of the applications found in the literature are concerned with self-powered sensors often relying on micro-generators and MEMS technologies.

An application of regeneration concepts to road vehicles is presented in this paper. Energy regeneration devices have in fact recently been introduced in the automotive industry. Most of these devices accumulate energy otherwise dissipated during braking and then use it to power vehicle electric loads or to accelerate the vehicle (KERS-Kinetic Energy Recovery System). A feasibility study for applying an electromagnetic vibration-driven generator (EVDG) to a road vehicle suspension system is instead presented in this study (Figure 1). A preliminary analysis has thus been carried out in order to assess the capabilities of the device to recover energy from road induced vibrations of the vehicle unsprung masses. Basic concepts for maximize the harvested energy are exposed and physical constraints due to the device practical realization are illustrated.

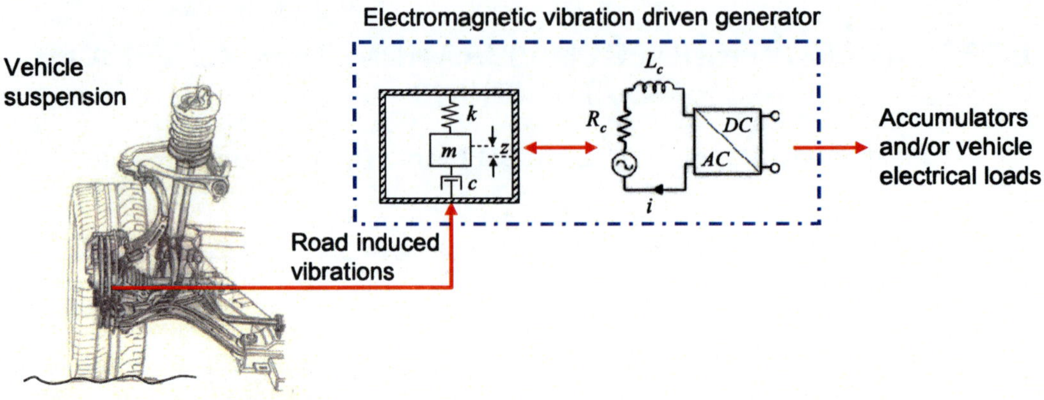

Figure 1: Scheme of an EVDG applied to a vehicle suspension.

ELECTROMAGNETIC VIBRATION DRIVEN GENERATOR

In this section the basic theory of resonant electromagnetic vibration driven generators is illustrated.
The basic vibration driven generator (VDG) consists of a seismic mass, m, connected through a spring, k, to a rigid housing. As the housing is vibrated, a relative motion between the seismic mass and the housing occurs. The mechanical kinetic energy of the moving mass, m, is transformed into electrical energy by having the mass move a magnet relative to a coil. This could be obtained with either a moving a magnet or a moving coil configuration. In practice a moving magnet configuration is simpler to fabricate since electrical connection to the mass are avoided. In any case, voltage is induced in the coil due to a varying flux linkage, with the resultant currents causing forces opposing the relative motion between the magnet and the coil.
Thus, the motion of the VDG can be described through the mass-spring-damper system shown in Figure 2.

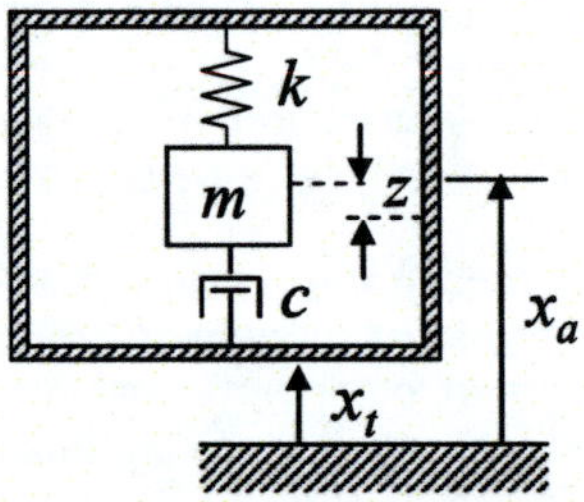

Figure 2: Scheme of the VDG.

The basic equation of motion of the mass m relative to the housing is therefore given by:

$$m\ddot{z} + c\dot{z} + kz = -m\ddot{x}_t \quad (1)$$

where z is the relative displacement between the mass and the housing (i.e. between the magnet and the coil), x_t is the vibration imposed to the housing and c is an equivalent damping coefficient accounting for electromagnetic damping (the conversion of mechanical into electrical energy damps the mass) and parasitic damping due to air resistance and material loss.
As it can seen, the VDG is a an inertial generator, i.e. it only needs to be anchored to a moving body to generate electric power and the damper, c, represents the energy extraction mechanism.
This generator is intended to operate in resonance conditions and for optimum energy extraction should be designed in order to have the natural frequency of the VDG synchronized with the one of the intended application environment (i.e. the frequency of the vibration applied to the housing). In fact, assuming that the generator is driven by an harmonic excitation $x_t = X_t sin(\omega t)$, the average power dissipated, P_d, within the damper (i.e. the power extracted by the transduction mechanism) is ([3],[4],[5]):

$$P_d = \frac{m\zeta X_t^2 \omega_c^3 \omega^3}{\left(1-\omega_c^2\right)^2 + \left(2\zeta\omega_c\right)^2} \quad (2)$$

where $\zeta=c/(2m\omega_n)$ is the damping ratio, $\omega_c=\omega/\omega_n$ and ω_n is the natural frequency of the VDG. Thus, maximum power dissipation within the generator occurs when the device is operated at ω_n (i.e. ω_c=1) and in this case it is given by:

$$P_d = \frac{mX_t^2\omega_n^3}{4\zeta} \tag{3}$$

Eq. (3) suggests that the dissipated power increases:

- linearly with the mass m;
- with the cube of the natural frequency ω_n;
- with the square of the amplitude X_t;
- when the damping ratio ζ approaches to zero.

This occurs only if the source of motion is capable of supplying infinite power, the relative displacement is not limited and no parasitic damping is present into the system. In practice, these conditions are not realizable. In particular, reducing the damping ratio increases the displacement of the mass. Thus, the damping ratio must be high enough to prevent the mass displacement exciding the limit Z_{max}. If the inertial mass displacement is limited, eq. (3) becomes ([4],[5]):

$$P_d = \frac{1}{2}m\omega_n^3 X_t^2\left(\frac{Z_{\max}}{X_t}\right) \tag{4}$$

and the damping ratio maximizing the dissipated power is given by ([4]):

$$\zeta_{opt} = \frac{1}{2\omega_c}\sqrt{\omega_c^4\left(\frac{X_t}{Z_{\max}}\right)^2 - \left(1-\omega_c^2\right)^2} \tag{5}$$

The generator damping factor may be further constrained by unwanted parasitic damping, c_p. In fact the maximum power that can be extracted by the transduction mechanism can be calculating only by including the parasitic damping ratio ζ_p:

$$P_e = \frac{m\zeta_e X_t^2\omega_n^3}{4\left(\zeta_e+\zeta_p\right)^2} \tag{6}$$

Maximum power is delivered to the electrical domain when $\zeta_e=\zeta_p$, i.e. when damping arising from the electrical domain (ζ_e) equals the mechanical losses (ζ_p). In this case, eq. (6), becomes:

$$P_e = \frac{mX_t^2\omega_n^3}{16\zeta_e} \tag{7}$$

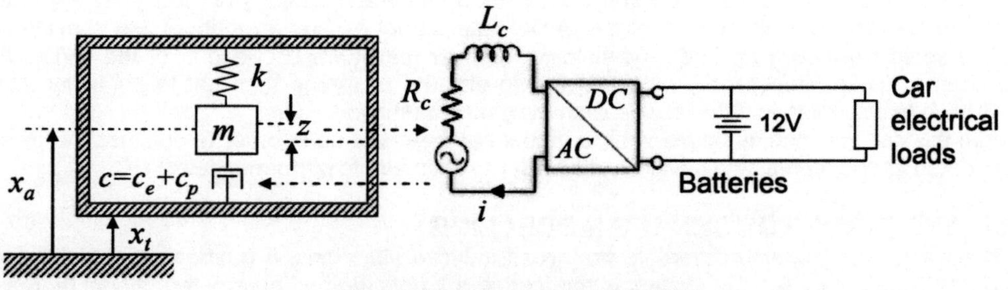

Figure 3: Scheme of the EVDG.

The damping due to the electromagnetic transduction c_e can be estimated as ([5]):

$$c_e = \frac{(NlB)^2}{R_c + L_c j\omega} \tag{8}$$

where N is the number of turns in the generator coil, l is the length of the coil, R_c, L_c are the resistance and the inductance of the coil (see Figure 3) and B is the flux density to which it is subjected to.
In the following, the application of the previously exposed criteria to a vehicle suspension is presented. Physical constraints for the resonant electromagnetic vibration driven generators design are illustrated and an estimation of the expected performances of the device is carried out considering a simplified vehicle model (quarter car vehicle model).

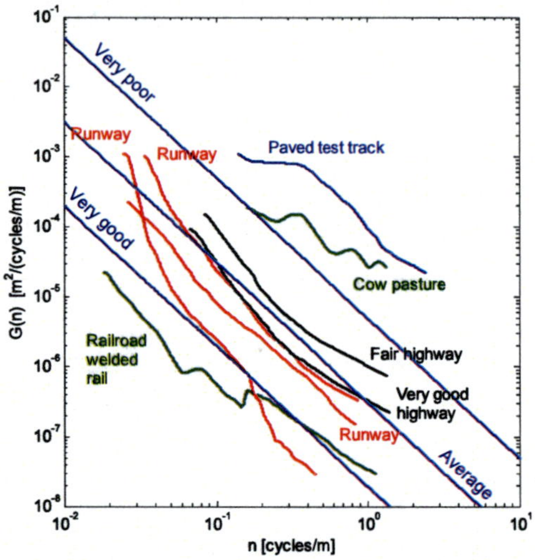

Figure 4: PDS of various road roughness ([11]).

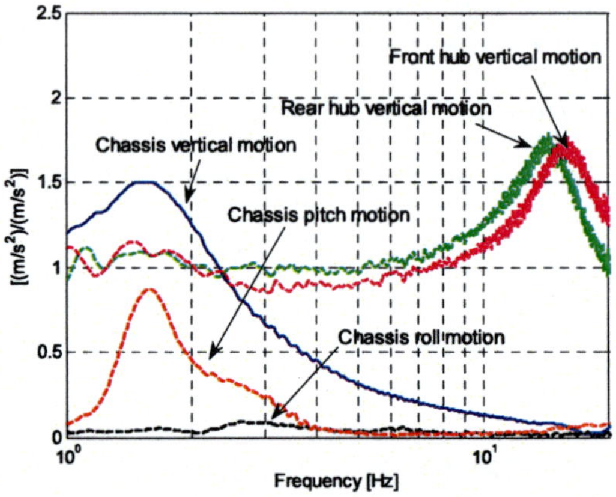

Figure 5: Amplitude of the experimental FRFs of a passenger car.

APPLICATION OVERVIEW

The intended application for the electromagnetic vibration driven generator presented in this paper is a road vehicle suspension. As known, due to road roughness, vibrations are transmitted from the ground to the vehicle hubs (unsprung masses) and chassis (sprung mass). Typical roads provide a broadband random excitation characterized by a displacement power spectral density (PSD) presenting an amplitude decreasing with the spatial frequency, n (see Figure 4, [11]), while passenger cars generally present resonance peaks between 1 and 2.5Hz due to the vehicle chassis vertical, pitch and roll motions and resonance peaks in the range 10-15Hz associated with the tires vertical motion ([10], see Figure 5). In theory, the EVDG can be attached either to the chassis or to the unsprung masses of the vehicle and its natural frequency can be synchronized either with the ones of the sprung or of the unsprung masses. Although the most suitable synchronism frequency seems to be the natural frequency of the sprung mass being road induced vibration amplitudes higher at low frequency, in practice, vibrations transmitted to the chassis are filtered by the vehicle suspensions. Thus the EGVD must be anchored to the unsprung masses, where only a slight resonance peak associated with the chassis natural frequencies can be observed (see Figure 5). These considerations have led to synchronize the natural frequency of the EVDG with the one of the unsprung mass. The seismic mass m of the EVDG thus cannot exceed a weight of 3kg to avoid negative effect on the vehicle handling and ride dynamics. Moreover maximum oscillations of the EVDG must be limited in order to avoid interferences with the sprung and the unsprung masses. In particular, maximum oscillations have been limited to 1/3 of suspension maximum deflection.
Starting from this assumptions, a simple vehicle model has been set up in order to optimize the parameters of the considered energy harvesting device and in order to estimate its performances.

VEHICLE AND ROAD ROUGHNESS MODELLING

In order to estimate the power recoverable from road induced vibrations, a quarter car vehicle model has been used ([12], see Figure 6). The system is reduced to a 2 dofs lumped parameters model representative of one of the four corners of the vehicle. The model thus includes the sprung mass, m, and the tire unsprung mass, m_t. The two masses are connected between them and with the ground through spring-damper elements respectively representing suspension and tire stiffness and damping.

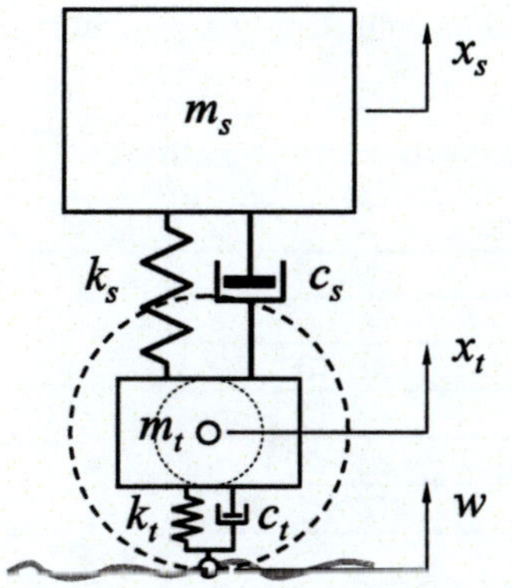

Figure 6: Quarter car vehicle model.

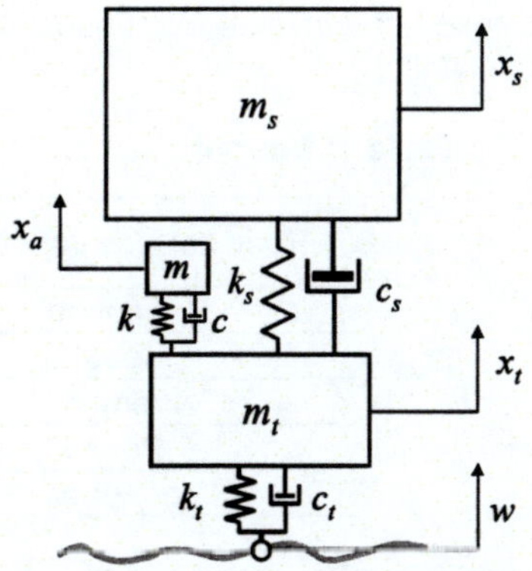

Figure 7: Quarter car vehicle model+EGDV.

The parameters of the quarter car vehicle model have been obtained from previous studies ([13]). Figure 8 shows the Frequency Response Functions (FRFs) between the vertical acceleration of the sprung/unsprung mass and the vertical acceleration imposed by the road profile. Numerical results (Figure 8.b) are compared with the measurements (Figure 8.a) carried out with an instrumented vehicle tested on a four post test bench. During the experimental tests, eight accelerometers were placed in correspondence of the four vehicle hubs and of the four attachments between the suspensions and the vehicle chassis. Imposed displacements were measured as well In particular, the FRFs referred to the front left corner of the tested vehicle are shown in Figure 8.a. a good agreement between the quarter car vehicle model and the experimental measurements can be noticed.

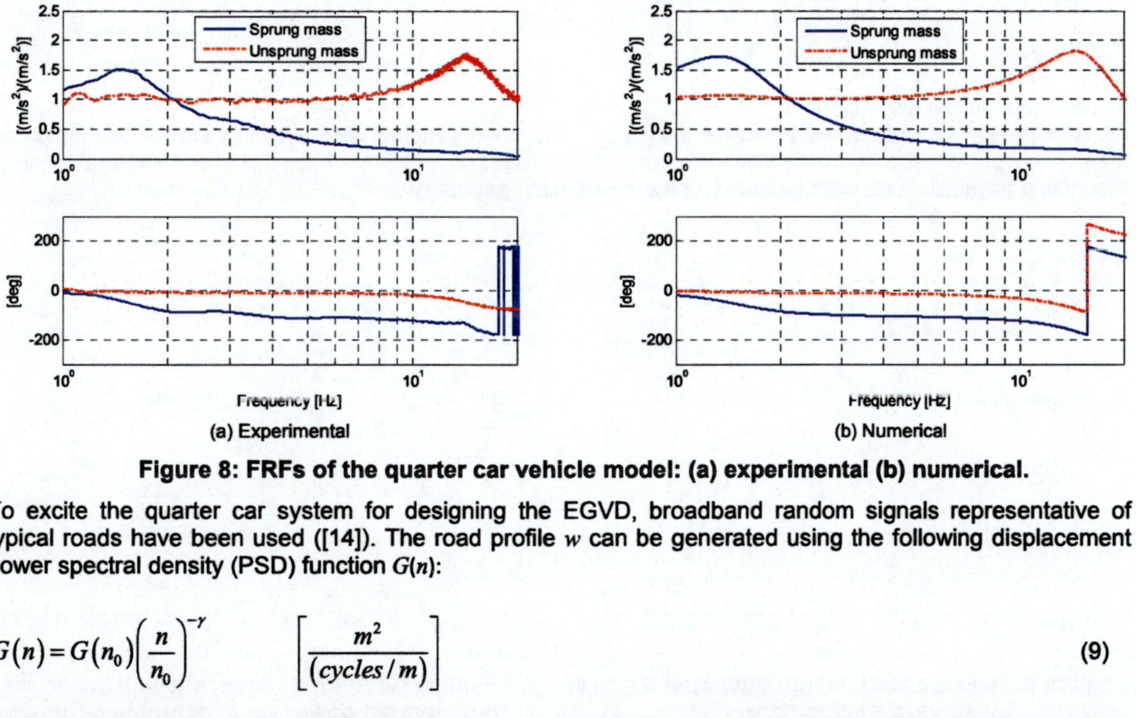

Figure 8: FRFs of the quarter car vehicle model: (a) experimental (b) numerical.

To excite the quarter car system for designing the EGVD, broadband random signals representative of typical roads have been used ([14]). The road profile w can be generated using the following displacement power spectral density (PSD) function $G(n)$:

$$G(n) = G(n_0)\left(\frac{n}{n_0}\right)^{-\gamma} \qquad \left[\frac{m^2}{(cycles/m)}\right] \tag{9}$$

where n is the spatial frequency (cycles/m), n_0 is the reference spatial frequency, while γ is a fitting constant describing the severity of the road roughness. The ISO suggested values for these latter parameters are n_0=0.1 cycles/m and γ=2 ([14]). The ISO classified roads into five classes depending on surface conditions. Classification of roads and recommended parameters are reported in Table 1, while the lower bounds for *very good*, *average* and *very poor* road profiles are depicted in Figure 4.
According to the ISO classification, several road profile have been generated with a frequency content from 0Hz to 30Hz. In fact, being the spatial frequency n given by f/V, where f is the vibration frequency and V is

the vehicle speed, the inverse fast Fourier transform can be applied to determine the road surface height in the time domain ([15]).

Table 1: Roughness coefficient ranges for different types of road surface.

Road surface condition	$G(n_0)$	
	Range	Mean
Very good	$2\text{-}8{\cdot}10^{-6}$	$4{\cdot}10^{-6}$
Good	$8\text{-}32{\cdot}10^{-6}$	$16{\cdot}10^{-6}$
Average	$32\text{-}128{\cdot}10^{-6}$	$64{\cdot}10^{-6}$
Poor	$128\text{-}512{\cdot}10^{-6}$	$256{\cdot}10^{-6}$
Very poor	$512\text{-}2048{\cdot}10^{-6}$	$1024{\cdot}10^{-6}$

The equations of motion of the quarter car vehicle model when the EGDV is introduced (Figure 7) can be written as:

$$\begin{cases} m_s\ddot{x}_s + c_s(\dot{x}_s - \dot{x}_t) + k_s(x_s - x_t) = 0 \\ m\ddot{x}_a + c(\dot{x}_a - \dot{x}_t) + k(x_a - x_t) = 0 \\ m_t\ddot{x}_t - c_s(\dot{x}_s - \dot{x}_t) - k_s(x_s - x_t) - c(\dot{x}_a - \dot{x}_t) - k(x_a - x_t) + c_t(\dot{x}_t - \dot{w}) + k_s(x_t - w) = 0 \end{cases} \tag{10}$$

where k_s/k_t are suspension/tire stiffness, c_s/c_t are the suspension/tire damping and x_s/x_t are the sprung/unsprung mass vertical displacements.

As it can be seen the EGVD has been anchored to the unsprung mass.

Note that, the second of equations (10) is equivalent to eq. (1) if the following change of coordinates is applied:

$$x_a = z + x_t \tag{11}$$

being x_a the vertical displacement.

As anticipated, the natural frequency of the EVDG has been synchronized with the one of the unsprung mass. The parameters of the device have then been tuned in order to maximize the harvested energy, respecting physical constraints imposed by the suspension geometry.

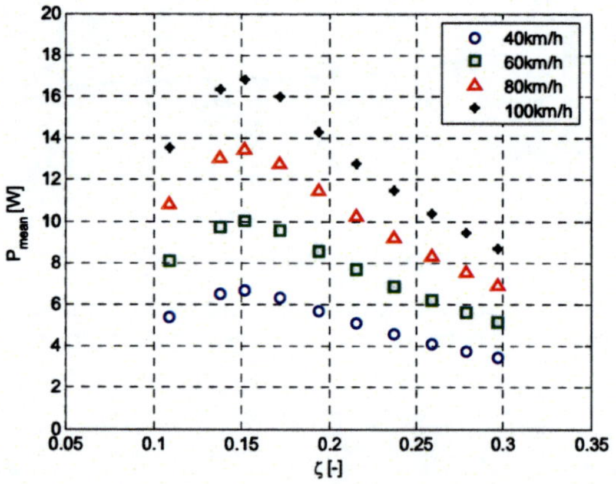

Figure 9: Average road: mean regenerated power vs. damping ratio.

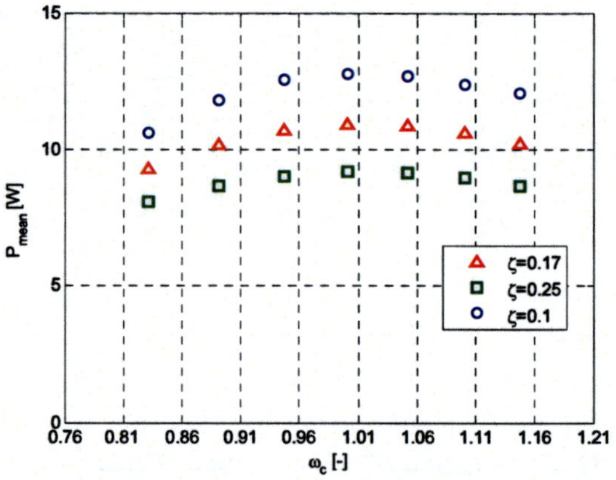

Figure 10: Average road, speed 80km/h: mean regenerated power vs. synchronism frequency.

NUMERICAL ANALYSIS OF THE EXPECTED PERFROMANCES

In order to evaluate the effective feasibility and the effective performances of the presented device, simulations with different road profiles and vehicle speeds have been performed. Simulations have allowed a direct verification of the regeneration criteria previously exposed.

In particular, the choice of the EGVD damping ratio ζ has been investigated through a parametric analysis. As anticipated, ζ in fact results from the compromise between energy harvesting (low damping) and working space requirements (high damping to prevent excessive vibration amplitudes of the seismic mass m, see Figure 11).

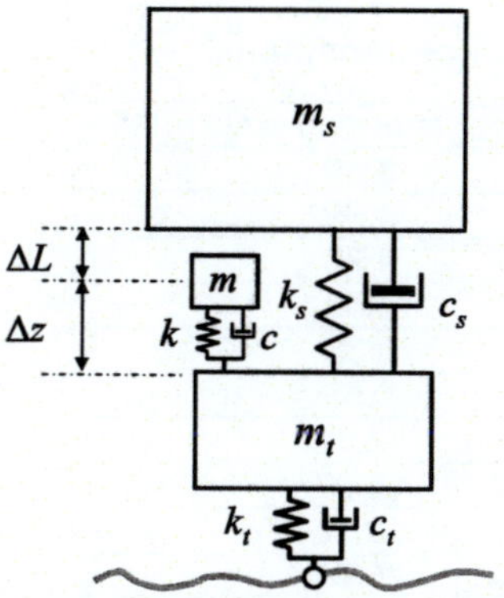

Figure 11: Suspension working space.

More in detail, Figure 10 depicts the variation of the mean value of regenerated power under different condition of vehicle speed and damping ratio. An average road profile, corresponding to the one of a typical main road, is considered. It must be pointed out that the road profile has been generated using the mean value of the range reported in Table 1 for the displacement PSD. As it can be seen, energy harvesting is maximized when ζ=0.15. A similar result is achieved if a different road profile is considered.

In order to verify that the value of the damping ratio maximizing the energy recovery would not lead to excessive vibration amplitudes of the EVDG, simulations considering poor and very poor road profiles have been carried out. In particular, Figure 12 shows the effect of the damping ratio on the r.m.s. (blue squares), minimum and maximum (upper and lower limits of the vertical bars) values of the ratio between the distance of the seismic mass m from the sprung mass, ΔL (see Figure 11), and the maximum available working space ΔL_{max}, while in Figure 13 the r.m.s. (blue squares), minimum and maximum (upper and lower limits of the vertical bars) values of the vibration amplitude of the VDG, Δz (see Figure 11), normalized by its maximum value (1/3 of the suspension maximum deflection) are depicted as a function of ζ. Results are referred to a poor road profile (poor minor road) run at the constant speed of 60km/h. A severe test has thus been simulated. It can be noticed that, in order to avoid interferences between the EGVD seismic mass and the suspension sprung/unsprung mass, the damping ratio ζ must be lower limited to 0.17, even if this value does not maximize the harvested energy. It must be pointed out that the most critical condition is concerned with the EGVD vibration amplitude Δz for any value of the damping ratio.

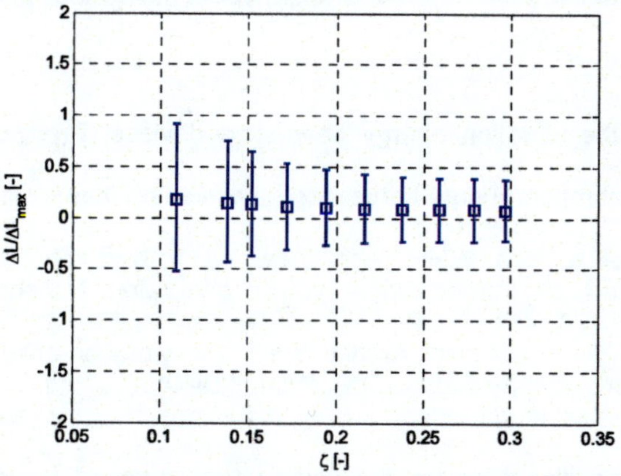

Figure 12: Poor road, speed 60km/h: EGVD-sprung mass relative displacement vs. damping ratio.

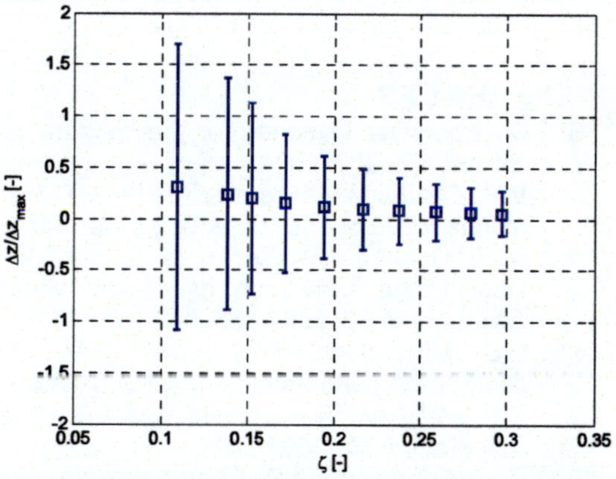

Figure 13: Poor road, speed 60km/h: EGVD-unsprung mass relative displacement vs. damping ratio.

Moreover, since in a suspension system several parameters are expected to change their value (e.g. vehicle mass varying with loads, tire stiffness changing with pressure, temperature and wear, etc.) or are difficult to estimate (e.g. suspension damping and compliance), the influence of errors in the synchronism between the EGVD and the sprung mass natural frequency has been analyzed. As it can be seen in Figure 10, reductions of about 15% in the mean regenerated power correspond to errors of 20% on the synchronism frequency. As expected, the negative effect of a synchronism error increases as the damping ratio decreases (Figure 10).

The capabilities of a linear permanent magnet alternator of recovering energy from road induced vibrations are summarized in Table 2 under different road conditions and vehicle speeds. The mean power regenerated by a single corner is reported when a damping factor ζ=0.17 is considered.

Under the hypothesis of having an energy harvesting device equipping each suspension, an average power of 16W can be recovered when travelling on a good quality highway, an average power of 50W can be recovered when travelling on a typical main road and an average power of 100W can be recovered when travelling on a minor road. Considering that the electric consumption of a vehicle ranges from 300W to

500W, it can be concluded that the proposed energy harvesting device seems to be promising, even considering that, during urban driving, paved surfaces are often encountered, this allowing to double (or more) the recovered energy (see Figure 4).

Table 2: Regenerated power for different road surface conditions.

Road surface condition	Vehicle speed [km/h]	Mean regenerated power [W]
Good	80	3.19
	100	3.99
	120	4.78
	140	5.58
Average	40	6.36
	60	9.55
	80	12.79
	100	15.97
Poor	20	12.75
	40	25.51
	60	37.98

CONCLUDING REMARKS

The present study has shown the possible use of a linear permanent magnet alternator on a road vehicle suspension in order to regenerate the kinetic energy produced by road induced vibrations of the unsprung masses (energy harvesting) which is usually dissipated by the suspension dampers. The regenerated energy can be transformed into electrical energy and used to supply electrical load on the vehicle.

In order to estimate the performances of the energy harvesting device, a quarter car vehicle model has been used. A parametric analysis involving different road profiles and vehicle speeds has been carried out in order to tune the structural parameters (damping factor and stiffness) of the energy harvesting device. Simulation results have been critically analyzed in order to evaluate the effective performances of the system considering physical constraints, such as the limited available working space.

Simulations stated that a mean power of 50W can be regenerated when travelling on typical main roads and of 100W when running on poor roads.

REFERENCES

[1] M. Pereyma: Overview of the modern state of the vibration energy harvesting devices, Proc. of MemsTech, 2007.

[2] V.R. Challa, M.G. Prasad, Y. Shi, F.T. Fisher: A vibration energy harvesting device with bidirectional resonance frequency tunability, Smart Mater. Struct., 17, 2008, 1-10.

[3] M. El-Hami, P. Glynne-Jones, E. James, S.P. Beeby, N.M. White, A.D. Brown, J.N. Ross, M. Hill: Design and fabrication of a new vibration-based electromechanical power generator, Sensors Actuators, A 92, 2001 335–342.

[4] P.D. Mitchenson, T.C. Green, E.M. Yeatman, A.S. Holmes: Architectures for vibration-driven micropower generators, Journal of Microelectromechanical Systems, 13(3), 2004, 429-440.

[5] C.B. Williams, R.B. Yates: Analysis of a micro-electric generator for microsystems, Sensor and Actuators, A 52, 1996, 8-11.

[6] R. Amirtharajah, A.P. Chandrakasan: Self-powered signal processing using vibration-based power generation, IEEE Journal of Solid State Circuits, 33, 1998, 687-695.

[7] P. Glynne-Jones, S.P. Beeby, N.M. White: A nover thick-film piezoelectric micro-generator, Smart Mater. Struct., 10, 2001, 850-852.

[8] S. Menninger, J.O. Mur-Miranda, R. Amirtharajah, A.P. Chandrakasan, J.H. Lang: Vibration-to-electric energy conversion, IEEE Trans. VLSI Syst., 9, 2001, 64-76.

[9] N.G. Stephen: On energy harvesting from ambient vibrations, Journal of Sound and Vibrations, 2006, 193, 409-425.

[10] R.A. Williams: Automotive active suspensions Part 1: basic principles, Proc. Instn. Mech. Engrs., 1997, 211 Part D, 415-426.

[11] E. Sevin, W.E. Pilkey: Optimum shock and vibration isolation, The Shock and Vibration Information Centre, United States Department of Defence, 1971.

[12] D.A. Crolla: Vehicle dynamics: theory into practice, Proc. Inst. Mech. Engrs, Part D: Journal of Automobile Engineering, 1996, 210, 83–94.

[13] F. Cheli, A. Costantini, N. Porciani, F. Resta, E. Sabbioni: A controlled semi-active suspension system for high performance vehicles, Proc. of ATA-ICC Conference, 2004.

[14] ISO 8608: Mechanical vibration - Road surface profile - Reporting of measured data, 1995.

[15] D. Cebon, D.E. Newland: The artificial generation of road surface topography by the inverse FFT method, Proc. of the 8th IAVSD Symposium, Cambridge, Massachussets, 1984, pp. 29–42.

Proceedings of the IMAC-XXVIII
February 1–4, 2010, Jacksonville, Florida USA

Development of MEMS-based Piezoelectric Vibration Energy Harvesters

Kasyap, A.[1], Phipps, A.[2], Nishida, T.[2], Sheplak, M.[3], Cattafesta, L.[3]
[1] Mechanical Analyst, AdaptivEnergy Llc., Hampton, VA
[2] Department of Electrical and Computer Engineering, University of Florida
[3] Department of Mechanical and Aerospace Engineering, University of Florida

NOMENCLATURE

a_0	Base acceleration	M_m	Effective mass
f_{ext}	Excitation frequency	C_{ms}	Short circuit mechanical compliance
U	Tip velocity	f_n	Natural frequency
I	Current across the PZT	d_m	Effective piezoelectric coefficient
F	Excitation force	ϕ	Transduction factor
V	Voltage across the PZT	κ^2	Effective coupling coefficient
ω	Frequency in rad/s	$\tan\delta$	Dielectric loss tangent
ζ	Damping ratio	C_{ef}	Free electrical capacitance
Q	Quality factor	C_{eb}	Blocked electrical capacitance

ABSTRACT

In this paper, the development of a first generation MEMS-based piezoelectric energy harvester capable of converting ambient vibrations into storable electrical energy is presented. The energy harvester is designed using a validated analytical electromechanical Lumped Element Model (LEM) that accurately predicts the behavior of a piezoelectric composite structure. The MEMS device is fabricated using standard sol gel PZT and conventional surface and bulk micro processing techniques. It consists of a piezoelectric composite cantilever beam (*Si/SiO$_2$/Ti/Pt/PZT/Pt/Au*) with a proof mass at one end. A prototype device packaged in a 5 mm^2 area produces 0.98 μW rms power into an optimal resistive load when excited with an acceleration of 1 m/s^2 at its resonant frequency of 129 Hz. Although the model predicts the general behavior of the device accurately, knowledge of the overall system damping is critical to accurately predict the power output, and therefore individual dissipation mechanisms in the system must be investigated. This effort lays the foundation for future development of MEMS piezoelectric energy harvester arrays as a potential power solution for self sustaining wireless embedded systems. The electromechanical model further enables intelligent and optimal design of these energy harvesters for specific applications minimizing prototype test runs.

INTRODUCTION

With ever-growing energy consumption, renewable sources that constitute the major portion of our current energy supply could soon be exhausted. Consequently, research in the past few decades has focused on using alternate energy resources, such as solar, wind, vibration, thermal, etc. Both solar and thermal energy harvesting have been demonstrated at microscales and are an existing MEMS technology. Mechanical energy is a commonly accessible resource and is particularly useful in applications where light and thermal energy are not readily available. Numerous meso-scale energy harvesters have been developed in the past using different transduction mechanisms. Examples include rotary generators, electromagnetic generators [1], dielectric elastomers with compliant electrodes [2], and electrostatic energy harvesters [3].

Of interest is piezoelectric energy harvesting that can potentially produce much higher power densities and has been previously investigated for harvesting human ambulation [4] for wearable electronics. Other piezoelectric technologies have demonstrated energy harvesting using impact [5], fluid flow [6], cantilevers [7-8], stacks, PVDF

T. Proulx (ed.), *Structural Dynamics and Renewable Energy, Volume 1*, Conference Proceedings of the Society for Experimental Mechanics Series 10, DOI 10.1007/978-1-4419-9716-6_8, © The Society for Experimental Mechanics, Inc. 2011

polymer films and tapered beams. Key applications for piezoelectric micro power generators include operating as a power source for embedded systems, wireless sensor networks, supplementing battery storage devices, etc. Ambient vibration sources are commonly available in household appliances, machinery equipment, HVAC ducts and vehicles, such as ground transport and aircraft [9].

The development of self-powered systems requires self contained power generators that possess an inherent mechanism to extract power from the ambient environment for their operation. Furthermore, advancements in MEMS technology have enabled implantation of micro devices into various host structures that remain remote and isolated from the outside world. Using conventional batteries as a power supply to these embedded systems limits their lifetime and prevents deployment of many such devices in a system. The ever-reducing size of CMOS circuitry and correspondingly lower power consumption have also provided immense opportunities to design and build micro power generators that can be ideally integrated with CMOS. Therefore, developing a micro-scale self-contained power supply offers great potential for applications in remote systems.

Initially, thick-film piezoelectric layers have been imprinted on micromachined silicon substrates to form the desired structure with thicknesses ranging up to 100 *μm* [10]. Zinc oxide has been the most commonly used piezoelectric material for MEMS applications until Lead Zirconate Titanate ($Pb(Zr_x, Ti_{1-x})O_3$ or PZT) gained popularity. PZT materials exhibit high dielectric constants and electromechanical coupling, when *x*, the percentage composition of Zr in PZT ranges between 0.52 and 0.55 [11]. PZT has been extensively studied and used lately due to its excellent electromechanical coupling and piezoelectric properties. Piezoelectric thin films in micro systems are used in a wide variety of applications such as micro actuators, micro mirrors, micropumps, microphones, micro accelerometers, and fiber bulk wave acoustic resonators, etc. Piezoelectric micro power generators have also been developed recently that employ the d_{33} mode to extract power [12].

ELECTROMECHANICAL LUMPED ELEMENT MODEL

The main objective of this paper is to report on the design, fabrication, and characterization of micro piezoelectric cantilevers for vibration energy harvesting from an ambient environment, subject to fixed input parameters and constraints. A simple schematic of the piezoelectric composite beam energy harvester is shown in Figure 1. In the figure, a_o is the input acceleration, f_{ext} is the excitation frequency of the base vibration, and V is the resulting voltage from the piezoceramic.

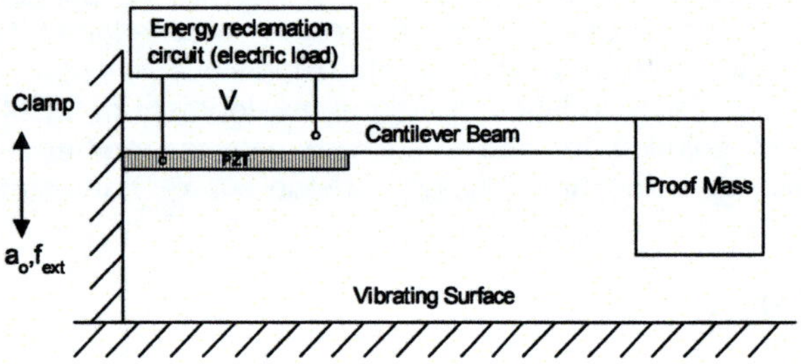

Figure 1: Schematic of a piezoelectric cantilever beam subject to source vibration.

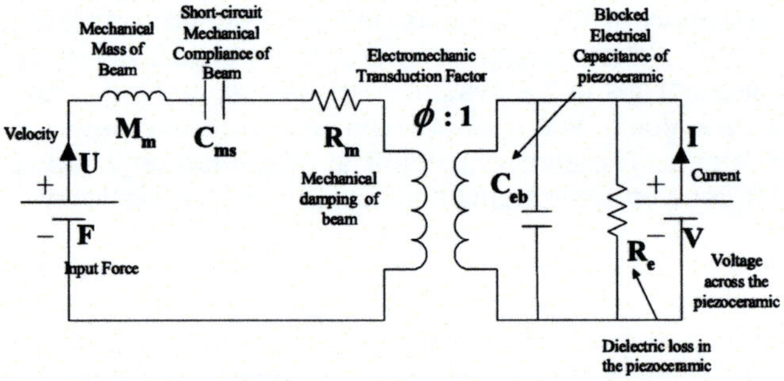

Figure 2: Overall equivalent electromechanical circuit representation of a piezoelectric composite beam.

The device consists of a cantilever shim with an attached piezoelectric layer and a proof mass. The device, when directly attached to a vibrating surface, converts the input base acceleration into an effective inertial force at the tip that deflects the beam, thereby inducing mechanical strain in the piezoelectric layer. This strain produces an effective voltage in the layer that is converted into usable power when connected to a power extraction circuit. The composite beam is analyzed using first-principles based composite Euler-Bernoulli beam theory, and an equivalent electromechanical circuit model is formulated using the lumped element model (LEM) technique [13]. The lumped element modeling technique is effective in analyzing and designing coupled energy systems using equivalent circuit elements to represent the device behavior. Furthermore, it enables complete system level simulation using existing circuit analysis tools. Equation 1 governs the coupled LEM system shown in Figure 2. Here, *F* is the effective force applied to the structure, *U* is the resulting tip velocity, *V* and *I* are the output voltage and current generated at the ends of the piezoelectric material. All of the parameters are obtained by lumping the energy at the tip using the relative motion of the tip with respect to the clamp/base [13].

$$\begin{bmatrix} U \\ I \end{bmatrix} = \begin{bmatrix} j\omega C_{ms} & j\omega d_m \\ j\omega d_m & j\omega C_{ef} \end{bmatrix} \begin{bmatrix} F \\ V \end{bmatrix}. \quad (1)$$

where $C_{ms} = \left(U / j\omega F\right)\big|_{V=0}$ is the short circuit compliance, $C_{ef} = \left(I / j\omega V\right)\big|_{F=0}$ is the free electrical capacitance, and $d_m = \left(U / j\omega V\right)\big|_{F=0} = \left(I / j\omega F\right)\big|_{V=0}$ is an effective reciprocal piezoelectric constant.

DEVICE FABRICATION

The test structures are fabricated using conventional surface and bulk micro processing techniques. The first three major steps employ the PZT layer deposition and patterning process at the Army Research laboratory (ARL). ARL's standard process involves three masks for depositing and patterning the PZT layer along with its electrodes. First, the base SOI wafer is RCA cleaned to remove any surface particles and impurities. Silicon dioxide (0.1 µm) is deposited on the wafer using the Plasma Enhanced Chemical Vapor Deposition (PECVD) technique. This layer is necessary to prevent any lead diffusion into the silicon overlayer during the PZT anneal step. Any lead out diffusion will immensely affect the performance of the PZT to the extent that it will lose its piezoelectric property. Then, Titanium (Ti, 20 nm) and Platinum (Pt, 200 nm) are sputter deposited on the whole wafer to form the bottom electrode. A Titanium seed layer is provided to ensure good adhesion between Pt and the substrate. Next, sol gel PZT is deposited using a 3-step process involving spin, bake and anneal steps. The sol gel is initially spun at a certain rotational speed in a dedicated deposition spinner. Then, the wafer is baked on a hot plate for a short interval usually, 50-60 seconds. For higher PZT thicknesses, multiple spin and bake steps are carried out before the final anneal step is done at a much higher temperature in a rapid thermal anneal (RTA) furnace. Multiple spin and bake steps are adopted to maintain consistent PZT properties even at higher thickness. Consequently, the top electrode for the PZT is patterned, and then platinum is deposited and patterned using the lift-off photolithography technique. Next, openings are patterned on the devices to provide access to the bottom electrode. The exposed PZT in those areas is wet etched revealing the bottom Pt underneath. The final step in this process involves patterning the electrode area. The exposed PZT and bottom Pt/Ti are physically etched using ion milling process resulting in the PZT features along with top and bottom electrodes with provisions for bond pads. The mask shapes are indicated in the process flow shown in Figure 3.

After the PZT process is completed, the wafers are returned for the device release steps carried out at UF. Before performing the release, Gold (Au) is deposited on the wafer and then etched using a Potassium Iodide solution while protecting the bond pads that are patterned previously. Next, the release trenches are patterned for the beams. This is followed by wet etching the pad oxide with BOE that is deposited in the first step. Consequently, the 12 µm Si overlayer is removed using DRIE with the STS tool. The etching from the top stops at the buried oxide layer (BOX). Finally, etch areas are patterned on the backside using back to front alignment. The exposed silicon is removed using the DRIE technique and stopped at the Box layer. Then, the 0.4 µm BOX layer is etched out using BOE and the remaining photoresist is stripped using a solvent clean with acetone, eventually releasing the cantilever devices. The individual dies are then mounted in a specially designed Lucite package using conductive Ag epoxy. The bond pads are connected to the copper posts with gold wires using the wire-bonder. In areas where wire-bonding is difficult, silver epoxy is used to create the contacts. The package shown in Figure 5 has a threaded hole in the bottom to be mounted on a vibrating shaker for characterization.

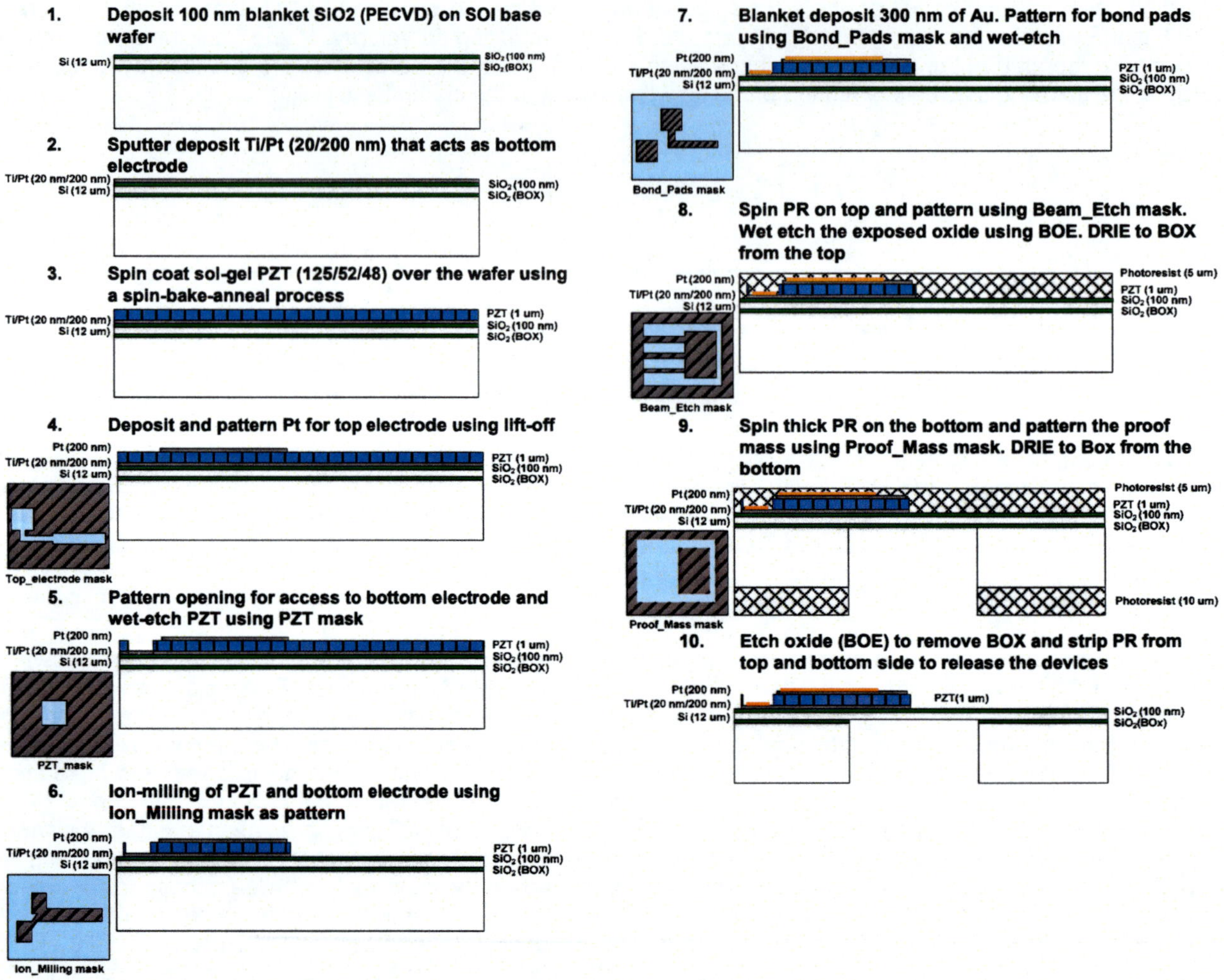

Figure 3: Device fabrication process flow for piezoelectric micro power generators.

The material properties and dimensions of the packaged PZT-EH-07 device are listed in Figure 4 along with a photograph of the actual MEMS device.

PZT-EH-07

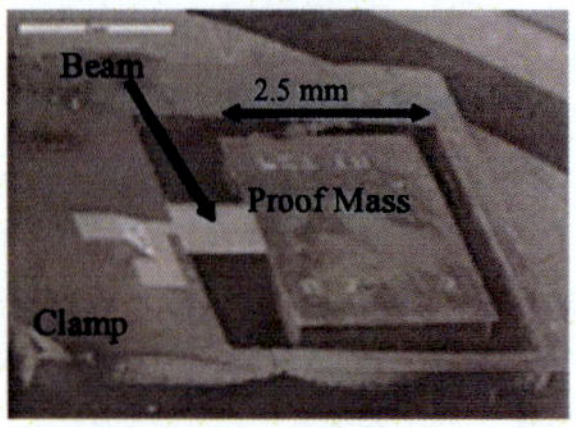

Length of beam	1 mm
Width of beam	1 mm
Thickness of beam	12 μm
Length of PZT	1 mm
Width of PZT	1 mm
Thickness of PZT	1 μm
Length of proof mass	2.5 mm
Width of proof mass	4 mm
Thickness of proof mass	500 μm

Figure 4: Device dimensions and photograph for PZT-EH-07.

EXPERIMENTAL SETUP AND CHARACTERIZATION

The experimental setup used to characterize the PZT energy harvesters is shown in Figure 5. The packaged MEMS PZT device is rigidly mounted on a Bruel & Kjaer mini shaker (type 4810) using a 1" 10-32 threaded screw with a locking nut. The setup is mounted underneath an Olympus BX60 optical microscope to perform velocity measurements using a Polytec (PSV 300) scanning laser vibrometer with a microscope adapter (OFV-074). A signal is generated using a waveform generator to drive the shaker through a B&K power amplifier (type 2718).

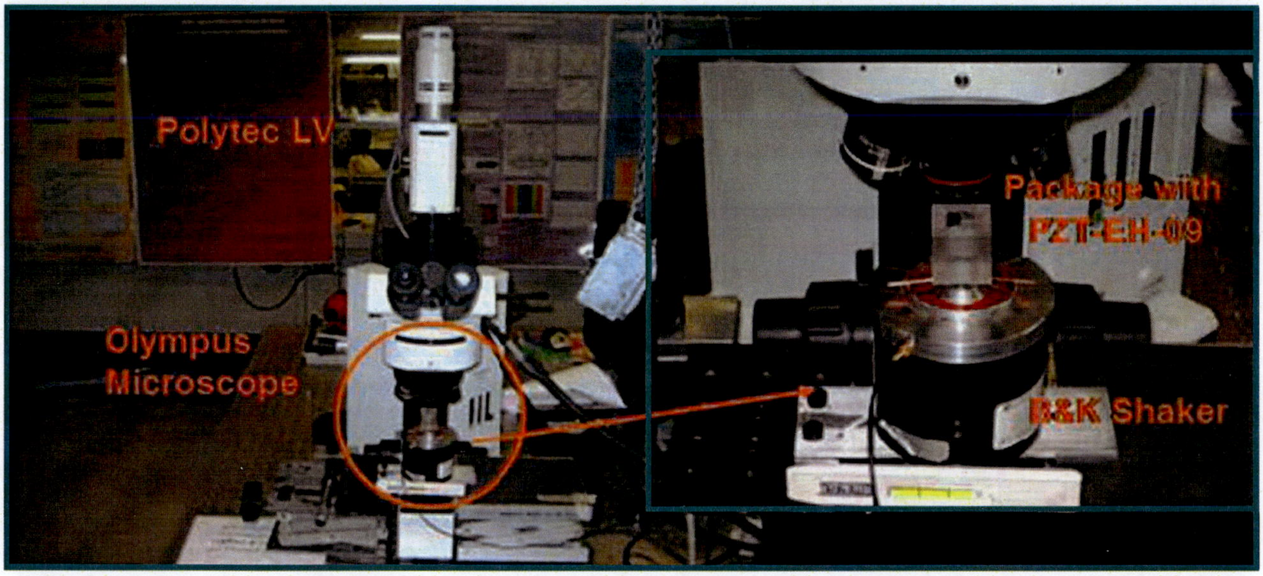

Figure 5: Experimental setup for device characterization

For electrical characterization, blocked impedance measurements are first carried out using an Agilent Impedance Analyzer across the electrodes of the piezo layer to obtain the effective blocked capacitance (C_{eb}) and dielectric loss tangent (*tanδ*). Using the thickness of the PZT layer $1.02 \pm 0.1\ \mu m$, the overall dielectric permittivity and the dielectric loss tangent are estimated to be 943 ± 18 and $(1.41 \pm 0.46) \times 10^{-2}$. For mechanical and electromechanical characterization, the input base acceleration to the device is measured at the clamp of the device and the tip deflections are recorded by the vibrometer. The effective piezoelectric coefficient (d_m), defined as the static tip deflection for an applied voltage to the PZT, is obtained using the low frequency asymptotic dc response of the device when excited using an ac voltage. Next, for the open circuit voltage response, a known periodic chirp acceleration signal is applied to the device mechanically through the dynamic shaker, and the resulting voltage is measured across the PZT layer. Finally, a sinusoidal signal is generated using the waveform generator at the measured resonance frequency of each device to mechanically excite the device via the shaker. A low vibration level of 0.1 *g* is used operate the device within the linear regime to validate the model and avoid geometric nonlinearities. Furthermore, excessive deflections at resonance can potentially result in device failure. Furthermore, small deflections are required to validate the model within the linear regime. The resulting output voltage at the PZT is measured across various resistive loads (1 *kΩ* - 750 *kΩ*) to observe the voltage and power characteristics.

RESULTS AND DISCUSSION

An extraction algorithm is used to obtain all the lumped element parameters required to represent the circuit [25]. The device responses obtained in the characterization experiments conducted above are modeled using frequency response equations that model the device behavior. These equations are applied to different frequency regions such as "*low frequency region*" and near "*resonance*" and fitted with an effective set of LEM parameters. This approach simplifies the expressions facilitating easy parameter extraction via the method of least squares.

All of the extracted LEM parameters using measured responses are summarized in Table 2 for two PZT-EH-07 devices and compared with model predictions. The short circuit mechanical parameters such as M_m, C_{ms} and F_n match very well with theory (< 5%), thereby validating the beam model for current designs. Furthermore, the electrical parameters, C_{eb} and $tan\delta$ also match well with the model. However, the measured effective piezoelectric coefficient d_m is considerably smaller than the value predicted in the model as discussed below. Consequently, other related piezoelectric parameters such as the coupling and transduction factors do not match well with the model.

	Model	PZT-EH-07-02	PZT-EH-07-03
M_m (mg)	4.71	4.68	4.72
C_{ms} (m/N)	33.71 x 10^{-2}	33.76 x 10^{-2}	32.30 x 10^{-2}
ζ	0.01*	0.12 x 10^{-2}	0.05 x 10^{-2}
f_n (Hz)	126.30	126.60	128.80
d_m (μm/V)	-1.91	-1.25	-1.66
ϕ (μN/V)	5.66	3.70	5.14
κ^2	9.19 x 10^{-4}	3.79 x 10^{-4}	6.87 x 10^{-4}
C_{eb} (nF)	11.84	12.19	12.41
tanδ	0.02*	1.94 x 10^{-2}	0.99 x 10^{-2}

* assumed values in the model

Table 2: Comparison between theory and experiments for PZT-EH-07

The excessive uncertainties in these parameters and the additional parasitic capacitances may have resulted in the large differences. Factors such as residual stresses, non-uniformity in piezoelectric properties on the wafer and further poling of the PZT may play an important role in finally determining the actual piezoelectric parameters. In addition, the uncertainties associated with the fabrication process for each batch of devices coupled with the dimensional tolerances of the structures need to be considered for the differences in measurements. Furthermore, characterization of many devices of the same type is required to understand the electromechanical behavior and obtain better repeatability of measurements between devices. Many improvements in the fabrication process and device characterization have also been proposed [25].

Dissipation	Quality Factor
Air Flow	536
Supports	5.71 x 10^6
Thermoelastic	56200
Effective Q	531
Effective ζ	9.41 x 10^{-4}

Table 3: Effective damping for PZT-EH-07

From Table 2, it can be noted that although the mechanical LEM parameters are reasonably close to the model, the mechanical damping measured is significantly lower than assumed. In the LEM, an assumed value for the damping ratio, ζ=0.01 is used based on the measured damping for conventional cantilever devices. However, the MEMS devices produce a much higher quality factor as confirmed by the low damping ratios. Common dissipation mechanisms such as air damping, thermoelastic damping, support losses, etc. need to be investigated in some detail to estimate overall damping in the device. Empirical relations for some of these dissipation mechanisms have been formulated to calculate individual losses in the device and understand their scaling [25]. The quality factors for different loss mechanisms are calculated using the empirical relations and presented in Table 3. The newly calculated effective damping ratio is of the same order of magnitude as the extracted value from experiments. Further experiments and a more detailed investigation of the dissipation mechanisms are required to better understand and accurately predict the resonant behavior of these devices. This estimate also confirms that the assumed value for ζ does not apply for these devices. From the calculated quality factors, it appears that the air flow damping is the dominant mechanism, followed by thermoelastic damping. The air damping can be minimized by performing the characterization under vacuum, in which case the overall material damping will approach the thermoelastic limit.

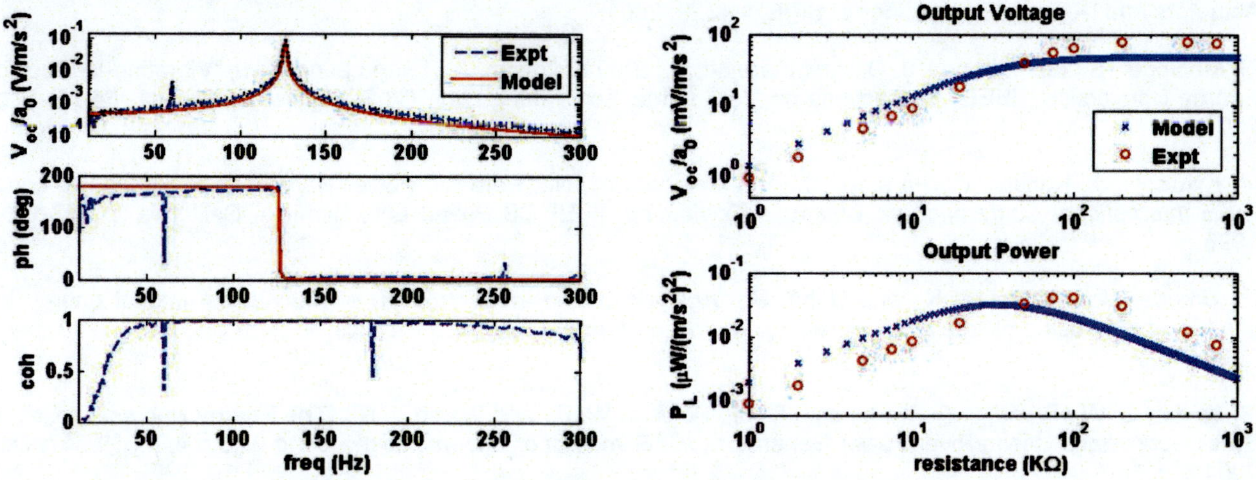

Figure 6: Open circuit voltage frequency response and output power for PZT-EH-07-03.

The measured open circuit voltage response and output power results across resistive loads are shown in Figure 6 for PZT-EH-07-03. Although the model fitted the various individual frequency responses well, the output power at resonance does not match well with model. The maximum power generated by the device at its resonance is 0.98 µW for a 1 m/s^2 vibration input. First, the most critical parameters determining the output power, damping ratio and the resonant frequency using the model are obtained from the extraction algorithm. The extracted resonant frequency does not match the resonant frequency where the actual data are measured. The predicted power output from the model is based on the extracted LEM parameters. In addition, due to very high *Q's* observed, the frequency resolution may have to be improved considerably to excite the device at its true resonance to obtain peak output power. As expected, the output voltage increases with increasing output resistance and finally approaches the open circuit voltage. On the other hand, the output rms power increases and reaches a maximum for an optimal output load and then decreases with increasing resistance. The model predicts the trend in the output power to a certain extent, but these fits can be further improved by implementing earlier proposed enhancements in the model, primarily in estimating system damping.

CONCLUSIONS

The development of a first generation MEMS-based piezoelectric energy harvester capable of converting ambient vibrations into storable electrical energy has been successfully demonstrated. The MEMS piezoelectric device is fabricated using standard sol gel PZT and conventional surface and bulk micro processing techniques. The developed prototype device produces 0.98 µW rms power into an optimal resistive load when excited with an acceleration of 1 m/s^2 at its resonant frequency of 129 Hz. The LEM predicts the device response with reasonable accuracy for all characterization experiments. Significantly high Q's are measured for the MEMS device which emphasizes the need to understand the individual loss mechanisms. The disadvantage with a higher Q resonant energy harvesting device is that a minor change in either the source vibration frequency or the device resonance will result in a significant drop in power generation. This work, coupled with a more detailed study on system damping and power extraction circuits, provides the required design tools for future development of MEMS piezoelectric energy harvester arrays as a potential power solution for self sustaining wireless embedded systems.

REFERENCES

[1] El-hami, M., Glynne-Jones, P., White, N. M., Hill, M., Beeby, S., James, E., Brown, A. D., and Ross, J. N., "Design and Fabrication of a New Vibration-based Electromechanical Power Generator," Sensors and Actuators A, p 335-342, 2001.

[2] Pelrine, R., Kornbluh, R., Eckerle, J., Jeuck, P., Oh, S., Pei, Q., and Stanford, S., "Dielectric Elastomers: Generator Mode Fundamentals and Applications", Smart Structures and Materials : Electroactive Polymer Actuators and Devices, Proceedings of SPIE, v 4329, 2001.

[3] Meninger, S., Mur-Miranda, J. O., Amirtharajah, R., Chandrakasan, A. P., and Lang, J.H., "Vibration-to-Electric Energy Conversion," IEEE Transactions on Very Large Scale Integration (VLSI) Systems, v9 , n1, 64-76, Feb 2001.

[4] Kymissis, J., Kendall, C., Paradiso, J., and Gershenfeld, N., "Parasitic Power Harvesting in Shoes", *Second IEEE International Conference on Wearable Computing*, IEEE CS Press, Los Alamitos, California, p 132-139, 1998.

[5] Umeda, M., Nakamura, K., and Ueha, S., "Analysis of the transformation of mechanical impact energy to electric energy using piezoelectric vibrator", Japanese Journal of Applied Physics, v 35, n 5B, 3267-3273, May 1996.

[6] Taylor, G. W., Burns, J. R., Kammann, S. M., Powers, W. B., and Welsh,T. R., "The Energy Harvesting Eel: A Small Subsurface Ocean/River Power Generator", *IEEE Journal of Oceanic Engineering* v 26 n 4, p 539-547, Oct 2001.

[7] Kasyap, A., Lim, J-S., Johnson, D., Horowitz, S., Nishida, T., Ngo, K., Sheplak, M., and Cattafesta, L., "Energy Reclamation from a Vibrating Piezoceramic Composite Beam," 9th International Congress on Sound and Vibration (ICSV9), Orlando, FL, Jul 2002.

[8] Roundy, S. and Wright, P.K., "A Piezoelectric Vibration Based Generator for Wireless Electronics", Smart Materials and Structures, n13, 1131-1142, 2004.

[9] Beeby, S. P., Tudor, M. J., and White, N. M., "Energy harvesting vibration sources for Microsystems applications", Measurement Science and Technology, v 17, R175-R195, 2006.

[10] Barth, P. W., Pourahmadi, F., Mayer, R., Poydock, J., and Petersen, K., "A monolithic silicon accelerometer with integral air damping and overrange protection," in *Tech. Dig. Solid-State Sensors and Actuators Workshop,* Hilton Head Island, SC, pp. 35–38, Jun 1988.

[11] Wang, Z-J., Maeda, R., and Kikuchi, K., "Preparation and characterization of sol-gel derived PZT thin films for micro actuators", Symposium on Design, Test and Microfabrication of MEMS and MOEMS, Proceedings of SPIE, v 3680, n 11, 948-955, 1999.

[12] Jeon, Y.B., Sood, R., Jeong, J.-h., and Kim, S.-G, "MEMS power generator with transverse mode thin film PZT", Sensors and Actuators A, n122, 16-22, 2005.

[13] Kasyap, A., "Development of MEMS-Based Piezoelectric Cantilever Arrays for Vibrational Energy Harvesting", PhD Dissertation, Department of Mechanical and Aerospace Engineering, University of Florida, 2007.

Proceedings of the IMAC-XXVIII
February 1–4, 2010, Jacksonville, Florida USA

ANALYSIS OF SWITCHING SYSTEMS USING AVERAGING METHODS FOR PIEZOSTRUCTURES

WOON K. KIM, ANDREW J. KURDILA, BRAD A. BUTRYM
VIRGINIAN POLYTECHNIC INSTITUTE AND STATE UNIVERSITY
BLACKSBURG, VA 24060

NOMENCLATURE

A	Cross-sectional area
A_p	Total electrode area
$\mathbf{A}_i(t)$	State matrix
b	Width of beam
$\mathbf{b}$	Electrical coupling vector
$B_p(x)$	Ectro-mechanical coupling parameter.
$\mathbf{B}_i(t)$	Input matrix
$\mathbf{c},(c_{11})$	Young's modulus vector, (component)
C	Capacitance
C_p^S	Inherent capacitance of PZT layers.
$\mathbf{d}$	PZT coefficient matrix
$\mathbf{D}$	Dielectric displacement matrix
$\mathbf{e},(e_{31})$	PZT stress constant matrix, (component)
$\mathbf{E}, E_3$	Electric field vector, (component)
$f(x_i,t)$	Discrete applied loads at x_i
$\mathbf{f}$	Force vector
$\mathbf{F}_i(t)$	Disturbance vector
h, h_p	Thicknesses of shim metal and PZT layer
$EI(x)$	Effective bending stiffness
I	2nd moment of area
$\mathbf{K}$	Stiffness matrix
$\mathbf{K}^*$	Normalized diagonal stiffness matrix
L	Length of beam
L_u	Linear differential operator for spatial variables
L_φ	Electrical potential operator
$\mathbf{M}$	Mass matrix
$\mathbf{M}^*$	Normalized diagonal mass matrix
$\mathbf{r}(t)$	Modal amplitude vector
q	Charge
Q_m	1st moment of area of piezoelectric layers,
$\mathbf{S}$	Mechanical strain vector
T	Kinetic energy or period time
$\mathbf{T}$	Mechanical stress vector
$\mathbf{u}(t)$	State input vector
U	Potential energy
$V, \tilde{V}$	Voltage, volume
$\mathbf{w},(w)$	Transverse displacement vector,(component)
W_{nc}	Non-conservative virtual work
$\mathbf{x}(t)$	State vector
z	Distance from neutural axis of the beam
$\boldsymbol{\varepsilon},(\varepsilon_{33})$	Dielectric permittivity matrix, (component)
ρ	Total density
ρ_s, ρ_p	Densities of the substrate and PZT.
φ	Electrical potential
$\chi_p(x)$	Characteristic function
$\boldsymbol{\psi}_r, \psi_{ri}$	Assumed mode shape
$\boldsymbol{\psi}_v, \psi_{vi}$	Assumed mode shapes of electrical field
$(\ )^{(2)}$	$=\partial^2/\partial x^2$

Superscripts

T	Transpose of a matrix
$\mathbf{T}$	Values taken at constant stress
$\mathbf{S}$	Values taken at constant strain
$\mathbf{E}$	Values taken at constant electric field

T. Proulx (ed.), *Structural Dynamics and Renewable Energy, Volume 1*, Conference Proceedings of the Society for Experimental Mechanics Series 10, DOI 10.1007/978-1-4419-9716-6_9, © The Society for Experimental Mechanics, Inc. 2011

Subscript

ij Strain (or stress) is applied in the j-axis and the normal direction of the electrode is i-axis.

ABSTRACT

This paper derives an analytical framework based on hybrid and switched system theory for the representation of some semi-actively shunted piezoelectric systems. In the case in which the gate driving frequency is a pulse-width modulated (PWM) signal, it is possible to obtain an approximation of the dynamic response of the shunted system by averaging the governing hybrid system equations. In contrast to some recent discretely switched (capacitive) shunted piezostructures, the semi-actively shunted piezostructure can be modulated continuously over the "adaptive bandwidth" by varying the duty cycle of the drive PWM signal. The adaptive bandwidth of the semi-actively shunted piezoelectric system is qualitatively similar to that of a corresponding discretely switched (capacitive) shunted piezostructure. The theoretical considerations in the paper are validated and verified via experiments.

INTRODUCTION

Techniques for achieving vibration absorption via "intelligent" or "smart" or "mechatronic" actuation have evolved from passive, to active, and most recently to *semi-active* methodologies. Passive techniques are often praised for their robustness and are typically the methods of first resort in design [1, 2]. Historically speaking, active vibration attenuation arose due to the need to improve performance over passive methods. This field is now mature enough that the foundations can be found in numerous textbooks, for example [3] gives a good account of the fundamentals. Semi-active vibration attenuation is a technique that resembles both passive and active methods. *A semi-active vibration attenuation system is characterized by the synthesis of low-power (usually switched) electronics that are intrinsically and physically integrated with the electromechanical system.* One of the goals of semi-active control techniques is to improve performance over passive approaches while reducing the energy requirements often associated with active control approaches. Hence, there is an emphasis on simple control methods that are implementable via energy efficient switching strategies.

Most frequently, semi-active systems with piezoelectric materials consist of passive shunt systems and active control systems. Piezoelectric materials, in conjunction with shunt circuitry, have been used to provide improved vibration attenuation compared to other passive control approaches. Hollkamp [4] studied a self tuning shunt circuit that tracks a particular mode by adjusting inductance and resistance. This strategy adapts the resonant frequencies of shunt circuit and the target structure. Clark [5, 6] and Corr et al [7-9] developed vibration suppression techniques using a piezoceramic actuator coupled to a switching system. Semi-active switching systems including capacitive shunt circuits are typically used to vary the effective stiffness of transducers used for vibration attenuation. This results in a broader bandwidth of resonant frequency shifts. Davis and Lesieutre [10] developed an adaptive transducer controlled by electromechanical switches such as relays with capacitive ladder shunt circuits. They showed that a finite number of shunt capacitive switches controlled the broadband discrete notch frequency shifts. Kurdila et al. [11] created a fast-switching topology by replacing multiple switches with a single PWM controlled switch. In this paper, they proposed a simpler "averaging method" for analyzing this type of switching piezostructure.

The progress in semi-active vibration attenuation has been closely coupled to the development of switching strategies to reclaim energy from ambient vibrations. Ottman et al [12, 13] developed a pulse-width modulated switching converter. The goal of the design was to maximize power transfer from the piezoelectric system to the energy storage device. According to the authors, a 400% power efficiency increase over the non-PWM switching system was achieved. Lesieutre et al [14] exploited electrical damping effects to enhance the energy conversion from mechanical energy to electrical energy.

This paper considers analytical methodology which employs a fast-switched, capacitively shunted system. The switching topology effectively induces modal responses that can be adaptively tuned via modulation of the duty cycle. The switching system modulates capacitance and is used to vary the effective stiffness. This strategy enables amplitude attenuation over a broader bandwidth. This system can be designed such that it provides a continuous range of operation by use of the switching subsystem. The present study introduces an enhanced averaging analysis of the switching shunt circuit for the piezoelectric structure for finding a predictive mathematical model of a PWM switching system. This method is used to approximate the solution which is not readily available for time-variant systems. Analysis or simulation of the averaged approximation is much faster and simpler than the analytical solution of the original systems.

LINEAR PIEZOSTRUCTURAL SYSTEM MODELING

The governing equations for a linear piezostructural system are readily available in the literature [3, 15, 16]. The governing equations are derived from well-known principles of linear piezoelectricity and linear electrical networks. The equations are time invariant as long as the topology of the network is time invariant, and we invoke the usual assumptions of linearity of the electrical components and transducer elements. With the introduction of idealized switches, the network topology changes abruptly in time, and the right hand side of the equations are *discontinuous*. In this paper an averaging method is used to approximate the discontinuous system as a continuous system. Switching between various electronic network topologies can be modeled with the introduction of *hybrid or switched systems*. In this case, the dynamics can be represented in terms of a collection of discrete switching states and continuous states.

In this section, the bimorph (symmetric) cantilevered structure is considered. The two piezoelectric layers are of uniform thickness. For these structures, it is assumed that the adhesion to the substrate shim metal is perfect. Also, the shim metal between piezoelectric layers is perfectly conducting. The voltage varies linearly along the piezo element layer thickness, where the voltage is distributed along the z –axis (31 mode of operation). We use Hamilton's Extended Principle (HEP) for piezoelectricity to derive the closed form solution of the governing equations. Shown in Figure 1 is the schematic of the bimorph beam that is considered in this study.

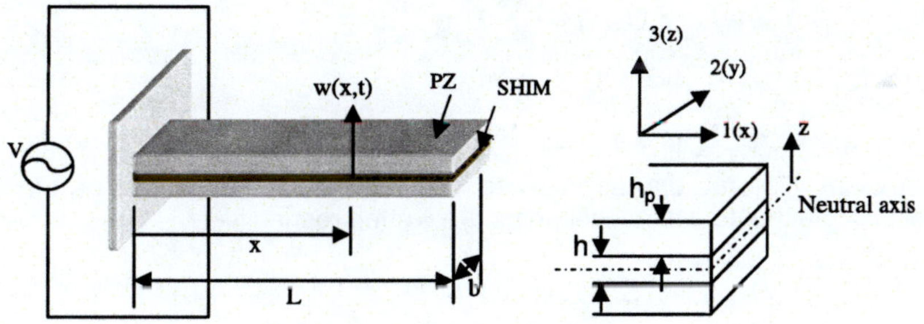

Figure 1. The bimorph-cantilevered structure symmetrically covered with piezoelectric layers.

Hamiltonian's principle can be expressed in integral form over the time interval t_1 to t_2. This principle is developed for linear piezoelectricity in the literature [1, 15, 17, 18].

$$H \triangleq \int_{t_1}^{t_2} \delta L dt + \int_{t_1}^{t_2} \delta W_{nc} dt = 0 \qquad (1)$$

$L = T - U$ is the Lagrangian of dynamic systems. We will now introduce the constitutive law for linear piezoelectricity. The relationship between mechanical deformation and electrical charge is written as [19]

$$\begin{bmatrix} \mathbf{T} \\ \mathbf{D} \end{bmatrix} = \begin{bmatrix} \mathbf{c}^{\mathbf{E}} & -\mathbf{e}^{T} \\ \mathbf{e} & \boldsymbol{\varepsilon}^{\mathbf{S}} \end{bmatrix} \begin{bmatrix} \mathbf{E} \\ \mathbf{S} \end{bmatrix} \qquad (2)$$

$\boldsymbol{\varepsilon}^{\mathrm{S}}$ is expressed in terms of $\mathbf{d}$ and $\boldsymbol{\varepsilon}^{\mathrm{T}}$

$$\boldsymbol{\varepsilon}^{\mathrm{S}} = \boldsymbol{\varepsilon}^{\mathrm{T}} - \mathbf{e}\mathbf{d}^{T} = \boldsymbol{\varepsilon}^{\mathrm{T}} - \mathbf{d}\mathbf{c}^{\mathrm{E}}\mathbf{d}^{T} \tag{3}$$

The kinetic and potential energies in the piezostructure are defined as,

$$T = \frac{1}{2}\int_V \rho \dot{\mathbf{w}}^T \dot{\mathbf{w}} d\tilde{V} \tag{4}$$

$$U = \frac{1}{2}\int_V \mathbf{E}^T \mathbf{D} d\tilde{V} - \frac{1}{2}\int_V \mathbf{S}^T \mathbf{T} \tag{5}$$

The non-conservative work term includes external electrical and mechanical loads. The virtual work is expressed as,

$$\delta W_{nc} = \sum_{i=1}^{N_f} \delta \mathbf{w}(x_i, t)\mathbf{f}(x_i, t) - \sum_{j=1}^{N_f} \mathbf{E}_j \mathbf{q}_j \tag{6}$$

After the integration over time and by parts, the Hamiltonian formulation for the equations of motion can be described by substituting Equations (4) ~ (6) into Equation (1),

$$\begin{aligned} H &= \int_{t_1}^{t_2} \left[\delta(T-U) + \delta W_{nc}\right] dt \\ &= \int_{t_1}^{t_2} \Bigg[-\int_V \rho \delta \dot{\mathbf{w}}^T \dot{\mathbf{w}} d\tilde{V}\, dt + \int_V \delta \mathbf{E}^T \left(\boldsymbol{\varepsilon}\mathbf{E} + \mathbf{e}^T \mathbf{S}\right) d\tilde{V} + \int_V \delta \mathbf{S}^T \left(\mathbf{e}\mathbf{E} - \mathbf{c}\mathbf{S}\right) d\tilde{V} \\ &\quad + \sum_{i=1}^{N_f} \delta \mathbf{w}(x_i, t)\mathbf{f}(x_i, t) - \sum_{j=1}^{N_q} \mathbf{E}_j \mathbf{q}_j \Bigg] dt \end{aligned} \tag{7}$$

The standard differential operators, L_u and L_φ, are introduced as shown below. The reader is referred to Hagood [1] and Muriuki and Clark [20] for the appropriate and detailed descriptions of these operators. Strain-displacement and electrical field-potential relationships are written respectively,

$$\mathbf{S} = L_u \mathbf{w}(x,t) \tag{8}$$

$$\mathbf{E} = L_\varphi \boldsymbol{\varphi}(x,t) \tag{9}$$

Analytic computations of composite structures, which are usually formulated as partial differential equations, require special mode shapes and time-dependent modal amplitudes. This is due to the complexity of the partial differential equations. If the mode shapes are given, modal amplitude can be expressed as a set of ordinary differential equations. This makes the numerical approximation of the switching system possible. Numerical approximation begins with modal expansion of the transverse displacement and the electrical potential;

$$\mathbf{w}(x,t) = \boldsymbol{\psi}_r(x)\mathbf{r}(t) = \sum \psi_{ri} r_i(t) \tag{10}$$

$$\boldsymbol{\varphi}(x,t) = \boldsymbol{\psi}_v(x)\mathbf{V}(t) = \sum \psi_{vi} V_i(t) \tag{11}$$

where i is i^{th} mode. Substituting Equations (10) and (11) into Equations (14) and (15) respectively, we have

$$\mathbf{S}(x,t) = \mathbf{N}_{\mathrm{r}}(x)\mathbf{r}(t) = \sum N_{ri} r_i(t) \tag{12}$$

$$\mathbf{E}(x,t)=\mathbf{N}_{\mathbf{v}}(x)\mathbf{V}(t)=\sum N_{vi}V_i(t) \tag{13}$$

where

$$\mathbf{N}_{\mathbf{r}}(x)=L_u\boldsymbol{\psi}_{\mathbf{r}}(x) \text{ and } \mathbf{N}_{\mathbf{v}}(x)=L_\varphi\boldsymbol{\psi}_{\mathbf{v}}(x)$$

In the case of the simple beam model (Euler – Bernoulli assumption), the elongation is in one direction along the x axis (or 1 axis). Strain distribution is

$$S=-z\frac{\partial^2 w}{\partial x^2} \tag{14}$$

The strain differential operator (L_u) can be defined as $-z\frac{\partial^2}{\partial x^2}$. Also, linear variation of the electric field through the PZT material thickness is given as

$$E=-\nabla\varphi(x,t)=-\nabla V/h_p \tag{15}$$

The equations of motion for the electro-mechanically coupled structure, which is a simple cantilever beam, are now presented. The kinetic energy of the piezostructure is

$$T=\int_0^L \rho(x)A\dot{w}(x)^2dx \tag{16}$$

$$\rho(x)=hb\rho_s+2\chi_p(x)h_pb\rho_p \tag{17}$$

$$\chi_p(x)=\begin{cases}1, & x\in \mathit{Patch}\\ 0, & \mathit{Otherwise}\end{cases} \tag{18}$$

The potential energy is given by

$$\begin{aligned}U&=\frac{1}{2}\int_0^L dx\int_A\left(\varepsilon_{33}E_3^2-2w^{(2)}ze_{31}E_3-c_{11}w^{(2)2}z^2\right)dA\\ &=\frac{1}{2}\int_0^L\left(\varepsilon_{33}E_3^2A_p-2w^{(2)}Q_me_{31}E_3-EI(x)w^{(2)2}\right)dx\end{aligned} \tag{19}$$

where $A_p=\int_A dA$, $Q_m=\int_A zdA$ and $EI(x)=c_sI^{(S)}+\chi_pc_{11}^EI^{(B)}+\chi_pc_{11}^EI^{(T)}$.

Therefore, the Hamiltonian formulation is taken

$$\begin{aligned}H&=\int_{t_1}^{t_2}\left[\delta(T-U)+\delta W_{nc}\right]dt\\ &=\int_{t_1}^{t_2}dt\int_0^L\left(-\rho A\ddot{w}-V(t)B_p(x)^{(2)}-\left(EI(x)w^{(2)}\right)^{(2)}+f_i\right)\delta w dx\\ &\quad+\int_{t_1}^{t_2}\delta V(t)\left[C_p^SV(t)+\int_0^L B_p(x)w^{(2)}dx-\sum_j^{N_q}N_{vj}q_j\right]dt+B.C.\\ &=0\end{aligned} \tag{20}$$

The coefficients are given by

$$C_p^S = \varepsilon_{33}^S \frac{A_p}{h_p}, \qquad B_p(x) = \frac{e_{31}}{h_p} Q_m \chi_p(x) \text{ and } \qquad Q_m = \frac{b}{2}\left[\left(\frac{h}{2}+h_p\right)^2 - \left(\frac{h}{2}\right)^2\right].$$

Two important strong form governing equations are derived. One is for the piezoelectric actuator equation and the other is for the sensor equation:

$$\rho A\ddot{w} + \left(EI(x)w^{(2)}\right)^{(2)} = V(t)B_p(x)^{(2)} + f \tag{21}$$

$$q = \sum_{j=1}^{N_q} N_{vj} q_j = C_p^S V(t) + \int_0^L B_p(x) w^{(2)} dx \tag{22}$$

AVERAGING ANALYSIS FOR THE SWITCHING SHUNT SYSTEM

The averaging method has been extensively studied for analyzing complicated switching behavior in power electronics and nonlinear vibrations [21, 22]. Recently, this approach was applied to piezoelectric switching systems in Kurdila et al. [11]. In order to analyze these systems, it is essential to cast the original linear time variant system model into a continuous and time-invariant one. In this section, the capacitively switched shunt model will be developed. The system considered is a natural generalization of the slower relay-switched systems, such as those found in Davis et al [10], to a fast-switching system. This section describes a prototype of a vibration absorber or energy harvester that can shift the notch frequencies among a finite-fixed collection of settings. In the special case, the dynamics can be represented in terms of a collection of multi-discrete switching states or a single-fixed switching state as shown Figures 2 and 3. Figure 2 is a compact representation of the linear piezoelectric system and a discrete switching system. The total capacitance in this system is attained via a collection of parallel capacitors that are engaged by actuating a finite number of discrete switches. In contrast, a fast-switching configuration is depicted in Figure 3. In this configuration, total capacitance of the shunt circuit is changed by using a switching cycle that is modulated very rapidly. This results in a system that (at least approximately) exhibits continuous variation in total capacitance.

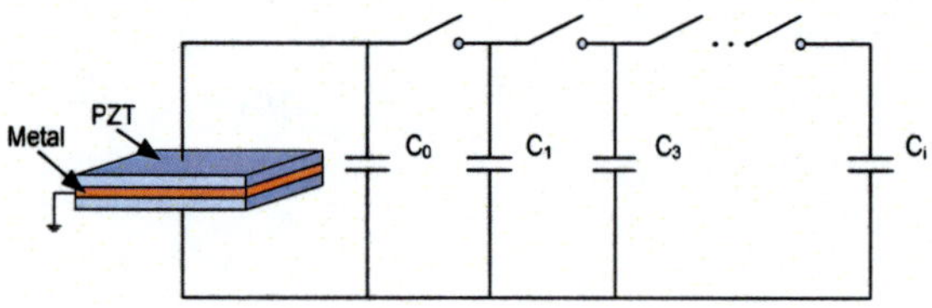

Figure 2. Multi-discrete capacitive shunt system.

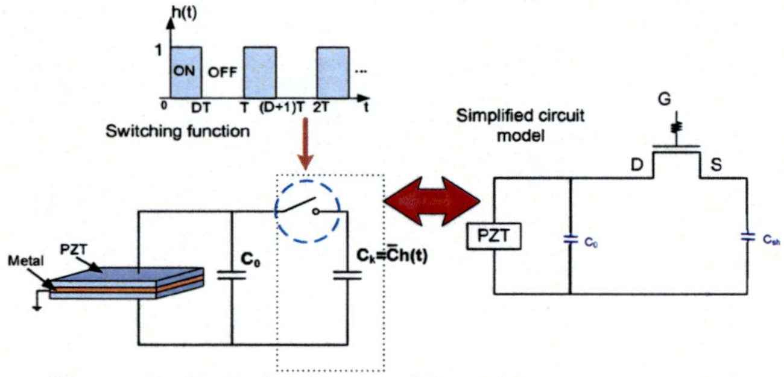

Figure 3. Switched capacitive shunt system and switching function

Specifically, we consider the fast switching system illustrated in Figure 3. Simply, the charges are a function of the sum of capacitances. Recall that charge and voltage are related by

$$q_{sh} = CV \tag{23}$$

According to KCV's law, the sum of the charge (q) from the PZT structure and charge (q_{sh}) from the shunt capacitance is

$$q_{sh} + q = 0 \tag{24}$$

The voltage across the shunt capacitance is taken by substituting Equation (24) into Equation (22), we have

$$V(t) = -\frac{\int_0^L B_p(x)\frac{\partial^2 w}{\partial x^2}dx}{C_p^S + C(t)} \tag{25}$$

where $C(t)$ is time-variant capacitance. This value is determined by the switching function ($h(t)$), which is driven by a pulse-width-modulated (PWM) signal, and characterized by the period T and duty ratio D. The pulse function $h(t)$ is one when the switch is on and zero when the switch is off. This is simply illustrated in Figure 3.

The shunt capacitance of the switching circuit is represented by the sum of the inherent capacitance of piezoelectricity, initial capacitance, and a single fixed capacitance.

$$C(t) = C_P^S + C_0 + \bar{C}h(t) \tag{26}$$

For an easy approach to the numerical analysis, strong form equations may be transformed into weak form equations. The weak form equation of the actuator is yielded by substituting Equation (25) into Equation (21), multiplying by test function ψ_r and integrating over the beam length,

$$\int_0^L \rho A\frac{\partial^2 w}{\partial t^2}\psi_{ri}dx + \int_0^L EI(x)\frac{\partial^2 w}{\partial x^2}\frac{\partial^2 \psi_{ri}}{\partial x^2}dx + \frac{\int_0^L B_p(x)\frac{\partial w^2}{\partial x^2}dx \cdot \int_0^L B_p(x)\frac{\partial^2 \psi_{ri}}{\partial x^2}dx}{C_p^S + C(t)} = \int_0^L f(x_i,t)\psi_{ri}dx \tag{27}$$

Now, we have a set of ordinary differential equations with respect to temporal amplitude $\mathbf{r}(t)$:

$$\mathbf{M}\ddot{\mathbf{r}}(t) + (\mathbf{K} + \Delta\mathbf{K})\mathbf{r}(t) = \mathbf{f} \tag{28}$$

where,

$$\mathbf{M} = \int_0^L \rho A\psi_{ri}\psi_{rj}dx$$

$$\mathbf{K} = \int_0^L EI(x)\cdot\psi_{ri}^{(2)}\psi_{rj}^{(2)}dx$$

$$\Delta\mathbf{K}(t) = \frac{\int_0^L B_p(x)\psi_{ri}^{(2)}dx \cdot \int_0^L B_p(x)\psi_{rj}^{(2)}dx}{C_p^S + C(t)} = \begin{cases} \dfrac{\mathbf{b}\cdot\mathbf{b}^T}{C_p^S + C_0 + \sum_{i=1} C_i} & : \text{For multi-switching system} \\ \dfrac{\mathbf{b}\cdot\mathbf{b}^T}{C_p^S + C_0 + \bar{C}h(t)} & : \text{For single switching system} \end{cases}$$

$$\mathbf{b} = \int_0^L B_p(x)\psi_{ri}^{(2)}dx$$

$$\mathbf{f} = \int_0^L f(x_i,t)\psi_{ri}dx$$

Switching topology and State-space representation

One of the objectives is to determine how to represent a piecewise affine system as smooth and linear. The dynamic system we consider introduces switching topology. Since the switching topology may change abruptly in time, the dynamics of the switched systems are mathematically and computationally complex. For this reason, we need the simplified but more accurate models for the physical representations. Our objective in this section is to develop the analytical techniques and tools for the design and analysis of the systems by adopting the averaging method. This method begins with state-space representation written as first order differential equations.

Under the mild assumptions that state vector $\mathbf{x}(t)$ is smooth and continuous during the switching interval and the transition, the switched piezoelectric system provides the piecewise linear system [23]:

$$\dot{\mathbf{x}}(t) = \mathbf{A}_i(t)\mathbf{x}(t) + \mathbf{B}_i(t)\mathbf{u}(t) + \mathbf{F}_i, \qquad i = 1,2 \tag{29}$$

After substitution of Equation (28) into (29), we have the state-space matrix form for the switching system for the averaging analysis.

$$\dot{\mathbf{x}} = \mathbf{A}_i\mathbf{x} + \mathbf{F}_i(t), \qquad i = 1,2 \tag{30}$$

where

$$\mathbf{A}_i(t) = \begin{bmatrix} \mathbf{0} & \mathbf{I} \\ -\mathbf{M}^{-1}\left(\mathbf{K} + \dfrac{\mathbf{b}\cdot\mathbf{b}^T}{C_p^S + C_0 + \bar{C}h(t)}\right) & \mathbf{0} \end{bmatrix} \tag{31}$$

It should be noted that the input matrix is zero in this special case. Now let's consider the switching interval. The ideal switch, which is driven by the PWM signal, is simply described by the switching period T and duty ratio, as illustrated in Figure 3. The state form of the dynamic model is then written as

$$\dot{\mathbf{x}}(t) = \begin{cases} \mathbf{A}_1(t)\mathbf{x}(t) + \mathbf{F}_1(t) & t \in \left[(n-1)T \quad \{(n-1)+D\}T\right] \\ \mathbf{A}_2(t)\mathbf{x}(t) + \mathbf{F}_2(t) & t \in \left[\{(n-1)+D\}T \quad nT\right] \end{cases}, \; n = 1,2,\cdots \tag{32}$$

where

$$\mathbf{A}_1(t) = \begin{bmatrix} \mathbf{0} & \mathbf{I} \\ -\mathbf{M}^{-1}\left(\mathbf{K} + \dfrac{\mathbf{b}\cdot\mathbf{b}^T}{C_p^S + C_0 + \bar{C}}\right) & \mathbf{0} \end{bmatrix} \tag{33}$$

$$\mathbf{A}_2(t) = \begin{bmatrix} \mathbf{0} & \mathbf{I} \\ -\mathbf{M}^{-1}\left(\mathbf{K} + \dfrac{\mathbf{b}\cdot\mathbf{b}^T}{C_p^S + C_0}\right) & \mathbf{0} \end{bmatrix} \tag{34}$$

Theoretical background of the averaging method

The averaging method will now be introduced for application to a typical switched linear piezostructural system. This approach is useful in that it can simplify the study of a switched system. Generally speaking, it converts a time-varying linear system into a simpler time-invariant system. In order to establish a continuous approximation from the original time-variant system, we introduce the time scaling factor ε. One common standard form for the averaging is given in Equation (35) [24, 25],

$$\frac{dx(t)}{dt}=\varepsilon f(t,x(t)),\ x(t_0)=x_0 \text{ and } 0<\varepsilon\ll 1 \tag{35}$$

with $f(t,x)$ continuous as a function of t and x. Under the assumption that $f(t,x)$ is T-periodic, the time average of Equation (35) is defined to be

$$f^0(\cdot)\triangleq\frac{1}{T}\int_0^T f(s,\cdot)ds \tag{36}$$

Then, Equation (36) is transformed into the approximate equations which have the form,

$$\dot{y}=\varepsilon f^0(y),\ y(t_0)=x_0 \tag{37}$$

Averaging analysis to switching shunt systems

The switched equation can be cast in a form that is suitable for the classical averaging technique by introducing a time scaling factor. The time scaling is written as

$$t=\varepsilon\tau \tag{38}$$

The parameter ε relates time scale and the number of switches. Therefore, the periodic time scale function is $t\in[0\quad NT]$ if $\tau\in[0\quad N]$. The switching function in rescaled time is defined as

$$\hat{h}(\tau)\triangleq h(\varepsilon\tau)=h(T\cdot\tau)=\begin{bmatrix}1, & \tau\in[(n-1), & n-1+D]\\ 0, & \tau\in[(n-1+D), & n]\end{bmatrix} \tag{39}$$

where $n=1,2,3,\cdots,N$. N is the number of switches. In order to apply the averaging method, we need to use Equation (38). In the time domain, the general averaging equation is represented as

$$\mathbf{x}_{avg}(t)=\frac{1}{T}\int_0^T\mathbf{x}(t) \tag{40}$$

$\mathbf{y}(\tau)$ is the averaging variable in the time scale. Therefore,

$$\begin{aligned}
\mathbf{y}(\tau) &= \frac{1}{TN}\int_0^{TN} \mathbf{x}(t)dt \\
&= \frac{1}{TN}\int_0^{TN} \mathbf{x}(T\tau)(Td\tau) \\
&= \frac{1}{N}\int_0^{N} \hat{\mathbf{x}}(\tau)d(\tau) \\
&\triangleq \hat{\mathbf{x}}_{avg}(\tau)
\end{aligned} \tag{41}$$

Another time-variant variable is $\dot{\mathbf{x}}(t)$ in left-hand side of the state-space representation. By using the chain rule,

$$\frac{d(\cdot)}{dt} = \frac{d(\cdot)}{d\tau}\frac{d\tau}{dt} = \frac{1}{\varepsilon}\frac{d(\cdot)}{d\tau} \tag{42}$$

$\dot{x}(t)$ in the scaled time can be rewritten as

$$\begin{aligned}
\mathbf{y}'(\tau) \triangleq \frac{d\mathbf{y}(\tau)}{d\tau} &= \frac{1}{TN}\int_0^{TN} \frac{d\mathbf{x}}{dt}(t)dt \\
&= \frac{1}{TN}\int_0^{TN} \frac{d\mathbf{x}(T\tau)}{d\tau}\frac{d\tau}{Td\tau}(Td\tau) \\
&= \frac{1}{T}\frac{1}{N}\int_0^{N} \frac{d\hat{\mathbf{x}}(\tau)}{d\tau}d\tau \\
&= \frac{1}{T}\hat{\mathbf{x}}'_{avg}(\tau)
\end{aligned} \tag{43}$$

where $\hat{\mathbf{x}}'_{avg}(\tau) = \frac{1}{N}\int_0^{N} \frac{d\hat{\mathbf{x}}(\tau)}{d\tau}d\tau$. And,

$$\begin{aligned}
\hat{\mathbf{A}}_{avg} &= \frac{1}{N}\int_0^{N} \mathbf{A}_i(\hat{h}(\tau))d\tau \\
&= \frac{1}{N}\left\{\sum_{n=1}^{N}\int_{n-1}^{n-1+D} \mathbf{A}_1(\hat{h}(\tau))d\tau + \sum_{n=1}^{N}\int_{n-1+D}^{n} \mathbf{A}_2(\hat{h}(\tau))d\tau\right\} \\
&= \frac{1}{N}\left\{\sum_{n=1}^{N}\begin{bmatrix} \mathbf{0} & \mathbf{I} \\ -\mathbf{M}^{-1}\left(\mathbf{K} + \frac{\mathbf{b}\cdot\mathbf{b}^T}{C_p^S + C_0 + \bar{C}}D\right) & -\mathbf{M}^{-1}\mathbf{Q} \end{bmatrix} + \sum_{n=1}^{N}\begin{bmatrix} \mathbf{0} & \mathbf{I} \\ -\mathbf{M}^{-1}\left(\mathbf{K} + \frac{\mathbf{b}\cdot\mathbf{b}^T}{C_p^S + C_0}(1-D)\right) & -\mathbf{M}^{-1}\mathbf{Q} \end{bmatrix}\right\} \\
&= \hat{\mathbf{A}}_{1avg}(D) + \hat{\mathbf{A}}_{2avg}(D)
\end{aligned} \tag{44}$$

where

$$\hat{\mathbf{A}}_{1avg} = \begin{bmatrix} \mathbf{0} & \mathbf{I} \\ -\mathbf{M}^{-1}\left(\mathbf{K} + \dfrac{\mathbf{b}\cdot\mathbf{b}^T}{C_p^S + C_0 + \bar{C}} D\right) & -\mathbf{M}^{-1}\mathbf{Q} \end{bmatrix}$$

$$\hat{\mathbf{A}}_{2avg} = \begin{bmatrix} \mathbf{0} & \mathbf{I} \\ -\mathbf{M}^{-1}\left(\mathbf{K} + \dfrac{\mathbf{b}\cdot\mathbf{b}^T}{C_p^S + C_0}(1-D)\right) & -\mathbf{M}^{-1}\mathbf{Q} \end{bmatrix}$$

and D is duty ratio of the modulation of PWM signal. Therefore, the averaged equations of motion are given as

$$\mathbf{y}'(\tau) = \varepsilon\left[\left\{\mathbf{A}_{1avg}(D) + A_{2avg}(D)\right\}\mathbf{y}(\tau) + \mathbf{F}(\tau)\right] \tag{45}$$

The next step is that we need to convert back to the domain of t. It is easy to see that $y(\tau) = \hat{x}_{avg}(\tau)$ transfers to:

$$\begin{aligned} \mathbf{y}(\tau) &= \hat{\mathbf{x}}_{avg}(\tau) \\ &= \mathbf{x}_{avg}(\varepsilon\tau) \\ &= \mathbf{x}_{avg}(t) \\ &\triangleq \mathbf{y}(t) \end{aligned} \tag{46}$$

and $y'(\tau)$ is

$$\begin{aligned} \mathbf{y}'(\tau) &= \frac{1}{T}\hat{\mathbf{x}}'_{avg}(\tau) = \frac{\varepsilon}{NT}\int_0^{NT} \frac{d\mathbf{x}_{avg}(\varepsilon\tau)}{d\varepsilon\tau} \\ &= \varepsilon\mathbf{x}_{avg}(t) \\ &\triangleq \varepsilon\dot{\mathbf{y}}(t) \end{aligned} \tag{47}$$

as defined above, $\mathbf{x}_{avg}(t)$ is $\dfrac{1}{NT}\displaystyle\int_0^{NT} \frac{d\mathbf{x}_{avg}(t)}{dt}$, which is the averaging over the time period $t \in [0 \quad NT]$. Substituting Equation (46) and Equation (47), the average model is rewritten as

$$\frac{d}{dt}\mathbf{y}(t) = \left[\mathbf{A}_{1avg} + \mathbf{A}_{2avg}\right]\mathbf{y}(t) + \mathbf{F}(t) \tag{48}$$

If the mass is invariant over the switching period, the dynamic equation with respect to the averaging variable $\mathbf{y}(t)$ can be written as

$$\mathbf{M}\ddot{\mathbf{y}}(t) + (\mathbf{K} + D\frac{\mathbf{b}\cdot\mathbf{b}^T}{C_p^S + C_0 + \bar{C}} + (1-D)\frac{\mathbf{b}\cdot\mathbf{b}^T}{C_p^S + C_0})\mathbf{y}(t) = \mathbf{f} \tag{49}$$

The above equation shows that the stiffness is a function of the duty cycle and capacitance. It is easy to see that natural frequency can be shifted by controlling duty cycle operated by the PWM-driven control input. It also implies that the semi-active switching system will provide (approximately) continuous natural frequency shifts.

Let Φ_r and η_r be the modal vector and modal amplitude at the rth mode respectively. The modal expansion of I-modes is given by

$$\mathbf{y}(t) = \boldsymbol{\Phi}\boldsymbol{\eta} = \sum_{r=1}^{I} \Phi_r \eta_r \tag{50}$$

Substitution Equation (50) into Equation (49), one finds,

$$\mathbf{M}^{*}\ddot{\boldsymbol{\eta}}\ (t) + \mathbf{K}^{*}\boldsymbol{\eta}\ (t) = \mathbf{f}^{*} \tag{51}$$

where

$$\mathbf{M}^{*} = \boldsymbol{\Phi}^{T}\mathbf{M}\boldsymbol{\Phi}$$

$$\mathbf{K}^{*} = \boldsymbol{\Phi}^{T}(\mathbf{K} + D\frac{\mathbf{b}\cdot\mathbf{b}^{T}}{C_p^S + C_0 + \overline{C}} + (1-D)\frac{\mathbf{b}\cdot\mathbf{b}^{T}}{C_p^S + C_0})\boldsymbol{\Phi}$$

$$= \mathbf{k}^{*} + D\frac{\boldsymbol{\theta}\ \cdot\boldsymbol{\theta}^{T}}{C_p^S + C_0 + \overline{C}} + (1-D)\frac{\boldsymbol{\theta}\ \cdot\boldsymbol{\theta}^{T}}{C_p^S + C_0}$$

$$\mathbf{k}^{*} = \boldsymbol{\Phi}^{T}\mathbf{K}\boldsymbol{\Phi}$$

$$\boldsymbol{\theta} = \boldsymbol{\Phi}^{T}\mathbf{b}$$

$$\mathbf{f}^{*} = \boldsymbol{\Phi}^{T}\mathbf{f}$$

From the modal expansion of the dynamic equation, we obtain the shifted natural frequency range. The effective frequency ω_r^{eff} in the averaged equations is written as

$$\omega_r^{eff} = \sqrt{\left(k_r^{*} + D\frac{\theta_r\cdot\theta_r^{T}}{C_p^S + C_0 + \overline{C}} + (1-D)\frac{\theta_r\cdot\theta_r^{T}}{C_p^S + C_0}\right)\Big/ M_r^{*}}\ , \qquad r = 1,2,\ldots \tag{52}$$

Equation (52) is readily interpreted as the ratio between natural frequency shift and natural frequency at closed-circuit condition.

$$\frac{\omega_r^{eff}}{\omega_r^{E}} = \sqrt{\left(k_r^{*} + D\frac{\theta_r\cdot\theta_r^{T}}{C_p^S + C_0 + \overline{C}} + (1-D)\frac{\theta_r\cdot\theta_r^{T}}{C_p^S + C_0}\right)\Big/\left(k_r^{*} + \frac{\theta_r\cdot\theta_r^{T}}{C_p^S + C_0 + \overline{C}}\right)} \tag{53}$$

where ω_r^{E} are the resonant frequency at the closed-circuit condition.

EXPERIMENTAL SET UP

The schematic of the experimental setup is shown in Figure 4. The piezoelectric specimen, T226-A4-503X, is used for the bimorph. The detailed material properties are described in Table 1. The boundary conditions are clamped and free. A laser vibrometer (OFV 303 Head sensor and OFV3001 controller) is used to measure the velocity of the tip at the free end. A force transducer (PCB 2008C02) was used to measure the input signal generated by an electromagnetic shaker. The SIGLAB was used to analyze the signal.

The electrodes on the top and bottom surfaces of the piezoelectric layers are connected to the electric circuits. For the capacitive shunt tests, two different types of switching systems: "Discrete Capacitor (DC) Configuration and "MOSFET Fast Switched (MFS) Configuration" are used. The DC configuration as illustrated in Figure 2 is a capacitance array operated by several switches. The capacitors are in parallel connection between switches. It is easy to see that the capacitance is maximized when all of the switches are closed. For the single switching test, MOSFET IRF 820 was used. The control input is a PWM signal. This signal is generated by a function generator (HP3341A).

Table 1. Material properties of beam and piezoceramics: T226-A4-503X (source :www.piezo.com)

Beam	Density(ρ_s)	8700 kg/m^3
	Effective length(L)	50×10^{-3}m
	Effective width(b)	31.8×10^{-3}m
	Thickness(h)	0.1266×10^{-3}m
	Elastic modulus(c_S)	95 Gpa
Piezoceramic	Density (ρ_p)	7800 kg/m^3
	d constant (d_{31})	-190e-12 m/V
	Zero strain capacitance (C_P^T)	50nF
	Dielectric (ε_{33}^T)	1800 ε^0
	Coupling coefficient (k_{31})	0.35
	Effective length(L)	50×10^{-3}m
	Effective width(b)	31.8×10^{-3}m
	Effective thickness (h_p)	0.2667×10^{-3}m
	Elastic modulus (c_{11}^E)	66 Gpa
Shunt capacitance	Initial shunted capacitance (C_0)	330 pF
	Shunted capacitance ($\overline{C}$)	480 nF

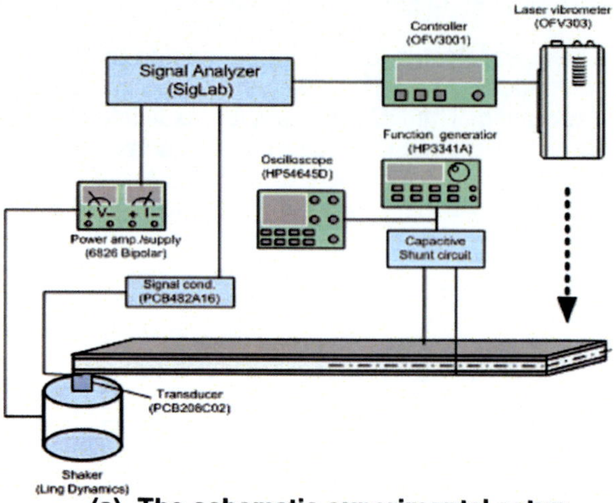

(a) The schematic experimental setup

(b) Bimorph beam

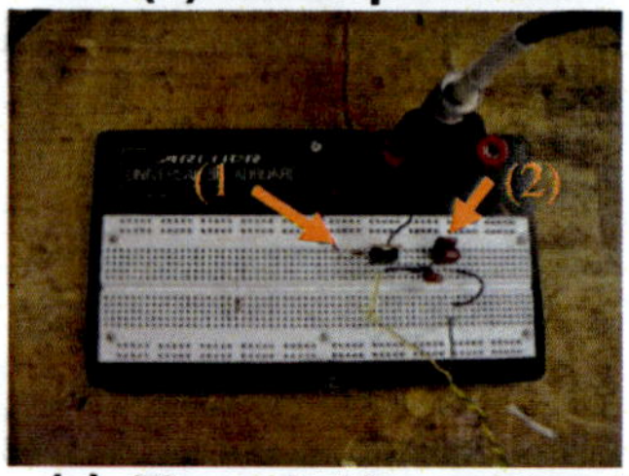

(c) The switching system:(1) MOSFT 820, (2) Single-fixed capacitance.

Figure 4. Experimental set up in bimorph cantilever beam

RESULTS AND DISCUSSION

The present study has two goals. The first is to obtain continuously varying natural frequencies by integrating a PWM-driven single-fixed switch with piezostructures. The second goal is to determine the analytic solution of the dynamic model for the time-varying switching system by introducing an averaging method. We also conducted an experiment to verify and validate the analytic solution. It should be noted that we only consider the first mode in this paper.

Figures 5 and 6 depict the most general results of the experiment. On the vertical axis the normalized effective frequency ratio $\omega_r^{eff} / \omega_r^E$ is plotted, where ω_r^{eff} is the resonant frequency of the shunted bimorph and ω_r^E is the natural frequency of the elastic short circuit bimorph. The normalized effective frequency ratio measures the frequency range over which the resonant frequency of the shunted bimorph may be shifted. The value of this ratio varies from $\omega_r^{eff} / \omega_r^E = 1$, to roughly $\omega_r^{eff} / \omega_r^E = 1.05$, which means that the frequency range over which the resonant frequency can be changed is roughly 5% of the short circuit frequency.

Figure 5 illustrates the ratio of the effective natural frequency with respect to the normalized discrete capacitances (C^{eff} / C_p). C^{eff} is the sum of the ladder capacitances and C_p is the initial capacitance. In this graph, the effective range of capacitance values is found between 0.01 and 10 times the initial capacitance. The data points marked in blue (*) in Figure 5 correspond to measurements made in the DC configuration. As noted in the figure, the data for the DC configuration is in good agreement with the theoretical prediction of Equation(28). This is also consistent with the results in [10].

Figure 6 suggests that we have an alternative to selecting among a fixed set of pre-selected design points in creating a resonance. The green (◊) represent the MFS configuration. The data in Figure 6 corresponding to the MFS configuration shows that the same range of effective frequencies is achieved as in the DC configuration, but the effective frequency of the shunted piezoelectric is controlled by the duty cycle D which can be modulated (nearly) continuously. Moreover, the data for the MFS Configuration are in good agreement with the analysis represented in Equation(53).

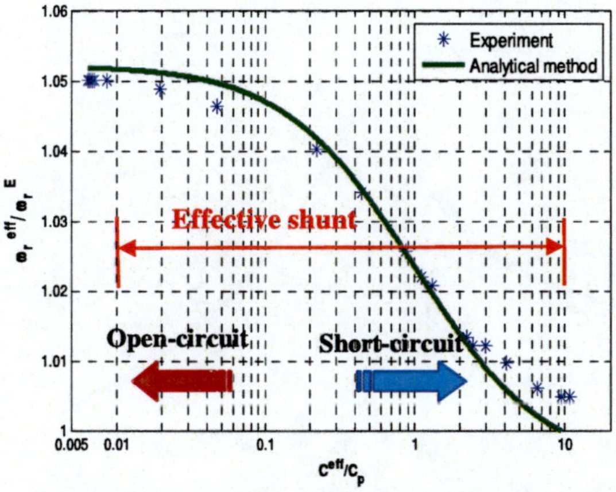

Figure 5. Normalized effective frequency versus (effective) shunt capacitance for DC configuration at 1st mode.

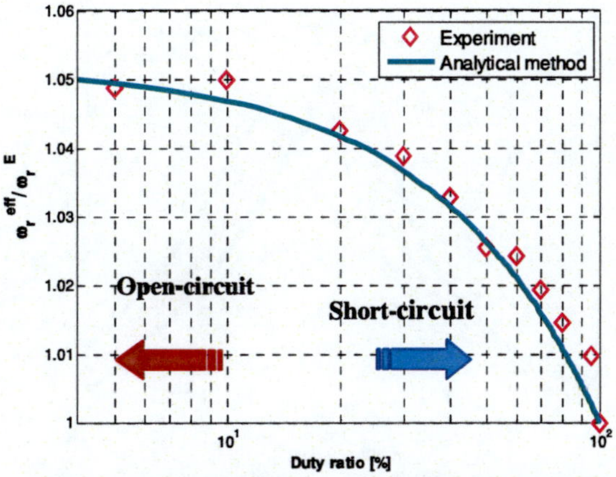

Figure 6. The comparison of experimental results and analytical solution at 1st mode: Normalized effective frequency versus Duty ratio in MFS configuration.

The following graphs show modal amplitude variation in the MFS configuration with respect to duty cycle and switching frequency (fs). It is easy to observe that the switching frequency is one of the important factors to ensure the continuous and equal variation of natural frequencies with duty cycle. Figures 7(a) and 7(b) show good performance of the variation of the modal amplitude of the transfer function when switching frequencies are between 100 Hz and 200Hz. In this range of switching frequencies, the natural frequency shifts (almost) equally vary when duty ratio changes.

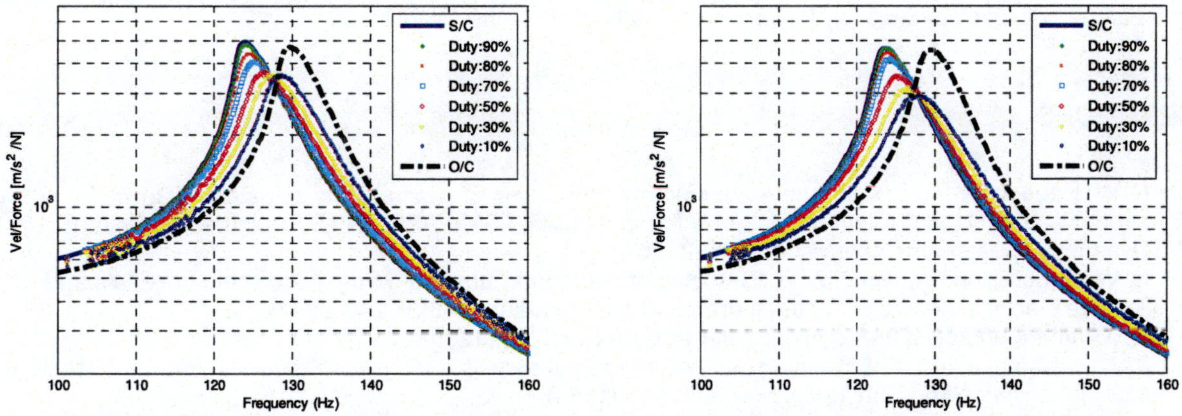

(a) Switching frequency (fs) = 100Hz **(b) Switching frequency (fs) = 200Hz**

Figure 7. Effective range of PWM switching frequency at 1st mode:

Unlike the previous Figures 7(a) and 7(b), Figures 8(a) and 8(b) show that it is not possible to generalize this observation beyond a range of effective frequencies. Figure 8(a) shows that when very low switching frequency is used (50Hz), the results are very noisy. Figure 8(b) shows that at high values of switching frequency (600Hz), all of the values of frequency for different duty cycles tend to accumulate near the short circuit condition.

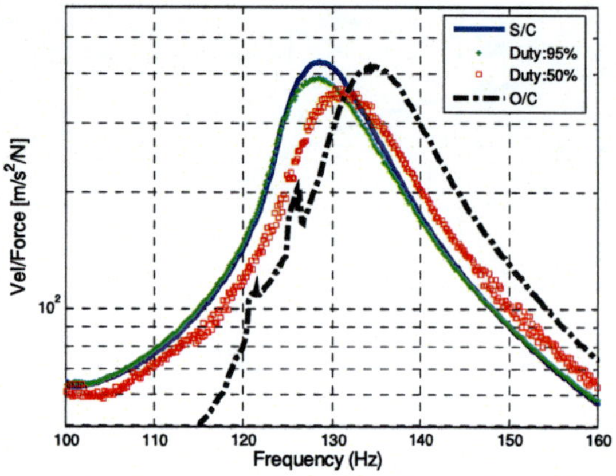

(a) Switching frequency (fs) = 50Hz

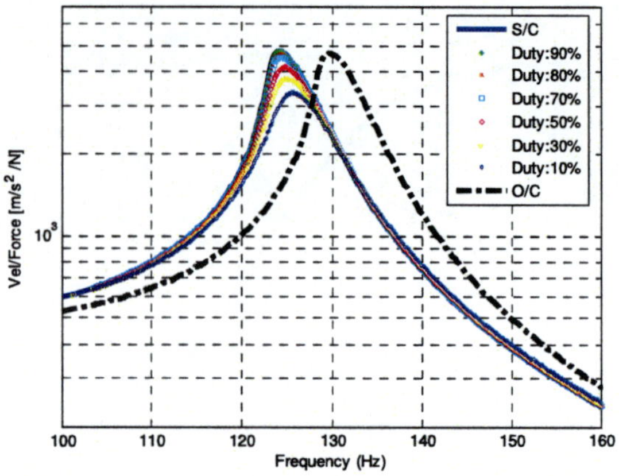

(b) Switching frequency (fs) = 600Hz

Figure 8. Effective range of PWM switching frequency at 1st mode

CONCLUSIONS

This paper has summarized a model for a type of semi-active piezostructural system that represents the dynamics in terms of hybrid system theory. It is shown that for the special case in which the gate-driving signal is restricted to a pulse-width modulated (PWM) signal, the hybrid system equations can be averaged. The averaged equations provide a simple, accurate estimate of the dynamics for some semi-actively shunted piezoelectric bimorph systems. The experimental study shows that the first mode of a semi-actively shunted piezoelectric bimorph can be adjusted continuously by varying the duty cycle of the PWM input.

REFERENCES

[1] N. W. Hagood, W. H. Chung, and A. von Flotow, "Modelling of piezoelectric actuator dynamics for active structural control," Collection of Technical Papers - AIAA/ASME/ASCE/AHS Structures, Structural Dynamics & Materials Conference. 2242-2256.

[2] N. W. Hagood, and A. von Flotow, "Damping of structural vibrations with piezoelectric materials and passive electrical networks," Journal of Sound and Vibration, 146(2), 243-68 (1991).

[3] D. J. Inman, [Vibration with control] John Wiley & Sons Ltd., (2006).

[4] J. J. Hollkamp, and T. F. Starchville Jr, "Self-tuning piezoelectric vibration absorber," Journal of Intelligent Material Systems and Structures, 5(4), 559-566 (1994).

[5] W. W. Clark, "Semi-active vibration control with piezoelectric materials as variable stiffness actuators," Proceedings of SPIE - The International Society for Optical Engineering, 3672, 123-130 (1999).

[6] W. W. Clark, "Vibration control with state-switched piezoelectric materials," Journal of Intelligent Material Systems and Structures, 11(4), 263-271 (2000).

[7] L. R. Corr, and W. W. Clark, "Comparison of low frequency piezoceramic shunt techniques for structural damping," Proc. SPIE - Int. Soc. Opt. Eng. (USA). 4331, 262-72.

[8] L. R. Corr, and W. W. Clark, "A novel semi-active multi-modal vibration control law for a piezoceramic actuator," Journal of Vibration and Acoustics, Transactions of the ASME, 125(2), 214-222 (2003).

[9] L. R. Corr, and W. W. Clark, "Similarities between variable stiffness springs and piezoceramic switching shunts," AIAA Journal, 44(11), 2797-800 (2006).

[10] C. L. Davis, and G. A. Lesieutre, "Actively tuned solid-state vibration absorber using capacitive shunting of piezoelectric stiffness," Journal of Sound and Vibration, 232(3), 601-617 (2000).

[11] A. J. Kurdila, G. A. Lesieutre , X. Zhanga *et al.*, "Averaging Analysis of State-Switched Piezoelectric Structural Systems," Proceedings of SPIE, Vol. 5760, 413-422 (2005).

[12] G. K. Ottman, H. F. Hofmann, A. C. Bhatt *et al.*, “Adaptive piezoelectric energy harvesting circuit for wireless remote power supply,” IEEE Transactions on Power Electronics, 17(5), 669-676 (2002).

[13] G. K. Ottman, H. F. Hofmann, and G. A. Lesieutre, “Optimized piezoelectric energy harvesting circuit using step-down converter in discontinuous conduction mode,” IEEE Transactions on Power Electronics, 18(2), 696-703 (2003).

[14] G. A. Lesieutre, G. K. Ottman, and H. F. Hofmann, “Damping as a result of piezoelectric energy harvesting,” Journal of Sound and Vibration, 269(3-5), 991-1001 (2004).

[15] R. R. Craig, and A. J. Kurdila, [Fundamentals of structural dynamics] John Wiley, (2006).

[16] D. J. Leo, [Engineering analysis of smart material systems] John Wiley & Sons, (2007).

[17] A. Preumont, [Mechatronics : dynamics of electromechanical and piezoelectric systems] Springer, Dordrecht :(2006).

[18] R. C. Smith, -, [Smart material systems : model development] Society for Industrial and Applied Mathematics, Philadelphia :(2005).

[19] “IEEE STANDARD ON PIEZOELECTRICITY,” American National Standards Institute, Standards, var paging-var paging (1987).

[20] M. G. Muriuki, and W. W. Clark, “Analysis of a technique for tuning a cantilever beam resonator using shunt switching,” Smart Materials and Structures, 16(5), 1527-1533 (2007).

[21] R. W. Erickson, and D. Marksimovic, [Fundamentals of Power Electronics] Kluwer academic publishers, (2004).

[22] J. Sun, and H. Grotstollen, "Averaged modelling of switching power converters: reformulation and theoretical basis," PESC '92 Record. 23rd Annual IEEE Power Electronics Specialists Conference (Cat. No.92CH3163-3). 1165-72.

[23] A. J. Kurdila, W. W. Clark, W. Wang *et al.*, “Stability of a class of real-time switched piezoelectric shunts,” Journal of Intelligent Material Systems and Structures, 13(2-3), 107-116 (2002).

[24] P. T. Krein, J. Bentsman, R. M. Bass *et al.*, “On the use of averaging for the analysis of power electronic systems,” IEEE Transactions on Power Electronics, 5(2), 182-90 (1990).

[25] J. A. Sanders, Verhulst, F., [Averaging methods in nonlinear dynamical systems] New York: Springer-Verlag, (1985).

Proceedings of the IMAC-XXVIII
February 1–4, 2010, Jacksonville, Florida USA
©2010 Society for Experimental Mechanics Inc.

In-field testing of a steel wind turbine tower

Marco Molinari[1], Matteo Pozzi[1], Daniele Zonta[1] and Lorenzo Battisti[1]

[1] Department of Structural and Mechanical Engineering, University of Trento,
Via Mesiano 77, 38100 Trento, Italy

ABSTRACT
The strong drive to exploit wind energy has recently led to consideration of new types of location for wind turbines, including mountain regions. A major concern for wind farm installation in these sites is the long-term reliability of the support structures. In a flexible steel tower, a combination of vortex shedding and gusting caused fatigue cracks at the base joint. To identify the conditions critical to the development of this phenomenon, a two-blade down-wind turbine was thoroughly investigated in an in-field experimental campaign. This turbine features a 13m diameter rotor of rated power 11kW, mounted on an 18m tubular steel tower. In operation, the blades rotate at a fixed rate of 2 Hz. The tower, instrumented with 15 accelerometers, was first dynamically characterized in the absence of significant wind. Next, its spectral response to wind excitation was identified both in operation and with the rotor at rest. The outcome of the experiment suggests that the vulnerability to fatigue of this model of turbine is very sensitive to its modal behaviour, this in turn depending on the mechanical admittance of the foundations.

1. INTRODUCTION

The strong drive to exploit wind energy has recently led to consideration of new types of location for wind turbines, including mountain regions. A major concern for wind farm installation in these sites is related to the long-term reliability of the supporting structures. In a survey of recent technical literature, we see growing interest by the scientific community in the dynamic behaviour of windmills, reflecting the massive economic investment by EU and US governments in this field [1]. Especially in the last three years, many technical reports feature methods for modal testing or structural health monitoring (SHM) as applied to wind turbines, while most major structural engineering conferences, including this one, more and more often include in their program special sessions dedicated to this topic [2, 3]. However, most of the work focuses on the dynamic behaviour either of the rotor blades [4] or of the single blades seen as elements independent of the rest of the structure [5,6]. Other studies concentrate on characterization of the action induced by wind on the rotor system, see for instance [7]. On the other hand, few papers examine the effect of rotor dynamics on the supporting tower (we mention the work by Lynch [8] and also [7] and [9]). It appears that current research is driven by the concern to prevent failure modes expected in the blade/rotor system, which is evidently seen as the most critical part of the plant.

Contrary to this general impression, we observe that wind-induced vibration can severely damage high lightweight masts: recent failures of lighting towers have raised questions as to the robustness and safety of similar existing structures [10]. Particularly for flexible steel towers, a combination of vortex shedding and natural wind gusts induces fatigue cracking, mainly at the base joint; small cracks gradually develop due to wind-induced cyclic loads and eventually the cracks with highest stress concentration propagate, occasionally resulting in a catastrophic failure. The vulnerability of steel joints to fatigue is often coupled with corrosion due to aging of the structure: as corrosion reduces the thickness of metal connections, the likelihood of a failure grows dramatically with the age of the structure. Predicting the strength of these structures is difficult, for a number of reasons: first, because cracks can initiate at different points, including the base flange-to-column weld, the handhole detail and the anchor rods [11]; second, the response depends on the complex interaction between wind action and dynamic vibration. More specifically, in the steel mast supporting the turbine, the aerodynamic vibration directly induced by wind is always coupled with that produced by the rotor. The latter effect is even more critical when the rotor mass become eccentric as a result of blade icing, or of an exceptional blade failure.

T. Proulx (ed.), *Structural Dynamics and Renewable Energy, Volume 1*, Conference Proceedings of the Society for Experimental Mechanics Series 10, DOI 10.1007/978-1-4419-9716-6_10, © The Society for Experimental Mechanics, Inc. 2011

For flexible steel structures, such as featuring hollow thin steel sections, fatigue is mainly related to dynamic amplification when ambient vibrations are close to some resonance frequencies, and particularly the locking phenomena slightly extend the resonance frequency range. Without correct design of the welding between the elements forming the pipe, and especially between the pipe and the base flange, additional local bending moments can cause fatigue cracks, reducing the operational life of the base flange and consequently of the whole mast. If early warnings are missed, fatigue collapse is particularly insidious and difficult to predict. Among the various SHM techniques, methods based on the detection of variations in dynamic modal parameters offer a cheap and effective way to cope with fatigue damage.

In this contribution we report the ongoing investigation of this approach in a specific case study: this is a prototype low-rise windmill, conceived for effective widespread application in mountain regions, and currently operating in the experimental wind farm in Trento, Italy. With the aim of predicting the in-field fatigue damage evolution in the supporting tower, the dynamic behaviour of this prototype was studied in detail via an experimental modal analysis, including the response to operational winds. In the next section of the paper we give the reader an overview of the ongoing pilot project and describe the specific windmill selected as a case study for this investigation. Section 3 illustrates the wind field characteristics, observed at site during two years continuous monitoring of wind action. Details of the dynamic experiment carried out on the supporting tower are reported in Section 4, while the outcomes of the dynamic characterization are reported and discussed in Section 5. Finally some concluding remarks are made at the end of the paper.

a) 

b) 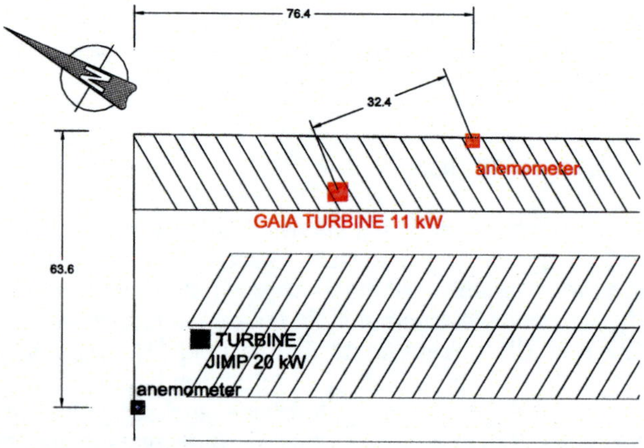

Figure 1. Overall view (a) and plan view (b) of Trento Experimental Wind Farm.

2. DESCRIPTION OF THE WIND FARM AND THE WIND TURBINE INVESTIGATED

The project targets the monitoring of a pilot wind turbine at the Trento Experimental Wind Farm [12,13], located at the bottom the Adige valley, north of the town of Trento, at altitude 200m asl. The objective of this farm is to create a reference in research on wind energy deployment in Italy and Europe. It provides the necessary equipment for analysis and comparison of structural and functional characteristics of mini and micro wind turbines. In particular, the extensive use of such turbines, their technical and "economic" efficiencies, their environmental and acoustic impact and the problems during installation and removal are experimentally assessed in the farm.

To date three small turbines are operating in the wind farm: a three blade upwind turbine named JIMP20 (rated power 20kW, tower height 18m, rotor diameter 8m), a two bladed downwind turbine, named GAIA WIND, (rated power 11kW, tower height 18m, diameter of the rotor 13m) plus a three bladed upwind wind micro turbine ZEPHIR (rated power 1kW, tower height 9m, diameter of the rotor 1.8m). Apart from the instrumentation used in the specific test reported below in Section 4, all wind turbines and the masts are set up with a standard set of sensors, which acquire fluid-dynamic, functional and structural data. The data is recorded by an acquisition system at the wind farm and then sent to the Turbomachinery Laboratory of the University of Trento to be post-processed. Two dedicated wind measurement masts are placed according IEC standards [14] near each of the larger wind turbines.

These mount sonic anemometers at two heights, 9m and 18m, allowing wind turbulence to be fully characterized. Because of the wide use of this type of rotor in the European energy industry, the two bladed GAIA windmill, shown in Fig. 1, has been chosen as a case study for this investigation. Due to the favorable mechanical layout of the GAIA turbine, this machine has been heavily instrumented to achieve good structural identification: one 2D accelerometer has been placed in the nacelle, one in the blade tip, and a strain gauge bridge captures flap and edge bending moments of the blade. Another strain gauge measures the mechanical torque in the low speed shaft. This experimental set up has been improved by a set of sensors. We expect the aerodynamic interaction of the blades with the upstream standing tower to cause large fluid-dynamic effects and stress levels in these components, with increased fatigue on both. Another set of sensors is dedicated to energy measurements.

The tower sustaining the GAIA turbine is 18m tall overall and consists of a 6 mm thick tapered hollow steel pipe, of circular section, having an external diameter varying between 938 mm, at the base, and 410 mm, at the top. It is built in three flanged tubular segments, connected with bolts. Each conical pipe segment is obtained by rolling and welding steel sheets of length 3.025 m; six of these elements were joined in pairs by full-section welding in the factory, thus obtaining 3 pieces 6.05 m long; end flanges are welded at the ends of the three elements, then joined in the field by preloaded high-strength bolts. The bottom plate was connected to a concrete base size 4.5 m square and 2 m thick. Layered sand, gravel and lime are present in the subsoil. The overall mass of the wind turbine, including blades, onboard gear box and alternator, was estimated in 940 kg.

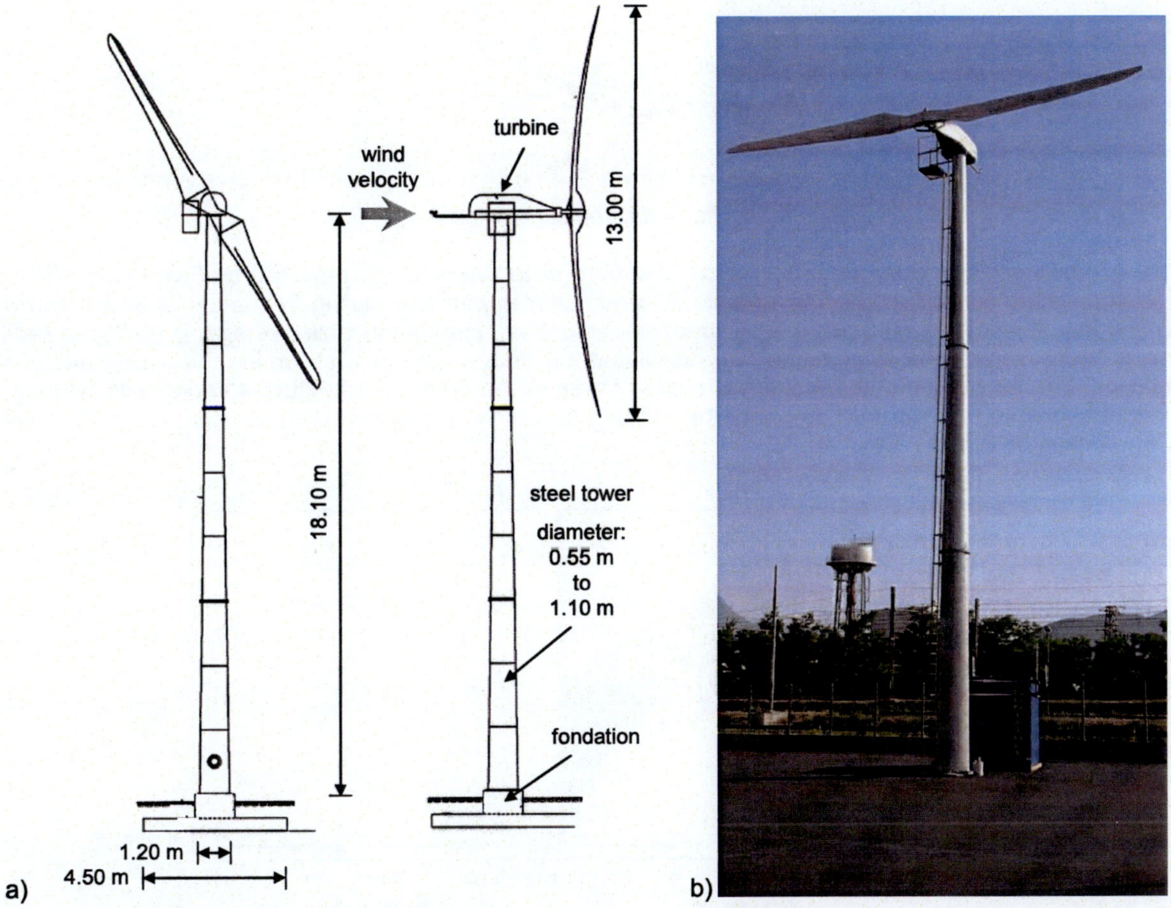

Figure 2. Scheme (a) and (b) picture of one of the GAIA wind turbines at the Experimental Wind Farm in Trento.

3. CHARACTERIZATION OF THE WIND FIELD

The test field of Trento was conceived according to standard IEC 62400-12 [15], and all measurements and corrections are taken according to this standard. The area is subject to thermal wind regimes, and, located in a typical complex terrain, exhibits very irregular winds with high levels of turbulence. This condition leads to intermittent operation of the wind turbines characterized by several hundred starts and stops per day. The irregularity of the wind causes widespread wind shear. As an example, Figure 3 reports ten different wind speed vertical profiles, all observed at an average speed U_{18} of approximately 10 m/s, recorded at level 18 m. The graph clearly highlights the variability of the wind profile in the operation space of the rotor, from 12 to 24 m. at the same average wind speed. This observation is of paramount importance to correctly characterize the dynamic response of the turbine to wind action.

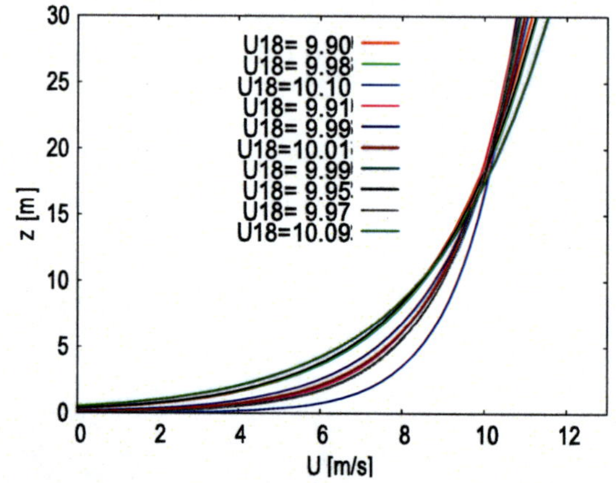

Figure 3. Wind shear patterns measured at the GAIA dedicated wind mast.

Figure 4 shows on the left in green the typical spectrum of turbulence *S(f)*, obtained by Fast Fourier Transform (FFT), of the time series recorded by a sonic 3D anemometer with a sampling frequency of 50 Hz and plotted against a logarithmic frequency scale. The spectrum, smoothed using a movable average, is plotted in red in the same graph. Based on the same signal, we calculated the power spectrum, assuming the hypothesis of frozen turbulence. The resulting power spectrum is plotted in red in the right box of Figure 4, along with Kolmogorov's condition, shown in blue, in order to verify the aliasing condition. The spectrum is consistent with the standard curve proposed by ESDU [16].

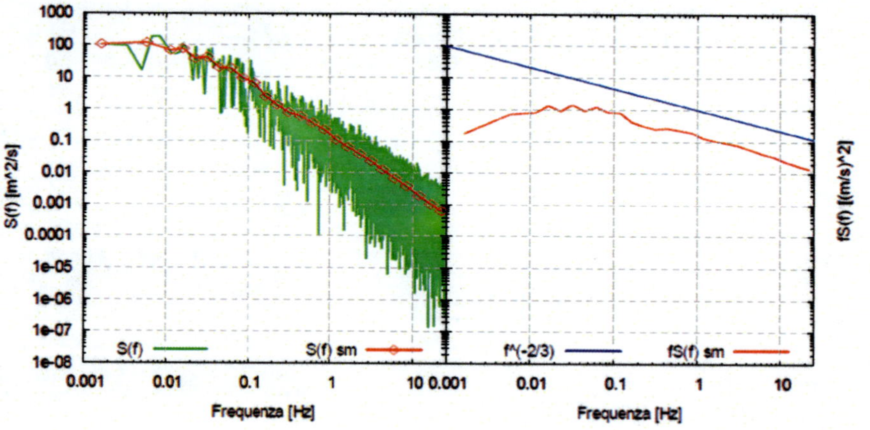

Figure 4. Representative Fourier spectrum (left) and power spectrum (right) of wind in the test field of Trento. Left side: raw data in green, averaged data in red. Right side: experimental power spectrum in red and Kolgomorov's condition in blue.

4. IN-FIELD DYNAMIC EXPERIMENT

4.1. Tests and instrumentation overview

To characterize the dynamic behaviour of the GAIA turbine, and specifically of its supporting tower, both Shock Hammer Tests (SHT) and Ambient Vibration Tests (AVT) were performed. SHT aimed to identify the frequency response function, and therefore the modal properties of the structure, by controlling the input force, in absence of wind and environmental noise. The purpose of AVT was to investigate the response of the mast in windy conditions; for this reason, data acquisition was repeated at various wind speeds and with the rotor blades both locked and free-rotating. To record the structural response, we used 13 piezoelectric accelerometers, models PCB 393C and PCB 393B12, as illustrated in Figure 5: five pairs of sensors were applied to the tower equally spaced at a vertical pitch of 3.025 m, oriented in two horizontal orthogonal directions x and y; the remaining three were arranged vertically above the foundations, in order to detect possible base rotation during vibration. Given the high torsion stiffness of the hollow tower, confirmed by preliminary numerical analysis, we did not need to use additional accelerometers to acquire torsion data. In all tests, the data sampling rate was fixed at 400Hz.

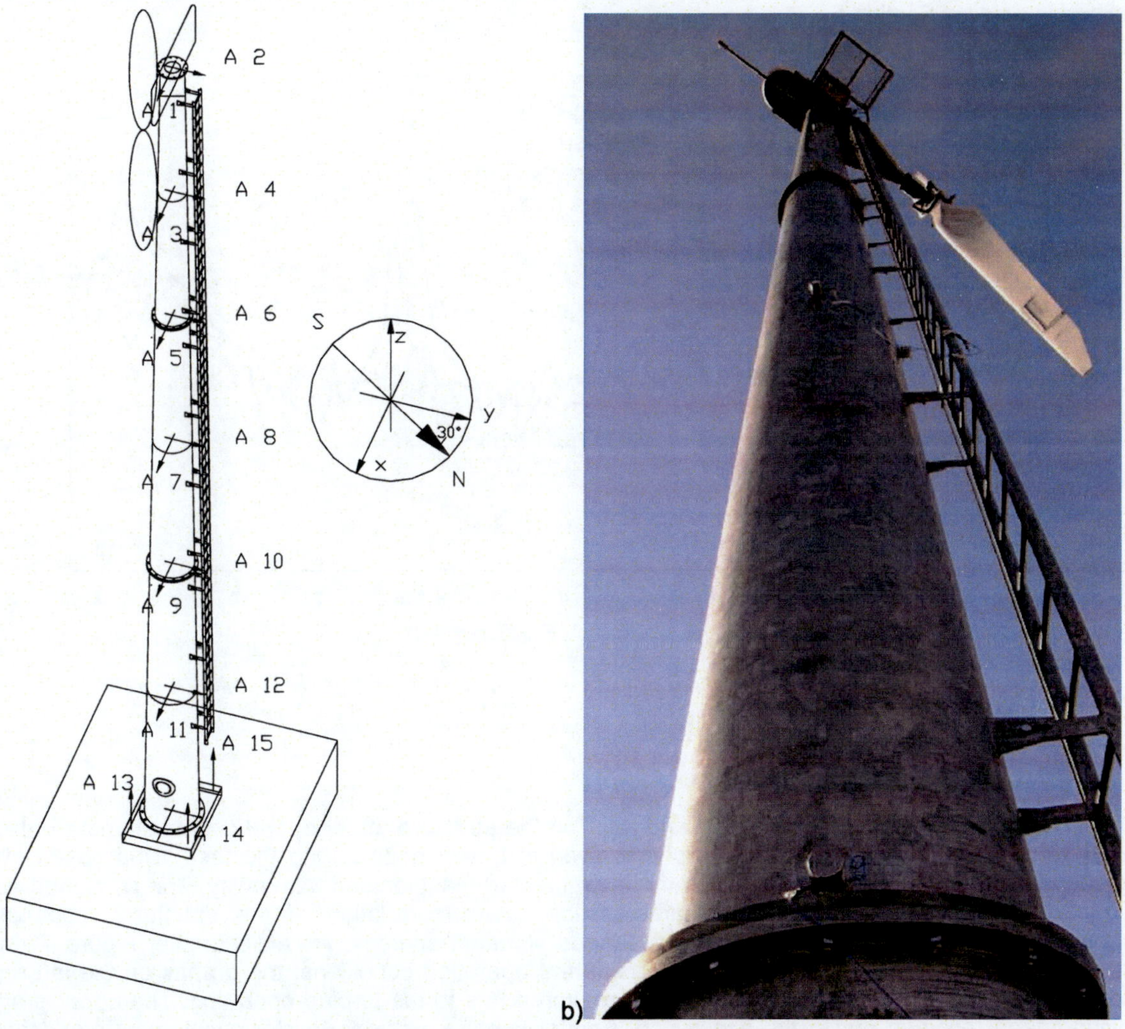

a) b)

Figure 5. Sketch of the supporting tower with accelerometer location and identification (a); detail view of the tower showing the accelerometers installed during the dynamic experiment (b).

4.2. Shock hammer tests

Because of the eccentricity between the rotor and tower axis, evidently the modal properties of the wind tower change with the angle of rotation of the nacelle. During the shock tests, the nacelle was therefore temporarily locked with the rotor blade axis in the x direction and fixed to the tower. In these tests both input and output signals in terms of acceleration are acquired, thus allowing direct computation of the Frequency Response Function (FRF) of the system. Impact tests were performed using a PCB 086D50 instrumented sledgehammer with an extra-soft tip characterized by a cut-off frequency of 80 Hz.

The tests were repeated a number of times choosing different excitation points: impact forces were applied at the two connection flanges of the tower, located at level 6.05 m and 12.10 m above the ground, in both x and y directions. These two flanges were found to be the only points on the tower stiff enough to safely receive the strong sledgehammer blow necessary to excite the structure. The impact energy delivered was in any case limited, in order to avoid saturation of the acceleration signals acquired, but still sufficient to excite the seven lowest modes, corresponding to first, second and third flexural modes in both directions. Sample experimental FRFs are reported in Figure 6, for accelerometer 1, 2, 5 and 6 and for excitation points close to sensors 5 (x direction) and 6 (y direction). For convenience, the graph also indicates the model frequency extracted, as explained in detail in Section 5.

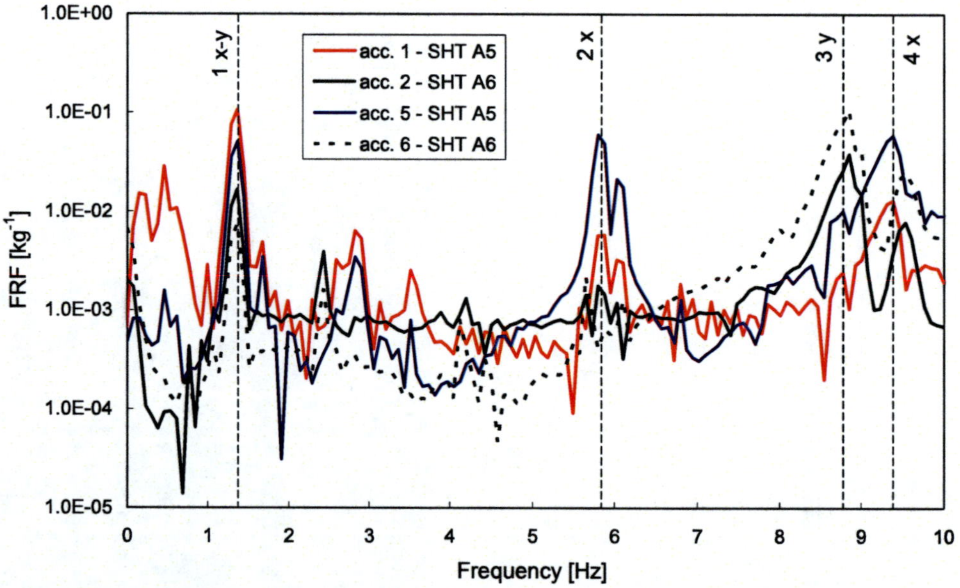

Figure 6. FRF relevant to accelerometers 5 and 6 in tests SHT #A5 and SHT #A6, elevation 12.1 m on the ground, x and y directions, respectively.

4.2. Ambient Vibration Tests

AVTs aimed to explore the effects of wind and rotor on the structure under operation conditions; for this reason, contrary to SHT, in this case the nacelle was allowed to freely rotate during the test. Wind speed at different heights and wind direction were acquired at the instrumented mast located 32m away. The combined acquisition of structural vibration and wind field characteristics allows correlation of the structural response to the environmental action and identification of the various vibration sources. As an example, Figure 7 reports the outcomes of two acquisitions, performed under different operation conditions: the graphs on the left report data recorded with the turbine parked, while the right column refers to the turbine operating. The upper graphs show the wind speed recorded during the acquisition. Fourier spectra are reported for accelerometers located at the tower tip (sensors 1 and 2) and at the flange at level 12.1m (sensors 5 and 6).

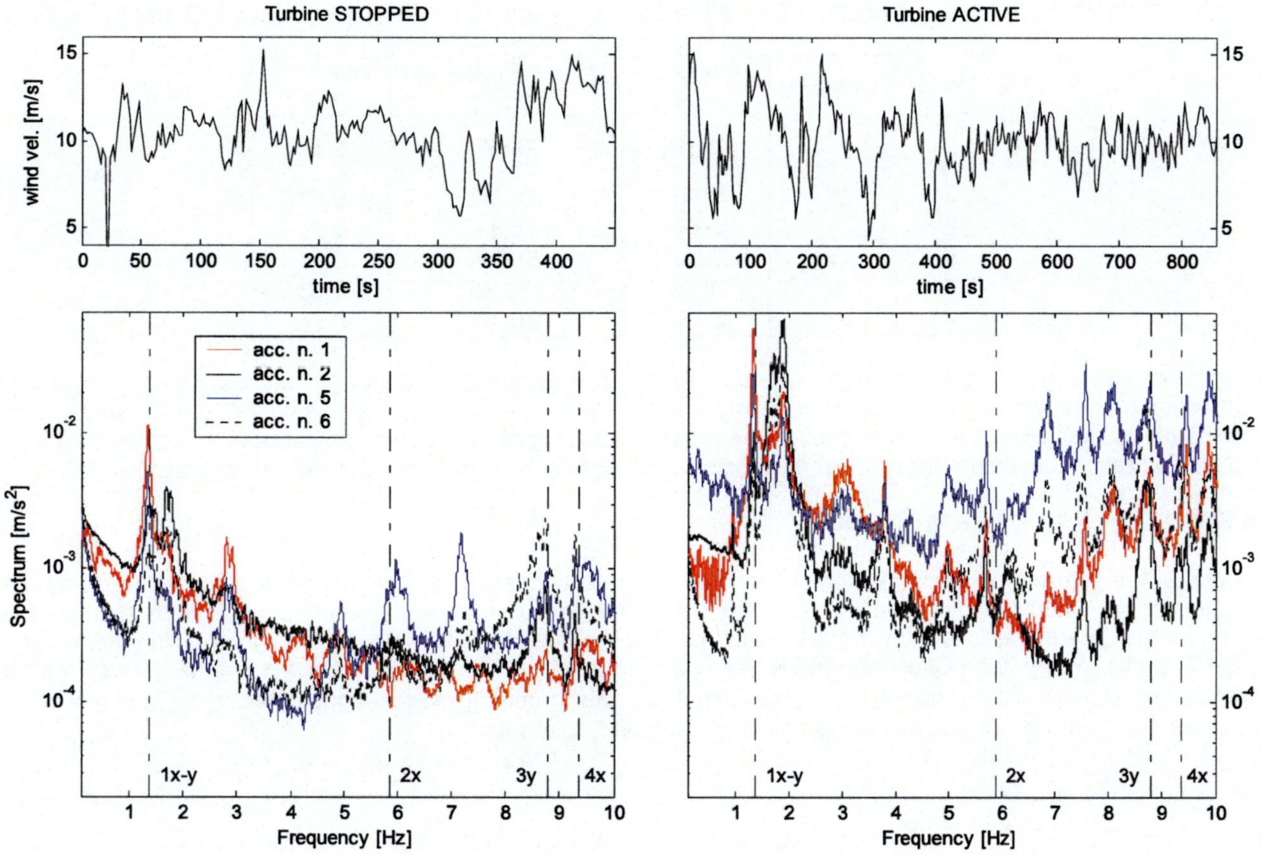

Figure 7. Wind speed (upper graphs) and FFT of the AVTs (lower graphs), for the condition with rotor parked (right column) and active (left column).

Comparison between the graphs highlights that rotor and blade movement strongly influences the frequency components. The fundamental contribution of the 2-blade rotor motion is evident at about 1.9 Hz, consistent with its fixed rotation speed of 58 rpm, but even higher harmonic components can be seen.

The FFT vibration analysis in the x direction reveals a remarkable difference between the two operating modes of the rotor, parked and rotating. When parked, the main turbine components excited by the wind are the tower and the blades, these behaving as independent structures. Both will generate vortex-induced vibration due to the vortices shed alternately from their opposite sides. Over the *Re* range of this application, eddies are shed continuously from each side of the tower, forming rows of vortices. The frequency of shed is given by the Strouhal number (*St*), the dimensionless number which compares the frequency of vibration of the forcing actions on the body with its characteristic frequency [17]. The frequency of vortex shedding f_s as a function of height from the ground *z* is given by:

$$f_s(z) = St\frac{V(z)}{d(z)} \tag{1}$$

Where *d*(z) is the diameter of the tower and *V*(z) is the wind speed. Since the tower diameter decreases with height while the velocity increases according to the wind shear, the frequency and the intensity of the vortices are expected to increase slightly from bottom to top of the tower. This is actually what is noticed in the left side of figure 6, where a pronounced peak at a frequency ranging from 1.3 to 1.8 Hz occurs, the highest frequencies recorded at the top accelerometers. Although this pattern is modified due to the wind turbulence and structure aero elasticity, this range of frequencies interferes with the first natural frequency of the tower resulting in the series of resonance peaks. The smaller blade chord size determines further peaks at higher frequencies, at around 6 Hz. Higher frequency harmonics depend on complex fluid-structure interaction.

Table 1. Modal data extract from SHT A5 (mode 1, 2, 4 and 6) and SHT A6 (modes 1, 3 and 5).

Mode	Direction	Frequency [Hz]	Angular Frequency [rad/s]	Damping [%]
1x 1y	x y	1.36	8.57	0.58
2	x	5.86	36.84	0.53
3	y	8.79	55.21	0.75
4	x	9.37	58.89	0.76
5	y	25.20	158.35	0.53
6	x	26.20	164.60	0.24

A more complex pattern arises from the vibration analysis of the rotating mode. This situation shows again the interaction of the tower vortices, but far more important is the 1P blade frequency behind the rotor around 1.9 Hz. The blade passes twice in a rotation through the tower wake. This causes a periodic deficit in thrust resulting in an induced vibration that determines further harmonics at 2P (3.8 Hz), 4P (5.7 Hz) and higher frequencies.

5. RESULTS OF MODAL EXTRACTION

Basically, the symmetrical configuration of the tower respect to axis x (the direction of the nacelle) suggests that modal shapes should appear in two separate sets, one with dominant x- and the other with dominant y-components. Therefore, to identify the x-direction modes, we used the FRFs generated by impacting the tower at point 5; and similarly, for y-direction modes, we used those of point 6. Modal extraction was performed via the circle fitting technique [18], assuming a proportional viscous damping model. Consequently the receptance α_{jk} measured at point *j* for a force applied at point *k* at frequency ω reads:

$$\alpha_{jk}(\omega)=\sum_{r=1}^{N}\frac{\varphi_{jr}\varphi_{kr}}{\omega_r^2+2i\xi_r\omega_r\omega-\omega^2} \tag{2}$$

where ω_r, ξ_r, φ_{jr} indicate the frequency, damping rate and *j*-th modal component for the *r*-th mode shape, respectively. The main results of the extraction are listed in Table 1, while mode shapes in directions x and y are graphically plotted in separate graphs in Figure 8.

As expected, the two fundamental bending modes, 1-x and 1-y, are highly coupled, to the point that their resonance peaks are indistinguishable in the FRF. It is also worthy of note that modes 2 and 3, both in x direction but well separated in frequency, apparently have similar shapes. The analysis carried out using a numerical model highlights that in fact the two modes exhibit a very different behaviour of the rotor: in mode 2 (f_2=5.86Hz) the rotor vibrates in phase with the tower tip, while in mode 3 (f=9.37Hz) it moves in counter-phase.

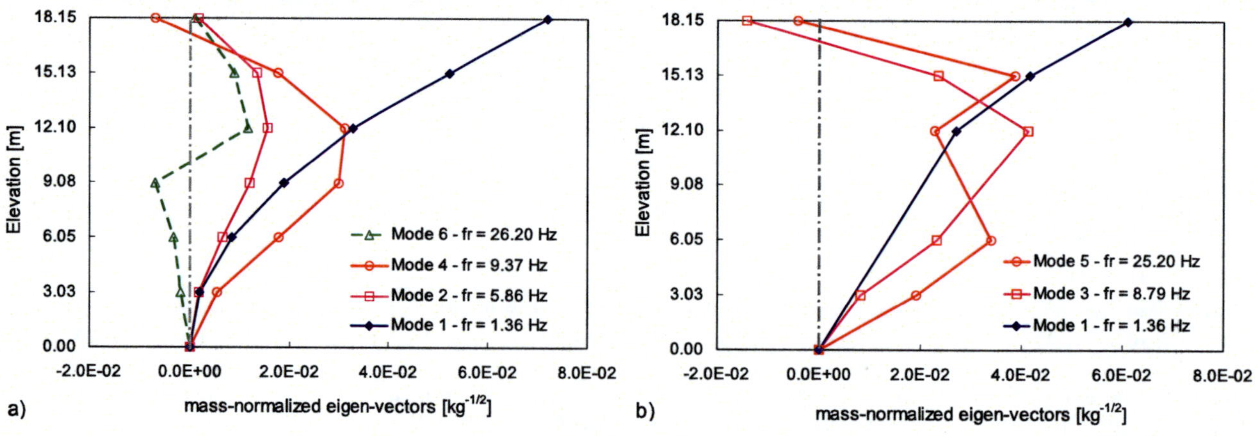

Figure 8. Experimental mode-shapes: (a) x direction, derived SHT A5, (b) y direction, derived SHT A6.

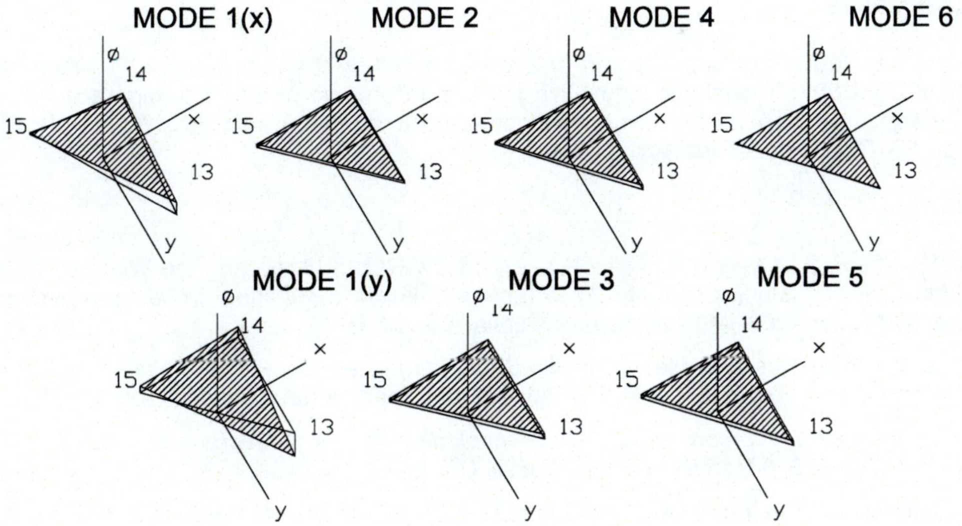

Figure 9. Base joint displacements, as identified by SHT A6.

Figure 9 shows detail of the modal components recorded at the base of the tower. It is observed that in most modes the rotational components are basically negligible. The only significant exceptions are the two fundamental modes: in these cases it appears that the soil-foundation interaction plays an important role in the dynamic response of the tower, and should be taken into account in the design of new turbines as well as in the assessment of existing ones. For instance, in the specific case, the first two modes, with frequency of 1.35 Hz, are very close to the vortex shedding frequency range f_s, thus it is clear that careful estimation of the natural frequencies at the design phase is critical to prevent unwanted effects on the supporting tower.

Damping is also critical in correct evaluation of fatigue-inducing conditions. As shown in table 1, the values of damping ratio ξ identified are relatively low, in the order of 0.5%, for all the frequencies. However, before using this information as a general result, we note that here modal extraction is based on low amplitude vibration tests, while energy dissipation in operation is expected to be higher.

6. CONCLUSIONS

We present the experimental dynamic behaviour of a prototype windmill under operational conditions. As far as the supporting tower is concerned, we observed that the vibration amplitude at different points depends on many linked factors: the modal resonances of the structure, the excitation produced by the rotor and that related to the vortices. In this work, we attempt to identify experimentally all these different contributions, to better understand how they interact. To do so, the windmill was subjected to modal analysis in a non-windy state, and the derived FRFs were qualitatively compared with the response spectra in windy conditions. Particularly the effect of blade rotation was investigated first with the rotor parked, then comparing this acquired data with the active rotor response.

We present a preliminary interpretation of the various spectral components ; specifically, we clearly identified the contributions of the vortex shedding effect and of the harmonics due to rotor motion. These findings are expected to be validated by future research. In the next step of the project, the behaviour of the turbine under windy conditions will be compared with that predicted by an aero-elastic model able to predict the shedding even when the rotor is active. Moreover the wind flow data recorded by a high-frequency anemometer will be analyzed to better de-couple the contributions. This information will allow better understanding of the fatigue-induced damage in the mast and so predict its operational life.

ACKNOWLEDGEMENTS

The experiment presented in this paper is part of a work funded by the Autonomous Province of Trento (PAT) within the project CRS2007 "Tecnologie innovative per il monitoraggio di torri e turbine eoliche installate in siti complessi". The experimental wind farm was developed and is operated within the framework of PAT FaStFal 2007-2010 and EU COST MP0702 research projects.

REFERENCES

[1] Butterfield S., Sheng S., Oyague F., "Wind Energy's New Role in Supplying the World's Energy: What Role will Structural Health Monitoring Play?", *Structural Health Monitoring 2009*, proceedings of the 7th International Workshop on SHM, Standford University, CA, 2009.

[2] *IMAC-XXVII: a Conference and Exposition on Structural Dynamics*. Proceedings of the 27th International Modal Analysis Conference, Orlando, Fl, US, Society for Experimental Mechanics-SEM, 2009.

[3] Proc. SPIE, *Sensors and Smart Structures Technologies for Civil, Mechanical, and Aerospace Systems 2009*, Vol. 7292, 729215, 9 March 2009, San Diego, CA, 2009.

[4] Turner A., Graver T.W., Rumsey M.A. and Mendez A., "Comparison of the Three Different Approaches for Damage Detection in the Part of the Composite Wind Turbine Blade", *Structural Health Monitoring 2009*, proceedings of the 7th International Workshop on SHM, Standford University, CA, 2009.

[5] Frankenstein B., Schubert L., Weihnacht B., Schulze E., Ebert C., Friedmann H. and Grahnert T. "Development of Condition Monitoring Systems for Rotor Blades of Windmills", *Structural Health Monitoring 2009*, proceedings of the 7th International Workshop on SHM, Standford University, CA, 2009.

[6] Luczak M., Mevel L., Ostachowicz W. and Martyniuk K. "Comparison of the Three Different Approaches for Damage Detection in the Part of the Composite Wind Turbine Blade", *Structural Health Monitoring 2009*, proceedings of the 7th International Workshop on SHM, Standford University, CA, 2009.

[7] Klinkov M. and Fritzen C.-P. "On Field Wind Load Reconstruction for a 5MW Onshore Wind Energy Plant" , *Structural Health Monitoring 2009*, proceedings of the 7th International Workshop on SHM, Standford University, CA, 2009.

[8] Rolfes R., Zerbst S., Haake G., Reetz J. and Lynch J.P. "Integral SHM-System for Offshore Wind Turbines Using Smart Wireless Sensors", *Structural Health Monitoring 2007*, proceedings of the 6th International Workshop on SHM, Standford University, CA, 2007.

[9] Fritzen C. P., Kraemer P., and Klinkov M. "Structural Health Monitoring of Off-Shore Wind Energy Plants", *Structural Health Monitoring 2008*, proceedings of the 4th European Workshop on SHM, CA, 2008.

[10] Drewry, M.A. and Georgiou, G.A. "A review of NDT techniques for wind turbines", NDT.net, 49(3), 2007.

[11] Connor R.J. and Hodgson I.C. "Field Instrumentation and Testing of High-Mast Lighting Towers in the State of Iowa", Draft Final Report, Iowa Department of Transportation Office of Bridges and Structures, 2006.

[12] Web site: http://www.eolicotrento.ing.unitn.it/uk/index.htm

[13] Battisti L., Hansen M.O.L. and Soraperra G. "Aeroelastic simulations of an iced MW-Class wind turbine rotor" Proceedings of the VII BOREAS Conference, 7-8 March, Saarisalka, Finland, 2005.

[14] International Electro-technical Commission (IEC), *IEC 61400-1 Ed.3 CD. 2. Revision. Wind Turbines. Part 1: Design Requirements, International Electrotechnical Commission*, Geneva, Switzerland. 2005.

[15] International Electro-technical Commission (IEC), *Wind turbines - part 12-1: Power performance measurements of electricity producing wind turbines*, 2005.

[16] ESDU, *85020 G Characteristics of atmospheric turbulence near the ground. Part II: single point data for strong winds (neutral atmosphere)*. Boston: British Library, 1985.

[17] Schlichting H., Gertsten K. *Greanzschicht Theorie*, Springer, 1997.

[18] Ewins D.J. *Modal Testing*, 2nd ed., Wiley, 2001.

Proceedings of the IMAC-XXVIII
February 1–4, 2010, Jacksonville, Florida USA

Full-scale modal wind turbine tests: comparing shaker excitation with wind excitation

Richard Osgood[(1)], Gunjit Bir[(1)], Heena Mutha[(1)],
Bart Peeters[(2)], Marcin Luczak[(2)], Gert Sablon[(2)]

[(1)]National Renewable Energy Laboratory, National Wind Technology Center – MS 3911, 1617 Cole Blvd., Golden, CO 80401, USA
[(2)]LMS International, Interleuvenlaan 68, B-3001 Leuven, Belgium, bart.peeters@lms.be

ABSTRACT

The test facilities at the National Wind Technology Center (NWTC) of the National Renewable Energy Laboratory (NREL) include a three-bladed Controls Advanced Research Turbine (CART3). The CART3 is used to test new control schemes and equipment for reducing loads on wind turbine components. As wind turbines become lighter and more flexible to reduce costs, novel control mechanisms are necessary to stop high winds from damaging the turbine. However, wind turbines must also be designed to capture the maximum amount of energy from the wind, so engineers must devise new ways of achieving this while controlling wind loads that would cause the turbines to fatigue quickly. New control mechanisms and computer codes can help the wind turbine shed some loads in extreme or very turbulent winds. The special configuration of the CART3 allows researchers to analyze these diverse control schemes.

This paper reports on the initial results of a major full-scale modal testing campaign to validate and refine simulation models. One model is a tailored multi-body dynamic simulation model that will be used to develop an advanced controller designed to optimize power and minimize structural loads. Researchers would also like to tune Finite Element Models of the blades, nacelle and tower assembly to predict the higher order rotating modes of the wind turbine for a range of inflow conditions.

The paper will discuss an Experimental Modal Analysis approach where the wind turbine in parked condition is excited by shakers connected with cables. This approach will be compared to Operational Modal Analysis where the same structure is subjected to wind excitation without the shakers activated. These tests and data analyses will provide experience and increase confidence in the approach used for future tests in rotating conditions.

1 INTRODUCTION

During operation, wind turbines, like many engineering structures, are subject to dynamic loads, because the aerodynamic and gravitational loads vary with time as the turbine rotates [1]. The optimal design of a turbine is dependent on reducing vibrations, which can be done by accurately identifying turbine's natural frequencies and mode shapes [2]. By determining the modes of the wind turbine, it can be ensured that the turbine's operational conditions preclude resonant frequencies, thereby minimizing dynamic loads and lengthening the life of the turbine [2][3]. In order to develop accurate structural models, extensive experimental testing must be done to refine and validate computer simulations. Modal testing, an experimental method for measuring the modes (natural frequencies, mode shapes, and damping ratios) of a structure, is a common method for validating a structural model as modes are readily computed from models and define the dynamics of the turbine over the operating speed of the rotor [1].

In addition, dynamic loads can be reduced through controls, which actively damp component vibration. Wind turbine control can increase power production while reducing fatigue on the turbine parts. In order to accurately design controls, the turbine system model must accurately predict the dynamics for a variety of operating conditions [4][5]. Poorly designed controls can result in control loop interference, causing the turbine to vibrate unstably, introducing large damaging loads. Modal testing validates the inputs and plant parameters used to develop and test the controls model. The more insight a modal test provides, the more accurate a model becomes, and the better the final structure operates. This modal investigation will be used to validate an analytical structural model and the chosen inputs used for a turbine control model.

Figure 1: (Left) Test facilities at NREL's National Wind Technology Center (NWTC), Boulder, Colorado. (Right) CART3 – Controls Advanced Research Turbine: 3-bladed – 600 kW Wind Turbine (40m rotor diameter & hub-height).

In a traditional Experimental Modal Analysis (EMA), measured input forces are applied by shakers connected to cables which excite the structure [6][7][8][9]. In this method, both input and output signals are measured to collect Frequency Response Functions (FRFs). By fitting a parametric model through the FRFs, the modal parameters of the structure are extracted. This method provides an approximation of the static (i.e. parked rotor) system modes, which are used as a starting point for tuning the structural and control models. However, in this method, the experimentally determined vibrations are not the turbine's pure modes [9]. Mode shapes are changed by boundary conditions, and shaker excitation ignores the ambient loads acting on the turbine during operation: the wind. Wind excitation has sufficient energy to excite the fundamental system modes, however, the input forces are not measured and therefore, FRFs cannot be computed. In these cases the deterministic knowledge of the input is replaced by the assumption that the input is white noise and Operational Modal Analysis (OMA) techniques are applied to extract the modal parameters from output-only measurements. OMA allows the investigator to collect modes that are relevant to the operation of the machine. A review of OMA methods can be found in [10] and some recent developments are discussed in [11]. The original application of OMA (although that name was not in use at that time) to wind turbines can be found in [12][13].

Modal tests can be conducted with the turbine in a parked position or in rotational operation. Standstill modes are useful for validating and refining models, however there is no inclusion of the rotational effects that dominate modes [9]. Modal characteristics vary between a parked and rotating turbine, so including rotational effects is important for computing modal parameters [3][6]. Rotational testing has been conducted since the 1980s. Carne et al. [3][6][12][13] focused on vertical-axis wind turbines (VAWTs), which do not deal with the changes in gravitational forces acting on the rotating structure or the bending moments acting out of the plane of rotation – challenges that rotating horizontal axis turbines face. Horizontal-axis wind turbines (HAWTs) have been tested with a variety of excitation methods [9].

This investigation describes a series of full modal tests conducted on a highly modified Westinghouse model WWG-0600 wind turbine known as the CART3 (three-bladed Controls Advanced Research Turbine; see Figure 1). Standard EMA test methods are employed to identify the parked-rotor or static wind turbine modes of vibration; hydraulic shakers are used to excite the turbine in the fore-aft and side-to-side directions (i.e. along the turbine's wind axis and across the turbine's wind axis). A secondary test using wind excitation to collect an ambient response is also conducted. OMA is applied to this wind excitation data. A new distributable data acquisition system (DAS) is used to collect both static and operational modes of vibration. This new next-generation DAS allowed for measurements to be collected in the wind turbine hub, nacelle, and tower in stark contrast with the old DAS, which was centrally located in the test shed at the base of the wind turbine tower.

2 STRUCTURE AND DATA ACQUISITION

2.1 Wind Turbine Description

The three-bladed rigid-hub upwind rotor of CART3 incorporates independent full span pitch of each blade to control aerodynamic torque. The drive train employs advanced power electronics which allows for variable speed operation unlike the original constant speed synchronous generator. The rotor diameter is approximately 41 m and operates over a speed range of 10-43 RPM, and produces rated power of 600 kW for a wind speed of 18 m/s. The hub height of the turbine is 36.6 m. The tower base has a diameter of 4.42 m, which linearly tapers to 2.18 m at a height of 9.20 m from ground. From 9.20 m to the nacelle, the tower has a constant 2.18 m diameter. The drive train has a low-speed brake located between the main shaft and gearbox and a high-speed brake at the end of the generator shaft. The instrumented CART3 turbine is shown in Figure 2 (Left). As shown in the figure, this test has blade one aligned vertically with the yaw axis and blades two and three oriented 120° about either side of it.

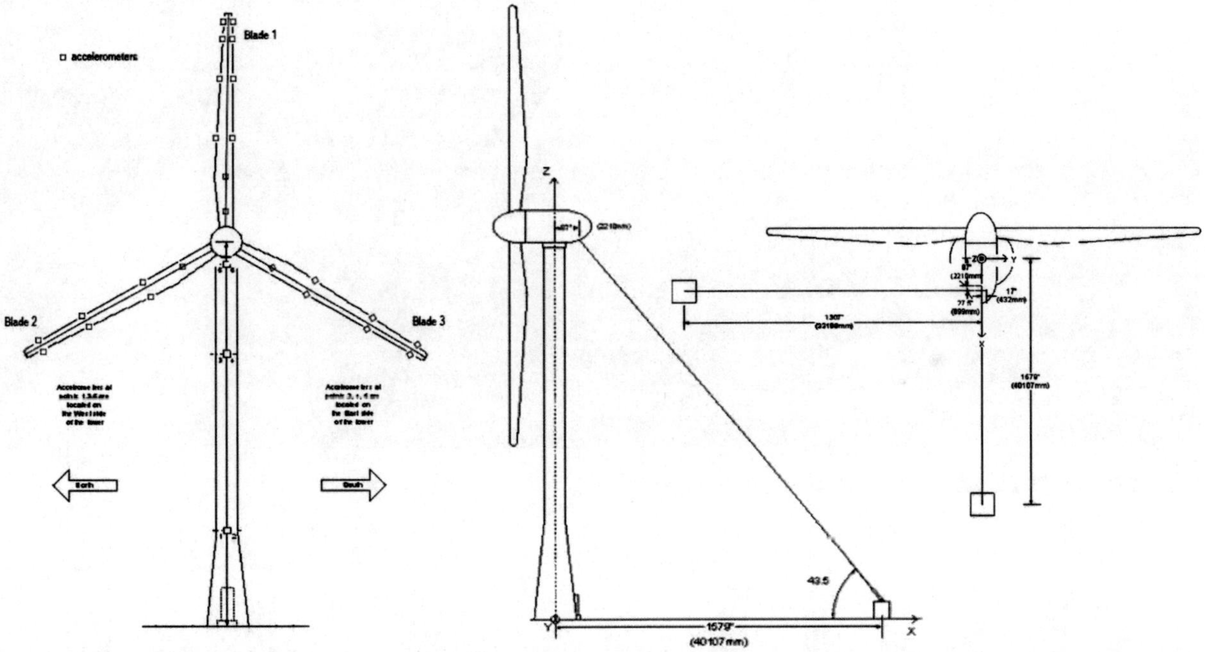

Figure 2: (Left) CART-3 turbine node map. Accelerometers were located along the tower wall, blade edges, and nacelle for modal characterization. (Right) Hydraulic shaker set-up, side and top view. One steel cable was connected to the nacelle and another to the tower.

2.2 Approach and Test Setup

The initial test was conducted similar to the CART2 test done in 2002 [7], using a multiple-input/multiple-output (MIMO) procedure to measure the modes. Like the CART2 [7], the turbine was excited in the fore-aft and side-to-side directions utilizing two independent hydraulic actuators. The fore-aft actuator was connected to the downwind side (East) of the nacelle while the side-to-side actuator was connected along the tower's South face, as shown in Figure 2 (Right). Tests were conducted running one actuator at a time. These different orientations were used to put energy into the nacelle pitching and yawing modes. The tower side-to-side excitation put energy in the rotor's in-plane and asymmetric bending modes. The tower fore-aft excitation put energy into the rotor's symmetric and asymmetric out-of-plane bending modes. In order to determine modes excited at low energy, the turbine was parked in the wind, and excited by the ambient wind.

2.3 Test Equipment and Instrumentation

Figure 2 (Left) provides a schematic of the setup for the modal test. Force transducers were attached to the nacelle and tower to measure the force applied to the turbine by the hydraulic ram. These sensors were attached to the ram by a high-strength, tensioned steel-cable (Figure 3). Biaxial and triaxial accelerometers were configured from single-axis accelerometers. Twenty-four biaxial accelerometers (48 channels) were used to map the rotor's in-plane and out-of-plane bending modes. The blade accelerometers were located along the leading and trailing edges, as well as along the chord line of the blade near its root. One axis was aligned to measure flap (out-of-plane bending) and the other lag (in-plane bending) motion. Blade one was equipped with more accelerometers to determine higher-order mode shapes (Figure 4). Five triaxial accelerometers (15 channels) were used to map bending in the drive train (nacelle). The accelerometers were oriented to capture the fore-aft, side-to-side, and the pitching motion of the nacelle (Figure 5 – Right). Finally, six biaxial accelerometers (12-channels) were used to map the bending modes of the tower. These sensors were placed along the upwind and downwind sides; collecting data in the fore-aft and side-to-side directions (Figure 5 – Left). The coordinate systems and orientations comply with the International Electrotechnical Commission (IEC) convention for the geometry of the blades, tower, and nacelle, and are shown in Figure 4 and Figure 5.

Figure 3: CART3 excitation by hydraulic shaker.

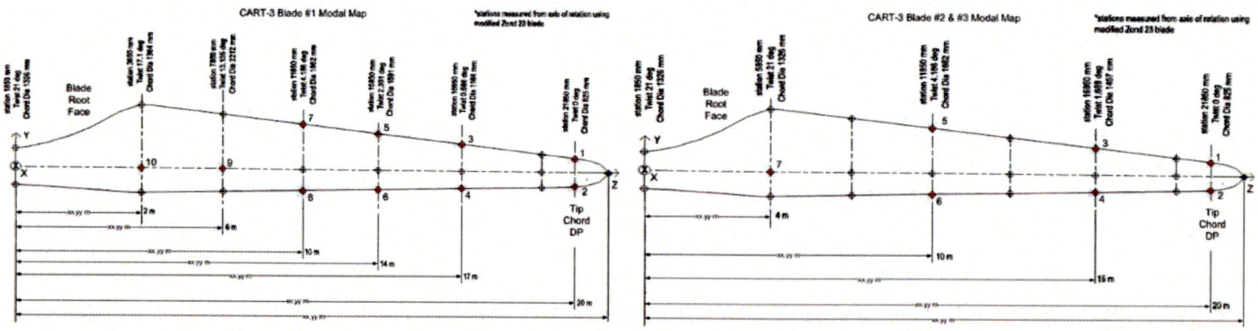

Figure 4: Blade one, two, and three node map. Blade one was instrumented with ten biaxial accelerometers, while two and three were instrumented with seven. Locations were approximated using the station information from a modified Zond 23 m blade design.

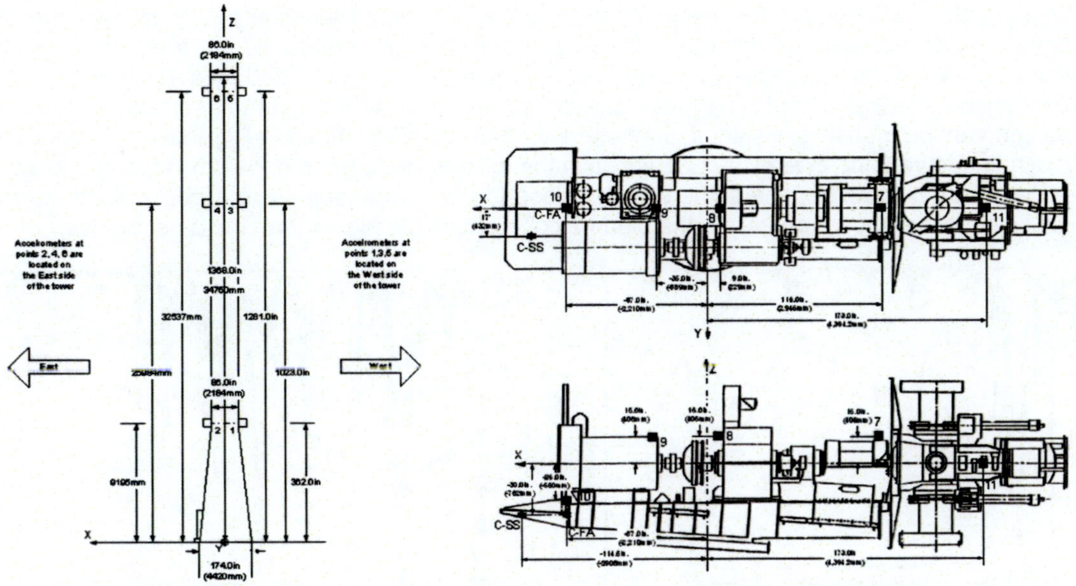

Figure 5: (Left) Tower node map. Six biaxial accelerometers were located along the height of the turbine. (Right) Nacelle node map, top view and side views. Five triaxial accelerometers were located along the central axis.

In total, seventy-five channels of acceleration were measured to map the modal response of the wind turbine system. In addition, two driving points were collected using four channels; two channels of force and two channels of acceleration. A driving point is the collocated measurement of the input excitation force and response acceleration in the direction of the applied force. Therefore, the modal test system collected 79 channels of data for the modal test. The LMS SCADAS Mobile (Figure 6) data acquisition system (DAS) was used to collect, reduce, and save the processed modal test data on a laptop PC running LMS Test.Lab software [14][15].

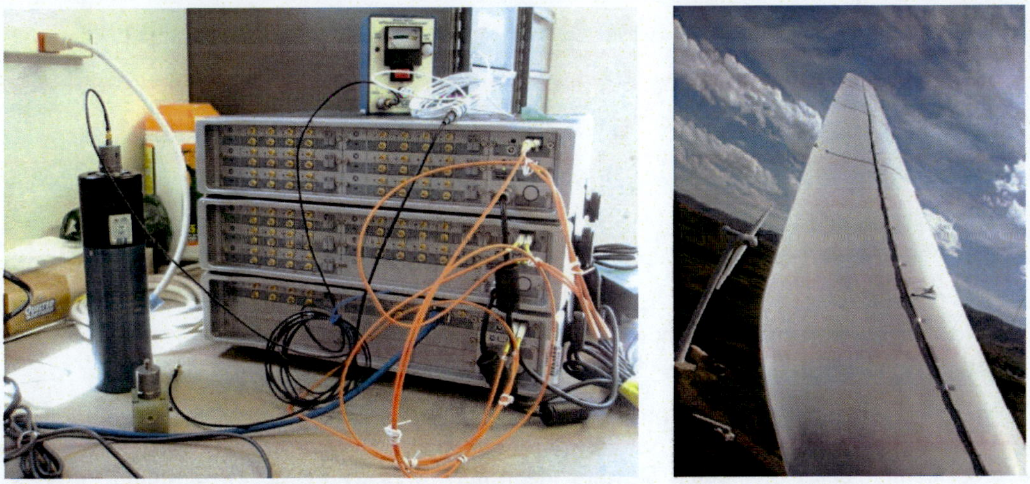

Figure 6: (Left) LMS Scadas Mobile data acquisition system. (Right) blade accelerometers.

3 EXPERIMENTAL MODAL ANALYSIS – SHAKER EXCITATION

A random signal was sent to the shaker. A Hanning window was applied to the excitation and response signals. The sampling frequency was 50 Hz and the block size 4096, yielding a frequency resolution of the spectra, FRFs and coherences of 0.0122 Hz. Ten averages were used in the computation of the measurement functions. Therefore the measurement time was 819.2 s. The driving point FRFs and corresponding coherences are shown in Figure 7. A quality indicator for the measurements is the coherence function that combines the effects of leakage, noise and non-linearity on the FRFs. When performing high channel count measurements, browsing through all coherence functions is rather cumbersome. A synthetic view is provided by averaging the coherences over certain frequency ranges and plotting these at each sensor location in a geometry display. A relative quality assessment can then be made: for instance, the coherences of close and structurally related locations should not differ too much. In Figure 8 the averaged coherences in the full frequency band (0.5 – 25 Hz) and a low-frequency band (0.5 – 5 Hz) are plotted. In general the sensors close to the force inputs have a better signal-to-noise ratio and hence a higher coherence. Apparently, the blade 2 coherences are higher than the other blades.

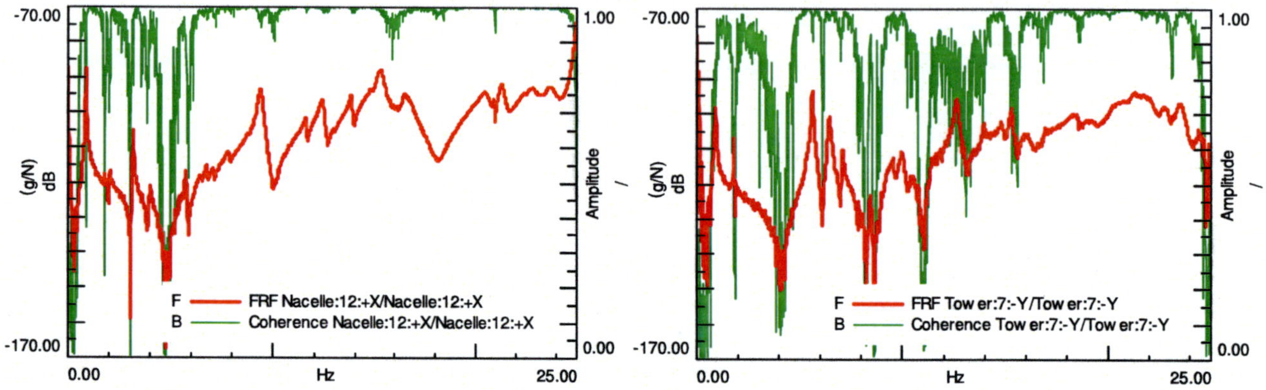

Figure 7: Driving point FRFs and coherences. (Left) nacelle fore-aft excitation; (Right) tower side-to-side excitation.

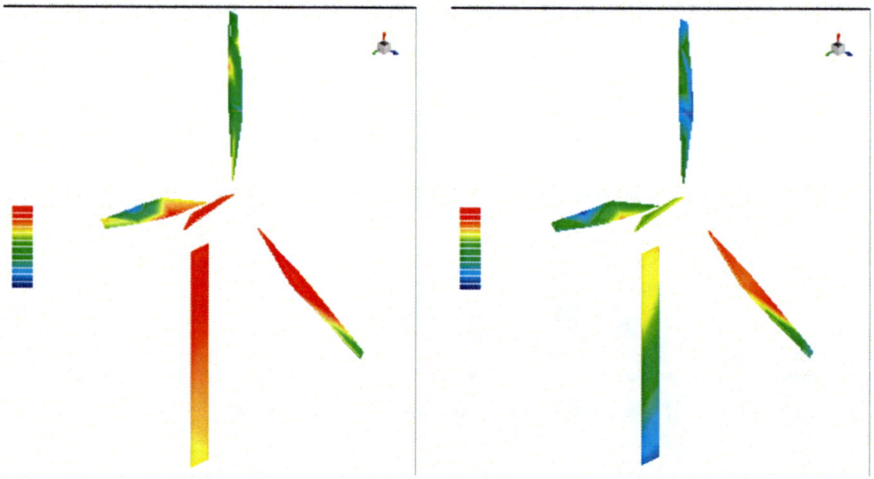

Figure 8: Tower excitation averaged coherences in a geometry display. (Left) full frequency band 0.5 – 25 Hz; (Right) low-frequency band (0.5 – 5 Hz). The color scale varies between 0 (blue) and 1 (red).

After the data validation, the PolyMAX modal parameter estimation algorithm [15][16] was applied to the measured FRFs. Both shaking tests were carried out sequentially, but processing both datasets simultaneously yielded the most complete list of modes. Despite the relatively noisy FRFs, the structural modes are clearly identified in the PolyMAX stabilization diagram (Figure 10 – Left). A stabilization diagram shows the modes that are independently calculated at different model orders in the system identification process. Experience on a very large range of problems shows that in such analysis, the pole values of the "physical" eigenmodes always appear at a nearly identical frequency, while mathematical poles tend to scatter around the frequency range. Fifteen modes were found below 10 Hz. Their frequencies and damping ratios are listed in Table 1. The first nine mode shapes are represented in Figure 9. The FRFs can be synthesized from the identified modes and compared with the measured FRFs. Figure 10 (Right) shows the comparison between the sum of the measured FRFs and the sum of the synthesized FRFs. The good agreement indicates that most important dynamics have been identified from the data.

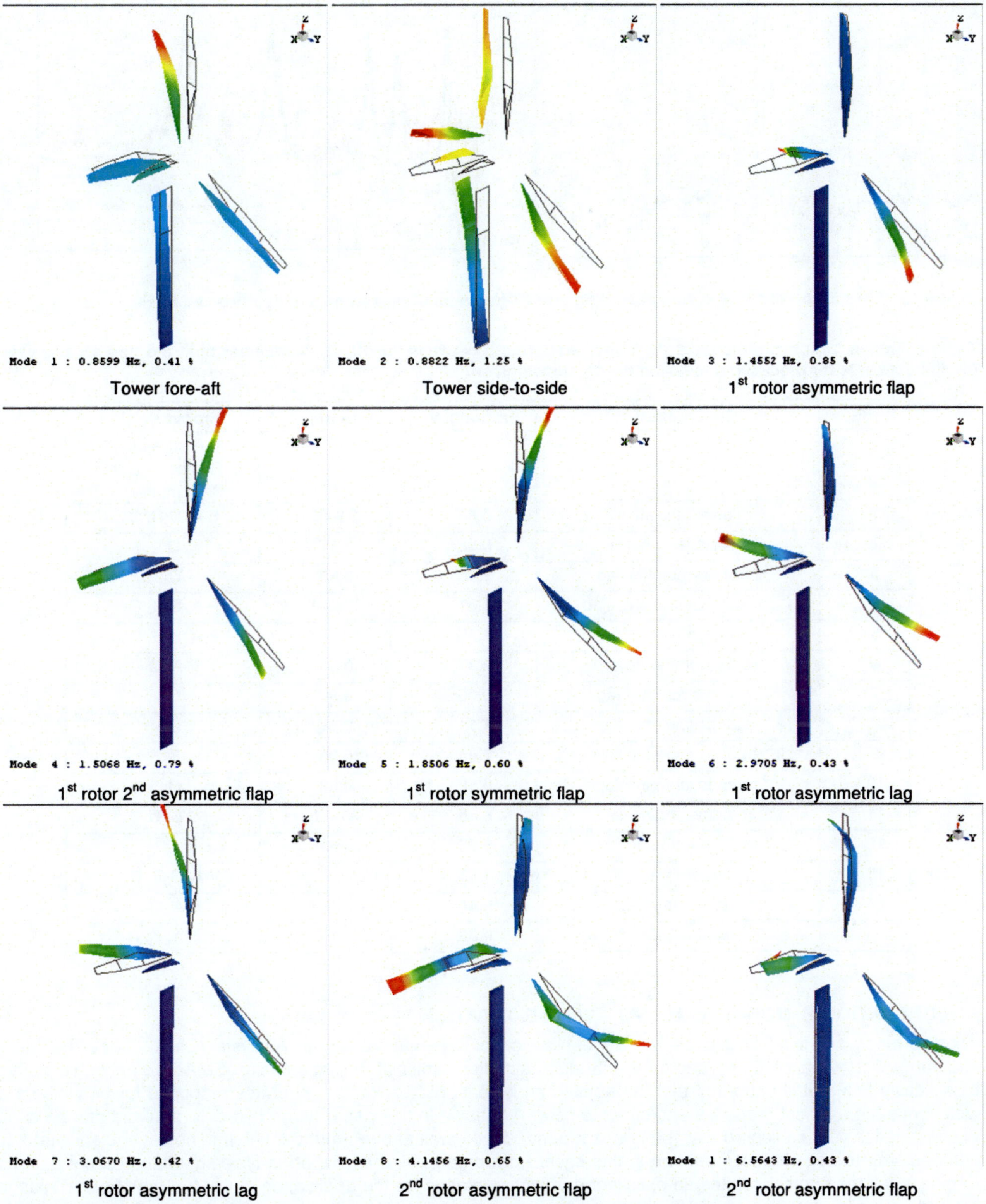

Figure 9: EMA-PolyMAX mode shapes. The color scale indicates the relative amplitudes.

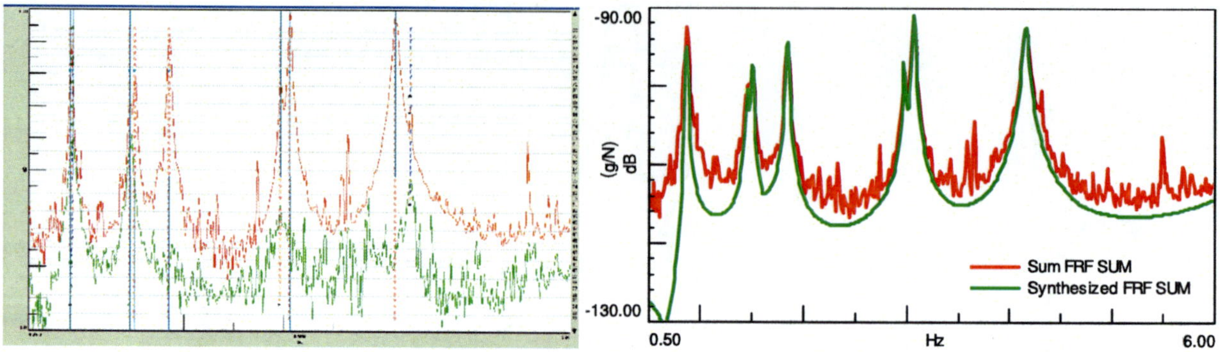

Figure 10: (Left) EMA-PolyMAX stabilization diagram; (Right) sum of measured and synthesized FRFs.

Table 1: Modes from hydraulic tests. 15 modes were found between 0.8-10 Hz. Combining both the fore-aft and side-to-side data sets for processing revealed the 15 modes given.

Mode	Description	Frequency	Damping	Processing
#		Hz	%	F-A: fore-aft S-S: side-to-side C: Combined
1	Tower fore-aft	0.86	.24-.38	F-A, S-S, C
2	Tower side-to-side	0.88	1.12	C
3	1st rotor asymmetric flap	1.45	0.88	C
4	1st rotor 2nd asymmetric flap	1.51	.35-.61	F-A, S-S, C
5	1st rotor symmetric flap	1.85	.55-.57	F-A, S-S, C
6	1st rotor asymmetric lag	2.97	0.33	F-A, C
7	1st rotor asymmetric lag	3.06	0.38	C
8	2nd rotor asymmetric flap	4.14	0.65	C
9	2nd rotor asymmetric flap	6.44	0.25	C
10	2nd rotor asymmetric flap	6.56	0.53	C
11	2nd rotor symmetric flap	6.89	.31-.46	S-S, C
12	2nd rotor asymmetric flap	6.96	0.65	C
13	2nd tower fore-aft	7.16	.51-.77	F-A, S-S, C
14	3rd asymmetric flap	9.44	.45-.87	F-A, C
15	3rd asymmetric flap	9.89	1.1-1.21	F-A, C

4 OPERATIONAL MODAL ANALYSIS – WIND EXCITATION

Wind turbines are ideal structures for the application of Operational Modal Analysis (OMA). They are large structures which makes it challenging to introduce sufficient artificial excitation in an EMA approach. To obtain high-quality FRFs, it is important that the applied and measured excitation generates response levels which are well above the ambient response and/or noise floor. As shown in this paper, large hydraulic shakers can be used in combination with cables so that the forces are introduced well above ground level. It is however clear that it is cumbersome to use shaker excitation in this case. Fortunately, wind excitation is typically often present at wind turbine locations and constitutes an excellent ambient excitation for the application of OMA. An additional reason for performing in-operation test is that the eigenfrequencies of wind turbines depend on the rotation speed of the blades due to centrifugal and Coriolis forces and the effect of tension stiffening. In this paper, OMA will be applied with the rotor in parked conditions, but at a later stage also rotating conditions will be experimentally investigated.

The wind excitation data used in this paper originates from a preliminary limited test in which power and cross spectra were acquired under wind excitation (Figure 11), but without storing the time histories. Therefore, the ideal leakage-free pre-processing of the time data into so-called half spectra could not be performed. Moreover, the use of an exponential window in the pre-processing stage is an excellent "de-noising" tool that could not be applied here. More details about the advantages of the use of half spectra (or correlograms) in comparison with the more classical spectra (or periodograms) can be found in [11]. In the near future, more extensive operational time-domain data will be acquired. Nevertheless, the present data already yielded very interesting results.

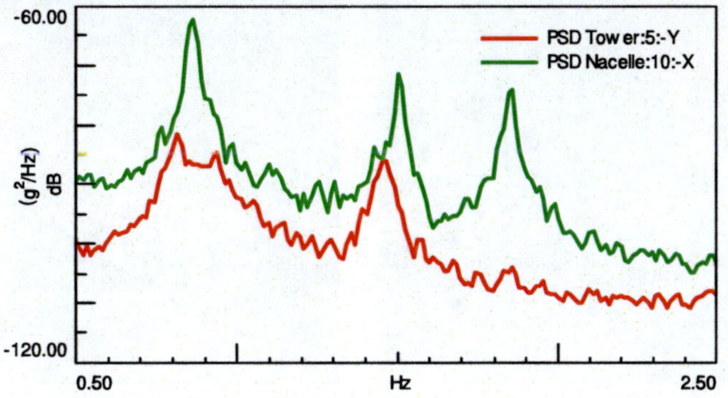

Figure 11: Lower-frequency part of wind turbine response spectra due to wind loading.

Figure 12 compares the stabilization diagrams of two modal parameter estimation methods available in Test.Lab [15]: Operational PolyMAX and Stochastic Subspace Identification (SSI). PolyMAX is known to yield very clear stabilization diagrams, but identifies modes which are heavily affected by noise with negative damping. These modes are not shown in the stabilization diagrams. Reason for the noise influence on the modes is the non-ideal pre-processing of the data. The SSI method does not suffer from this. From Figure 12 (Right), it is clear that more modes are identified though the diagram is less clear. Note that the SSI method is a time-domain method, but it has been implemented in such a way that also data reduced to power and cross spectra can be used. The MAC values in Figure 13 (Left) indicate that both Operational PolyMAX and SSI mode shapes are in excellent agreement. Also the correspondence between mode shapes identified using shaker excitation (EMA) and wind excitation (OMA) is excellent (Figure 13 – Right). The modes that were not identified using OMA (modes 2 and 6, see Figure 9 for the shapes) are indeed modes that are not well excited by the wind.

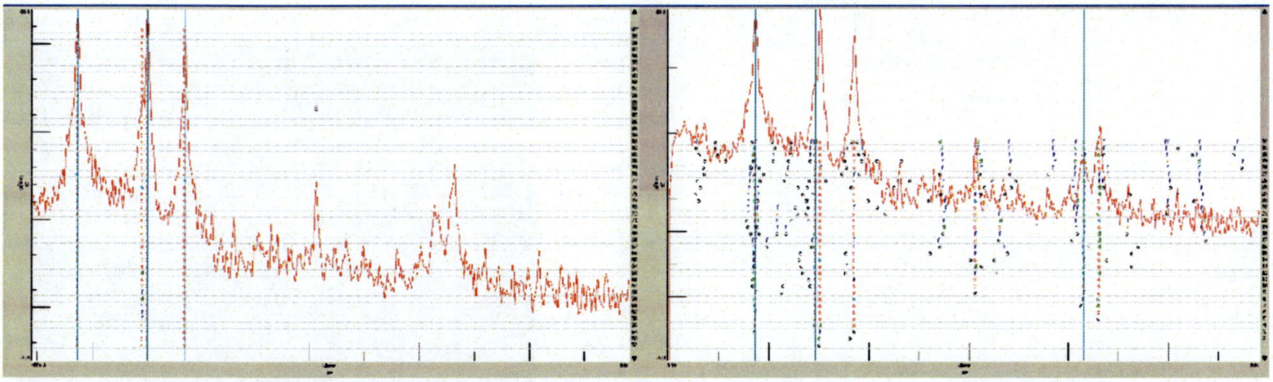

Figure 12: Operational PolyMAX (Left) versus Stochastic Subspace Identification (Right) stabilization diagram.

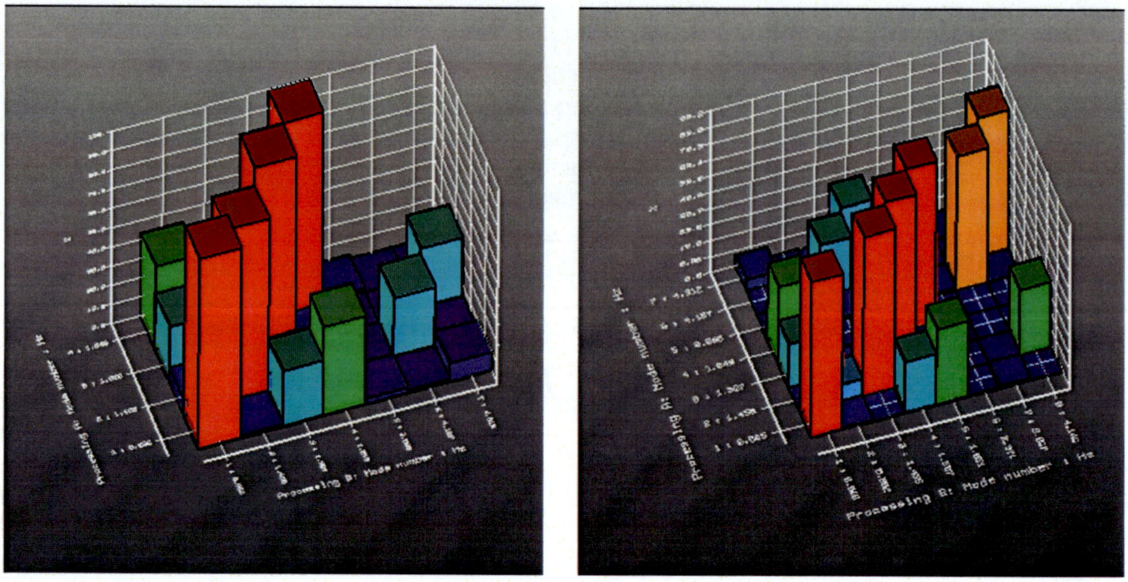

Figure 13: (Left) MAC values between Operational PolyMAX and SSI mode shapes. (Right) MAC values between shaker excitation (EMA / PolyMAX) and wind excitation (OMA / SSI) mode shapes.

5 FURTHER USE OF EXPERIMENTAL MODAL PARAMETERS

The Campbell diagram given in Figure 14 (Left) compares the experimentally measured modes (See Figure 9 and Table 1) to the CART3 operating frequencies. The harmonics of the rotor frequency for different rotor speeds are given by the black lines, 1P, 2P, 3P, etc. where P is the given rotor frequency. At the rated operation, the rotor speed is 41.7 rpm. From this figure, it can be observed that the flap modes and second tower fore-aft modes coincide with the rotor harmonics of 2P, 6P and 10P at this rated speed. However, since the modal test does not include aerodynamic damping, operating data is needed to determine if the indicated natural frequencies pose a resonance hazard to the operation of the turbine. Note that in reality, all rotor frequencies should show a slight increase as the rotor speed is increased. This is not shown in Figure 14 because here the idea is to get a preliminary estimate of the cross-over frequencies and hence possible resonance points just based on measured frequencies at zero rotor speed conditions.

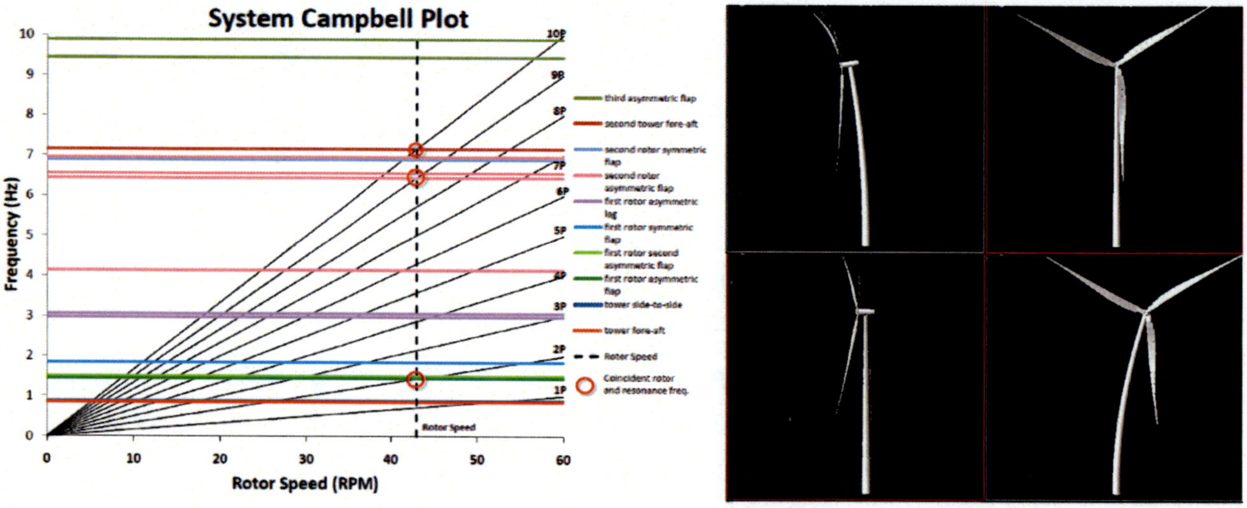

Figure 14: (Left) Campbell plot for comparing rotor harmonics (black lines) with structural resonances (colored lines). (Right) Examples of simulated full turbine system modes for a 3-bladed parked rotor wind turbine.

The modes determined from the hydraulic test differed significantly from the modes computed from the initial multi-body dynamic simulation (MBS) model of CART3 (Figure 14 – Right). Therefore, the experimental results proved useful in tuning the numerical model. The results of the comparison between test and tuned model can be found in Table 2. The first and second fundamental bending modes and the first rotor flap mode extracted from the modal data were determined within 1% of the model's prediction. Modes 4 and 5 representing the other first rotor flap modes were predicted within 3-4% of the experimentally found modes. The MBS model, mode 6, with drive-train twist and first symmetric lag, was not found experimentally. This is likely due to the location of tower side-to-side excitation not being sufficiently offset from the tower and nacelle yaw axis. Model modes 7 and 8, the first rotor asymmetric lag responses, were predicted within 3% of the experimental results. Model mode 9, the first emergent second rotor asymmetric flap mode, is predicted within a 4% difference with the experiment. Model mode 10 however was predicated with a 30.8% difference. This is likely due to a wrong pairing of experimental and model modes. Experimental modes 9-11, second rotor asymmetric flap shapes, were not found by the model. Likewise, model mode 11, a first tower torsion mode at 5.867 Hz was not found from the modal data extraction.

Table 2: Correlation of experimental modes with model predicted modes. The low fundamental modes yield a percent difference of < 5%, indicating high correlation. The experiment does not determine torsion or twist modes. The last mode predicted by the model shows a 30% difference from the experiment.

Exper. Mode	Exper. Nat Freq [Hz]	Model Mode	Model Nat Freq [Hz]	% Diff	Shape
1	0.86	1	0.868	0.930	Tower fore-aft
2	0.88	2	0.882	0.227	Tower side-to-side
3	1.45	3	1.45	0.000	1st rotor asymmetric flap
4	1.51	4	1.556	3.046	1st rotor 2nd asymmetric flap
5	1.85	5	1.931	4.378	1st rotor symmetric flap
6	2.97	7	3.047	2.593	1st rotor asymmetric lag
7	3.06	8	3.113	1.732	1st rotor asymmetric lag
8	4.14	9	3.98	3.865	2nd rotor asymmetric flap
12	6.96	10	4.813	30.848	2nd rotor asymmetric flap

6 DISCUSSION AND CONCLUSION

This paper reported on the initial results of a major full-scale modal testing campaign on the CART3 wind turbine to validate and refine simulation models. The paper compared an Experimental Modal Analysis approach, where the wind turbine in parked condition is excited by shakers connected with cables, with an Operational Modal Analysis approach, where the same structure is subjected to wind excitation. Hydraulic testing with fore-aft and side-to-side shaking revealed 15 modes below 10 Hz. Both shaking tests were carried out sequentially, but processing both datasets simultaneously yielded the most complete list of modes. Simultaneously exciting side-to-side and fore-aft in hydraulic testing seems more ideal. This is considered as an option in future.

Though, the use of wind excitation data in an OMA approach was only preliminary investigated, already very promising results were obtained: the identified OMA modes agreed very well with their EMA counterparts. In near future, more extensive wind excitation time-domain data will be acquired, allowing a more detailed OMA. Also, higher levels of wind excitation data are desired. The observed low amplitude of mean wind speed recorded indicated that there was insufficient low frequency excitation of the wind turbine structure. Usually wind excitation provides greater amounts of low frequency excitation when compared to the available low frequency excitation energy supplied by the hydraulic test system.

Comparing the multi-body simulation model with the shaker experiment modes revealed close agreement for many modes. The lower modes were predicted very well by the model. Due to limitations in resolution afforded by the modal map, the test was unable to identify the rotor twist and torsion modes of the blade, and there was poor correlation for higher second rotor flap modes. In order to interpret higher order system modes, more measurement points will be required.

The next step for this test is to conduct the experiment with the turbine operating. Rotation of the rotor changes the loads and damping forces acting on the turbine, thus changing the boundary conditions on the turbine. Identifying modes from operating conditions allows a further refinement of the structural and control models of the turbine. This test however, has provided the opportunity to develop test procedures and data analysis methods which will be used to plan modal surveys for multi-megawatt large-scale wind turbines. Two of which (a 1.5 and 2.3 MW turbine) are currently being installed at the National Wind Technology Center.

ACKNOWLEDGEMENTS

This work was supported by the Department of Energy Office of Science through the Summer Undergraduate Laboratory Internship Program and the Office of Education Programs at the National Renewable Energy Laboratory. The authors would like to thank the following individuals: Garth Johnson, Scott Wilde, Don Baker, Lucas Adams, Paul Fleming, John Hiatt, Gordon Green, Michael Salowitz, Linda Lung.

REFERENCES

[1] D.T. Griffith, T.G. Carne, and J.A. Paquette. Modal testing for validation of blade models, *Wind Engineering*, **32**(2):91-102, 2008.

[2] G. Bir and J. Jonkman. Modal dynamics of large wind turbines with different support structures, *Proc. of the ASME 27th International Conference on Offshore Mechanics and Arctic Engineering*, Estoril, Portugal, Jun. 15-20, 2008.

[3] T.G. Carne, D.W. Lobitz, A.R. Nord, and R.A. Watson. *Finite Element Analysis and Modal Testing of a Rotating Wind Turbine*, SAND82-0345, Sandia National Laboratories, Albuquerque, NM, Oct. 1982.

[4] A.D. Wright, L.J. Fingersh, and K.A. Stol. Designing and Testing Controls to Mitigate Tower Dynamic Loads in the Controls Advanced Research Turbine, *Proc. of the 45th AIAA Aerospace Sciences Meeting and Exhibit, Wind Energy Symposium*, Reno, Nevada, Jan 8-11, 2007.

[5] A. Wright and M. Balas. Design of state-space-based control algorithms for wind turbine speed regulation, *Proc. of the 21st American Society of Mechanical Engineers (ASME) Wind Energy Symposium*, Reno, Nevada, Jan. 14-17, 2002.

[6] J.P. Lauffer, T.G. Carne, and T.D. Ashwill. *Modal Testing in the Design Evaluation of Wind Turbines*, SAND87-2461, Sandia National Laboratories, Albuquerque, NM, Apr. 1988.

[7] R.M. Osgood. *Dynamic Characterization Testing of Wind Turbines*, NREL Technical report NREL/TP-500-30070, May 2001.

[8] R.M. Osgood, H. G. McFarland, and G.L. Johnson. *Full System Modal Survey Test Results of the Controls Advanced Research Turbine (CART)*, NREL Internal Letter Report, National Renewable Energy Laboratory, Golden, CO, Feb. 2002.

[9] M.H. Hansen, K. Thomsen, and P. Fuglsang. Two methods for estimating aeroelastic damping of operational wind turbine modes from experiments, *Wind Energy*, **9**:179-191, Jan. 2006.

[10] B. Peeters and G. De Roeck. Stochastic system identification for operational modal analysis: a review, *ASME Journal of Dynamic Systems, Measurement, and Control*, **123**(4):659-667, 2001.

[11] B. Peeters, H. Van der Auweraer, F. Vanhollebeke, and P. Guillaume. Operational modal analysis for estimating the dynamic properties of a stadium structure during a football game, *Shock and Vibration*, **14**(4):283-303, 2007.

[12] G.H. James, T.G. Carne, and P.S. Veers. Damping measurements using operational data, *Transactions of the ASME*, **118**:190-193, Aug. 1996.

[13] G.H. James, T.G. Carne, J.P. Lauffer, and A.R. Nord. Modal testing using natural excitation, *Proc. of the 10th International Modal Analysis Conference*, San Diego, CA, 1992.

[14] LMS International. *LMS SCADAS Mobile Data Acquisition Front-end*, Breda, The Netherlands, www.lmsintl.com, 2009.

[15] LMS International. *LMS Test.Lab Structures*, Leuven, Belgium, www.lmsintl.com, 2009.

[16] B. Peeters, H. Van der Auweraer, P. Guillaume, and J. Leuridan. The PolyMAX frequency-domain method: a new standard for modal parameter estimation? *Shock and Vibration*, **11**:395-409, 2004.

Proceedings of the IMAC-XXVIII
February 1–4, 2010, Jacksonville, Florida USA
©2010 Society for Experimental Mechanics Inc.

Damage Detection in Wind Turbine Blade Panels Using Three Different SHM Techniques

Marcin Luczak[1,a], Bart Peeters [1,b], Michael Döhler [2,c], Laurent Mevel [2,d], Wieslaw Ostachowicz[3,e], Pawel Malinowski[3,f], Tomasz Wandowski [3,g], and Kim Branner [4,h]

[1]*LMS International, Interleuvenlaan 68, B–3001 Leuven, Belgium*

[2] *INRIA, Centre Rennes – Bretagne Atlantique, Campus de Beaulieu, F–35042 Rennes, France*

[3] *Institute of Fluid Flow Machinery, Polish Academy of Sciences, ul. Fiszera 14, 80 952 Gdansk, Poland*

[4] *Wind Energy Division, Risø National Laboratory for Sustainable Energy, Technical University of Denmark, Building 118, DK–4000 Roskilde, Denmark*

[a]*marcin.luczak@lmsintl.com*, [b]*bart.peeters@lmsintl.com*, [c]*michael.dohler@irisa.fr*, [d]*lmevel@irisa.fr*, [e] *wieslaw@imp.gda.pl*, [f] *pmalinowski@imp.gda.pl*, [g]*tomaszw@imp.gda.pl* [h] *kibr@risoe.dtu.dk*

Abstract.

A comparison of three different damage detection methods is made on three nominally identical glass reinforced composite panels, similar to the load carrying laminate in a wind turbine blade. Sensor data were recorded in the healthy state and after the introduction of damage by means of a four–point bending quasi–static test. Acceleration sensors, PZT transducers and the piezoelectric excited Lamb waves were used for the measurements of the panels. All three methods are based on the comparison of the healthy and damaged structure. The first method is statistical covariance–driven damage detection using a subspace–based algorithm, where one damage indicator for all three panels was computed. The second method is based on PZT transducers and the A0 mode of Lamb waves propagating in the panel, making use of the reflection of the signal at damage in the panel. The third method is based on the estimation of modal parameters of the intact and damaged panel using pLSCF and following their deviations. The results from these three damage detection methods are compared and discussed.

1. INTRODUCTION

The paper presents the results of the damage detection investigation in the multilayer E–glass wind turbine blade composite material with three SHM methods. The objects of the investigation were three nominally identical plates A, B and C. The nominal dimension were 20×320×320 mm (see Figure 2). Due to non repeatable manufacturing process dimension variability reaches ±0.7%. Damage was introduced into plates by means of four–point bending quasi–static test. Plate A was loaded with the force of 230 kN, plate B 210 kN and plate C 220 kN. The scope and results of each method is presented in following sections.

T. Proulx (ed.), *Structural Dynamics and Renewable Energy, Volume 1*, Conference Proceedings of the Society for Experimental Mechanics Series 10, DOI 10.1007/978-1-4419-9716-6_12, © The Society for Experimental Mechanics, Inc. 2011

2. SUBSPACE-BASED IDENTIFICATION AND DAMAGE DETECTION

2.1 Subspace-based Covariance Driven Identification Algorithm

We consider a linear multi-variable output-only system described by a discrete-time state space model:

$$\begin{cases} X_{k+1} = FX_k + V_{k+1} \\ Y_k = HX_k \end{cases} \tag{1}$$

where state X and observed output Y, at each time sample k, have dimensions m and r respectively. The state noise V is assumed to be stationary, unmeasured Gaussian white noise with zero mean. For system identification we construct the Hankel matrix

$$\mathrm{H}_{p+1,q} \stackrel{\text{def}}{=} \begin{pmatrix} R_0 & R_1 & \cdots & R_{q-1} \\ R_1 & R_2 & \cdots & R_q \\ \vdots & \vdots & & \vdots \\ R_p & R_{p+1} & \cdots & R_{p+q-1} \end{pmatrix} \stackrel{\text{def}}{=} Hank(R_i) \tag{2}$$

containing the correlations $R_i = \mathrm{E}(Y_k Y_{k-i}^T)$, use the well-known factorization property of $\mathrm{H}_{p+1,q}$ into observability and controllability matrix and recover H and F from the observability matrix. The eigenstructure (λ, Φ_λ) of the system (Eq. 1) and a corresponding modal parameter θ are retrieved from

$$\det(F - \lambda I) = 0,\ (F - \lambda I)\varphi_\lambda = 0,\ \Phi_\lambda \stackrel{\text{def}}{=} H\varphi_\lambda,\ \theta \stackrel{\text{def}}{=} \begin{pmatrix} \Lambda \\ \mathrm{vec}\ \Phi \end{pmatrix} \tag{3}$$

where Λ is the vector whose elements are the λ's and Φ is the matrix whose columns are the Φ_λ's. The natural frequencies and damping values of the system (Eq. 1) are computed from the λ's, see also [1].
The outputs Y_k for each of the plates were measured at 49 points during four setups: at 14 points in the first three setups and at 7 in the 4th setup. The excitation is assumed to be different for each of the four setups due to slight changes in the environmental conditions, but stationary during each measurement. Hence the state noise V in (Eq. 1) has a different variance for all the four setups and the measurements are normalised with respect to a change in the excitation. For this normalizing procedure, one additional reference sensor common to all the setups was available and with its help the excitation of the different setups can be normalized to a common excitation level. A merged matrix $\mathrm{H}_{p+1,q}$ containing measurement information from all the four setups is obtained according to the merging procedure described in [1], [15].

2.2 Damage Detection

For our subspace-based damage detection method a statistical comparison is made by defining a decision variable evaluating the system state [2]. This decision variable is an asymptotically centered Gaussian variable when the system is near the reference/healthy state ($\theta = \theta_0$) and non-centered in case of a change in modal parameters ($\theta \neq \theta_0$), which can be determined with an appropriate χ^2 test. We compute our decision variable ζ_n as the residual:

$$\zeta_n \stackrel{\text{def}}{=} \sqrt{n}\,\mathrm{vec}(S^T \hat{\mathrm{H}}_{p+1,q}) \text{ with } \hat{\mathrm{H}}_{p+1,q} = Hank(\hat{R}_i),\ \hat{R}_i = \frac{1}{n}\sum_{k=1}^{n} Y_k Y_{k-i}^T \tag{4}$$

where n is the sample length, S is the left kernel of the Hankel matrix at the reference state and $\hat{\mathrm{H}}_{p+1,q}$ is the Hankel matrix at the actual state. With the residual covariance Σ at the reference state, the global χ^2 test statistics built on the residual boils down to [2, 3]

$$\chi_n^2 = \zeta_n{}^T \Sigma^{-1} \zeta_n \tag{5}$$

As there are three references in our example (plates A, B and C in intact state, resp. Hankel matrices $\hat{\mathrm{H}}^{(\mathrm{A})}_{p+1,q}, \hat{\mathrm{H}}^{(\mathrm{B})}_{p+1,q}, \hat{\mathrm{H}}^{(\mathrm{C})}_{p+1,q}$), we create a residual ζ_n robust to all the plates by computing the left kernel S on the juxtaposed matrix $\left(\hat{\mathrm{H}}^{(\mathrm{A})}_{p+1,q} \;\; \hat{\mathrm{H}}^{(\mathrm{B})}_{p+1,q} \;\; \hat{\mathrm{H}}^{(\mathrm{C})}_{p+1,q}\right)$, inspired by the temperature change rejection approach in [3]. The damage detection is then done by computing the appropriate χ^2 test (Eq. 5) and comparing it to a threshold.

2.3 Numerical Results

2.3.1 Identification

With the subspace-based identification algorithm of Section 2.1 we computed the natural frequencies (*f*), damping ratios (*d*) and mode shapes of the three plates in intact and damaged state using the modal analysis toolbox COSMAD [16, 17]. Nine modes were recovered (see Table 1, Figure 1) in the frequency range of interest [0 – 2000 Hz]. As the damage to plate A led to considerably lower frequencies, three further modes were detected there (mode 10: 1326 Hz, 3.7 %; mode 11: 1489 Hz, 1.6 %; mode 12: 1858 Hz, 1.3 %).
The mode shapes were also computed on the intact samples B and C and especially the first modes (1, 2, 3, 4 and 7) turned out to be very similar, as it can be seen in a MAC comparison in Figure 2. This indicates that the other modes are more affected by differences in the manufacturing process of the plates, or that the statistical uncertainties carry more weight for the higher modes. Besides that, mode 6 was difficult to extract as it seems to be weakly excited and hence the uncertainty on the mode shape estimates is quite big, resulting in a low MAC value in the comparison.

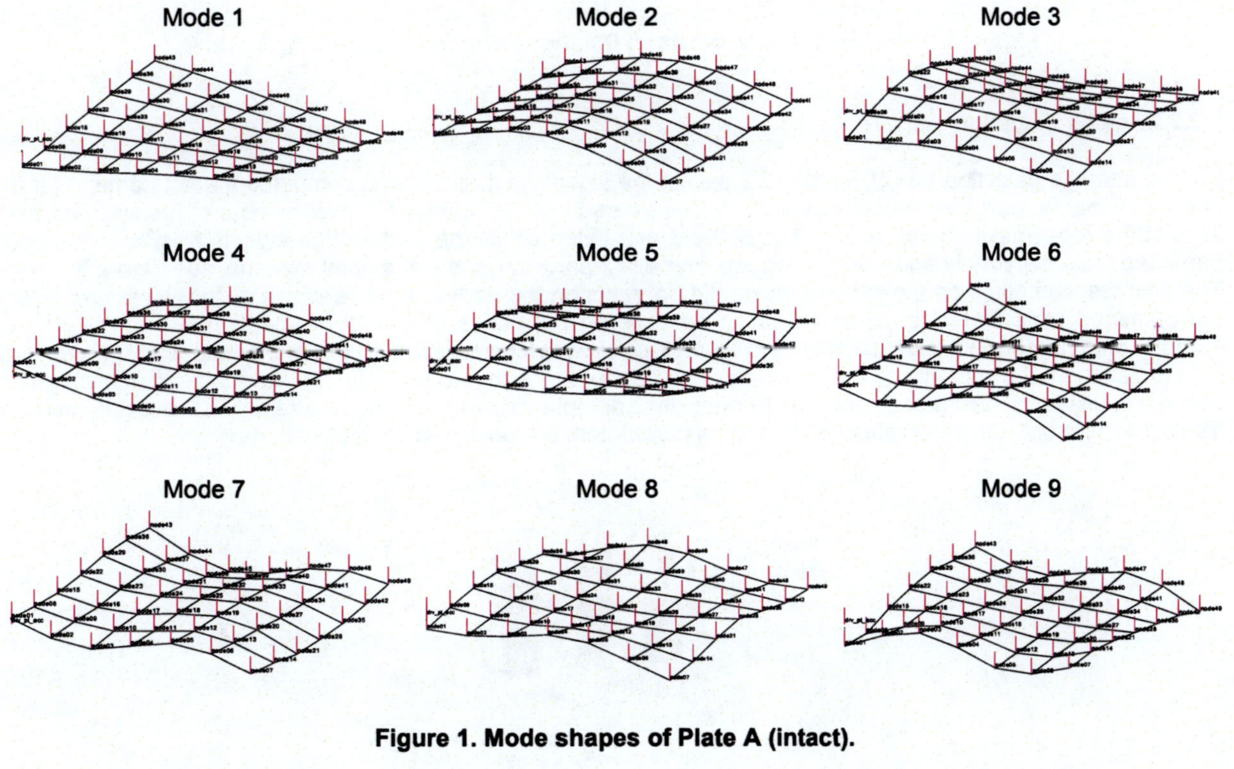

Figure 1. Mode shapes of Plate A (intact).

Mode	A intact		A damaged		B intact		B damaged		C intact		C damaged	
	f (Hz)	*d* (%)	*f* (Hz)	*d* (%)	*f* (Hz)	*d* (%)	*f* (Hz)	*d* (%)	*f* (Hz)	*d* (%)	*f* (Hz)	*d* (%)
1	350	2.2	257	1.1	357	2.1	358	1.0	359	1.7	359	2.4
2	559	2.5	387	1.1	564	1.5	538	1.0	554	3.5	545	3.7
3	818	3.3	528	2.1	788	2.6	784	0.6	784	3.3	835	3.0
4	917	1.8	654	1.5	919	1.9	957	1.4	929	2.7	928	2.4
5	1081	1.1	729	1.1	1102	3.3	1123	0.9	1097	1.9	1106	1.6
6	1246	5.7	-	-	1292	5.8	1295	3.8	1270	5.0	1242	2.5
7	1472	1.5	874	1.3	1536	2.4	1515	1.9	1506	2.4	1554	2.5
8	1849	1.3	1108	1.3	1846	2.5	1844	1.1	1850	2.1	1835	1.3
9	1934	2.5	1273	1.4	1962	3.6	1927	1.7	1946	1.7	1891	1.9

Table 1. Natural frequencies and damping coefficients.

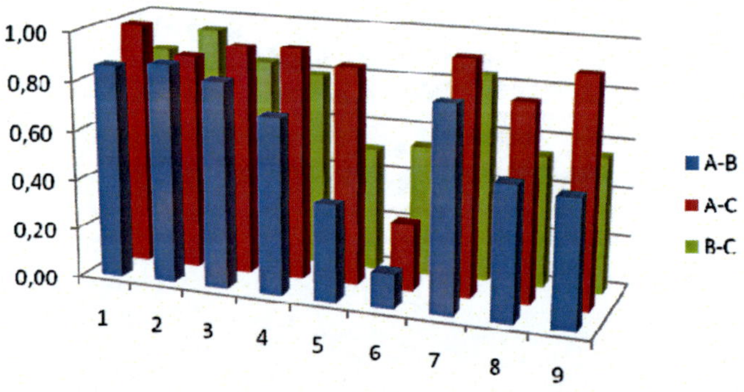

Figure 2. MAC comparison between the mode shapes of plates A, B and C.

2.3.2 Damage detection

For the damage detection test of Section 2.2 we cut the signals in 12 parts and computed the left kernel S for the decision variable ζ_n on three Hankel matrices – one for each plate – containing healthy data of the first part only. To test the robustness, the χ^2 values (Eq. 5) were calculated on all the parts of the signals from the intact and damaged state separately and finally averaged over all 12 parts to get a meaningful average, see Figure 3.
The joint test computed on the three samples did not react on the plates in the healthy states, but reacted well in the damaged state and was hence indicating damage on the plates. It should be mentioned that the χ^2 value means computed on the three samples in Figure 3 cannot be compared to evaluate the extent of the damage. Due to its inherent stochastic nature, their variance has to be taken into account. Moreover, the χ^2 test value is a pondered distance, taking into account frequencies, damping ratios and mode shapes. Its exact relation to Table 1, where only selected values of the modal parameters are observed, is not trivial.

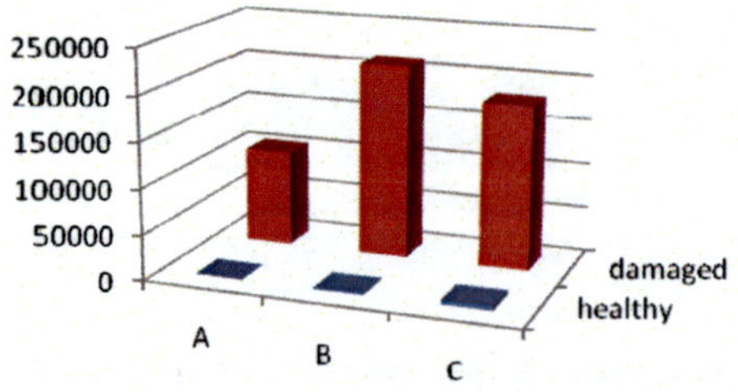

Figure 3. χ^2 damage detection test on intact and damaged plates.

3. DAMAGE DETECTION USING PIEZO–EXCITED ELASTIC WAVES

In order to perform damage detection piezoelectric transducers were used to excite and register elastic waves in considered specimen. Propagating waves interact with material inhomogeneities and can be used for Structural Health Monitoring purposes, particularly in this case for damage detection. In this study even approximate size of introduced damage was unknown, therefore high frequency (short wavelength) waves were used to ensure sensitivity. Piezoelectric transducers were excited and then responses from their electrodes were gathered using integrated generation/acquisition device (see Figure 4). It conducts generation and amplification of the signals to drive the piezos. In registration path it collects the signals, refines them and sends to a PC via USB. The concept of this device was born at the Department of Mechanics of Intelligent Structures (Institute of Fluid–Flow Machinery, Polish Academy of Sciences). Device setting (see Figure 4) allow to choose various types of excitation. In case of this research a narrow band tone burst excitation was chosen to reduce wave dispersion – 5 cycles of Hanning window modulated sine.

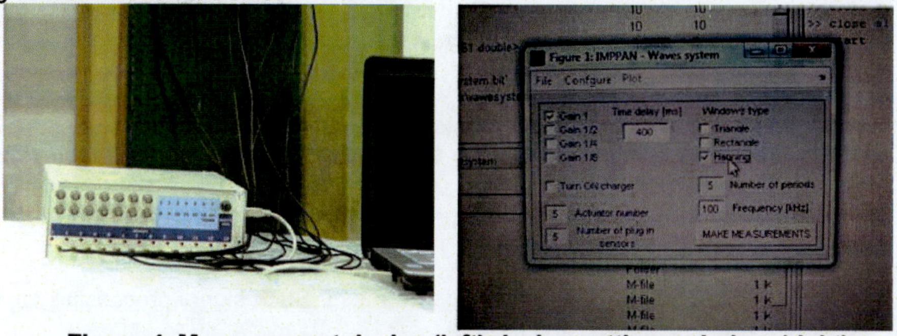

Figure 4. Measurement device (left) device settings window (right).

Transducer network was designed to obtain as much information about the specimen condition as possible. Twelve transducers were distributed on the whole specimen surface (see Figure 5). However they could not be placed uniformly due to the fact that four–point bending quasi–static test was performed on this specimen to introduce damage. Excitation was applied to each transducer from configuration while registration was realised in the rest of transducers. In result 12^2–12=132 signals were obtained for intact plate and the same number of signals was obtained from measurement for damaged specimen.

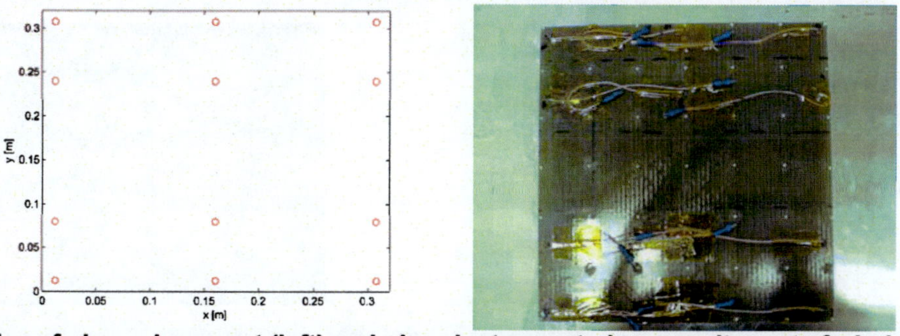

Figure 5. Idea of piezo placement (left) and piezo instrumented composite part of wind turbine blade (right). Black dashed lines represent the location of steel cylinders used in the four point bending test.

Measured signals were processed with special signal processing algorithm. The surface of monitored specimen was covered with a uniform mesh of points Pi=(xi , yi) [4]. Points separation was chosen to be dependent of the excitation signal

$$s = A c_g \frac{N}{f_c} \tag{6}$$

where cg – Lamb wave group velocity, N – number of sine cycles in excitation, fc – excitation central frequency, A – ratio to be chosen. In order to create an image of the specimen condition signals differences were calculated

$$B_{jk}(t) = B^{d}_{jk}(t) - B^{r}_{jk}(t), \tag{7}$$

where $B^{r}_{jk}(t)$ – signals from intact sample, $B^{d}_{jk}(t)$ – signals from investigated sample, i, j denote generating and receiving transducer, respectively. Distances between wave generating transducer to mesh point |Tj Pi| and from this point to wave receiving transducer |Pi Rk| were calculated and used to cut out a part of the $B_{jk}(t)$ signal. The cut out part has a length of l=s/cg and is centred in

$$t^{i}_{jk} = \frac{|T_j P_i| + |P_i R_k|}{c_g}. \tag{8}$$

Let's denote this part of the signal as Fn, this signal is discrete so the index takes values n=1,2,...,N, N depends on the length l. The signals are mapped into point Pi by summing signal power from all the Tj and Rk pairs:

$$M(P_i) = \sum_{j=1}^{J} \sum_{k=1}^{K} \sum_{n=1}^{N} F_n^2 . \tag{9}$$

This procedure is repeated for all point P_i in considered mesh. Such signal processing approach causes that the $M(P_i)$ lies on an ellipsis which loci are T_j and R_k [5].

In the conducted experiment A=0.1 was chosen (Eq. 6). It gave a good balance between computational speed and resolution of the mapping. Results for two considered frequencies (100 and 110 kHz) are depicted on Figure 6. Obtained results were normalised to the maximum value. Colour scale is from blue – minimum to red – maximum. Conducted mapping procedure indicated that the greatest differences between damaged an intact sample are in its lower half (see Figure 6). This suggests that damage could occur in this area. However it should be underlined that the difference could be also a result of transducer debonding cased by four–point bending test. In order to ensure this is not the reason a transducer self–testing procedure ought to be incorporated in the detection procedure.

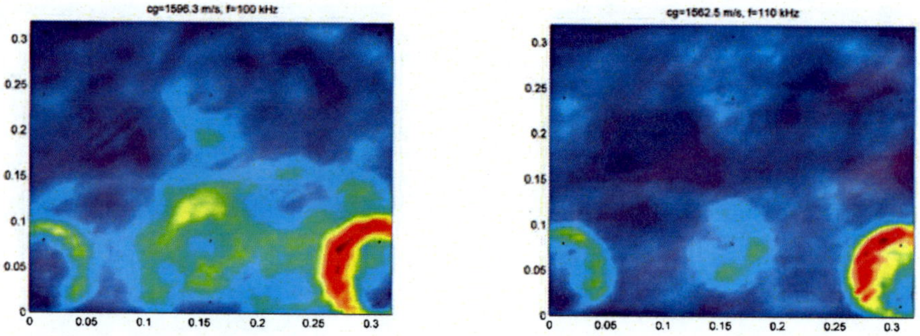

Figure 6. Results of damage detection using Lamb wave propagation for two central frequencies 100 (left) and 110 kHz (right).

4. POLYREFERENCE LSCF (PolyMAX) METHOD

4.1 Theoretical background

The Polyreference LSCF (PolyMAX) analysis method is based on Input/Output data in Frequency Domain (compare with Subspace method based on Output data in Time Domain).This method is used for estimation and comparison of the modal models identified for intact (reference) and damaged structure. Just like the Frequency-domain direct parameter identification method (FDPI) [6, 7,8] the LMS PolyMAX method uses measured frequency response function (FRF's) as primary data. Time-domain methods, such as the polyreference LSCE method [9] typically require impulse responses (the inverse Fourier transforms of the FRF's) as primary data. In the LMS PolyMAX method, following so-called right matrix-fraction model is assumed to represent the measured FRF's:

$$[H(\omega)]=\sum_{r=0}^{p} z^r[\beta_r]\left(\sum_{r=0}^{p} z^r[\alpha_r]\right)^{-1} \tag{10}$$

where $[H(\omega)]\in C^{lxm}$ is the matrix containing the FRF's between all m inputs and all l outputs; $[\beta_r]\in R^{lxm}$ the numerator matrix of polynomial are coefficients; the denominator matrix polynomial coefficients $[\alpha_r]\in R^{mxm}$ and is the model order. It should be noted that the so-called –domain model (i.e. a frequency-domain model that is derived from a discrete-time model) is used in Eq. 10, with:

$$z = e^{-j\omega\Delta t} \tag{11}$$

where Δt is the sampling time.

Basically, the unknown model coefficients $[\alpha_r]$ and $[\beta_r]$ are being found as the Least-Squares solution of these equations (after linearization). More details about this procedure can be found in [10, 11]

Once the denominator coefficients $[\alpha_r]$are determined, the poles and modal participation factors are retrieved as the eigenvalues and eigenvectors of their companion matrix:

$$\begin{pmatrix} 0 & I & \dots & 0 & 0 \\ 0 & 0 & \dots & 0 & 0 \\ \dots & \dots & \dots & \dots & \dots \\ 0 & 0 & \dots & 0 & I \\ -[\alpha_0^T] & -[\alpha_1^T] & \dots & -[\alpha_{p-2}^T] & -[\alpha_{p-1}^T] \end{pmatrix} . V = V\Lambda \tag{12}$$

The modal participation factors are the last m rows of $V\in C^{mpxmp}$; the matrix $\Lambda\in C^{mpxmp}$ contains the (discrete time) poles $[e^{-\lambda_i\Delta t}]$on its diagonal. They are related to the eigenfrequencies $[\omega_i]$ [rad/s] and damping ratios $[\xi_i]$ as follows (•* denotes complex conjugate):

$$\lambda_i,\lambda_i^* = -\xi_i\omega_i \pm j\sqrt{1-\xi_i^2}\,\omega_i \tag{13}$$

This procedure is similar to what happens in the time-domain LSCE method and allows constructing a stabilization diagram for increasing model orders p and using stability criteria for eigenfrequencies, damping ratios and modal participation factors.

Although theoretically, the mode shapes could be derived from the model coefficients $[\alpha_r]$ and $[\beta_r]$, we proceed in a different way.

The mode shapes can be found by considering the so-called pole-residue model:

$$[H(\omega)]=\sum_{i=1}^{n}\frac{\{v_i\}<l_i^T>}{j\omega-\lambda_i}+\frac{\{v_i^*\}<l_i^H>}{j\omega-\lambda_i^*}-\frac{[LR]}{\omega^2}+[UR] \tag{14}$$

where n is the number of modes; $\bullet^H$ denotes complex conjugate transpose of a matrix; $\{v_i\}\in C^l$ are the mode shapes; $<l_i^T>\in C^m$ are the modal participation factors and λ_i are the poles (Eq. 13). $[LR],[Ur]\in R^{lxm}$ are respectively the lower and upper residuals modeling the influence of the out-of-band modes in the considered frequency band. The interpretation of the stabilization diagram yields a set of poles λ_f and corresponding participation factors $<l_i^T>$ Since the mode shapes $\{v_i\}$ and the lower and upper residuals are the only unknowns, they are readily obtained by solving (Eq. 14) in a linear least-squares sense. This second step is commonly called least-squares frequency-domain (LSFD) method [12, 13]. The same mode-shape estimation method is normally also used in conjunction with the time-domain LSCE method.

4.2 Experimental work

Data analysed with subspace based method described in Section 2 and PolyMAX was acquired from the same experimental setup build with LMS acquisition hardware and software. Main difference is that subspace algorithm works on output–only time domain signals while PolyMAX uses the frequency domain input/output FRFs. Number of natural frequencies and mode shapes identified from those two data sets are identical for the intact plates. In frequency domain analysis corresponding natural frequencies are slightly shifted towards higher values due to the influence of the shaker (see Table 1 and 2). This phenomenon is described in details in the [14]. Number of identical experiments followed by modal parameter estimation revealed existence of the scatter of identified frequencies values within consecutive tests of the same plate and in between plates A, B and C as well. Modal model parameters range caused by test data variability was statistically assessed on the dimensionless frequencies values and is presented on the Figure 7.

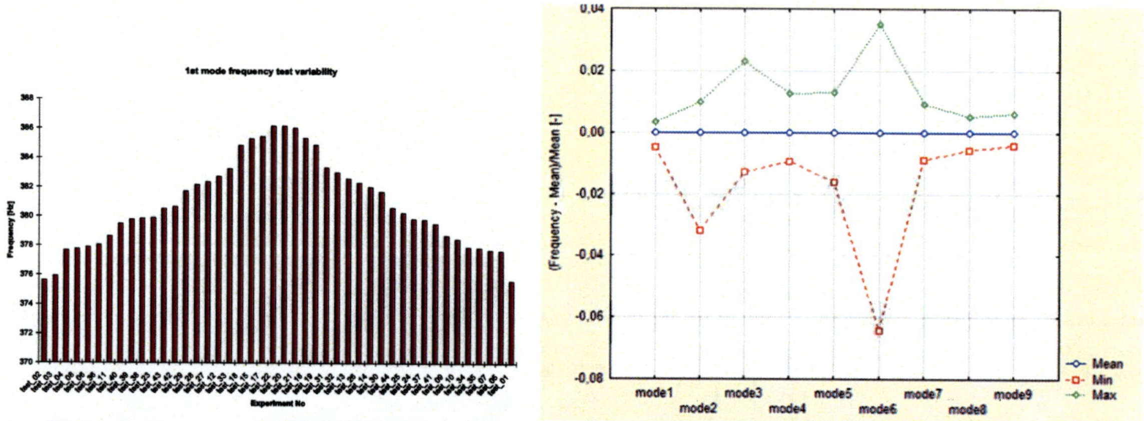

Figure 7. Test data variability leading to frequency values scatter. Left- 1st mode frq distribution, Right – dimensionless frequency differencies for all 9 modes

Intact sample	Damaged sample
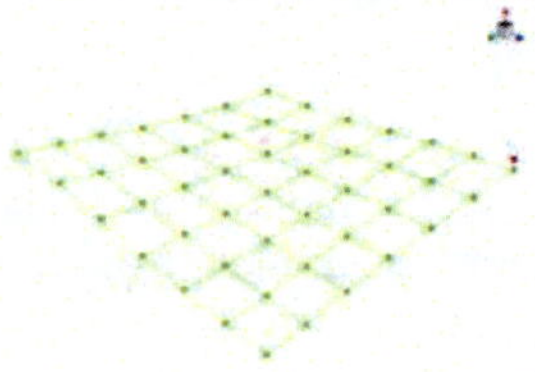	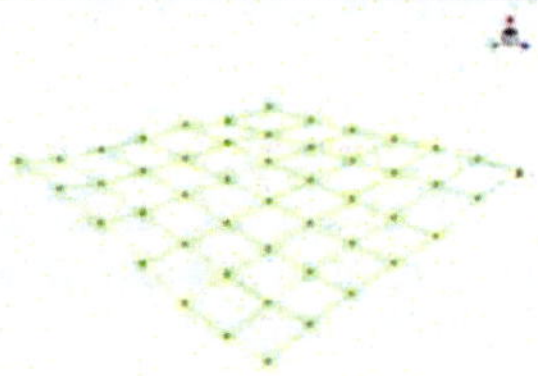
First mode 378 Hz	First mode 270 Hz

Table 2. Mode shape and natural frequencies.

Highest scatter observed for the sixth mode values is caused by relatively weak excitation of this mode. Damage was successfully detected in all three specimens. Example for plate A is presented in Table 2. Next to the decrease of natural frequency values caused by stiffness degradation due to delamination, fiber cracks and fiber–matrix debonding in all three plates appeared new modes. In the plate A with the largest damage three new modes were identified, in plate B there are two damage related modes and in plate C one mode.

A MAC comparison between the mode shapes of the plates obtained from the covariance-driven Stochastic Subspace Identification (SSI) from Section 2.3.1 and the PolyMAX method of this Section can be found in Figure 8. This shows that both methods lead to very close mode shapes estimates especially for the first modes. As the uncertainty on mode 6 is very high, the respective MAC value is low.

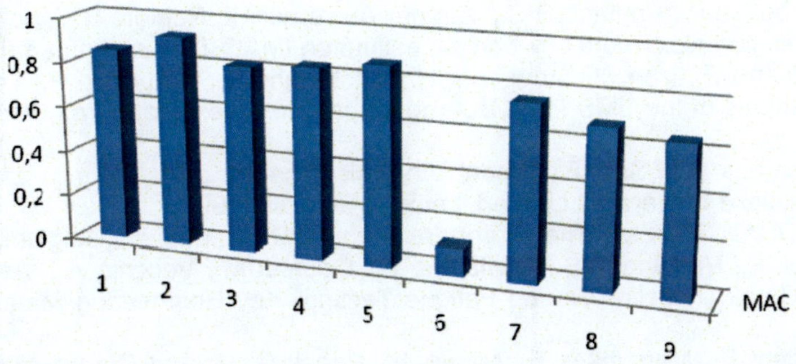

Figure 8. MAC comparison between mode shapes obtained using covariance-driven SSI and PolyMAX

5. CONCLUSIONS AND FURTHER RESEARCH

The comparison of three different damage detection methods in presence of test data variability for the E–glass composite material part of wind turbine blade was presented. All three methods proved their adequacy in detecting different levels of the damage. However the unexpected observation of increased natural frequency of some modes require more in–depth investigation. Therefore the scope of further research will focus on the precise identification of damage level and location by means of ultrasonic methods and radiography. Modal data will be processed with the application of the modal filter which is reported in many scientific papers as an indicator in order to differentiate between damage and intact state.

6. ACKNOWLEDGEMENTS

The authors gratefully acknowledge support for this research project "UNVICO–2" provided by the 6th EU FP Marie Curie Fellowships and K. Martyniuk from Wind Energy Division, Risø for her contribution.

REFERENCES

[1] Mevel, L., Basseville M., Benveniste A., and Goursat M. 2002. "Merging sensor data from multiple measurement setups for nonstationary subspace–based modal analysis," *J. Sound Vib.,* 249(4):719–741.

[2] Mevel, L., Goursat M., and Basseville M. 2003. "Stochastic subspace–based structural identification and damage detection and localization – Application to the Z24 bridge benchmark," *Mech. Syst. Signal Pr., Special issue on COST F3 Benchmarks,* 17(1):143–151.

[3] Balmès, É., Basseville M., F. Bourquin, Mevel, L., Nasser H., and Treyssède, F. 2008. "Merging sensor data from multiple temperature scenarios for vibration–based monitoring of civil structures," *Struct. Health Monit.,* 7(2):129–142.

[4] Michaels J. E., Croxford A. J., Wilcox P. D., Imaging algorithms for locating damage via in situ ultrasonic sensors, SAS 2008–IEEE Sensors Applications Symposium, Feb 12–14, 2008.

[5] Lu Y., Ye L. Su Z.: Crack identification in aluminium plates using Lamb wave signals of a PZT sensor network, Smart Materials and Structures 15: 839–849, 2006.

[6] Peeters B., Guillaume P., Van der Auweraer H., Cauberghe B., Verboven P., Leuridan J., Automotive and aerospace applications of the LMS PolyMAX modal parameter estimation method, IMAC 22, Dearborn (MI), USA, 2004.

[7] Lembregts F., Leuridan J., Zhang L., Kanda H. Multiple input modal analysis of frequency response functions based direct parameter identification, IMAC 4, 1986, Los Angeles (CA), USA.

[8] Lembregts F., Snoeys R., Leuridan J. Application and evaluation of multiple input modal pa-rameter estimation, International Journal of Analytical and Experimental Modal Analysis, 2(1), 19–31, 1987.

[9] Brown D.L., Allemang R.J., Zimmerman R., Mergreay M. Parameter estimation techniques for modal analysis, Society of Automotive Engineers, Paper No. 790221, 1979.

[10] Guillaume P., Verboven P., Vanlanduit S., Van der Auweraer H., Peeters B. A poly-reference implementation of the least-squares complex frequency-domain estimator, IMAC 21, Kis-simmee (FL), USA, 2003.
[11] Peeters B., Guillaume P., Van der Auweraer H., Cauberghe B., Verboven P., Leuridan J., Automotive and aerospace applications of the LMS PolyMAX modal parameter estimation method, IMAC 22, Dearborn (MI), USA, 2004.
[12] Heylen W., Lammens S., Sas P. Modal Analysis Theory and Testing, Department of Me-chanical Engineering, Katholieke Universiteit Leuven, Leuven, Belgium, 1995.
[13] LMS INTERNATIONAL, The LMS Theory and Background Book, Leuven, Belgium, 2000.
[14] Peeters B., Mevel, L., Vanlanduit S., Guillaume P., Goursat M., Vecchio A., Van der Auweraer H. "Online Vibration–based Crack Detection During Fatigue Testing" Key Engineering Materials Vols. 245–246 (2003) pp. 571–578
[15] Döhler M., Reynders E., Magalhães F., Mevel, L., Roeck G. D. and Cunha A. Pre- and post-identification merging for multi-setup OMA with covariance-driven SSI. In Proceedings of IMAC 28, Jacksonville, FL, 2010.
[16] Goursat M. and Mevel, L. An example of analysis of thermal effects on modal characteristics of a mechanical structure using Scilab. In S.-Y. Qin, B. Hu, S. Li, and C. Gomez, editors, Scilab Research, Development and Applications, pages 241–256. Tsinghua University Press - Springer, Beijing, China, 2005.
[17] Goursat M. and Mevel, L.. COSMAD: Identification and diagnosis for mechanical structures with Scilab. In Proceedings of the Multi-conference on Systems and Control, International Symposium on Computer-Aided Control Systems Design (CACSD), San Antonio, TX, US, 2008.

Proceedings of the IMAC-XXVIII
February 1–4, 2010, Jacksonville, Florida USA
©2010 Society for Experimental Mechanics Inc.

Force Estimation via Kalman Filtering for Wind Turbine Blade Control

Jonathan C. Berg[1] Dr. A. Keith Miller
Department of Mechanical Engineering
New Mexico Institute of Mining and Technology
801 Leroy Place, Socorro, NM 87801

ABSTRACT

Under simulation, the forces on a wind turbine blade are estimated in real time and then included in a control algorithm which seeks to manage loads and improve rotor performance. A Kalman filter, modified to include the input vector in the state vector, is employed to estimate the stochastic load state and full blade state from a small number of structural response measurements. Blade pitch and flap-like control surfaces are considered in a control scheme utilizing the estimated modal-equivalent forces. It is shown that the shape of the load profile can be regulated with such a controller, thereby reducing the undesirable effects of a wind gust.

1 BACKGROUND

In current technology, wind turbines are mostly passive structures. Control authority is limited to relatively slow moving actuators, and sensing is limited to what is required for essential control functions and electric grid integration. Research trends indicate that future generations will have active components which respond quickly to the wind environment, thereby improving efficiency and controlling loads [1].

Making the active components respond properly to the wind environment requires knowledge of either the local airflow conditions or the resulting aerodynamic forces on the blades. The first part of this paper proposes a method of estimating the forces which involves measuring the structural response and then using a modified Kalman filter to infer the stochastic wind load. The method allows a small number of sensors to produce a full state estimate. The second part of this paper discusses how the load estimate might be used to control blade pitch or an active aerodynamic device.

Input estimation of beam structures has been studied by C.K. Ma [2] using the method described by Tuan [3]. Their technique also uses the Kalman filter, but the excitation forces are estimated by a recursive least-squares algorithm that operates on information provided by the standard Kalman filter.

In regard to wind turbines, estimation of effective wind speed has been explored by X. Ma [4] and Østergaard [5]. Boukhezzar [6] uses the Kalman filter to estimate aerodynamic torque from the rotor speed and also demonstrates how to use the estimate for control purposes. Ehlers [7] looks at sensor selection and turbine state estimation for control purposes.

2 SYSTEM MODEL

The force estimation technique described in this paper begins with a linear structural model of the form

$$\mathbf{M}\ddot{\boldsymbol{\xi}} + \mathbf{C}\dot{\boldsymbol{\xi}} + \mathbf{K}\boldsymbol{\xi} = \mathbf{F}\mathbf{u} + \mathbf{F_d}\mathbf{w} \tag{1}$$

where $(\dot{\ })$ denotes the time derivative and $\mathbf{M}$, $\mathbf{C}$, and $\mathbf{K}$ are respectively the mass, damping, and stiffness matrices. The variable $\boldsymbol{\xi}$ is the vector of displacement degrees-of-freedom and is also called the physical vector. The variable $\mathbf{u}$ is the vector

[1]Work in partial fulfillment of completion of a Masters of Science Degree in Mechanical Engineering, NMT.

of applied forces and can also be considered the vector of control inputs. Disturbance inputs to the system are represented by the vector $\mathbf{w}$. The matrices $\mathbf{F}$ and $\mathbf{F_d}$ are included to facilitate mapping of the inputs to the system. Second order differential equations result naturally from mechanical systems, however, the following first order description is more convenient for analysis.

The state space representation of a dynamic system has the form

$$\dot{\mathbf{x}} = \mathbf{A}\mathbf{x} + \mathbf{B}\mathbf{u} + \mathbf{G}\mathbf{w} \tag{2}$$

where $\mathbf{x}$ is the vector of state variables, $\mathbf{u}$ is the control input vector, and $\mathbf{w}$ is the input disturbance (or process noise) vector. Matrices $\mathbf{A}$, $\mathbf{B}$, and $\mathbf{G}$ are respectively the state matrix, control input matrix, and input disturbance matrix. Physical measurements of the system are modeled as linear combinations of the state variables. The output equation is

$$\tilde{\mathbf{y}} = \mathbf{H}\mathbf{x} + \mathbf{v} \tag{3}$$

where $\tilde{\mathbf{y}}$ is the vector of measured outputs, $\mathbf{H}$ is the output matrix, and $\mathbf{v}$ is the sensor noise vector.

The state variable assignment for a given system is not unique. For the general mechanical system of equation (1), the most common choice is $\mathbf{x} = [\xi^T\ \dot{\xi}^T]^T$. The resulting system matrices are given in equation (4).

$$\mathbf{A} = \begin{bmatrix} \mathbf{0} & \mathbf{I} \\ -\mathbf{M}^{-1}\mathbf{K} & -\mathbf{M}^{-1}\mathbf{C} \end{bmatrix} \quad \mathbf{B} = \begin{bmatrix} \mathbf{0} \\ \mathbf{M}^{-1}\mathbf{F} \end{bmatrix} \quad \mathbf{G} = \begin{bmatrix} \mathbf{0} \\ \mathbf{M}^{-1}\mathbf{F_d} \end{bmatrix} \tag{4}$$

Equation (2) is continuous in time, meaning that the system progresses smoothly from one state to the next and the dynamics are described by differential equations. In certain situations such as simulation or control implementation, the system is sampled at regular time intervals, which makes a discrete-time description more convenient. The differential equations become recursive equations relating the state variables at integer multiples of the time step Δt. In the following notation, the subscript k refers to the kth time step, meaning $\mathbf{x}_k = \mathbf{x}(k\Delta t)$. Reference [8] shows that, when $\mathbf{A}$ and $\mathbf{B}$ are constant in time, the equations become

$$\mathbf{x}_{k+1} = \mathbf{\Phi}\mathbf{x}_k + \mathbf{\Gamma}\mathbf{u}_k + \mathbf{\Upsilon}\mathbf{w}_k \tag{5a}$$

$$\tilde{\mathbf{y}}_k = \mathbf{H}\mathbf{x}_k + \mathbf{v}_k \tag{5b}$$

where

$$\mathbf{\Phi} = e^{\mathbf{A}\Delta t} \qquad \mathbf{\Gamma} = \left[\int_0^{\Delta t} e^{\mathbf{A}t}dt\right]\mathbf{B} \qquad \mathbf{\Upsilon} = \left[\int_0^{\Delta t} e^{\mathbf{A}t}dt\right]\mathbf{G}$$

In practice, the discrete-time system matrices can be found using a numerical approach. The MATLAB® command `c2d()`, which is part of the Control System Toolbox, implements one numerical solution and is used in this work.

```
[Phi,Gamma]=c2d(A,B,DT)
[Phi,Upsilon]=c2d(A,G,DT)
```

3 THE KALMAN FILTER

The Kalman filter is a recursive filter that can smooth measurements and estimate the complete state of a dynamic system. Its formulation is based in probability theory and stochastic system modeling, making it a good choice for application to the highly variable wind environment. If the stochastic properties of the measurement noise $\mathbf{v}$ and process noise $\mathbf{w}$ are well known, state estimates can be considered optimal in the sense of least-square estimation error.

Revisiting the discrete-time state space formulation given by equation set (5), the noise vectors are now described using statistical descriptions. The notation $\mathbf{v}_k \sim N(\mathbf{0}, \mathbf{R}_k)$ indicates that $\mathbf{v}_k$ is a random vector with normal (Gaussian) distribution having zero mean and covariance matrix $\mathbf{R}_k$. It is assumed that $\mathbf{w}_k$ and $\mathbf{v}_k$ are both white-noise processes.

$$\mathbf{x}_{k+1} = \mathbf{\Phi}_k\mathbf{x}_k + \mathbf{\Gamma}_k\mathbf{u}_k + \mathbf{\Upsilon}_k\mathbf{w}_k, \quad \mathbf{w}_k \sim N(\mathbf{0}, \mathbf{Q}_k) \tag{6a}$$

$$\tilde{\mathbf{y}}_k = \mathbf{H}_k\mathbf{x}_k + \mathbf{v}_k, \quad \mathbf{v}_k \sim N(\mathbf{0}, \mathbf{R}_k) \tag{6b}$$

3.1 Discrete-Time Linear Kalman Filter

Crassidis and Junkins [8] provide a thorough discussion of various forms of the Kalman filter. Equation set (7) is the discrete-time linear Kalman filter. The matrix $\mathbf{K}_k$ is known as the Kalman gain matrix, and it serves to improve the model's state prediction with information from the output measurement $\tilde{\mathbf{y}}_k$. Defining the estimation error as the difference between the state estimate $\hat{\mathbf{x}}_k$ and the true state $\mathbf{x}_k$, the matrix $\mathbf{P}_k$ is the covariance of the estimation error.

$$\hat{\mathbf{x}}_{k+1} = \boldsymbol{\Phi}_k \hat{\mathbf{x}}_k + \boldsymbol{\Gamma}_k \mathbf{u}_k + \boldsymbol{\Phi}_k \mathbf{K}_k \left[\tilde{\mathbf{y}}_k - \mathbf{H}_k \hat{\mathbf{x}}_k\right] \tag{7a}$$

$$\mathbf{K}_k = \mathbf{P}_k \mathbf{H}_k^T \left[\mathbf{H}_k \mathbf{P}_k \mathbf{H}_k^T + \mathbf{R}_k\right]^{-1} \tag{7b}$$

$$\mathbf{P}_{k+1} = \boldsymbol{\Phi}_k \mathbf{P}_k \boldsymbol{\Phi}_k^T - \boldsymbol{\Phi}_k \mathbf{K}_k \mathbf{H}_k \mathbf{P}_k \boldsymbol{\Phi}_k^T + \boldsymbol{\Upsilon}_k \mathbf{Q}_k \boldsymbol{\Upsilon}_k^T \tag{7c}$$

3.2 Input Estimation

To estimate the input vector $\mathbf{u}$, one approach is to redesign the Kalman filter such that the input vector is included in the state vector. However, the dynamics relating $\dot{\mathbf{u}}$ to the state vector are likely unknown. This lack of knowledge can be modeled by setting $\dot{\mathbf{u}} = \mathbf{0}$ and then adding a random vector $\mathbf{z}$, thus making $\mathbf{u}$ constant except for model uncertainty. The discrete-time description becomes

$$\begin{bmatrix}\mathbf{x}_{k+1}\\ \mathbf{u}_{k+1}\end{bmatrix} = \begin{bmatrix}\boldsymbol{\Phi}_k & \boldsymbol{\Gamma}_k\\ \mathbf{0} & \mathbf{I}\end{bmatrix}\begin{bmatrix}\mathbf{x}_k\\ \mathbf{u}_k\end{bmatrix} + \begin{bmatrix}\boldsymbol{\Upsilon}_k & \mathbf{0}\\ \mathbf{0} & \Delta t\mathbf{I}\end{bmatrix}\begin{bmatrix}\mathbf{w}_k\\ \mathbf{z}_k\end{bmatrix}, \quad \begin{bmatrix}\mathbf{w}_k\\ \mathbf{z}_k\end{bmatrix} \sim N\left(\mathbf{0}, \begin{bmatrix}\mathbf{Q}_{\mathbf{w}k} & \mathbf{0}\\ \mathbf{0} & \mathbf{Q}_{\mathbf{z}k}\end{bmatrix}\right) \tag{8a}$$

$$\tilde{\mathbf{y}}_k = \begin{bmatrix}\mathbf{H}_k & \mathbf{0}\end{bmatrix}\begin{bmatrix}\mathbf{x}_k\\ \mathbf{u}_k\end{bmatrix} + \mathbf{v}_k, \quad \mathbf{v}_k \sim N(\mathbf{0}, \mathbf{R}_k) \tag{8b}$$

where Δt is the time step.

Thus, the extended Kalman filter for input estimation uses the following alterations to equation sets (6) and (7).

$$\begin{gathered}\boldsymbol{\Phi}_k^\star = \begin{bmatrix}\boldsymbol{\Phi}_k & \boldsymbol{\Gamma}_k\\ \mathbf{0} & \mathbf{I}\end{bmatrix}, \quad \boldsymbol{\Gamma}_k^\star = \mathbf{0}, \quad \boldsymbol{\Upsilon}_k^\star = \begin{bmatrix}\boldsymbol{\Upsilon}_k & \mathbf{0}\\ \mathbf{0} & \Delta t\mathbf{I}\end{bmatrix},\\ \mathbf{Q}_k^\star = \begin{bmatrix}\mathbf{Q}_{\mathbf{w}k} & \mathbf{0}\\ \mathbf{0} & \mathbf{Q}_{\mathbf{z}k}\end{bmatrix}, \quad \mathbf{H}_k^\star = \begin{bmatrix}\mathbf{H}_k & \mathbf{0}\end{bmatrix}\end{gathered} \tag{9}$$

4 SIMULATION SETUP AND RESULTS

4.1 Initial Testing

Before applying the technique to a full wind turbine simulation, a single blade was simulated by applying realistic forces to a non-rotating cantilever beam model. To produce realistic forces, the FAST wind turbine code [9] with AeroDyn subroutines [10] was employed to calculate blade forces on the Baseline 1.5 MW model provided with version 6.01 of FAST. Full-field turbulent wind input was generated by TurbSim [11] using the normal turbulence model, intensity level C, and Kaimal spectral model. A mean wind speed of 9 m/s was chosen, placing the turbine operating point in Region 2 (variable rotor speed with blade pitch held constant).

Aerodynamic forces calculated by AeroDyn were saved as text files and then loaded into MATLAB for testing of the estimation procedure. The force vector had 15 entries, corresponding to the 15 blade elements defined in the AeroDyn input file. Six separate wind files and responses were generated to provide a variety of load profiles and events.

The cantilever beam model in MATLAB was constructed from simple Euler-Bernoulli 2D beam elements. Mass and stiffness properties were taken from the 1.5 MW blade description and utilized along with modal damping to produce a tapered beam representative of the blade. The beam was likewise given 15 elements, and discrete-time simulation was carried out using the same time step as the pre-calculated force data.

The matrix $\mathbf{F}$ was defined to map element forces to nodal forces and moments. Process noise was not of great interest for this part of the testing, so the disturbance input mapping matrix $\mathbf{F}_\mathbf{d}$ was set equal to an identity matrix of appropriate size and

$\mathbf{w}_k$ was given small variance. Considering $\mathbf{Q}_{\mathbf{z}k}$, it was noted from equation (8a) that the statistical properties of $\mathbf{z}$ are related to those of the input vector. So, element forces from over 10 minutes of simulated turbine operation were taken to calculate the statistics of the input vector, specifically the covariance of $\mathbf{u}$. Testing the filter with $\mathbf{Q}_{\mathbf{z}k}$ equal to $\text{Cov}(\mathbf{u})$ produced input estimates that were deemed acceptable.

Using this setup, a study of various measurement types, sensor locations, and sensor noise levels was conducted. Measurement type and location were defined in the output matrix $\mathbf{H}_k$, and noise level was selected with the variable R. In summary, the parameters of the Kalman filter were

$$\mathbf{Q}_{\mathbf{w}k} = 1.0 \cdot \mathbf{I} \tag{10a}$$

$$\mathbf{Q}_{\mathbf{z}k} = \text{Cov}(\mathbf{u}) \tag{10b}$$

$$\mathbf{R}_{\mathbf{k}} = R \cdot \mathbf{I} \tag{10c}$$

Of all the measurement types that were studied, FAST can output bending moment most easily. The next section describes how the force estimation technique was applied with the full FAST model in the loop with a controller. Thus, a brief discussion of the bending moment results of the sensor study is included here.

At least two sensors were required to estimate both the shape and magnitude of the load profile. With $R = 10^{-4}$, estimation error was minimized when bending moment was measured at the blade root and at node 8 (53% span). Note that the noise level was relative to the output level, which was scaled so that the output magnitude was approximately equal to one. For bending moment, this scaling resulted in output units of 1×10^6 Newton-meters. Figure 1 provides a plot of force estimation error over time and a few snapshots of the true force profile (solid line) and estimated force profile (dotted line). The estimate is not perfect, but it does track the shape of the force profile.

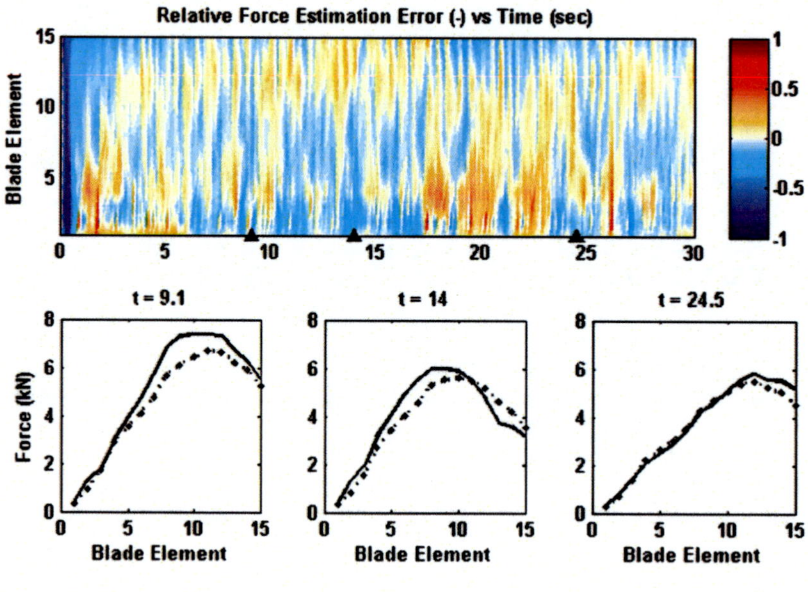

Figure 1: Top: force estimation error as a function of time and blade element. Bottom: the true force profile (solid) and estimated force profile (dotted) at three points in time.

4.2 Control Development

The technique was then applied inside MATLAB's Simulink environment with FAST included as an S-function block. With this configuration, the force estimate could be fed into a control algorithm while the simulation was running. The Kalman filter was implemented in MATLAB code as an S-function block.

It was found that a reduced-order model of the previous higher order beam model was useful for control purposes. First, smaller system matrices resulted in a faster Kalman filter. Second, the 15 estimated forces were reduced down to a more manageable set of inputs for the control algorithm.

The form of the reduced model was given by

$$\mathbf{M_r}\ddot{\boldsymbol{\eta}} + \mathbf{C_r}\dot{\boldsymbol{\eta}} + \mathbf{K_r}\boldsymbol{\eta} = \mathbf{u_r} + \mathbf{w_r} \tag{11}$$

The transformation from physical coordinates to the reduced set of DOFs was accomplished by first finding the real-valued eigenvectors $\mathbf{V}$ of the mechanical system equations. A reduced set of bending modes $\mathbf{V_r}$ was then formed by selecting a few columns from $\mathbf{V}$. The eigenvectors had been scaled such that the reduced mass matrix equaled the identity matrix and the reduced stiffness matrix was diagonal and contained the squared natural frequencies.

$$\boldsymbol{\xi} = \mathbf{V_r}\boldsymbol{\eta} \tag{12a}$$
$$\mathbf{M_r} = \mathbf{V_r}^T\mathbf{M}\mathbf{V_r} = \mathbf{I} \tag{12b}$$
$$\mathbf{K_r} = \mathbf{V_r}^T\mathbf{K}\mathbf{V_r} = \lceil \omega^2 \rfloor \tag{12c}$$
$$\mathbf{C_r} = \mathbf{V_r}^T\mathbf{C}\mathbf{V_r} = \lceil 2\zeta\omega \rfloor \tag{12d}$$
$$\mathbf{u_r} = \mathbf{V_r}^T\mathbf{F}\mathbf{u} \tag{12e}$$
$$\mathbf{w_r} = \mathbf{V_r}^T\mathbf{F_d}\mathbf{w} \tag{12f}$$

The state space form was then given by

$$\dot{\mathbf{x}}_\mathbf{r} = \mathbf{A_r}\mathbf{x_r} + \mathbf{B_r}\mathbf{u_r} + \mathbf{G_r}\mathbf{w_r} \tag{13a}$$
$$\tilde{\mathbf{y}} = \mathbf{H_r}\mathbf{x_r} + \mathbf{v} \tag{13b}$$

where $\mathbf{x_r} = [\boldsymbol{\eta}^T \; \dot{\boldsymbol{\eta}}^T]^T$ and

$$\mathbf{A_r} = \begin{bmatrix} \mathbf{0} & \mathbf{I} \\ -\mathbf{K_r} & -\mathbf{C_r} \end{bmatrix} \quad \mathbf{B_r} = \begin{bmatrix} \mathbf{0} \\ \mathbf{I} \end{bmatrix} \quad \mathbf{G_r} = \begin{bmatrix} \mathbf{0} \\ \mathbf{I} \end{bmatrix}$$
$$\mathbf{H_r} = \mathbf{H}\begin{bmatrix} \mathbf{V_r} & \mathbf{0} \\ \mathbf{0} & \mathbf{V_r} \end{bmatrix}$$

From this point, the procedure was the same as before. The continuous equations were converted to discrete time, and the Kalman filter's system matrices were augmented to estimate both the beam state and inputs as described previously.

To obtain the covariance of $\mathbf{w_r}$ in terms of $\mathbf{Q_w}$, the definition of covariance was applied to the random vector $\mathbf{w_r}$, resulting in

$$\mathbf{Q_{w_r}} = \mathbf{V_r}^T\mathbf{F_d}\mathbf{Q_w}\mathbf{F_d}^T\mathbf{V_r} \tag{14}$$

Likewise, the covariance of $\mathbf{u_r}$ was found to be

$$\mathrm{Cov}(\mathbf{u_r}) = \mathbf{V_r}^T\mathbf{F}\mathrm{Cov}(\mathbf{u})\mathbf{F}^T\mathbf{V_r} \tag{15}$$

In modal coordinates, the estimated input vector $\hat{\mathbf{u}}_\mathbf{r}$ can be compared directly to the true input vector $\mathbf{u_r}$. However, to compare results in physical coordinates, $\hat{\mathbf{u}}_\mathbf{r}$ must be related back to $\hat{\mathbf{u}}$. Although $\mathbf{V_r}^T$ is not square and can not be inverted exactly, the full set of eigenvectors is invertible. The resulting matrix $(\mathbf{V}^T)^{-1}$ is then reduced such that the remaining columns corresponded to the rows of $\hat{\mathbf{u}}_\mathbf{r}$.

$$\mathbf{F}\hat{\mathbf{u}} = [(\mathbf{V}^T)^{-1}]_\mathbf{r}\hat{\mathbf{u}}_\mathbf{r} \tag{16}$$

Thus, the estimated nodal forces $\mathbf{F}\hat{\mathbf{u}}$ computed by the above equation can be compared with the true nodal forces $\mathbf{F}\mathbf{u}$. Note that the columns of $(\mathbf{V}^T)^{-1}$ form a set of basis functions for the force profile. These basis functions have the unique property that when each is applied to the beam, the resulting static deformation is the corresponding mode shape. The first three force basis functions are plotted in Figure 2.

With the tools in place to estimate blade forces directly from the FAST outputs, a test case involving a wind gust was selected. In one particular wind input file, there was a point in the simulation where the forces at the blade tip dropped off drastically.

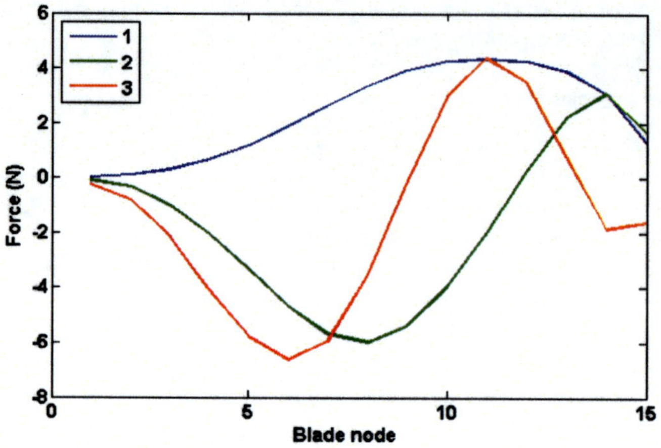

Figure 2: Force basis functions resulting from modal representation.

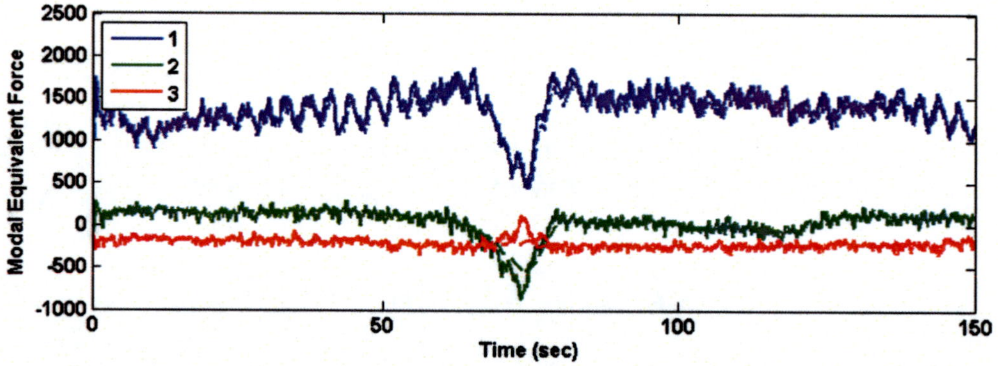

Figure 3: Modal equivalent forces showing a lull in the blade load. Dashed lines are the estimates.

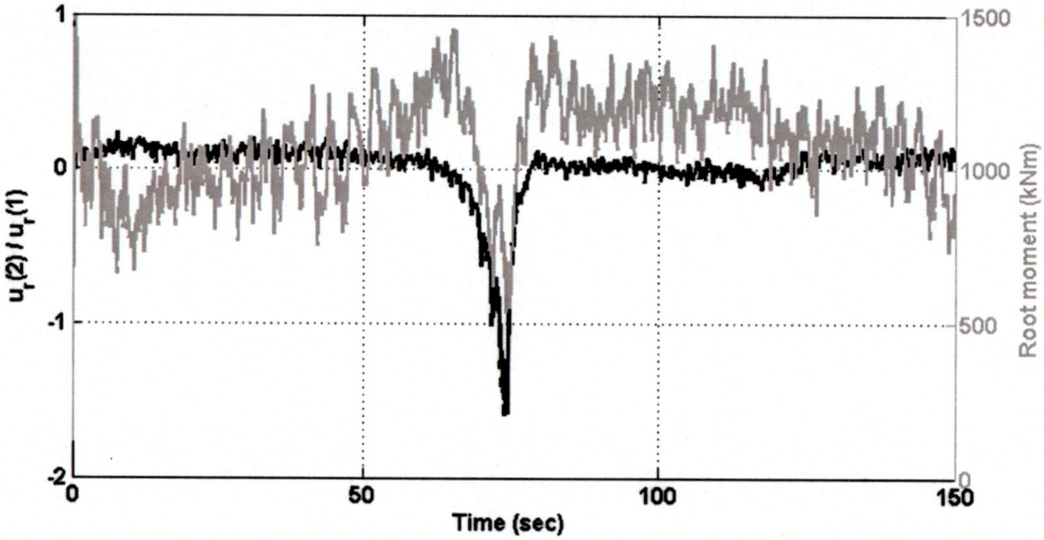

Figure 4: Ratio of the second modal force to the first. Root bending moment plotted for comparison.

A gust had caused the rotor to speed up and then the subsequent lull produced a drop in the tip load for a number of rotor revolutions. The lull can be seen in Figure 3 which plots the first three modal equivalent forces. The fact that the load profile had changed can be seen more clearly by plotting the ratio of the second modal force to the first (Figure 4).

If this ratio of the first and second modal forces were regulated by a controller, then the force profile would remain more constant. Having this control objective in mind, a simple proportional controller was defined which responded to this ratio whenever it fell below a threshold of 0.1. Equations (17) and (18) define the controller's input and output.

$$c_{in} = \hat{u}_{r,2}/\hat{u}_{r,1} - 0.1 \tag{17}$$

$$c_{out} = \begin{cases} K_P \cdot c_{in} & \text{for } c_{in} < 0 \\ 0 & \text{otherwise} \end{cases} \tag{18}$$

The control scheme was evaluated using two forms of actuation: blade pitch and a morphing flap device modeled as in reference [12]. In both cases, all three turbine blades were controlled collectively with the same command signal rather than independently. For blade pitch, the output of the load controller was added to the output of the standard pitch controller (which was a constant 2.6 degrees). Tuning the load controller resulted in the gain $K_P = -12 \cdot \pi/180$ (pitch command in radians). For the morphing flap, the tuned gain was $K_P = 30$ (output in degrees of flap deflection). Figures 5 and 6 show the results for each control device. The maximum pitch action, above the 2.6 degree nominal value, was 1.6 degrees. The maximum flap deflection was 4.9 degrees. The controller's effect on generated power can be seen in Figure 7.

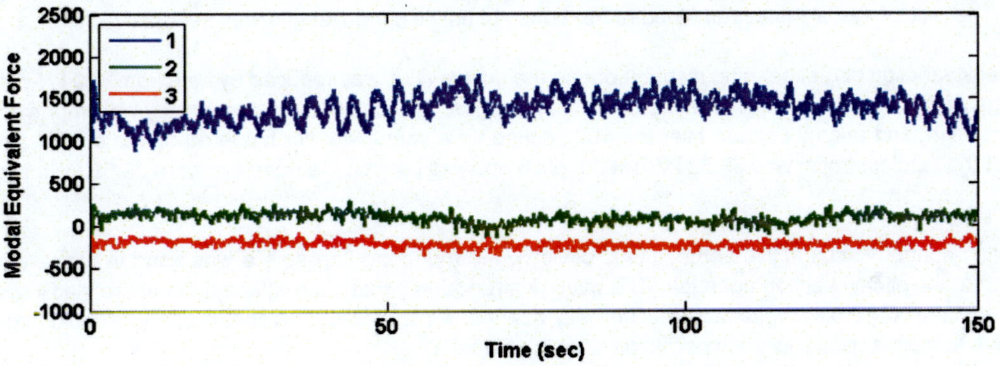

Figure 5: The lull in blade forces reduced by pitch action.

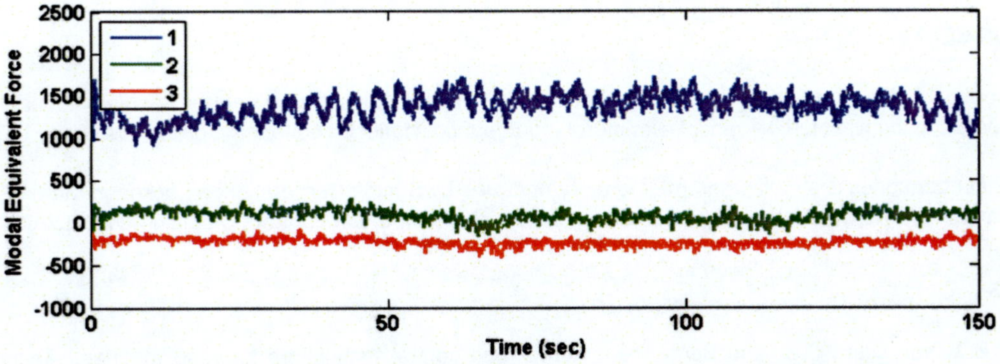

Figure 6: The lull in blade forces reduced by control of a morphing flap device.

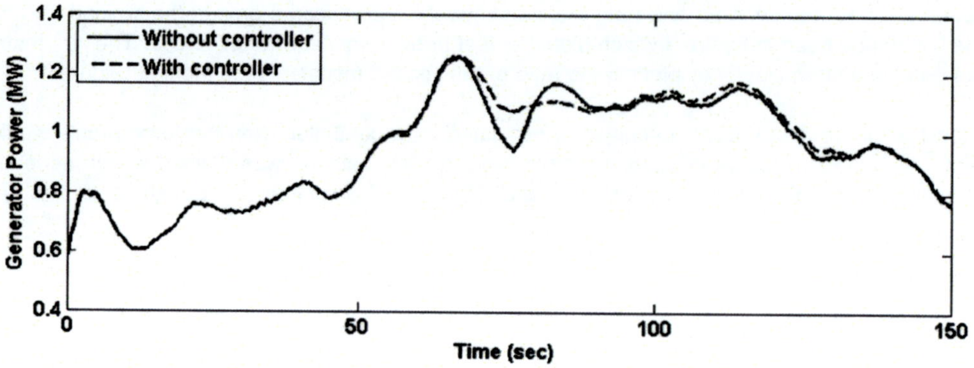

Figure 7: The controller's effect on generated power during the lull.

5 DISCUSSION

The purpose of this demonstration was to introduce the possibility of using blade force estimates in a control scheme. The approach taken here may not be the best solution when other factors are considered, and this controller has not been examined in terms of stability or robustness. However, the force estimation technique shows promise. Looking at Figure 4, the ratio of the modal forces provided more precise information than the root moment did by itself. Notice that the root moment increased before the lull, which would have set the controller going in the wrong direction if root moment were used to drive the controller. The ratio of modal forces remained rather constant until a control action was needed.

The peak-to-peak variation of root bending moment during the gust and lull was reduced by half, from 920 kN-m to 460 kN-m. Controlling large swings such as this is important for reducing fatigue. In addition, the power quality was improved by reducing the power fluctuation, and energy capture over the 150 second time frame was not hindered by the control action. Energy capture without the load controller was 39.38 kWh. With pitch and flap control, the numbers were 39.68 kWh and 39.78 kWh respectively.

In regard to pitch control versus active aerodynamic devices, the morphing flap device was used in this example in much the same way that the pitching action operates. The main difference was that pitch affected the entire blade span, while the morphing flap affected only the tip region where action was needed. A more advanced control scheme could target the effects of multiple active aerodynamic devices to specific parts of the blade.

Knowing the forces on each blade opens up the possibility of controlling directly for wind shear, yaw error, and pockets of turbulence hitting part of the rotor. These areas will be explored in future work.

ACKNOWLEDGMENTS

The authors would like to acknowledge the Wind & Water Power Technologies department of Sandia National Laboratories for providing access to the wind turbine model with morphing-flap modifications and for supporting this work.

Sandia National Laboratories is a multiprogram laboratory operated by Sandia Corporation, a Lockheed Martin Company, for the U.S. Department of Energy's National Nuclear Security Administration under contract DE-AC04-94AL85000.

REFERENCES

[1] Johnson, S.J., van Dam, C.P., and Berg, D.E. Active load control techniques for wind turbines. Technical Report SAND2008-4809, Sandia National Laboratories, August 2008.

[2] Ma, C.K., Chang, J.M., and Lin, D.C. Input forces estimation of beam structures by an inverse method. *Journal of Sound and Vibration*, 259(2):387–407, January 2003.

[3] Tuan, P., Ji, C., Fong, L., and Huang, W. An input estimation approach to on-line two-dimensional inverse heat conduction problems. *Numerical Heat Transfer, Part B: Fundamentals*, 29(3):345–363, April 1996.

[4] Ma, X., Poulsen, N.K., and Bindner, H. Estimation of wind speed in connection to a wind turbine. Technical Report IMM-1995-26, Informatics and Mathematical Modelling, Technical University of Denmark, DTU, December 1995.

[5] Østergaard, K.Z., Brath, P., and Stoustrup, J. Estimation of effective wind speed. *Journal of Physics: Conference Series*, 75, 2007.

[6] Boukhezzar, B., Siguerdidjane, H., and Hand, M. Nonlinear control of variable speed wind turbines for load reduction and power optimization. In *44th AIAA Aerospace Sciences Meeting and Exhibit*, pages 602–615, Reno, NV, January 2006.

[7] Ehlers, J., Diop, A., and Bindner, H. Sensor selection and state estimation for wind turbine controls. In *45th AIAA Aerospace Sciences Meeting and Exhibit*, Reno, NV, January 2007.

[8] Crassidis, J.L. and Junkins, J.L. *Optimal Estimation of Dynamic Systems*. Chapman & Hall/CRC, 2004.

[9] NWTC Design Codes. FAST by Jason Jonkman. http://wind.nrel.gov/designcodes/simulators/fast/, August 12, 2005. Accessed 21-May-2008.

[10] NWTC Design Codes. AeroDyn by Dr. David J. Laino. http://wind.nrel.gov/designcodes/simulators/aerodyn/, June 28, 2005.

[11] NWTC Design Codes. TurbSim by Neil Kelley, Bonnie Jonkman. http://wind.nrel.gov/designcodes/preprocessors/turbsim/, April 4, 2008. Accessed 28-May-2008.

[12] Wilson, D.G., Berg, D.E., Barone, M.F., Berg, J.C., Resor, B.R., and Lobitz, D.W. Active aerodynamic blade control design for load reduction on large wind turbines. In *European Wind Energy Conference & Exhibition*, March 2009.

Proceedings of the IMAC-XXVIII
February 1–4, 2010, Jacksonville, Florida USA
©2010 Society for Experimental Mechanics Inc.

Aspects of Operational Modal Analysis for Structures of Offshore Wind Energy Plants

Peter Kraemer and Claus-Peter Fritzen
University of Siegen, Department of Mechanical Engineering, Paul-Bonatz-Str. 9-11, 57068 Siegen, Germany, kraemer@imr.mb.uni-siegen.de, fritzen@imr.mb.uni-siegen.de

NOMENCLATURE

y	multivariate time series	b	deviation von zero mean
z	state vector	ε	model residuals
A_d	discrete state space matrix	δ, λ	eigenvalues
C_y	measurement matrix	ψ	eigenvectors
w, v	process, measurement noise	ω, ω_D	undamped, damped circular eigenfrequency
p	model-order	f	eigenfrequency
m	number of sensors	d	damping ratio
A	matrix with autoregressive coefficients	Φ	complex mode shapes matrix

ABSTRACT

Offshore wind energy plants (OWEP) are dynamic systems excited by stochastic loading due to wind, waves and operational interaction between the blades, machine and tower. For ensuring reliable operability of OWEP the knowledge of the dynamic characteristics are therefore of major importance. Furthermore, dynamic properties like eigenfrequencies and mode shapes can be used as damage sensitive features for structural health monitoring algorithms or model validation.

This paper starts with a discussion of the basic dynamic characteristics of the 5MW-OWEP (Multibrid M5000-2 with tripod foundation in Bremerhaven, Germany) and its environment by means of various measured signals. Subsequently, the results of an operational modal analysis (OMA) on this OWEP are presented. The extraction of the eigenfrequencies, damping ratios and mode shapes of the structure is performed automatically by means of Vector Autoregressive Models (ARV). The modal analysis algorithm is applied on measured acceleration signals from the tower over a period longer than seven month at different environmental and operational conditions (EOC). Finally, the results of the OMA are discussed with respect to some relationships between the changing EOC and variations of the dynamic properties of the structure.

INTRODUCTION: CONDITIONS FOR MONITORING OF OWEP

In order to perform online monitoring of OWEPs it is important to know the influences of the changing EOCs (wind velocity and wind direction, temperature, orientation of the nacelle, rotational speed of the blades and atmospheric conditions) on the dynamic behavior of the plant. Sometimes also the boundary conditions are changing (e.g. by ground erosion). For example, the wind velocity, the wind direction and the height of waves affect the excitation level of the structure and blades (e.g. at low wind speed only a few numbers of modes are excited). It is also known from bridges that a high wind velocity can cause changes in the eigenfrequencies [1] and a varying temperature can affect the dynamic behavior of structures [2, 3].

T. Proulx (ed.), *Structural Dynamics and Renewable Energy, Volume 1*, Conference Proceedings of the Society for Experimental Mechanics Series 10, DOI 10.1007/978-1-4419-9716-6_14, © The Society for Experimental Mechanics, Inc. 2011

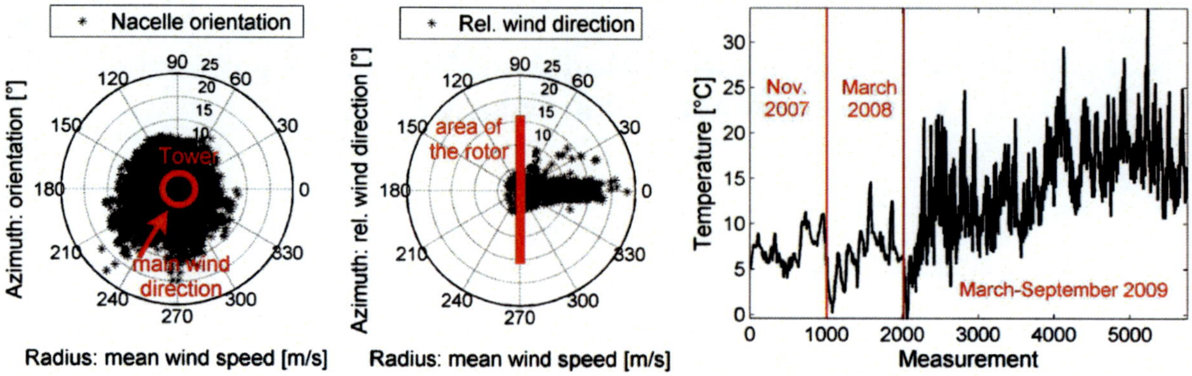

Figure 1: Changing EOC over a period of seven mounth

Since the structure of the OWEP is not perfectly symmetric along the tower axis (caused by transformers, pumps, etc. which are placed along the tower), the distribution of mass/mass moment of inertia changes due to the orientation of the nacelle, and with them also the dynamic properties of the system. Some operational and environmental conditions of the OWEP M5000-2 are displayed in Figure 1.

Figure 2 shows the response of one accelerometer placed on the top of the tower of the OWEP M5000-2. It can be seen that the probability density functions (PDF) of the measured sensor signal changes over the 5758 measurements. Thus the response of the structure can be classified as non-stationary.

The OWEP are not only stochastically excited by wind (Weibull-distributed) and waves (Pierson-Moskowitz-distributed). Also the rotational speed of the rotor Ω plays a major role in the structural excitation. Here the structure is periodically excited by the stall-effect appearing in the moment when one blade passes the tower. The mentioned periodic excitation can be amplified by additional flap moments generated by mass differences between the blades [4]. Figure 3 left explains which rotational speeds of the rotor excites the first eigenfrequency of the tower. The right part of Figure 3 presents the power spectrum densities (PSD) of three signals in the region of the first eigenfrequency of the tower. This figure supports the findings of the Campbell-diagram and shows that the structure can be stronger excited by its own operational condition as by a high wind speed v_w. Further periodical excitations occurs as an OWEP park-effect, when one plant stands in the slipstream of the other. This plant will be excited by the wind flow coming through the blades of the OWEPs standing in front of it.

Transient excitations of the structure and blades can occur by small wind speeds. In this case the rotation of the nacelle and the pitching of the blades in order to "find" better wind conditions are the only sources of excitation. Further transient excitation can be generated by sudden crosswind. Excitation of the lower frequencies of tower or blades by the drive train occurs especially by brake applications.

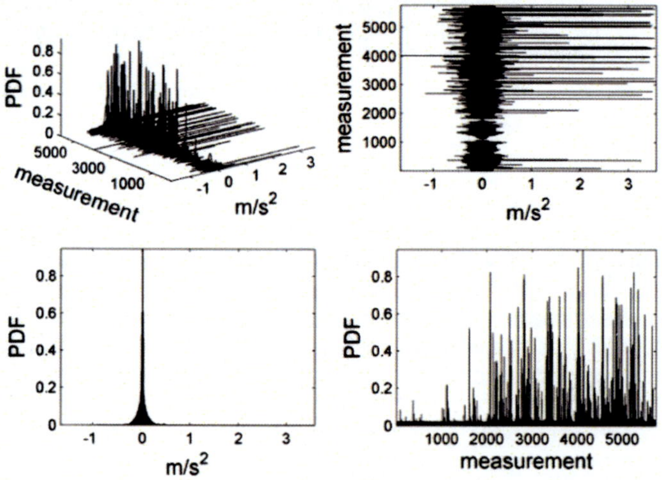

Figure 2: PDFs (in different views) of acceleration signal from channel 65 (see also Figure 5) over 5758 measurements

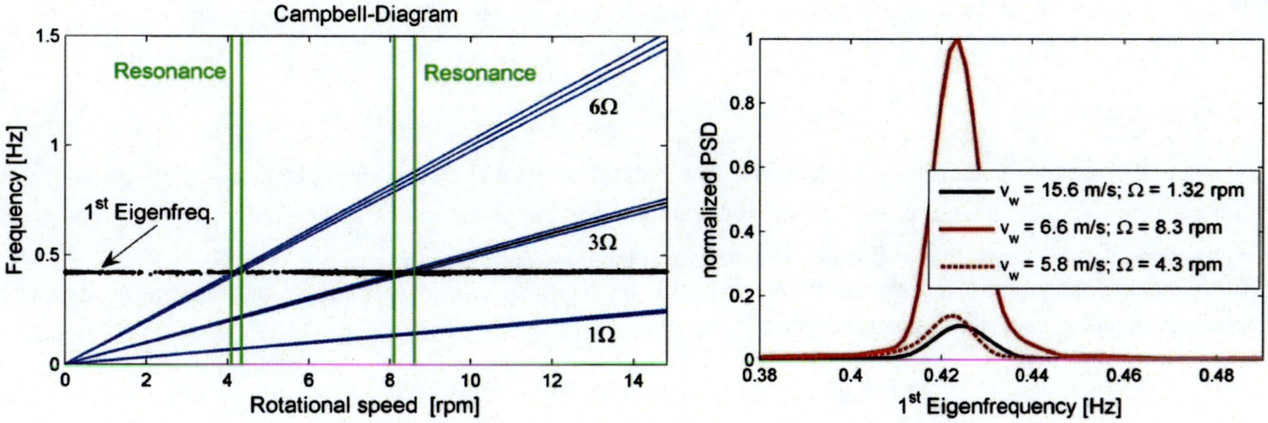

Figure 3: Campbell-diagram (left) and PSD for signals measured at different EOC (right)

OUTPUT-ONLY MODAL ANALYSIS WITH ARV-MODELLS

The modal analysis (MA) used on the measured data from M5000-2 is a part of an integrated monitoring system for OWEPs, which includes data selection, sensor fault identification and damage detection/localization [5-7]. The MA is used in this context to provide modal data for model updating and model based damage localization. The rough functionality of the hole system is presented in Figure 4.

Since the ARV coefficients are calculated in order to perform damage identification, these can be also used for MA without additional computational effort.

The applied MA uses the premise that the excitation of the structure is unknown but normally distributed (output-only system with white noise excitation). To satisfy this assumption it is avoided to apply the MA on data measured by strong transient excitation. In order to exclude such undesired data sets from the analysis, at first a multivariate outlier technique based on the Mahalanobis-distance with thresholds provided by means of the F-distribution is applied on the measured data [8].

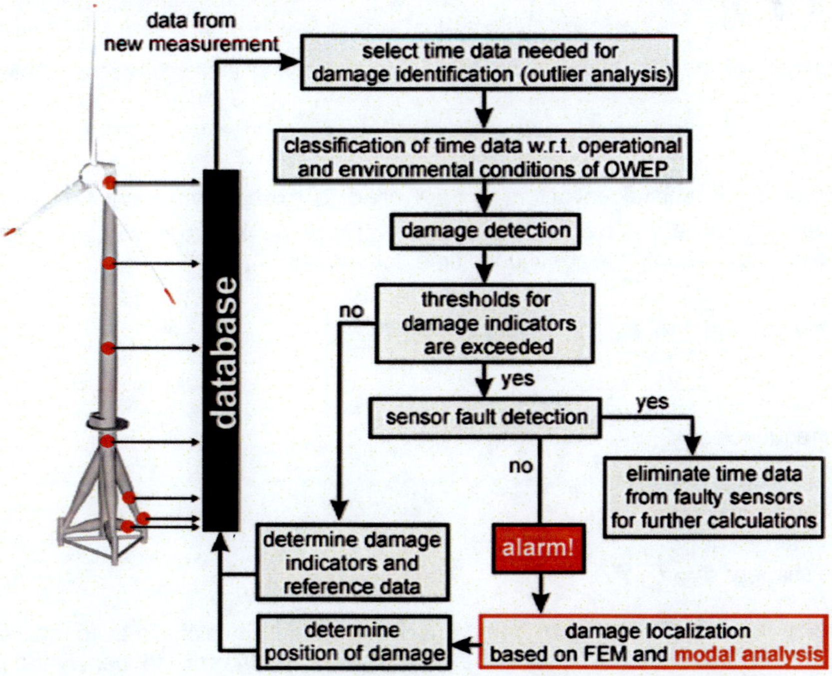

Figure 4: Concept for structural damage identification of OWEPs

The equations of motion for a stochastically excited linear system can be written in discrete form as:

$$\begin{aligned} z_{k+1} &= A_d z_k + w_k \\ y_k &= C_y z_k + v_k \end{aligned} \tag{1}$$

where z is the state vector and y contains the time data of m measured sensor signals. A_d is the discrete state space matrix, C_y the measurement matrix contained the position of the sensors, w and v are the process and measurement noise respectively. k is one time instant in the measured signals.

The simultaneously measured signals from eq.(1) can be modeled in one ARV-model of order p by means of multivariate statistical techniques as described in eq.(2).

$$y_k = b + \sum_{l=1}^{p} A_l y_{k-l} + \varepsilon_k \tag{2}$$

Here A is a matrix of dimension $m\text{x}mp$ containing the coefficients of the model calculated by means of a stepwise Least Square algorithm [9]. ε are the residuals between the measured time data and the model and b is a vector representing the deviations from the mean of zero of each time series in y.

The canonical form of the estimated discrete state space matrix A_d (dimension $mp\text{x}mp$) can be assembled by means of the coefficient matrices calculated by different model orders, see eq.(3). I is the identity matrix and 0 is the zero matrix of dimension $m\text{x}m$. The measurement matrix C_y of dimension $m\text{x}mp$ is assembled as shown in eq.(4).

$$A_d = \begin{bmatrix} A_1 & A_2 & \cdots & A_{p-1} & A_p \\ I & 0 & \cdots & 0 & 0 \\ \vdots & \vdots & \ddots & \vdots & \vdots \\ 0 & 0 & \cdots & I & 0 \end{bmatrix} \tag{3}$$

$$C_y = [0 \;\; 0 \;\; \cdots \;\; 0 \;\; I] \tag{4}$$

The dynamical properties of the structure can be calculated by solving the eigenvalue problem of the state space matrix.

$$(A_d - \delta_i I)\psi_i = 0 \tag{5}$$

The obtained eigenvalues δ and eigenvectors ψ are used to extract dynamical features of the structure like eigenfrequencies, damping ratios and mode shapes, see eq.(7). A pre-step for calculation of these features is the transformation of the eigenvalues into the continuous time domain, eq.(6).

$$\lambda_i = \frac{\ln(\delta_i)}{\Delta t} = -d_i\omega_i \pm j\underbrace{\sqrt{1-d_i^2}\,\omega_i}_{\omega_{D_i}}; \quad \Delta t = \text{time increment} \tag{6}$$

$$\begin{aligned} &\text{Damped eigenfrequencies: } f_i = \frac{\omega_{D_i}}{2\pi} = \frac{\text{Im}(\lambda_i)}{2\pi} \\ &\text{Damping ratios: } d_i = \frac{-\text{Re}(\lambda_i)}{|\lambda_i|} \\ &\text{Complex mode shapes: } \Phi = C_y \Psi \end{aligned} \tag{7}$$

A possibility to visualize the results of MA provides the well-known stability plot. To build this plot the measured time series are modeled by ARV-models of increasing maximal order of p. In order to identify the possible stable poles, the poles of one model of order $l-1$ will be compared with the poles of the next higher model of order l. The tolerances for the allowed variation of the poles are established by using following three criteria:

1. Criterion (stable frequencies): $$\text{ftol} = \frac{f_{l-1} - f_l}{f_{l-1}} 100\%$$
2. Criterion (stable damping ratios): $$\text{dtol} = \frac{d_{l-1} - d_l}{d_{l-1}} 100\%$$
3. Criterion (stable mode shapes): $$\text{MACtol} = \frac{\left(\boldsymbol{\Phi}_l^T \boldsymbol{\Phi}_{l-1}\right)^2}{\left\|\boldsymbol{\Phi}_l^T\right\|^2 \left\|\boldsymbol{\Phi}_{l-1}\right\|^2}$$

Practical choices of the tolerances for the stable poles are ftol = 0.1%, dtol = 5% and MACtol = 0.99 (MACtol = 1 represents the perfect matching between the mode shapes with l and $l-1$ poles).

RESULTS OF LONG TIME MONITORING OF THE DYNAMIC PROPERTIES OF OWEP M5000-2

The MA was applied on measured time data from the prototype OWEP M5000-2 of the company AREVA-Multibrid in Bremerhaven, Germany (see Figure 5, left). The sensor signals are provided by eight accelerometers used primarily for damage identification and positioned on the tower as shown in Figure 5, middle and right. The signals were measured simultaneously with a sample rate of 50Hz. The measurement time for one data set was ten minutes.
In order to perform long time monitoring 5758 data sets were considered. The first 2016 data sets were continuously measured in November 2007 (first 1008) and March 2008 (next 1008). The remaining 3742 data sets were measured between March and September 2009 (one data set per hour).
To identify the correct modes of the tower at first the results of the MA were compared with simulated modes by means of a FE-model. In order to avoid undesired effects from transient or periodical excitations one data set measured under constant orientation of the nacelle, wind speed and very low rotational speed of the blades was used (these EOCs are mentioned in Table 1). The angular orientation of the nacelle and the degree of freedoms which simulate the sensor positions in the FE-model correspond to the measured data, see Figure 5 right and Table 1.

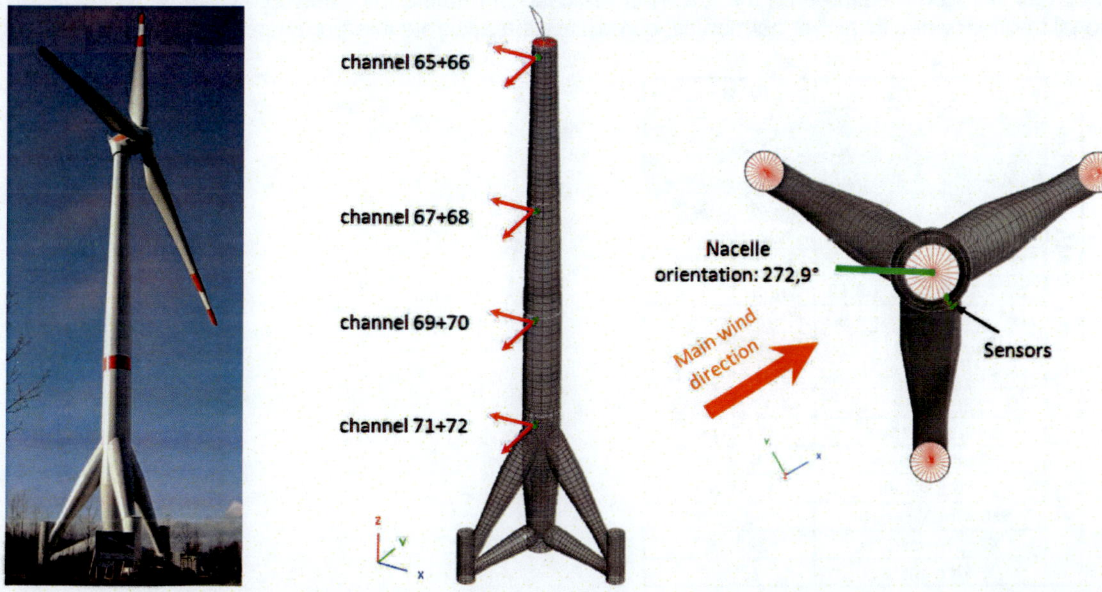

Figure 5: OWEP M5000-2 (left); sensor positions (middle); operational condition for reference MA (right)

The results of this "offline" MA are visualized in Figure 6 (stability plot) and Figure 7 (first four bending mode shapes). These results are very similar compared to results delivered by the MA with stochastic subspace identification method [10].

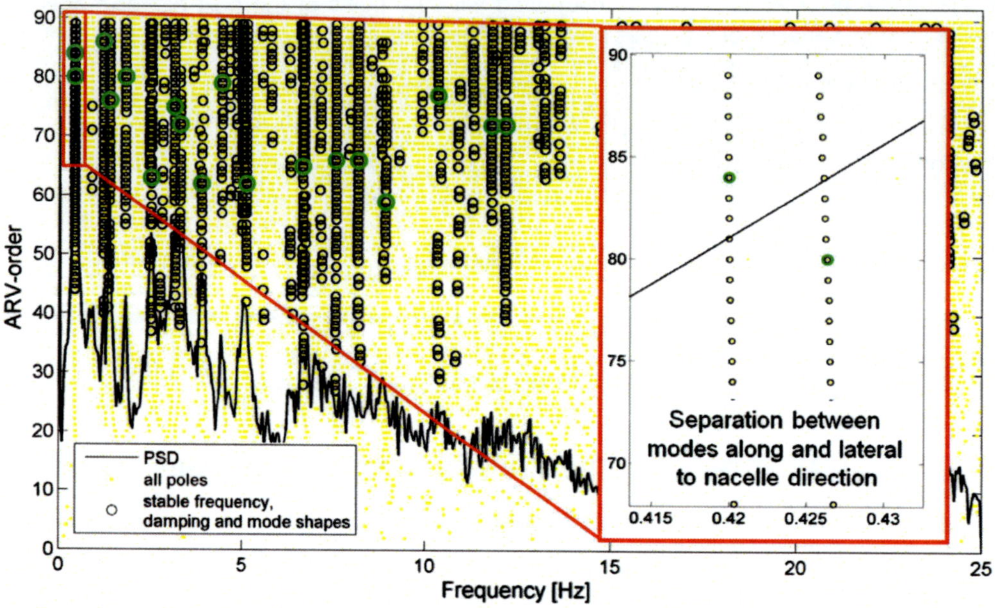

Figure 6: Stability plot

One important aspect by the calculation of the mode shapes is that the vibration direction depends strongly on the orientation of the nacelle. If the nacelle position changes, the vibration direction changes, too. So the bending modes appear always along or lateral to the nacelle orientation. Figure 6 shows exemplary that the first and the second modes arise lateral respectively along the nacelle direction and are very close to each other.

Table 1 (green marked) reveal the first four bending modes identified by MA and FEM. The sparse sensor network on the tower made the accurate identification of further modes difficult. Other stable poles displayed in Figure 6 come for e.g. from blades vibration measured on the tower. In the case of rotating rotor additional "stable" poles, coming from the rotational frequency and its higher harmonics, complicate the accurate interpretation of the real modes.

	FEM	Hz	MA	Hz	Diff. [%]	MAC[%]
Comparison: MA with FEM	1	0.43	1	0.42	3.15	100.0
	2	0.43	2	0.43	1.69	100.0
	3	3.25	9	3.13	3.90	99.6
	4	3.35	10	3.28	2.16	98.8
	5	5.89	16	5.79	1.76	15.7
	9	7.66	21	7.56	1.22	37.2
	10	8.09	22	8.18	-1.05	27.2
	13	8.90	23	8.80	1.09	93.6
	15	10.00	27	10.92	-8.49	96.9
	17	11.07	28	11.32	-2.21	95.0
	20	12.27	33	13.22	-7.20	95.5
EOC	Rotational speed				1,32 rpm	
	Temperature				5,60 °C	
	Wind speed				15,56 m/s	
	Rel. wind direction				-3,87 °	
	Nacelle position				272,90 °	
	Pitch angle				70,00 °	

Table 1: MA and FEM results by given EOC

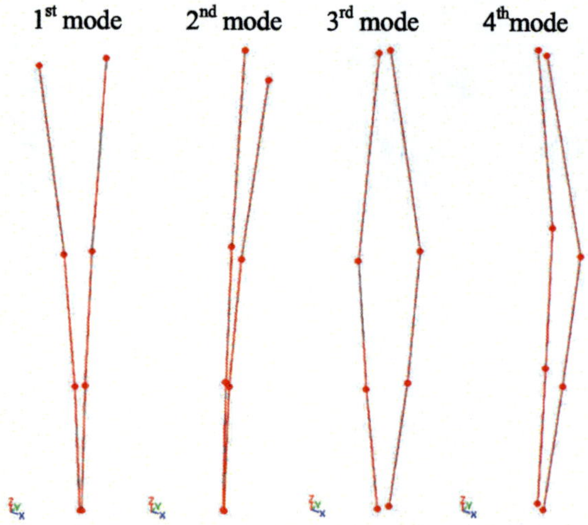

Figure 7: Mode shapes from MA

For long time monitoring the MA of the mentioned 5758 data sets is performed in an automatic way. Here the stable poles for the first four modes are monitored in the corresponding intervals (red lines in Figure 8). Also other poles from interaction of foundation piles and tower or blades vibration can be observed (blue dashed lines in Figure 8).

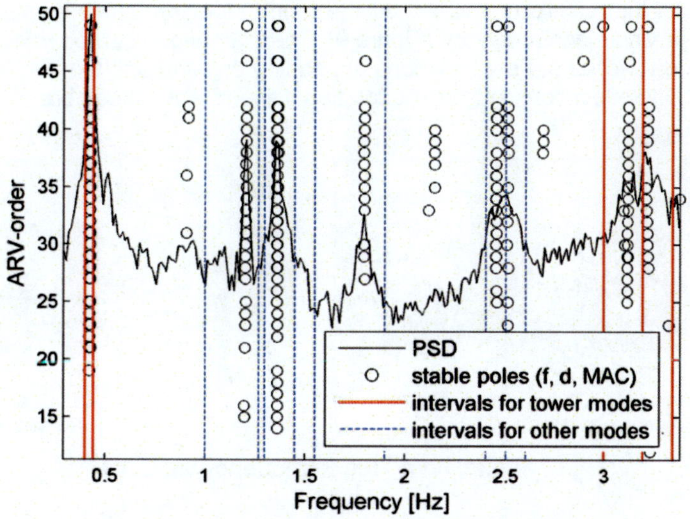

Figure 8: Frequency intervals for automated OMA

Since the first and the second mode are very close to each other and vary with EOCs, they are at first monitored together in the same frequency interval. However, the modes can be separated by means of the damping ratios. The vibration lateral to the nacelle direction are damped approx. 0,5% and the vibrations along the nacelle and in the wind direction are damped approx. 4% (probably as a result of the additional aerodynamic damping).

In Figure 9 each black dot represents the mean of the first and second eigenfrequencies in the frequency interval of 0.4-0.44Hz. To explain some variations of the averaged frequencies, they are plotted over important EOCs like wind speed, rotational speed of the rotor, temperature and orientation of the nacelle. It can be seen in Figure 9, that with growing wind velocity, especially between 6-12m/s, the eigenfrequencies are growing, too. This light trend comes from the second mode which vibrates along the nacelle orientation and in wind direction. After approx. 12m/s the effect of the wind load is kept constant by means of the pitching system.

The same trend can be observed by increasing the rotational speed of the rotor, especially higher than 9rpm. With higher temperatures the eigenfrequencies seem to decrease. Possible trends caused by the nacelle orientation are masked by variations of the eigenfrequencies due to changes of other EOCs.

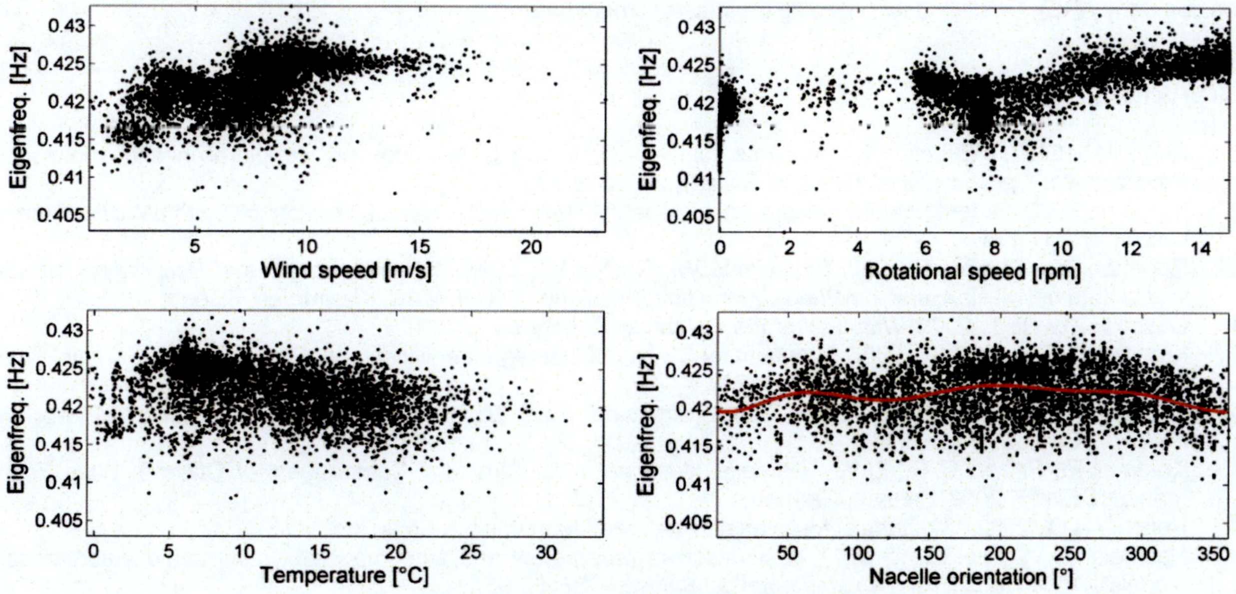

Figure 9: 1st Eigenfrequency vs. different EOC

In Figure 10 the influence of the nacelle orientation on the fourth eigenfrequency is much more significant than the influence on the first and second eigenfrequency (Figure 9). These changes can be referred to different stiffness and mass distributions along the tower axis. Few variations depending on wind and temperature can be also observed in the fourth eigenfrequency. The effect coming from the rotational speed is masked here by other influences.

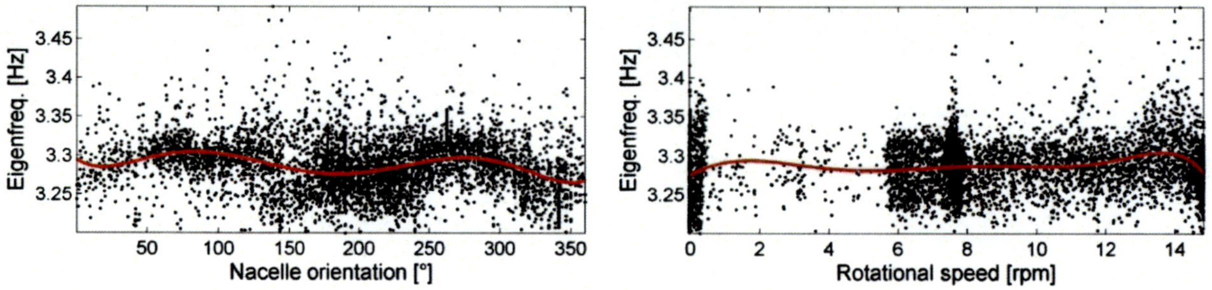

Figure 10: 4th Eigenfrequency vs. EOC

CONCLUSIONS

The industrial trend moves to bigger construction of WEP's and their global shift to offshore regions due to higher wind speeds. The dynamic behavior on sea of such huge vibration "joyful" structures of steel is of major importance in order to ensure a reliable design and operability of them. Furthermore, the extracted dynamic characteristics can be used as damage sensitive features for structural health monitoring algorithms.

The knowledge presented in this paper will be used to improve the reliability of the damage identification methods by changing EOC and to create more sophisticated models for a better understanding of the OWEP dynamics. Continued on the obtained findings it is expected that in the future more structural modes can be extracted followed by a better understanding of the interaction between the OWEP components.

ACKNOWLEDGEMENTS

The authors are grateful to the German Ministry of Economics for the financial support of the IMO-WIND project by funding of BMWi (grant no. 16INO327), to the Federal Institute for Materials Research and Testing (BAM VII-2) and to the company AREVA-Multibrid for providing the measurement data.

LITERATURE

1. Sohn H., 2007, "Effects of environmental and operational variability on structural health monitoring", *Philosophical Transactions of the Royal Society*, 365: 539-560.
2. Kullaa J., 2003, "Is temperature measurement essential in Structural Health Monitoring?", *4th IWSHM*, Stanford University: 717-724.
3. Rohrmann R., Bässler M., Said S., Schmid W., Rücker W., 1999, "Structural causes of temperature affected modal data of civil structures obtained by long time monitoring", *XVII IMAC*, Kissimmee, Florida: 1-6.
4. Gasch R., Twele J., 2002, Wind Power Plants. James & James.
5. Kraemer P., Fritzen C.-P., 2007, "Concept for Structural Damage Identification of Offshore Wind Energy Plants", *6th IWSHM*, Stanford University: 1881-1888.
6. Fritzen C.-P., Kraemer P., Klinkov M., 2008, "Structural Health Monitoring of Offshore Wind Energy Plants", *4th EWSHM*, Cracow, Poland: 3-21.
7. Kraemer P., Fritzen C.-P., 2008, "Damage Identification of Structural Components of Offshore Wind Energy Plants", *DEWEK 2008*, Bremen, Germany.
8. Rencher A.C., 2002, "Methods of Multivariate Analysis", John Wiley & Sons, Inc..
9. Neumaier A., Schneider T., 2001, "Estimation of parameters and eigenmodes of multivariate autoregressive models", *ACM Transactions on Mathematical Software*, 27: 58-65.
10. Peeters B., De Roeck G., 1999, "Reference-Based Stochastic Subspace Identification for Output-Only Modal Analysis", *MSSP*,12(6): 855-878.

Proceedings of the IMAC-XXVIII
February 1–4, 2010, Jacksonville, Florida USA
©2010 Society for Experimental Mechanics Inc.

Operational Modal Analysis of a Wind Turbine Mainframe using Crystal Clear SSI

Sven-Erik Rosenow
W2E Wind to Energy GmbH
Strandstraße 96, D-18055 Rostock, Germany, serosenow@wind-to-energy.de

Palle Andersen
Structural Vibration Solutions A/S
Niels Jernes Vej 10, DK-9220 Aalborg East, Denmark, pa@svibs.com

ABSTRACT

Experimental investigations of the dynamic behavior of wind turbine components as well as of the turbine's overall dynamics are important for the evaluation of applied prediction models. Operational Modal Analysis (OMA) was applied to identify resonant frequencies, corresponding mode shapes and damping values.
Experimental investigations were carried out at the main frame of a Fuhrländer AG 2.5 MW wind turbine with a rotor diameter and a hub height of each ca.100m. Structural responses were recorded at a stopped turbine excited by the stochastic wind only. The complex structure of the mainframe as well as its interaction with the tower and the turbine blades makes it non-trivial to perform the modal analysis.
In the paper the new Crystal Clear Stochastic Subspace Identification (CC-SSI) technique is applied. This technique produces very clear stabilization and easy automatic mode extraction even for complicated systems with a large number of modes as considered here.
At a wind turbine, there is usually a restriction in easily applicable measurement locations. Since time series were recorded only at the main frame of the turbine, overall mode shapes of the entire turbine were identified applying a combined identification using Finite Element Analysis (FEA).

1. INTRODUCTION

In order to avoid resonances in the variable speed range of a wind turbine, resonant frequencies of the entire turbine including substructure resonant frequencies as well as harmonic excitation must be known accurately. Whereas the harmonic excitation frequencies are multiples of the rotational speed and well known, resonant frequencies have to be calculated using a proper model or identified experimentally. A verification of calculated results by a measurement is the preferable approach.
The lowest resonant frequencies of the entire turbine and corresponding mode shapes, affected mainly by tower and blades, have to be considered with respect to the first and third order excitation (rotational frequency and blade passing frequency respectively) in order to avoid severe resonance problems during operation.
Resonant frequencies of the main frame have to be considered with respect to higher excitation orders resulting i.e. from the rotational frequency of the generator. Dynamic deflections of the main frame, potentially inducing undesirable loads of the drive train, have to be minimized.

Two aspects are important in order to obtain experimental data from a wind turbine. Firstly, how the turbine can be excited and, secondly, which measurement locations are accessible?
Easy applicable measurement locations can be found at the tower and at the nacelle, whereas locations at the rotor blades can only be used at a high expense. However, missing experimental rotor blade data are leading to uncertainties in the experimental identification procedure, caused by the fact that mode shapes cannot be identified completely and distinguishable.
In order to excite the entire wind turbine with sufficient frequency content, an artificial excitation using release lines is known but difficult to realize. The application of an excitation mass, as an alternative, is ineffective because of its restricted induced energy content. Therefore, the application of Operational Modal Analysis (OMA) using the stochastic excitation by the wind appeared to be a promising approach.

The experimental investigations described in this paper were carried out at a Fuhrländer AG 2.5 MW wind turbine (rotor diameter and hub height 100m), designed by W2E Wind to Energy GmbH (see Figure 1). Nacelle construction including main frame, gearbox, generator and transformer can be seen in Figure 2.

Figure 2: Nacelle construction with main frame

Figure 1: Fuhrländer 2.5 MW wind turbine, designed by W2E

2. EXPERIMENTAL INVESTIGATION

In using operational modal analysis the structure is excited by ambient vibrations resulting from the wind only. Excitations can be assumed to be random in space and time with small correlation length compared to the size of the wind turbine. Since fluctuations of the wind speed increases by increasing wave length, the white noise excitation level cannot be assumed to be constant over frequency. Accordingly, lower modes should be excited more intensive.

The applied measurement model is shown in Figure 3. In view of an easy sensor installation, only measurement locations at the main frame were used. Degrees of freedom that were not measured at the main frame could be

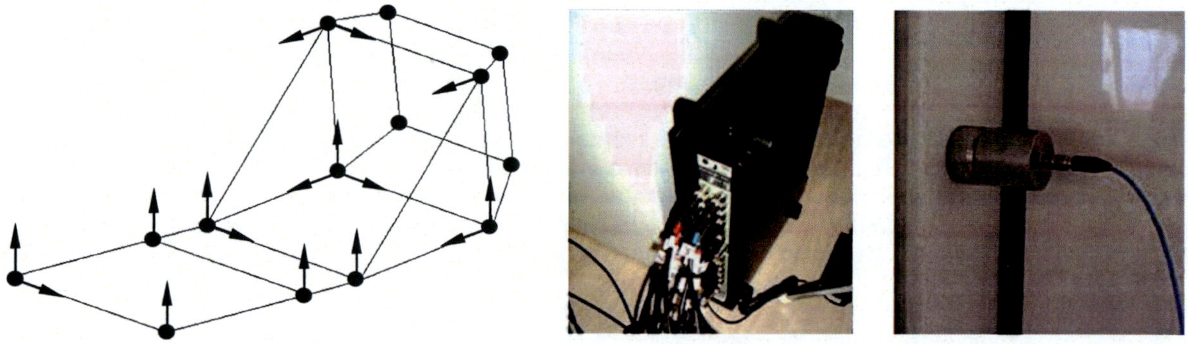

Figure 3: Measurement locations at the main frame, data acquisition system Dyn-X and seismic accelerometer

interpolated within the frequency range considered. To provide an identification of the mode shapes of the entire turbine based on the recorded main frame data, a Finite Element model was additionally generated.
During the measurement campaign the turbine was stopped and the rotor blades were pitched out of the wind (feathered position) The vibration responses were recorded over a period of 60 minutes, sampled with a frequency of 128Hz. For data acquisition a Dyn-X system (BRUEL&KJAER) with 2*24-bit AD-converter (dynamic range 160dB) in combination with seismic accelerometers (piezoelectric, sensitivity: 10V/g, PCB) were applied (see Figure 3).

3. IDENTIFICATION TECHNIQUE

Since the pioneering work of Overschee et al. [1] was presented a decade ago, there is no doubt that the Stochastic Subspace Identification (SSI) techniques are the most well-known and used parametric time domain estimators. Here in this paper a new generation of these techniques is applied. The SSI techniques fit discrete-time stochastic state space realizations with the following underlying model structure:

$$\begin{aligned} x_{k+1} &= Ax_k + v_{k+1} \\ y_k &= Cx_k \end{aligned} \tag{1}$$

where A is the state transition matrix including all the system dynamics. x_k is the internal state vector that holds the current state of the system at time instant k. v_{k+1} is the Gaussian white noise process that is driving the system. The observable output of the system is located in y_k and is obtained by a pre-multiplication of x_k with an observation matrix C.

Assuming that the system has been sampled using a sampling interval at T seconds, the continuous-time eigenvalue and mode shape ($\mu;\phi_\mu$) are given by the eigenstructure ($\lambda;\phi_\lambda$) of F:

$$\begin{aligned} e^{T\mu} &= \lambda \\ \phi_\mu &= C\phi_\lambda \end{aligned} \tag{2}$$

The natural frequency and damping are the obtained directly from μ.

Only knowing the output data y_k at the time instants $k = 1, \ldots, N$, the aim is to identify the eigenstructure ($\lambda;\phi_\lambda$) of the system (1) with Stochastic Subspace Identification algorithm called the Unweighted Principal Component algorithm, see Overschee et al. [1].

We choose p and q as index variables with $p + 1 \geq q$ that indicate the quality of the estimations, where larger p leads to better estimates, and the maximal system order ($\geq qr$ with the number of measurement channels r). Normally, we choose $p = q - 1$, but in the case of measurement noise $p = q - 1 + l$ should be chosen, where l is the order of the noise. Two data matrices of measured output are constructed:

$$Y_{p+1}^{+} = \begin{bmatrix} y_{q+1} & y_{q+2} & : & y_{N-p} \\ y_{q+2} & y_{q+3} & : & y_{N-p+1} \\ : & : & : & : \\ y_{q+p+1} & y_{q+p+2} & : & y_N \end{bmatrix}, \quad Y_q^{-} = \begin{bmatrix} y_q & y_{q+1} & : & y_{N-p-1} \\ y_{q-1} & y_q & : & y_{N-p-2} \\ : & : & : & : \\ y_1 & y_2 & : & y_{N-p-q} \end{bmatrix} \tag{3}$$

For the Unweighted Principal Component algorithm we build a weighted Hankel matrix from the above data matrices:

$$H_{p+1,q} = Y_{p+1}^{+} Y_q^{-T} (Y_q^{-} Y_q^{-T})^{-1} Y_q^{-} \tag{4}$$

As the computation of (4) usually is a large matrix the calculation is in practice based on QR decomposition of the data matrices, see Overschee et al. [1] for details. The weighted Hankel matrix of (4) can also be expressed in terms of the observability matrix $\boldsymbol{O}_{p+1}$ and a sequence of state vectors $\boldsymbol{X}_q$:

$$H_{p+1,q} = O_{p+1} X_q \quad , \quad O_{p+1} = \begin{bmatrix} C \\ CA \\ \vdots \\ CA^p \end{bmatrix} \tag{5}$$

The observability matrix $\boldsymbol{O}_{p+1}$ is obtained from an SVD of the matrix $\boldsymbol{H}_{p+1,q}$ and its truncation at the desired model order:

$$\begin{aligned} H_{p+1,q} &= U \Sigma V^T \\ &= \begin{bmatrix} U_1 & U_2 \end{bmatrix} \begin{bmatrix} \Sigma_1 & 0 \\ 0 & \Sigma_2 \end{bmatrix} \begin{bmatrix} V_1^T \\ V_2^T \end{bmatrix}, \\ O_{p+1} &= U_1 \Sigma_1^{1/2} \end{aligned} \tag{6}$$

The observation matrix $\boldsymbol{C}$ is then found in the first block-row of the observability matrix $\boldsymbol{O}_{p+1}$. The state-transition matrix $\boldsymbol{A}$ is obtained from the remaining block rows of $\boldsymbol{O}_{p+1}$, defined as

$$O_p^{\uparrow} = \begin{bmatrix} CA \\ CA^2 \\ \vdots \\ CA^p \end{bmatrix} = O_p A \tag{7}$$

by minimizing the norm of the difference $O_p^{\uparrow} - O_p A$ with respect to $\boldsymbol{A}$. In the ARTeMIS Extractor used for analysis in this paper the minimization is performed conditionally emphasizing a user defined number of most well-excited eigenvalues. The algorithm is called Crystal Clear SSI®, and the details are presented in Goursat et al. [2]. This algorithm results in much more clear stabilization than the usual unconditional minimization schemes by forcing almost all noise modes to have a natural frequency much higher than the Nyquist frequency and thus reducing their disturbing effects on the physical modes.

The Crystal Clear SSI® algorithm has been successfully applied to applications where conventional SSI algorithms usually have performance problems such as non-stationary short time series, e.g. measurements of launch of aerospace vehicles, see Goursat et al. [3].

4. RESULTS

4.1 PARAMETER IDENTIFICATION

As noted before, the software ARTeMIS Extractor was used for modal parameter identification. To extract resonant frequencies, corresponding mode shapes and damping values, a frequency domain technique (Enhanced Frequency Domain Decomposition, EFDD) was applied for a preliminary analysis as well as a time domain technique (Stochastic Subspace Identification, SSI [1]) for a detailed analysis. In case of the Stochastic Subspace Identification the new feature Crystal Clear SSI® was employed and compared to the standard algorithm.

Using the EFDD, a Singular Value Decomposition (SVD) of the Power Spectral Density (PSD) was carried out in a first step. Since the singular values near the resonant frequency are proportional to the PSD of a SDOF system, it was used as a starting point for modal parameter estimation [4], [5]. To reduce random and leakage errors in PSD estimation, especially for damping identification of the first modes (starting from 0.3Hz), the long recording times (60 minutes) were necessary [6], [7].
Singular values of PSD matrices for the lower frequency range are shown in Figure 4. Because of the spectral characteristic of ambient wind loading, especially low frequency modes of the turbine structure were well excited. Hence, singular values were showing their typical behavior and modal parameters could be identified easily. Problems occurred identifying the damping and separating the mode shapes of repeated modes using first and secondary singular values. Partially this is caused by the incompleteness of the measurement model, especially for the lower frequency range.

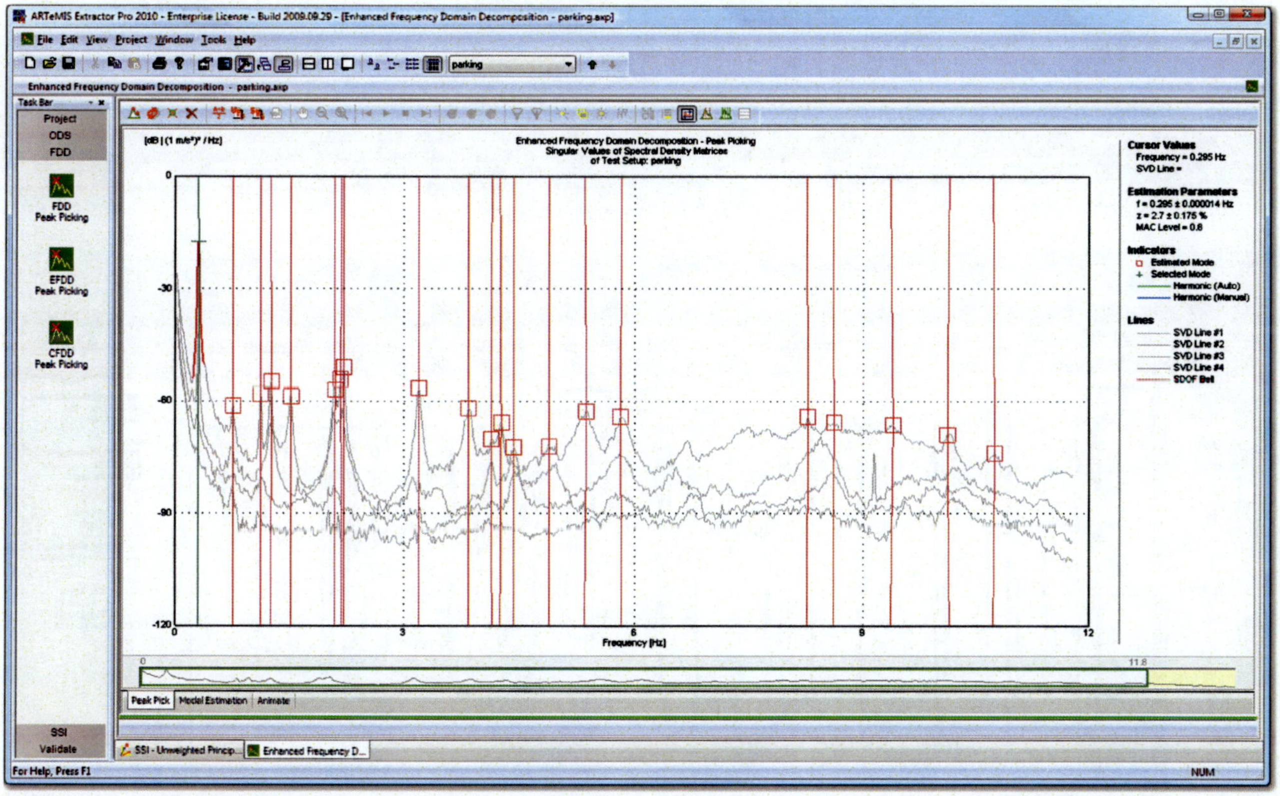

Figure 4: Singular values of spectral density matrices at the lower frequency range with selected spectral bell

Taking the advantage of the more sophisticated Stochastic Subspace Identification (SSI) [1], modal parameters were extracted in a detailed analysis. The SSI techniques rely on linear least squares estimation of the model using the raw measured time series. In the past this estimation was unconditionally. The new Crystal Clear SSI® estimation feature, described in detail above, allowed for a conditional estimation, emphasizing a user defined number of physical modes and suppressing the remaining parts of the information in the data.

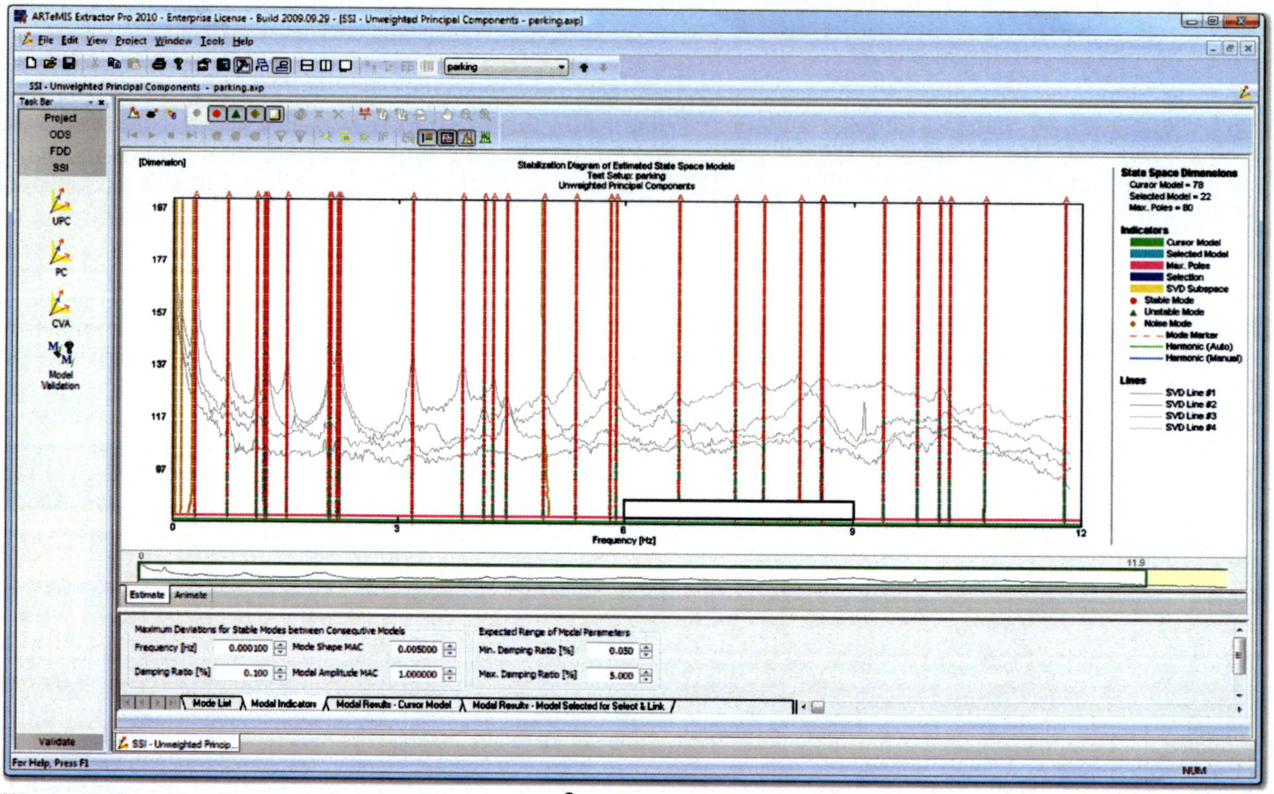

Figure 5: Stability diagram using Crystal Clear SSI® at the lower frequency range

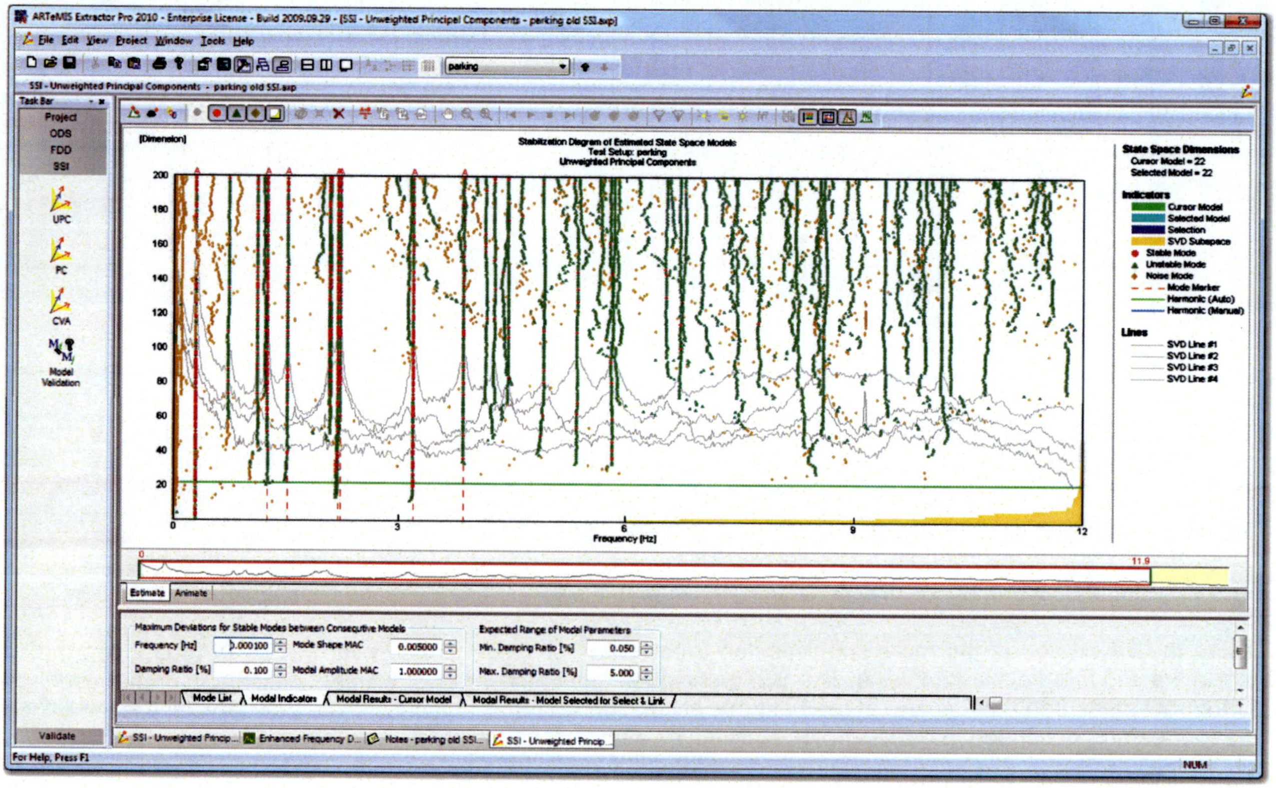

Figure 6: Stability diagram using standard SSI at the lower frequency range

Using this technique it is necessary to specify the maximum number of significant poles (eigenvalues) present in the measurements that should be estimated. The estimation algorithm focuses on the modes by having these poles, and any less significant noise poles are returned with a natural frequency estimate much higher than the Nyquist frequency, and a damping ratio of 100 %. This approach results in extremely clear stabilization of the number of modes specified and nearly no noise modes inside the visible frequency range. Due to the highly consistent estimation of the poles, the search for the optimal model order is less critical when using this new feature. By applying the Crystal Clear SSI® algorithm to the wind turbine data very stable poles occur and the modes could be identified easily with a high accuracy. The corresponding stabilization diagram is shown in Figure 5. Compared to the EFDD a higher number of modes could be identified.

Applying the standard SSI technique to the same dataset, differences in the stabilization diagram could be indicated very clearly as shown in Figure 6. In particular, less exited modes at higher frequencies and high number of modes in a narrow frequency range could be identified much better using the Crystal Clear SSI® algorithm compared to the standard procedure. Also the identification of repeated modes was excellent regardless of the spatial resolution of the measurement model (compared to EFDD). The separation of repeated modes can be seen in Figure 7.

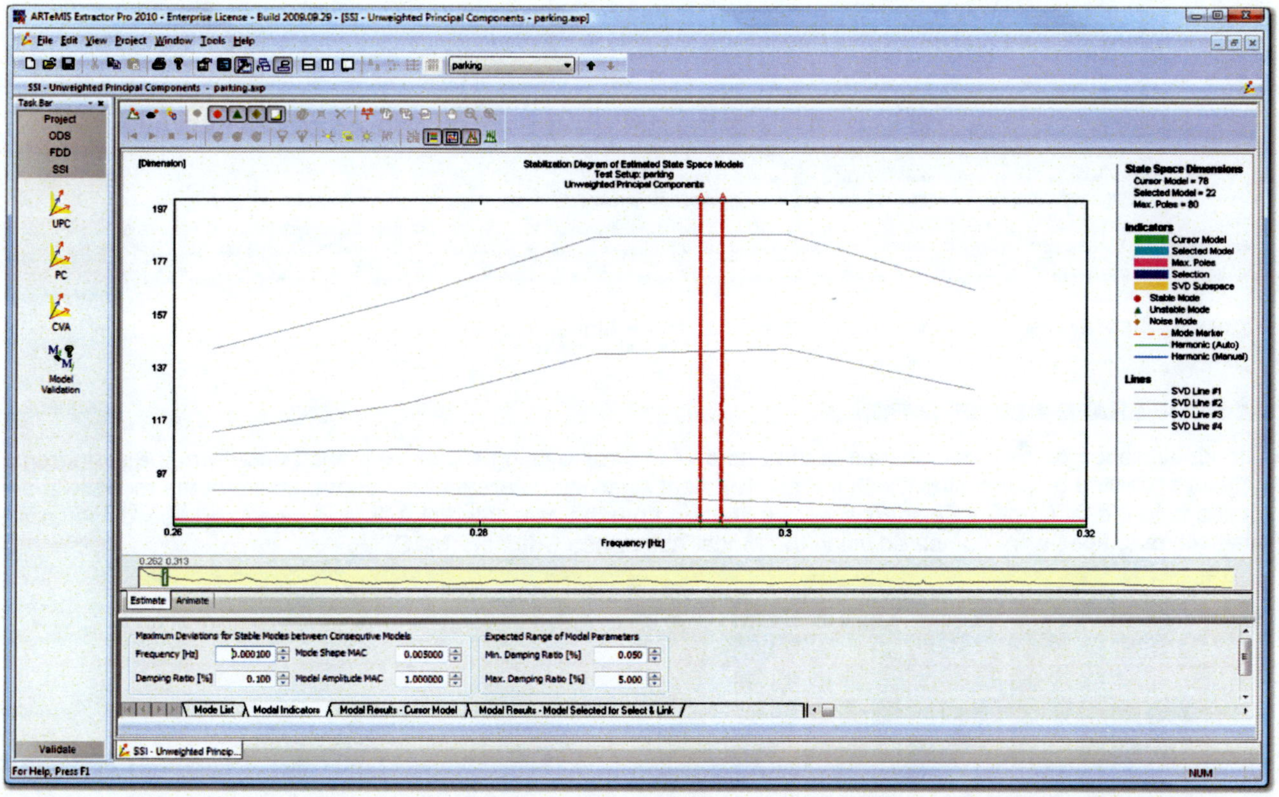

Figure 7: Separation and identification of repeated modes (first tower bending, perpendicular mode shapes)

As already mentioned, the fundamental resonant frequencies and corresponding mode shapes of the entire turbine are affected mainly by tower and blades and can be found in the lower frequency range. Because of the mass and stiffness properties, main frame resonant frequencies are located at higher frequencies. Unfortunately, the intensity of the natural excitation decreases with increasing frequency which results in less excited or weak modes with respect to the main frame. Using the new SSI algorithm a very clear stabilization diagram occurred, even for these modes as shown in Figure 8. The optimal model order could be selected and the model parameters were identified very easily.

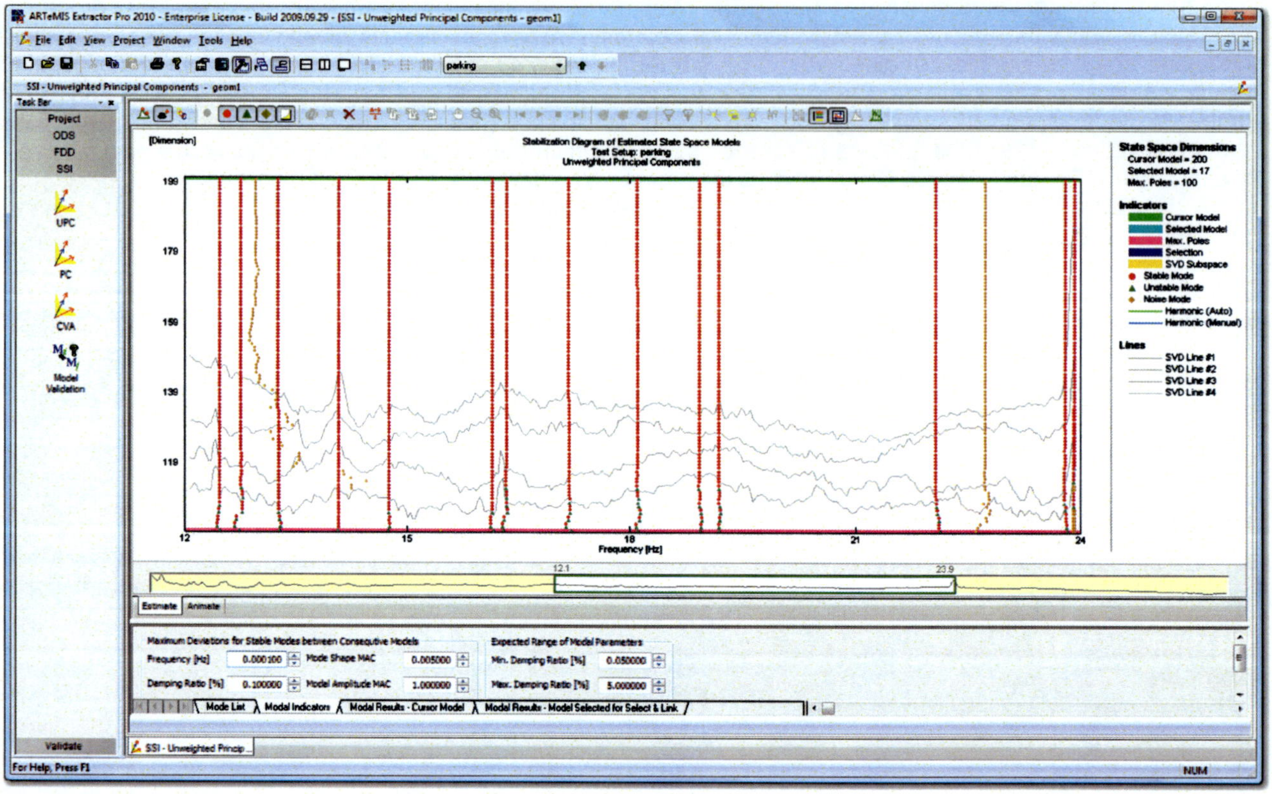

Figure 8: Stability diagram using Crystal Clear SSI® at the higher frequency range

4.2 MODE SHAPE IDENTIFICATION

The identification of the mode shapes of the entire wind turbine using the experimental data from the main frame only was realized by combining it with a Finite Element Analysis. Therefore, the contributions of the measurement locations to the different mode shapes were extracted from the calculated results and measured as well as calculated mode shapes were compared using the Modal Assurance Criterion (MAC).

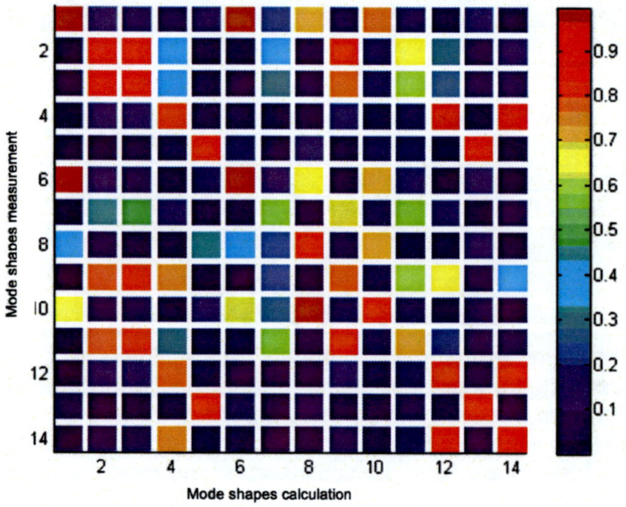

Figure 9: Application of Modal Assurance Criterion (MAC) to experimentally identified and calculated mode shapes

Calculated MAC values are shown in Figure 9 (remark: MAC value of two identical or corresponding mode shapes is equal 1). Using these mode correlation with reference to the expected frequency band (and additionally a "technical viewing") all resonant frequencies and corresponding mode shapes of the entire turbine in the lower frequency range could be identified. Exemplarily the correlation and mode shape identification for a flap wise blade bending mode is shown in Figure 10.

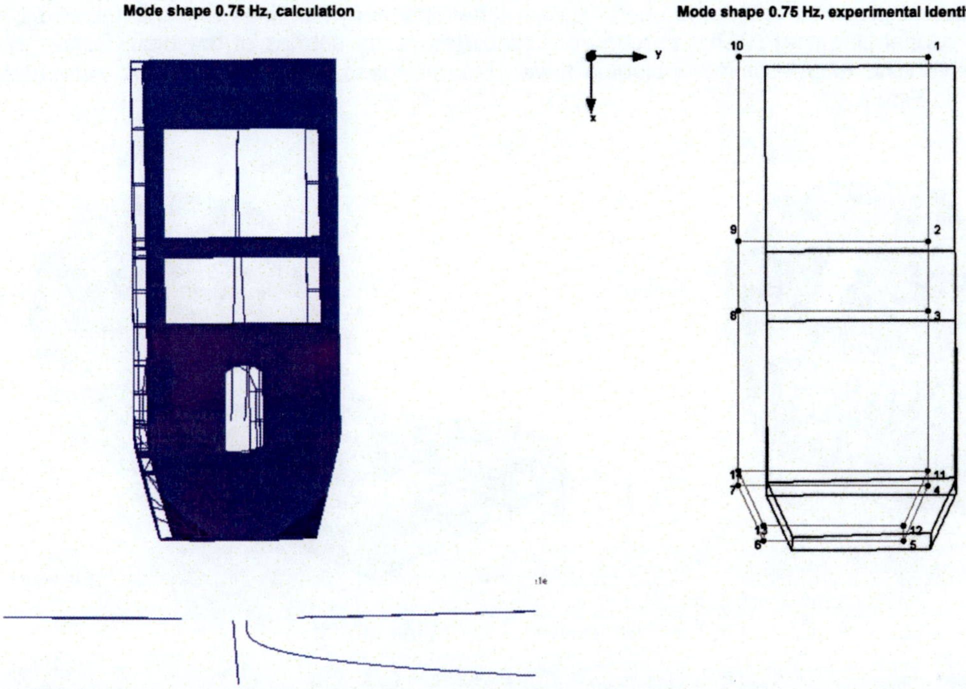

Figure 10: Mode shape identification using correlated modes: calculated flap wise (blades are pitched in feathered position) bending mode (left) and corresponding experimentally identified mode shape (right)

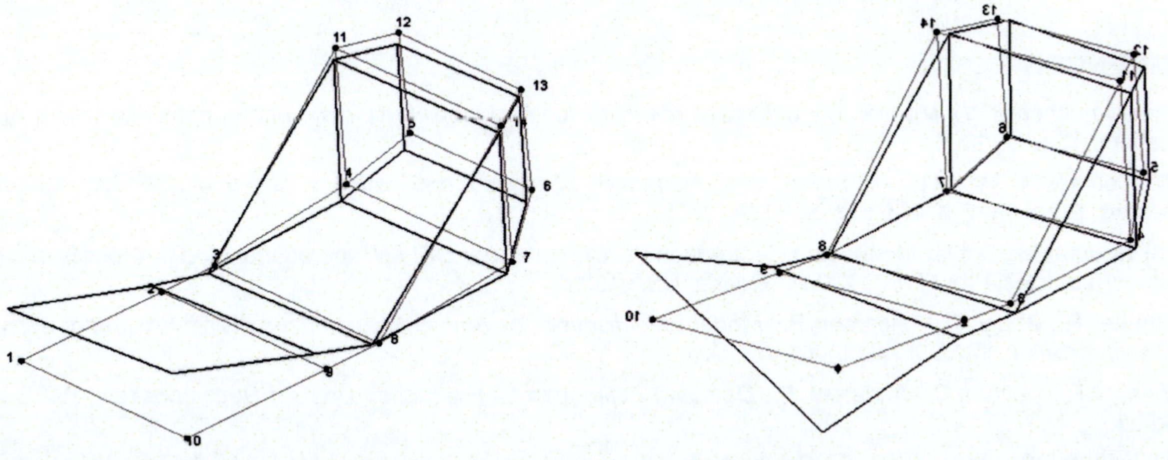

Figure 11: Experimentally identified mode shapes of the main frame, based on measurement data only

Unlike the mode shape identification of the entire turbine (using FEA) mode shapes of the main frame (or more precisely: mode shapes with a dominant main frame contribution) were extracted from the measured dataset only. Exemplarily, two fundamental mode shapes of the main frame are shown in Figure 11.

5. FURTHER INVESTIGATIONS

Based on the modal data obtained experimentally the generated prediction model could be evaluated with respect to the resonant frequencies. Particular attention was directed to the dynamic behaviour of the main frame. It is the intention to use the main frame model as a flexible body within a Flexible Multi-body Analysis for a load prediction of wind turbine components.

Structural as well as parametric uncertainties, leading to variations in the structural behaviour, have to be considered whenever a Finite Element prediction model is used. These variations were identified and minimized by a subsequent Computational Model Updating (CMU) procedure. Calculated mode shapes of the main frame, extracted from a comprehensive wind turbine model, including tower, blades, mass points for gearbox, generator, transformer etc., are shown in Figure 12.

Figure 12: Calculated mode shapes based on the comprehensive wind turbine model

The application of Operational Model Analysis was proven as an easily applicable tool for modal parameter estimation of wind turbines. The experimental data acquisition using the stochastic excitation from the wind as well as the very user friendly Crystal Clear SSI® parameter estimation technique are making this approach useful not only for scientific institutions but also for practical applications in the industry.

REFERENCES

[1] [1] Overschee P. V., Moor B. D., *Subspace Identification for Linear Systems*, Theory, Implementation, Applications, Kluwer, 1996.

[2] [2] Goursat M., Mevel L., *Algorithms for Covariance Subspace Identification: a Choice of Effective Implementations*, Proc. IMAC XXVII, Florida, USA, 2009.

[3] [3] Goursat M., Döhler M., Mevel L., Andersen P., *Crystal Clear SSI for Operational Modal Analysis of Aerospace Vehicles*, Proc. IMAC XXVIII, Florida, USA, 2010.

[4] Brinker R., Zhang L., Andersen P., *Modal Identification for Ambient Responses Using Frequency Domain Decomposition*, IMAC XVIII, 2000.

[5] Brinker R., Ventura C., Andersen P., *Damping Estimation by Frequency Domain Decomposition*, IMAC XIX, 2001.

[6] Uhlenbrock S., Rosenow S.-E., Schlottmann G., *Application of Operational Modal Analysis to Marine Structures*, IOMAC Workshop, 2006.

[7] Tamura Y., Yoshida A., Zhang L., Ito T., Nakata S., Sato K., *Examples of Modal Identification of Structures in Japan by FDD and MRD Techniques*, 1st IOMAC, 2005

Proceedings of the IMAC-XXVIII
February 1–4, 2010, Jacksonville, Florida USA
©2010 Society for Experimental Mechanics Inc.

Practical Aspects of Dynamic Substructuring in Wind Turbine Engineering

S.N. Voormeeren,* P.L.C. van der Valk and D.J. Rixen

Delft University of Technology, Faculty of Mechanical, Maritime and Materials Engineering
Department of Precision and Microsystem Engineering, section Engineering Dynamics
Mekelweg 2, 2628CD, Delft, The Netherlands
s.n.voormeeren@tudelft.nl

ABSTRACT

In modern day society concern is growing about the use of fossil fuels to meet our constantly rising energy demands. Although wind energy certainly has the potential to play a significant role in a sustainable future world energy supply, a number of challenges are still to be met in wind turbine technology. One of those challenges concerns the correct determination of dynamic loads caused by structural vibrations of the individual turbine components (such as rotor blades, gearbox and tower). Thorough understanding of these loads is a prerequisite to further increase the overall reliability of a wind turbine. Hence, improved insight in the component structural dynamics could eventually lead to more cost-effective wind turbines.
This paper addresses the use of dynamic substructuring (DS) as an analysis tool in wind turbine engineering. The benefits of a component-wise approach to structural dynamic analysis are illustrated, as well as a number of practical issues that need to be tackled for successful application of substructuring techniques in an engineering setting. Special attention is paid to the modeling of interfaces between components. The potential of the proposed approach is illustrated by a DS analysis on a Siemens SWT-2.3-93 turbine yaw system.

NOMENCLATURE

$\boldsymbol{M}$ – Mass matrix
$\boldsymbol{C}$ – Damping matrix
$\boldsymbol{K}$ – Stiffness matrix
$\boldsymbol{u}$ – Vector of degrees of freedom
$\boldsymbol{f}$ – External force vector
$\boldsymbol{g}$ – Connection force vector
$\boldsymbol{B}$ – Compatibility matrix (Boolean)
$\boldsymbol{L}$ – Localization matrix (Boolean)
$\boldsymbol{\lambda}$ – Vector of Lagrange multipliers
$\boldsymbol{R}$ – Reduction matrix
$\star^+$ – Generalized (pseudo) inverse

1 INTRODUCTION

At present there are few topics as heavily debated as "sustainability". On a daily basis the media are full of items on climate change, oil prices, CO_2 reductions, rising energy consumptions and so on. Regardless of one's opinion on the subject, a fact of the matter is that more sustainable ways of power generation need to be found simply because the currently used resources will some day be exhausted.

One of the more promising ways of generating "green" electricity on a large scale is provided by wind energy. As a result, the wind turbine industry has undergone a huge transition: from a small group of (mainly Danish) enthusiasts in the early 1980's, the modern wind power industry now has grown to a globalized multi billion dollar industry.[1] However, to enable

*This research is supported by Siemens Wind Power A/S
[1]From 2002 onwards, the wind power industry has seen an annual growth of no less than 25%.

T. Proulx (ed.), *Structural Dynamics and Renewable Energy, Volume 1*, Conference Proceedings of the Society for Experimental Mechanics Series 10, DOI 10.1007/978-1-4419-9716-6_16, © The Society for Experimental Mechanics, Inc. 2011

wind power to truly fulfill a significant role in a sustainable future energy supply, a number of technological challenges are still to be met. One of those challenges concerns the correct modeling and analysis of the structural dynamic behavior of the wind turbine.

1.1 Structural Dynamics in Wind Turbine Engineering

Naturally a wind turbine, with its large and relatively slender structure and the complex excitations, exhibits all kinds of structural dynamic behavior. The dynamic loading and structural vibrations sometimes can cause problems, from cracking blades, breaking gearboxes to "singing" towers. These problems have not been limited to a single manufacturer, but simply seem inherent to the structure of a modern wind turbine.

To cope with these dynamic effects, wind turbine manufacturers, research institutes and universities have developed many different aero-elastic codes [18]. These advanced codes are perfectly suited to analyze the global dynamics of a wind turbine, taking into account aerodynamic loads and coupling, possibly wave loads (for offshore turbines), and hence are commonly used for certification purposes.

Driven by today's highly competitive wind turbine market, manufacturers are searching for ways to optimize their turbine designs and hence save costs. An important way of achieving this is by reducing the total weight of turbine, by optimizing the design of each individual component. This causes a chain reaction of benefits as less material is used, transport and installation is made easier, a smaller foundation can be used and so on. On the downside, these optimized turbine designs generally introduce more flexibility to the structure. As a result, components start to exhibit local dynamic behavior, which can lead to increased component loading and decreased reliability. In some cases the local dynamic effects can interact with the global dynamics of the turbine, or vice versa. Thorough understanding of these dynamics is a prerequisite to further increase the overall reliability of a wind turbine. However, the aero-elastic models commonly used in wind turbine engineering are often incapable of predicting these local dynamic effects and their interaction with the global dynamics, due to their relatively few degrees of freedom and geometric simplifications.

Therefore, a need exists for more detailed structural dynamic analysis tools, without losing generality and versatility. In this paper we propose to use the paradigm of *dynamic substructuring* (DS) to fill this need.

1.2 Paper Outline

The remainder of the paper is organized as follows. The next section will introduce the concept of dynamic substructuring and discuss the details of substructure assembly. Although the basic principles of the DS methodology were established already some decades ago, implementing the dynamic substructuring approach in an industrial setting requires solving a number of practical issues. These issues and a number of solutions will be discussed in section 3. Section 4 thereafter presents a case study of the methodology on the yaw system of modern 2.3MW Siemens wind turbine. The paper is ended with some conclusions and recommendations in section 5.

2 INTRODUCTION TO DYNAMIC SUBSTRUCTURING

The theory of *dynamic substructuring* (DS) is about performing a dynamic analysis of a complex structure by dividing it into a number of smaller, less complex ones. These parts of the system are called substructures, subsystems or components, and their dynamic behavior is in general easier to determine than that of the complete system. When the dynamic properties of all the subsystems are known, DS techniques allow to construct the dynamic behavior of the complete system by coupling the subsystems together.

Performing the analysis of a structural system component-wise has some important advantages over global methods where the entire problem is handled at once:

- It allows the evaluation of the dynamical behavior of structures that are too large or complex to be analyzed as a whole. For experimental analysis this is true for large and complex systems such as aircrafts. For numerical models this holds when the number of degrees of freedom is such that solution techniques cannot find results in a reasonable time.

- By analyzing the subsystems, local dynamic behavior can be recognized more easily than when the entire system is analyzed. Thereby, DS allows identifying local problems and performing efficient local optimization.

- Dynamic substructuring gives the possibility to combine modeled parts (discretized or analytical) and experimentally identified components.

- It allows sharing and combining substructures from different project groups.

Dynamic substructuring methods have been long established; the first contributions in the literature date from over six decades ago [10, 23]. At the end of the 1960's the DS methodology saw rapid development with the rise of *component mode synthesis* methods [6, 15, 22], driven by the desire to reduce the complexity and size of computational structural dynamic models. Since then the methodology has seen many new developments, especially in the field of assembly of experimental component models, but the basic theory has remained the same. This basic theory of dynamic substructuring will be presented in this section, based on the discussion in [8].

2.1 Component Models and Interfacing

The starting point for the treatment of DS theory in this paper are the equations of motion in the physical domain. In this domain, the system is described by its mass, damping, and stiffness matrices as obtained from its mechanical and geometrical properties. Note however that the following discussion is also valid for substructure models in the frequency domain (where the component is seen through its frequency response functions) and the modal domain (where the dynamic behavior of a structure is interpreted as a combination of modal responses). The equations of motion of a discrete/discretized and linear(ized) dynamic subsystem s in the physical domain may be written as:

$$\boldsymbol{M}^{(s)}\ddot{\boldsymbol{u}}^{(s)}(t) + \boldsymbol{C}\dot{\boldsymbol{u}}^{(s)}(t) + \boldsymbol{K}^{(s)}\boldsymbol{u}^{(s)}(t) = \boldsymbol{f}^{(s)}(t) + \boldsymbol{g}^{(s)}(t) \tag{1}$$

Here $\boldsymbol{M}^{(s)}$, $\boldsymbol{C}^{(s)}$ and $\boldsymbol{K}^{(s)}$ are the mass, damping and stiffness matrices of substructure s, $\boldsymbol{u}^{(s)}$ denotes its vector of degrees of freedom (DoF), $\boldsymbol{f}^{(s)}$ is the external force vector and $\boldsymbol{g}^{(s)}$ is the vector of connecting forces with the other substructures. Suppose now that n substructure models of the form shown above are to be coupled. In order to simplify the notation, the equations of motion of these n substructures can be rewritten in a block diagonal format as:

$$\boldsymbol{M}\ddot{\boldsymbol{u}} + \boldsymbol{C}\dot{\boldsymbol{u}} + \boldsymbol{K}\boldsymbol{u} = \boldsymbol{f} + \boldsymbol{g} \tag{2}$$

With:

$$\begin{aligned}
\boldsymbol{M} &\triangleq \text{diag}\left(\boldsymbol{M}^{(1)}, \ldots, \boldsymbol{M}^{(n)}\right) = \begin{bmatrix} \boldsymbol{M}^{(1)} & \cdot & \cdot \\ \cdot & \ddots & \cdot \\ \cdot & \cdot & \boldsymbol{M}^{(n)} \end{bmatrix} \\
\boldsymbol{C} &\triangleq \text{diag}\left(\boldsymbol{C}^{(1)}, \ldots, \boldsymbol{C}^{(n)}\right) \\
\boldsymbol{K} &\triangleq \text{diag}\left(\boldsymbol{K}^{(1)}, \ldots, \boldsymbol{K}^{(n)}\right) \\
\boldsymbol{u} &\triangleq \begin{bmatrix} \boldsymbol{u}^{(1)} \\ \vdots \\ \boldsymbol{u}^{(n)} \end{bmatrix}, \quad \boldsymbol{f} \triangleq \begin{bmatrix} \boldsymbol{f}^{(1)} \\ \vdots \\ \boldsymbol{f}^{(n)} \end{bmatrix}, \quad \boldsymbol{g} \triangleq \begin{bmatrix} \boldsymbol{g}^{(1)} \\ \vdots \\ \boldsymbol{g}^{(n)} \end{bmatrix}
\end{aligned}$$

For the sake of simplicity, the explicit time dependence has been omitted here.

In order to actually establish the coupling of two or more substructures, two conditions must always be satisfied regardless of the domain and coupling method used:

1. Compatibility of the substructures' displacements at the interface is the so-called compatibility condition.
2. Force equilibrium on the substructures interface degrees of freedom is called the equilibrium condition.

Next, the compatibility condition can be expressed by:

$$\boldsymbol{B}\boldsymbol{u} = \boldsymbol{0} \tag{3}$$

The $\boldsymbol{B}$ matrix operates on the interface degrees of freedom and is a signed Boolean matrix if the interface degrees of freedom are matching (hence for conforming meshes on the interface). Note that in practice the substructures often do not originate from a partitioning of a global mesh but are meshed independently. In that case the interface compatibility is usually enforced through nodal collocation (see section 3.3.3), so that the compatibility condition can still be written as in (3) but now the matrix $\boldsymbol{B}$ is no longer Boolean. The subsequent discussion is valid both when $\boldsymbol{B}$ is Boolean or not. If $\boldsymbol{B}$ is a signed Boolean matrix, the compatibility condition states that any pair of matching interface degrees of freedom $u^{(k)}$ and $u^{(l)}$ must have the same displacement, i.e. $u^{(k)} - u^{(l)} = 0$. More details on the formulation of the Boolean matrix $\boldsymbol{B}$ can be found in the appendix.

The equilibrium condition is expressed by

$$\boldsymbol{L}^T \boldsymbol{g} = \boldsymbol{0} \tag{4}$$

where the matrix $\boldsymbol{L}$ is the Boolean matrix localizing the interface DoF of the substructures in the global set of DoF. The expression states that when the connection forces are summed, their resultant must be equal to zero, i.e. $g^{(k)} + g^{(l)} = 0$. More details can be found in the appendix.

The total system is now described by equations (2), (3) and (4):

$$\begin{cases} \boldsymbol{M}\ddot{\boldsymbol{u}} + \boldsymbol{C}\dot{\boldsymbol{u}} + \boldsymbol{K}\boldsymbol{u} = \boldsymbol{f} + \boldsymbol{g} \\ \boldsymbol{B}\boldsymbol{u} = \boldsymbol{0} \\ \boldsymbol{L}^T\boldsymbol{g} = \boldsymbol{0} \end{cases} \tag{5}$$

Note that the above equations describe the coupling between any number of substructures with any number of arbitrary couplings. Depending on whether one chooses displacement or forces as unknown at the interface, a primal or dual assembled system of equations is obtained, as discussed next.

2.2 Primal Substructure Assembly

In a primal assembly of the substructure models a unique set of interface degrees of freedom is defined and the interface forces are eliminated as unknowns using the interface equilibrium. This is how individual elements are classically assembled in finite element (FE) models. Mathematically this is obtained by stating that

$$\boldsymbol{u} = \boldsymbol{L}\boldsymbol{q}, \tag{6}$$

where $\boldsymbol{q}$ is the unique set of interface DoF for the system and $\boldsymbol{L}$ the Boolean matrix introduced earlier. Since (6) indicates that the substructure DoF are obtained from the unique set $\boldsymbol{q}$ it is obvious that the compatibility condition (3) is satisfied for any set $\boldsymbol{q}$, namely

$$\boldsymbol{B}\boldsymbol{u} = \boldsymbol{B}\boldsymbol{L}\boldsymbol{q} = \boldsymbol{0} \qquad \forall \boldsymbol{q}$$

Hence $\boldsymbol{L}$ actually represents the nullspace of $\boldsymbol{B}$, or vice versa:

$$\begin{cases} \boldsymbol{L} = \text{null}\,(\boldsymbol{B}) \\ \boldsymbol{B}^T = \text{null}\,(\boldsymbol{L}^T) \end{cases} \tag{7}$$

This is a very useful property when calculating the response of the coupled system, since in the assembly process only one Boolean matrix needs to be formulated. The construction of these Boolean matrices, as well as an explicit computation of the nullspaces, is discussed in more detail in the appendix.

Since the compatibility condition in (5) is satisfied by the choice of the unique set $\boldsymbol{q}$, the system is now described by:

$$\begin{cases} \boldsymbol{M}\boldsymbol{L}\ddot{\boldsymbol{q}} + \boldsymbol{C}\boldsymbol{L}\dot{\boldsymbol{q}} + \boldsymbol{K}\boldsymbol{L}\boldsymbol{q} = \boldsymbol{f} + \boldsymbol{g} \\ \boldsymbol{L}^T\boldsymbol{g} = \boldsymbol{0} \end{cases}$$

Premultiplication of the equilibrium equations by $\boldsymbol{L}^T$ and noting that according to the equilibrium condition $\boldsymbol{L}^T\boldsymbol{g}$ is equal to zero, the primal assembled system reduces to:

$$\tilde{\boldsymbol{M}}\ddot{\boldsymbol{q}} + \tilde{\boldsymbol{C}}\dot{\boldsymbol{q}} + \tilde{\boldsymbol{K}}\boldsymbol{q} = \tilde{\boldsymbol{f}} \tag{8}$$

with the primal assembled system matrices defined by:

$$\begin{cases} \tilde{\boldsymbol{M}} & \triangleq & \boldsymbol{L}^T\boldsymbol{M}\boldsymbol{L} \\ \tilde{\boldsymbol{C}} & \triangleq & \boldsymbol{L}^T\boldsymbol{C}\boldsymbol{L} \\ \tilde{\boldsymbol{K}} & \triangleq & \boldsymbol{L}^T\boldsymbol{K}\boldsymbol{L} \\ \tilde{\boldsymbol{f}} & \triangleq & \boldsymbol{L}^T\boldsymbol{f} \end{cases}$$

2.3 Dual Substructure Assembly

In a dual assembly of substructure models the full set of global DoF is retained, i.e. all interface DoF are present as many times as there are components connected to the corresponding node. From equation (5) the dual assembled system is obtained by satisfying a priori the interface equilibrium. This is obtained by choosing the interface forces in the form:

$$\boldsymbol{g} = -\boldsymbol{B}^T\boldsymbol{\lambda}$$

Here $\boldsymbol{\lambda}$ are Lagrange multipliers, corresponding physically to the interface force intensities. By choosing the interface forces in this form, they act in opposite directions for any pair of dual interface degrees of freedom, due to the construction of Boolean matrix $\boldsymbol{B}$. The equilibrium condition is thus written:

$$\boldsymbol{L}^T\boldsymbol{g} = -\boldsymbol{L}^T\boldsymbol{B}^T\boldsymbol{\lambda} = \boldsymbol{0}$$

Since it was shown that $\boldsymbol{L}^T$ was the nullspace of $\boldsymbol{B}^T$, see equation (7), this condition is always satisfied. Consequently, the system of equations (5) is now described by:

$$\begin{cases} \boldsymbol{M}\ddot{\boldsymbol{u}} + \boldsymbol{C}\dot{\boldsymbol{u}} + \boldsymbol{K}\boldsymbol{u} + \boldsymbol{B}^T\boldsymbol{\lambda} = \boldsymbol{f} \\ \boldsymbol{B}\boldsymbol{u} = \boldsymbol{0} \end{cases}$$

In matrix notation one finds the dual assembled system as:

$$\begin{bmatrix} \boldsymbol{M} & \boldsymbol{0} \\ \boldsymbol{0} & \boldsymbol{0} \end{bmatrix}\begin{bmatrix} \ddot{\boldsymbol{u}} \\ \ddot{\boldsymbol{\lambda}} \end{bmatrix} + \begin{bmatrix} \boldsymbol{C} & \boldsymbol{0} \\ \boldsymbol{0} & \boldsymbol{0} \end{bmatrix}\begin{bmatrix} \dot{\boldsymbol{u}} \\ \dot{\boldsymbol{\lambda}} \end{bmatrix} + \begin{bmatrix} \boldsymbol{K} & \boldsymbol{B} \\ \boldsymbol{B}^T & \boldsymbol{0} \end{bmatrix}\begin{bmatrix} \boldsymbol{u} \\ \boldsymbol{\lambda} \end{bmatrix} = \begin{bmatrix} \boldsymbol{f} \\ \boldsymbol{0} \end{bmatrix} \tag{9}$$

Now the general theory of dynamic substructuring has been outlined, the next section will discuss the use of such a methodology in an industrial setting.

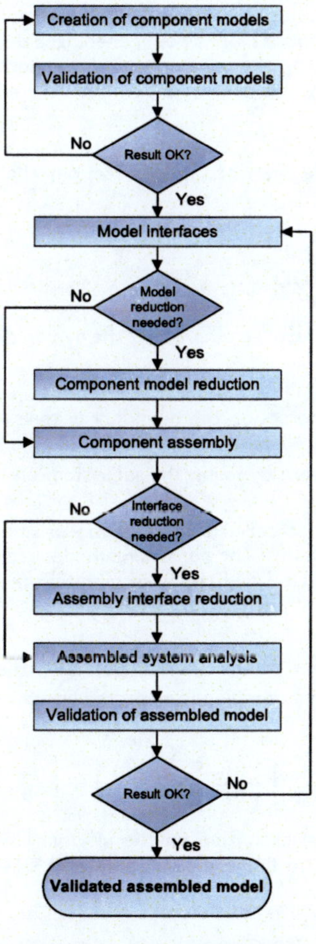

Figure 1: Stepwise DS analysis

3 DYNAMIC SUBSTRUCTURING IN AN INDUSTRIAL SETTING

As indicated in the introduction, the application of dynamic substructuring techniques to full sized industrial problems will require some practical issues to be solved. In this section we will discuss the steps that need to be taken in order to successfully apply dynamic substructuring to such problems and identify the issues that need to be resolved. Subsequently, some solutions will be presented to overcome these issues.

3.1 Basic Procedure

A full dynamic substructuring analysis comprises many aspects, such as model creation, interface modeling, possibly model reduction and model validation. This process can best be visualized in a flowchart indicating the individual steps in a DS analysis and the order in which they should be taken. Such a flowchart is depicted in figure 1.

The first step in the DS process already introduces some practical issues. Substructure models could be created in different software packages, or could even be measured FRFs. In order to enable a substructuring analysis at all, the individual component models need to be exported to one (flexible) software platform. This probably requires additional software which has to be able to import or export models from FE packages and read FRF data from for example universal files. Here we have chosen to use MatLab as the platform for the DS analysis; FE models and measured FRFs can be imported using additional toolboxes (such as FEMlink [3]).

3.2 Model Reduction

An important aspect of a DS analysis on an industrial problem is the reduction of substructure models. Since industrial FE models usually contain in the order of hundreds of thousands (or even millions) of DoF, model reduction is inevitable. When performing model reduction, substructures are represented in an approximated manner by transforming the full set of degrees of freedom to a reduced set of *generalized* DoF. These generalized degrees of freedom are usually a number of mode shape coordinates and interface DoF describing the behavior of the component. The process of applying model reduction to a component and subsequently assembling this component with other (reduced) components is called *component mode synthesis*.

Two model reduction/CMS methods will be described in this section: the well known Craig Bampton method and the relatively new Dual Craig-Bampton method.

3.2.1 THE CRAIG-BAMPTON METHOD

The Craig-Bampton method [6] is based on the observation that the dynamic behavior of a subsystem can be fully described in terms of two types of information:

- The static modes resulting from unit forces on the boundary degrees of freedom
- The internal vibration modes put forward by fixing the subsystem on its boundary with the neighboring subsystems

The Craig-Bampton method therefore uses so-called constraint modes and fixed vibration modes to form a (statically) complete reduction basis for the component [5]. The constraint modes represent the static response of the substructure due to a unit displacement at the interface DoF and the fixed vibration modes account for the dynamic behavior of the component

By splitting $\boldsymbol{u}^{(s)}$ into a set of boundary DoF $\boldsymbol{u}_b^{(s)}$ and internal DoF $\boldsymbol{u}_i^{(s)}$, the equations of motion of a single substructure from eq. (1) become (for the sake of simplicity damping is neglected and the superscript $^{(s)}$ to denote a single substructure is dropped):

$$\begin{bmatrix} \boldsymbol{M}_{bb} & \boldsymbol{M}_{bi} \\ \boldsymbol{M}_{ib} & \boldsymbol{M}_{ii} \end{bmatrix} \begin{bmatrix} \ddot{\boldsymbol{u}}_b \\ \ddot{\boldsymbol{u}}_i \end{bmatrix} + \begin{bmatrix} \boldsymbol{K}_{bb} & \boldsymbol{K}_{bi} \\ \boldsymbol{K}_{ib} & \boldsymbol{K}_{ii} \end{bmatrix} \begin{bmatrix} \boldsymbol{u}_b \\ \boldsymbol{u}_i \end{bmatrix} = \begin{bmatrix} \boldsymbol{f}_b \\ \boldsymbol{0} \end{bmatrix} + \begin{bmatrix} \boldsymbol{g}_b \\ \boldsymbol{0} \end{bmatrix}, \tag{10}$$

Here it is assumed that there is no external excitation on the internal DoF $\boldsymbol{u}_i$. By neglecting the contribution of the inertial forces, $\boldsymbol{u}_i$ can be condensed into $\boldsymbol{u}_b$ to find the constraint modes ($\boldsymbol{\Psi}_C$) as:

$$\boldsymbol{u}_i = -\boldsymbol{K}_{ii}^{-1}\boldsymbol{K}_{ib}\boldsymbol{u}_b = \boldsymbol{\Psi}_C\boldsymbol{u}_b \tag{11}$$

The reduction basis is now completed by adding fixed interface eigenmodes, these are computed by setting $\boldsymbol{u}_b = \boldsymbol{0}$ and substituting this into (10). Hence the modes are found from solving the following eigenproblem:

$$\boldsymbol{K}_{ii}\boldsymbol{u}_i = \omega^2\boldsymbol{M}_{ii}\boldsymbol{u}_i \tag{12}$$

From (12), only the first $m < n_i$ eigenfrequencies (ω_i) and mode shapes (ϕ_i) are taken and mass normalized, such that:

$$\boldsymbol{\Phi}_m = [\ \phi_1 \quad \phi_2 \quad \dots \quad \phi_m\ ]$$
$$\boldsymbol{\Phi}_m\boldsymbol{M}_{ii}\boldsymbol{\Phi}_m = \boldsymbol{I}$$
$$\boldsymbol{\Phi}_m^T\boldsymbol{K}_{ii}\boldsymbol{\Phi}_m = \text{diag}(\omega_1^2 \dots \omega_m^2) = \boldsymbol{\Omega}_m^2$$

The total set of degrees of freedom is reduced to a new set of DoF using the static constraint modes and fixed interface vibration modes. The Craig-Bampton reduction matrix can thus be expressed as:

$$\begin{bmatrix} \boldsymbol{u}_b \\ \boldsymbol{u}_i \end{bmatrix} = \begin{bmatrix} \boldsymbol{u}_b \\ \boldsymbol{\Psi}_c \boldsymbol{u}_b + \boldsymbol{\Phi}_m \boldsymbol{\eta} \end{bmatrix} = \begin{bmatrix} \boldsymbol{I} & \boldsymbol{0} \\ \boldsymbol{\Psi}_C & \boldsymbol{\Phi}_m \end{bmatrix} \begin{bmatrix} \boldsymbol{u}_b \\ \boldsymbol{\eta} \end{bmatrix} = \boldsymbol{R} \begin{bmatrix} \boldsymbol{u}_b \\ \boldsymbol{\eta} \end{bmatrix} \tag{13}$$

Using the reduction matrix $\boldsymbol{R}$ to reduce the original set of equations (10) then gives the reduced mass and stiffness matrices.

$$\tilde{\boldsymbol{K}} = \boldsymbol{R}^T \boldsymbol{K} \boldsymbol{R}$$
$$\tilde{\boldsymbol{M}} = \boldsymbol{R}^T \boldsymbol{M} \boldsymbol{R}$$

The reduction will generally decrease the number of DoF of the substructure model from several tens of thousands to some tens DoF, while maintaining an accurate description of the dynamic behavior within certain frequency limits. Since the original set of interface DoF is retained, the reduced substructures can easily be assembled with other (potentlally reduced) FE substructure models and are therefor also known as *superelements*.

The advantages of the Craig-Bampton reduction are the straightforward computation of the reduction basis and the easy assembly to other superelements and/or FE models. A disadvantage of Craig-Bampton reduced components is that if the interface of the substructure is changed, the entire reduction basis needs to be recomputed.

3.2.2 THE DUAL CRAIG-BAMPTON METHOD

As an alternative to the original Craig-Bampton method, the Dual Craig Bampton method was introduced in 2004 [21]. As the name suggests, this method is the dual counterpart of the Criag-Bampton method, that is, the substructure models are reduced and assembled in a dual manner. Again, two types of information are used for the reduction of the substructure models. In this case they are:

1. The free interface vibration modes of the structure to account for the dynamic behavior.
2. Residual flexibility modes to account for the static response of the structure when excited at its interface DoF.

The same ingredients for substructure reduction were already proposed by Rubin and MacNeal [22, 15], but there is an important difference between these methods methods and the Dual Craig-Bampton. Where Rubin and MacNeal transform the interface (connection) forces back to interface displacements to enable primal assembly of reduced structures (as discussed in section 2.2), the Dual Craig Bampton method keeps the interface forces as part of the new set of generalized DoF (as described in section 2.3). The substructures are assembled using the interface forces and thereby only enforce a weak interface compatibility. The method will be treated in detail below.

Let us start by writing the original set of DoF of a substructure as (again dropping the superscript $^{(s)}$ to denote a single substructure):

$$\boldsymbol{u} = \boldsymbol{u}_{\text{stat}} + \sum_{j=m+1}^{n} \boldsymbol{\theta}_j \boldsymbol{\eta}_j, \tag{14}$$

where the total response of the substructure $\boldsymbol{u}$ is represented in terms of the free vibration modes of the substructure and a static solution. Here m is the number of rigid body modes of the substructure. The static response can be expressed as

$$\boldsymbol{u}_{\text{stat}} = -\boldsymbol{K}^{+} \boldsymbol{b}^T \boldsymbol{\lambda}_l + \boldsymbol{\theta}_r \boldsymbol{\eta}_r, \tag{15}$$

where the first term describes the static flexible response to a unit force at the interface and the second part gives the contribution of the rigid body modes. In this expression $\boldsymbol{\theta}_r$ denotes the rigid body modes and $\boldsymbol{\eta}_r$ are the associated amplitudes. Furthermore, $\boldsymbol{b}$ is a local Boolean matrix locating the interface DoF of the substructure within its full set of DoF $\boldsymbol{u}$, and $\boldsymbol{\lambda}_l$ is a set of local (i.e. related to the associated substructure) Lagrange multipliers within the total set of multipliers in $\boldsymbol{\lambda}$.

In eq. (15) $\boldsymbol{K}^{+}$ is the generalized inverse of the stiffness matrix $\boldsymbol{K}$ and is thus a flexibility matrix. If the substructure is constrained such that it becomes statically determined then $\boldsymbol{K}^{+} = \boldsymbol{K}^{-1}$ and no rigid body modes exist ($m = 0$).

An approximation of the transformation of (14) is created by taking only the first k free interface vibration modes ($k << n$, with n the total number of original DoF).

$$\boldsymbol{u} \approx -\boldsymbol{K}^{+} \boldsymbol{b}^T \boldsymbol{\lambda}_l + \boldsymbol{\theta}_r \boldsymbol{\eta}_r + \boldsymbol{\theta}_f \boldsymbol{\eta}_f \tag{16}$$

In this approximation the flexibility associated to the free vibration modes in $\boldsymbol{\theta}_f$ is implicitly accounted for twice, since the spectral expansion of the the flexibility matrix is:

$$\boldsymbol{K}^{+} = \sum_{j=m+1}^{n} \frac{\boldsymbol{\theta}_j \boldsymbol{\theta}_j^T}{\omega_j^2} \tag{17}$$

To simplify the expressions of the reduced system one should therefore subtract the flexibility that is already accounted for in the free vibration modes from the generalized inverse of the stiffness matrix. As a result, the residual flexibility matrix is obtained:

$$\boldsymbol{G}_{\text{res}} = \boldsymbol{K}^{+} - \sum_{j=1}^{k} \frac{\boldsymbol{\theta}_j \boldsymbol{\theta}_j^T}{\omega_j^2} \tag{18}$$

The residual flexibility matrix has the following properties [21]:

$$\begin{aligned}
&\boldsymbol{G}_{\text{res}}^{(s)} = \boldsymbol{G}_{\text{res}}^{(s)^T} \\
&\boldsymbol{G}_{\text{res}}^{(s)^T} \boldsymbol{K}^{(s)} \boldsymbol{G}_{\text{res}}^{(s)} = \boldsymbol{G}_{\text{res}}^{(s)} \\
&\boldsymbol{\theta}_r^{(s)^T} \boldsymbol{K}^{(s)} \boldsymbol{G}_{\text{res}}^{(s)} = \boldsymbol{0} \\
&\boldsymbol{\theta}_r^{(s)^T} \boldsymbol{M}^{(s)} \boldsymbol{G}_{\text{res}}^{(s)} = \boldsymbol{0} \\
&\boldsymbol{\theta}_f^{(s)^T} \boldsymbol{M}^{(s)} \boldsymbol{G}_{\text{res}}^{(s)} = \boldsymbol{0} \\
&\boldsymbol{\theta}_f^{(s)^T} \boldsymbol{K}^{(s)} \boldsymbol{G}_{\text{res}}^{(s)} = \boldsymbol{0}
\end{aligned}$$

Substituting $\boldsymbol{G}_{\text{res}}$ for $\boldsymbol{K}^{+}$ in eq. (16) leads to the final approximation of the displacement field.

$$\boldsymbol{u} \approx -\boldsymbol{G}_{\text{res}} \boldsymbol{b}^T \boldsymbol{\lambda}_l + \boldsymbol{\theta}_r \boldsymbol{\eta}_r + \boldsymbol{\theta}_f \boldsymbol{\eta}_f \tag{19}$$

Recalling that the matrix $\boldsymbol{b}^T$ selects certain columns within the residual flexibility matrix, substituting these in (19) and rewriting the equation into a matrix-vector form leads to the reduction basis $\boldsymbol{T}_{DCB}$.

$$\begin{bmatrix} \boldsymbol{u} \\ \boldsymbol{\lambda}_l \end{bmatrix} = \boldsymbol{T}_{DCB} \begin{bmatrix} \boldsymbol{\eta}_r \\ \boldsymbol{\eta}_f \\ \boldsymbol{\lambda}_l \end{bmatrix} = \begin{bmatrix} \boldsymbol{\theta}_r & \boldsymbol{\theta}_f & -\boldsymbol{\Psi}_{\text{res}} \\ \boldsymbol{0} & \boldsymbol{0} & \boldsymbol{I} \end{bmatrix} \begin{bmatrix} \boldsymbol{\eta}_r \\ \boldsymbol{\eta}_f \\ \boldsymbol{\lambda}_l \end{bmatrix} \tag{20}$$

By projecting the substructure matrices onto the reduction basis $\boldsymbol{T}_{DCB}$, the reduced system is obtained:

$$\boldsymbol{T}_{DCB}^T \begin{bmatrix} \boldsymbol{M} & \boldsymbol{0} \\ \boldsymbol{0} & \boldsymbol{0} \end{bmatrix} \boldsymbol{T}_{DCB} \begin{bmatrix} \ddot{\boldsymbol{\eta}} \\ \ddot{\boldsymbol{\lambda}}_l \end{bmatrix} + \boldsymbol{T}_{DCB}{}^T \begin{bmatrix} \boldsymbol{K} & \boldsymbol{b}^T \\ \boldsymbol{b} & \boldsymbol{0} \end{bmatrix} \boldsymbol{T}_{DCB} \begin{bmatrix} \boldsymbol{\eta} \\ \boldsymbol{\lambda}_l \end{bmatrix} = \boldsymbol{T}_{DCB}^T \begin{bmatrix} \boldsymbol{f} \\ \boldsymbol{0} \end{bmatrix} + \boldsymbol{T}_{DCB}^T \begin{bmatrix} \boldsymbol{0} \\ \boldsymbol{u}_b \end{bmatrix} \tag{21}$$

Using the properties of the residual flexibility matrix the projection onto the reduction basis leads to the reduced matrices:

$$\tilde{\boldsymbol{M}} = \boldsymbol{T}_{DCB}^T \begin{bmatrix} \boldsymbol{M} & \boldsymbol{0} \\ \boldsymbol{0} & \boldsymbol{0} \end{bmatrix} \boldsymbol{T}_{DCB} = \begin{bmatrix} \boldsymbol{I} & \boldsymbol{0} \\ \boldsymbol{0} & \boldsymbol{M}_{\text{res}} \end{bmatrix}$$

$$\tilde{\boldsymbol{K}} = \boldsymbol{T}_{DCB}{}^T \begin{bmatrix} \boldsymbol{K} & \boldsymbol{b}^T \\ \boldsymbol{b} & \boldsymbol{0} \end{bmatrix} \boldsymbol{T}_{DCB} = \begin{bmatrix} \boldsymbol{\Omega}^2 & [\, \boldsymbol{R} \;\; \boldsymbol{\Theta} \,]^T \boldsymbol{b}^T \\ \boldsymbol{b}[\, \boldsymbol{R} \;\; \boldsymbol{\Theta} \,] & -\boldsymbol{F}_{\text{res}} \end{bmatrix}$$

$$\begin{aligned}
&\boldsymbol{M}_{\text{res}} = \boldsymbol{b} \boldsymbol{G}_{\text{res}} \boldsymbol{M} \boldsymbol{G}_{\text{res}} \boldsymbol{b}^T = \boldsymbol{\Psi}_{\text{res}}^T \boldsymbol{M} \boldsymbol{\Psi}_{\text{res}} \\
&\boldsymbol{F}_{\text{res}} = \boldsymbol{b} \boldsymbol{G}_{\text{res}} \boldsymbol{b}^T = \boldsymbol{b} \boldsymbol{\Psi}_{\text{res}}
\end{aligned}$$

Here $\boldsymbol{\Omega}^{(s)}$ is a square matrix with non-zero entries only on the diagonal, corresponding to the free interface eigenfrequencies (the "true" eigenfrequencies of the substructure).

One of the big advantages of reducing a substructure using the Dual Craig-Bampton method is that the reduction basis only slightly changes if other interface DoF are chose. One of the ingredients of the reduction basis is the set of free interface modes, these will not change if the interface DoF are changed; only the set of residual flexibility modes needs to be modified by deleting and/or adding columns containing static responses to the unit interface loads. Compared to the MacNeal-Rubin methods (using the same reduction basis as in (19)) the Dual Craig-Bampton provides more accurate results for a given order and has the advantage that the reduced matrices are nicely sparse. The downside of the method is that by replacing the interface DoF by interface force intensities, the assembly process is less straightforward and generally not implemented in commercial FE packages.

3.3 Interface Modeling & Assembly

Just as important as accurately reduced substructure models are accurate interface models. In many engineering applications interfaces will not just govern the compatibility between the different components, but will have a significant influence on the dynamic behavior of the total structure. One of the big challenges in dynamic substructuring is therefore creating accurate, but simple interface models. Complex interface models could lead to an increase in interface DoF, which will automatically lead to a decrease in a computational efficiency; a large number of interface DoF will lead to a large reduction basis. In this section a number of different interface modeling techniques will be discussed, starting with the simplest interface model: the rigid interface.

3.3.1 RIGID INTERFACES

In case an interface is located on a stiff part of the substructure, or is relatively small (and stiff) in comparison to the total substructure, one could approximate the behavior of the interface by a local rigid section with the larger flexible structure. This assumption will allow for a description of the interface displacements with six rigid motions only. The set of original interface DoF can hence be approximated by a set of only six interface DoF. This approximation can be described by a projection of the original (translational) boundary DoF on the six rigid body motions of the corresponding interface:

$$\begin{bmatrix} \boldsymbol{u}_{b,1} \\ \boldsymbol{u}_{b,2} \\ \vdots \\ \boldsymbol{u}_{b,n_i} \end{bmatrix} = \begin{bmatrix} \boldsymbol{T}_1 \\ \boldsymbol{T}_2 \\ \vdots \\ \boldsymbol{T}_{n_i} \end{bmatrix} \begin{bmatrix} q_x \\ q_y \\ q_z \\ q_\alpha \\ q_\beta \\ q_\gamma \end{bmatrix} \tag{22}$$

where:

$$\boldsymbol{T}_j = \begin{bmatrix} 1 & 0 & 0 & 0 & -d_{j,z} & d_{j,y} \\ 0 & 1 & 0 & d_{j,z} & 0 & -d_{j,x} \\ 0 & 0 & 1 & -d_{j,y} & d_{j,x} & 0 \end{bmatrix} \quad j = 1 \ldots n_i$$

$$\boldsymbol{d}_j = \begin{bmatrix} d_{j,x} \\ d_{j,y} \\ d_{j,z} \end{bmatrix} = \begin{bmatrix} x_j \\ y_j \\ z_j \end{bmatrix} - \begin{bmatrix} x_0 \\ y_0 \\ z_0 \end{bmatrix}$$

Here $\boldsymbol{u}_{b,j}$ is the vector of boundary DoF associated to an interface node j (with translational DoF only), $\boldsymbol{T}_j$ the corresponding transformation matrix and $\boldsymbol{d}_j$ the corresponding position vector with respect to a reference node $\boldsymbol{u}_0$. So, the boundary DoF ($\boldsymbol{u}_b$) are now described by six rigid motions, $\boldsymbol{q}_b$, as:

$$\begin{bmatrix} \boldsymbol{u}_b \\ \boldsymbol{u}_i \end{bmatrix} = \begin{bmatrix} \boldsymbol{T} & \boldsymbol{0} \\ \boldsymbol{0} & \boldsymbol{I} \end{bmatrix} \begin{bmatrix} \boldsymbol{q}_b \\ \boldsymbol{u}_i \end{bmatrix} = \boldsymbol{R}_r \begin{bmatrix} \boldsymbol{q}_b \\ \boldsymbol{u}_i \end{bmatrix} \tag{23}$$

By projecting the stiffness and mass matrix on $\boldsymbol{R}_r$, the stiffness and mass of the interface are condensed onto the single interface node. Rigidifying the interface will locally create an infinitely stiff section. Intuitively one can imagine that this will affect mostly the mode shapes in which this rigid section would normally deform, thereby leading to higher eigenfrequencies for these modes after rigidification. If a substructure has a large number of interfaces and/or interfaces take up a large portion of the substructure's surface, this approach will most likely not be desirable, since rigidification of the interfaces would then lead to a substantial increase of the stiffness of the entire structure. This approach could also be extended by including local interface modes to the basis given in (23) to account for some interface flexibility [16, 1, 9]

3.3.2 FLEXIBILITY AND DAMPING BETWEEN SUBSTRUCTURES

In many situations, the interface between two substructures is not perfect. Consider for example two components that have been connected by a bolt, a situation encountered very often in practice. Due to this connection some flexibility and/or damping is introduced on the interface that is not present in the separate components. Many other examples of connections are imaginable where some physics are added to the system simply through the coupling of components. The usual approach is to neglect these interface effects. However, this cannot always be done. Let us therefore investigate this issue in more detail. To this end, consider the coupling of two general substructures as depicted in figure 9. As before, one can write the equations of motion of the subsystems in block diagonal format as:

$$\boldsymbol{M}\ddot{\boldsymbol{u}} + \boldsymbol{C}\dot{\boldsymbol{u}} + \boldsymbol{K}\boldsymbol{u} = \boldsymbol{f} + \boldsymbol{g} \tag{24}$$

Also, since the springs exert equal forces to both substructures, the equilibrium condition still holds:

$$\boldsymbol{L}^T \boldsymbol{g} = \boldsymbol{0} \tag{25}$$

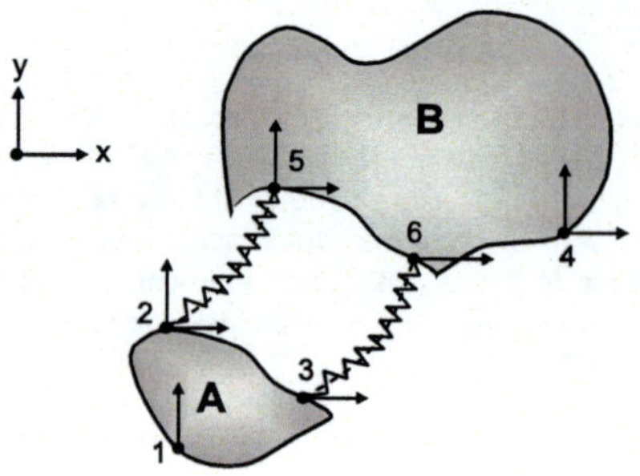

Figure 2: Coupling of two general substructures with stiffness on the interface.

However, due to the interface flexibility the compatibility condition no longer holds. Indeed, due to the flexibility the interface DoF are free to have a relative displacement. This means that two additional equations need to be obtained. One way to eliminate one unknown is to choose the interface forces as:

$$\boldsymbol{g} = -\boldsymbol{B}^T\boldsymbol{\lambda}$$

This way the interface forces in $\boldsymbol{g}$ are chosen such that, due to the construction of the Boolean matrix $\boldsymbol{B}$, the interface forces are always equal and opposite. As before, the Lagrange multipliers $\boldsymbol{\lambda}$ describe the force intensities. Hence, the equilibrium condition is always satisfied, which can be illustrated mathematically since $\boldsymbol{B}^T$ is in the nullspace of $\boldsymbol{L}^T$ and hence

$$\boldsymbol{L}^T\boldsymbol{g} = -\boldsymbol{L}^T\boldsymbol{B}^T\boldsymbol{\lambda} = \boldsymbol{0}$$

Now there is still one equation lacking to close the set of equations. However, we know that the spring on the interface behaves such that the interface force intensity can be written as:

$$\boldsymbol{\lambda} = \boldsymbol{K}_b\Delta\boldsymbol{u}_b$$

From the construction of the Boolean matrix $\boldsymbol{B}$ we also know that:

$$\Delta\boldsymbol{u}_b = \begin{bmatrix} u_2 - u_5 \\ u_3 - u_6 \end{bmatrix} = \boldsymbol{B}\boldsymbol{u}$$

Hence we can write for the Lagrange multipliers

$$\boldsymbol{\lambda} = \boldsymbol{K}_b\boldsymbol{B}\boldsymbol{u}$$

and subsequently for the connection forces:

$$\boldsymbol{g} = -\boldsymbol{B}^T\boldsymbol{K}_b\boldsymbol{B}\boldsymbol{u} \tag{26}$$

Inserting this expression for the connection forces $\boldsymbol{g}$ into the equations of motion of the subsystems in eq. (24) gives the assembled system as:

$$\boldsymbol{M}\ddot{\boldsymbol{u}} + \boldsymbol{C}\dot{\boldsymbol{u}} + \left(\boldsymbol{K} + \boldsymbol{B}^T\boldsymbol{K}_b\boldsymbol{B}\right)\boldsymbol{u} = \boldsymbol{f} \tag{27}$$

Note that there is no longer any choice whether to assemble the equations of motion in a dual or primal way; they are automatically assembled by the action of the interface spring. A primal formulation is not possible (since there are no longer redundant interface DoF) and a dual formulation would be trivial. Furthermore, note that "classic" assembly of systems with perfect connections can be regarded as a special case of the above situation, namely when $\boldsymbol{K}_b = \text{diag}(\infty, \infty)$. Then $\Delta\boldsymbol{u}_b = \boldsymbol{B}\boldsymbol{u} = \boldsymbol{0}$ and the compatibility condition indeed holds. It is interesting to know that the same formulation is found when one wants to enforce compatibility with a penalty method (not discussed here); the interface stiffness $\boldsymbol{K}_b$ is then the penalty.

Finally, note that the above is also true when (linear) damping is introduced at the interface. Suppose that in the system in figure 2, in addition to the interface stiffness $\boldsymbol{K}_b$, there is interface damping $\boldsymbol{C}_b$. One can then write for the interface force intensity:

$$\boldsymbol{\lambda} = \boldsymbol{K}_b\Delta\boldsymbol{u}_b + \boldsymbol{C}_b\Delta\dot{\boldsymbol{u}}_b$$

As before, this can be written as:

$$\boldsymbol{\lambda} = \boldsymbol{K}_b\boldsymbol{B}\boldsymbol{u} + \boldsymbol{C}_b\boldsymbol{B}\dot{\boldsymbol{u}}$$

Hence one can write the equations of motion of the connected systems, with (linear) stiffness and damping effects on the interface, as:

$$\boldsymbol{M}\ddot{\boldsymbol{u}} + \left(\boldsymbol{C} + \boldsymbol{B}^T\boldsymbol{C}_b\boldsymbol{B}\right)\dot{\boldsymbol{u}} + \left(\boldsymbol{K} + \boldsymbol{B}^T\boldsymbol{K}_b\boldsymbol{B}\right)\boldsymbol{u} = \boldsymbol{f}$$

The approach described above can be easily generalized to systems consisting of multiple substructures with multiple types of interfaces. The total $\boldsymbol{B}$ matrix can be partitioned into interfaces that are perfect (i.e. where the substructures are perfectly connected) and those where flexibility and/or damping is present between the interface DoF:

$$\boldsymbol{B} = \begin{bmatrix} \boldsymbol{B}_f \\ \boldsymbol{B}_p \end{bmatrix}$$

The subscripts $_f$ and $_p$ denote "flexible" and "perfect", respectively. The total system can then be described as:

$$\begin{cases} \boldsymbol{M}\ddot{\boldsymbol{u}} + \left(\boldsymbol{C} + \boldsymbol{B}_f^T\boldsymbol{C}_b\boldsymbol{B}_f\right)\dot{\boldsymbol{u}} + \left(\boldsymbol{K} + \boldsymbol{B}_f^T\boldsymbol{K}_b\boldsymbol{B}_f\right)\boldsymbol{u} = \boldsymbol{f} + \boldsymbol{g}_p \\ \boldsymbol{B}_p\boldsymbol{u} = \boldsymbol{0} \\ \boldsymbol{L}_p^T\boldsymbol{g}_p = \boldsymbol{0} \end{cases} \tag{28}$$

For the "perfect" interfaces a choice needs still to be made as to assemble the associated DoF in a primal or dual fashion. This is exactly done as described in sections 2.2 and 2.3.

3.3.3 NON-CONFORMING MESHES

One of the benefits of the DS approach is that it allows to combine substructures models created by different engineering groups. These models are often created without any knowledge of, or consideration, for the neighboring substructures, resulting in models with incompatible meshes. Since the models are meshed independently, the nodes at both sides of the interface are usually not collocated (i.e. at the same geometric position) and/or the models are meshed with different types of elements, leading to non-conforming meshes. Global geometric compatibility is usually not an issue, since the geometry of the substructures often originates from one large 3D CAD model.

One approach would be to re-mesh the substructure models, such that they become compatible. This leads to an additional computational step and hence reduces the overall efficiency of the DS strategy. A more efficient approach is to use the interpolation functions of the interface elements in order to enable an assembly of non-conforming substructure meshes [20]. In this paper the simple but effective *node collocation method* and its least square variant will be discussed.[2]

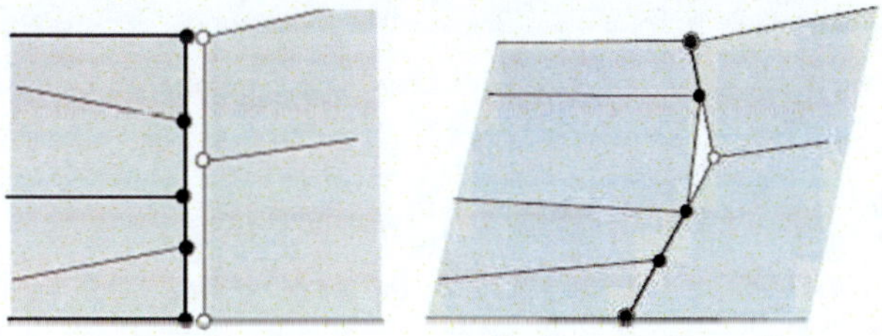

Figure 3: Non conforming meshes on the interface [20]

Node collocation method

Suppose two substructures need to be assembled, but the interfaces are not matching as depicted in figure 3. One option is to define an intermediate reference interface field $\boldsymbol{u}_b^{(ref)}$ and use the element shape functions of the substructures to interpolate and attach the nodes to the reference interface. This can be expressed as:

$$\boldsymbol{u}_b^{(s)} = \boldsymbol{D}^{(s)}\boldsymbol{u}_b^{(\text{ref})} \tag{29}$$

where $\boldsymbol{D}$ is the "collocation" matrix that needs to be computed for both substructures. A special case is obtained if the number of "reference nodes" is taken as the minimum of the number of nodes on each interface. In other words, taking the interface with the smallest number of nodes on the interface as the reference interface field:

$$n_b^{(ref)} \leqslant min\left(n_b^{(1)}, n_b^{(2)}\right) \tag{30}$$

[2]Note that in the last two decades, the assembly of structural models with non-conforming discretizations has become a research field on its own. An important contribution is the so-called Mortar element methods, as described in [4]. However, it is out of the scope of this work to treat such advanced methods.

The interface on the left is referred to as 1 and on the right as 2. From figure 3 it now becomes clear that $\boldsymbol{u}_b^{(2)}$ is the set of *master* interface nodes and $\boldsymbol{u}_b^{(1)}$ is the set of *slave* interface nodes. As a result, $\boldsymbol{D}^{(2)}$ becomes an identity matrix and only the collocation matrix of substructure 1 ($\boldsymbol{D}^{(1)}$) has to be computed. In the collocation method the matrix $\boldsymbol{D}^{(1)} = \boldsymbol{D}$ contains the values of the shape functions on the interface of substructure 2 at the locations of the interface nodes on substructure 1. This imposes that the nodes of substructure 2 remain on the interface of substructure 1. So:

$$\boldsymbol{u}_b^{(1)} = \boldsymbol{D}\boldsymbol{u}_b^{(2)} \tag{31}$$

The compatibility condition of (3) now transforms to:

$$\begin{bmatrix} -\boldsymbol{b}^{(1)} & \boldsymbol{D}\boldsymbol{b}^{(2)} \end{bmatrix} \begin{bmatrix} \boldsymbol{u}^{(1)} \\ \boldsymbol{u}^{(2)} \end{bmatrix} = \boldsymbol{B}\boldsymbol{u} = \boldsymbol{0} \tag{32}$$

The matrices denoted by $\boldsymbol{b}$ are local Boolean matrices acting on the set of boundary DoF within the total set of substructure DoF. Since $\boldsymbol{D}$ contains interpolation values between zero and one, the resulting matrix $\boldsymbol{B}$ will clearly no longer be a true Boolean matrix, although the part associated to substructure 1 will still be.

Discrete least squares compatibility

The interface constraint (29) together with condition (30) implicitly limits the behavior of the degrees of freedom on the sides of the interface that have more DoF then the number of reference DoF, thereby stiffening the interface behavior. This can also be seen for the condition in (32), that is when the coarsest side is chosen as reference. Equation (32) requires the nodes of the finest side of the interface to be exactly collocated with the interface on the coarse side as illustrated in figure 3. In a primal assembly eq. (32) would be satisfied by choosing $\boldsymbol{u}_b^{(2)}$ as the DoF in the global set, $\boldsymbol{u}_b^{(1)}$ being substituted using eq. (31). The collocation condition (29) or (32) however can lead to a severe stiffening of the interface model.

A way to render some flexibility to the interface is to relax the collocation condition. For that we look now at eq. (29) as an equation from which the reference DoF must be computed for arbitrary substructure DoF. Obviously, given condition (31), this is an overdetermined problem that can only be solved in a least square sense:

$$\frac{\partial}{\partial \boldsymbol{u}_{b,i}^{(\text{ref})}} \left(\left(\boldsymbol{u}_b^{(s)} - \boldsymbol{D}^{(s)} \boldsymbol{u}_{b,i}^{(\text{ref})} \right)^T \left(\boldsymbol{u}_b^{(s)} - \boldsymbol{D}^{(s)} \boldsymbol{u}_{b,i}^{(\text{ref})} \right) \right) = 0 \tag{33}$$

for:

$$i = 0 \dots n_b^{(\text{ref})} \qquad s = 1, 2$$

By solving (33) and, as before, choosing the interface with the smallest number of nodes as the reference interface, the compatibility condition of (3) is found in discrete least squares form as:

$$\begin{bmatrix} -\left((\boldsymbol{D}^T\boldsymbol{D})^{-1}\boldsymbol{D}^T\right)\boldsymbol{b}^{(1)} & \boldsymbol{b}^{(2)} \end{bmatrix} \begin{bmatrix} \boldsymbol{u}^{(1)} \\ \boldsymbol{u}^{(2)} \end{bmatrix} = \boldsymbol{B}\boldsymbol{u} = \boldsymbol{0} \tag{34}$$

Again, matrices denoted $\boldsymbol{b}$ are local boolean matrices acting on the set of boundary DoF within the total set of substructure DoF.

The number of constraints imposed by (34) is now equal to the number of DoF on the coarsest side, and not to the number of DoF of the finest side like in (31). As a matter of fact $\boldsymbol{u}_b^{(2)}$ can be computed for any arbitrary $\boldsymbol{u}_b^{(1)}$ so that if the interface would be assembled in a primal way one would keep all $\boldsymbol{u}_b^{(1)}$, $\boldsymbol{u}_b^{(2)}$ being eliminated by using (34). All $\boldsymbol{u}^{(1)}$ are independent but the DoF in $\boldsymbol{u}^{(2)}$ now should be such that the collocation conditions in eq. (31) are satisfied in a least square sense. The compatibility stated in (34) will therefore lead to a "best" fit, thus minimizing the interface incompatibility.

Both in the node collocation and in the discrete least square methods only local compatibility at nodes is considered. By doing so one disregards the compatibility error along the interface between the nodes, which leads to bad overall compatibility for non-uniform and highly incompatible meshes. Nonetheless, these methods methods are still used (also in many commercial software packages) since they are easy to implement and will in general not significantly alter the global dynamic behavior.

3.3.4 INTERFACE REDUCTION

Complex engineering structures, such as a modern wind turbine, commonly consist of a large number of (structural) components, consequently a large number of interfaces between these components exist. Not all interfaces can be assumed to behave rigidly as in section 3.3.1; the original set of interface DoF thus sometimes needs to be retained. If a component contains a large number of such interfaces, the number of interface DoF becomes unacceptably high. This is a problem especially when dealing with reduced substructure models, due to the size of the associated full (instead of sparse) reduction matrices. In this section two interface reduction methods will be presented in order to further reduce the total number of DoF.

Interface Reduction of Primal Assembled Systems

This first method for interface reduction is suited for the reduction of primal assembled substructures [7, 2, 25, 24]. This is (usually) the case when dealing with finite element models reduced in commercial software (i.e. *superelements*), such as Craig-Bampton reduced components.

Determining the interface behavior does not require detailed insight in the component's dynamic behavior; an accurate representation of the static behavior at the interface is often sufficient. A static condensation matrix of the substructure is therefore computed as (equal to Guyan's reduction [11]):

$$\begin{aligned} \boldsymbol{u}_i &= -\boldsymbol{K}_{ii}^{-1}\boldsymbol{K}_{ib}\boldsymbol{u}_b = \boldsymbol{\Psi}_c\boldsymbol{u}_b \\ \begin{bmatrix} \boldsymbol{u}_i \\ \boldsymbol{u}_b \end{bmatrix} &= \begin{bmatrix} \boldsymbol{\Psi}_c \\ \boldsymbol{I} \end{bmatrix} \boldsymbol{u}_b = \boldsymbol{R}\boldsymbol{u}_b \end{aligned} \tag{35}$$

Using the so obtained static constraint modes as a reduction basis, the entire substructure is condensed to the interface DoF, resulting in a generalized mass and stiffness matrix:

$$\boldsymbol{M}_{\text{int}}\ddot{\boldsymbol{u}}_b + \boldsymbol{K}_{\text{int}}\boldsymbol{u}_b = \boldsymbol{f}_b + \boldsymbol{g}_b \tag{36}$$

where

$$\begin{aligned} \boldsymbol{M}_{\text{int}} &= \boldsymbol{R}^T\boldsymbol{M}\boldsymbol{R} \\ \boldsymbol{K}_{\text{int}} &= \boldsymbol{R}^T\boldsymbol{K}\boldsymbol{R} \end{aligned}$$

An interface connects two substructures and hence its dynamic behavior can therefore not simply be described by a single (unassembled) substructure interface; it is dependent on all substructures participating in this interface. Recalling the primal assembly from section 2.2, the condensed stiffness and mass matrices can be assembled. In the case of assembly of two substructures, the equation would write:

$$\boldsymbol{M}_{\text{int}}^{(1+2)}\ddot{\boldsymbol{u}}_b + \boldsymbol{K}_{\text{int}}^{(1+2)}\boldsymbol{u}_b = \boldsymbol{f}_b^{(1+2)} \tag{37}$$

where:

$$\begin{aligned} \boldsymbol{M}_{\text{int}}^{(1+2)} &= \boldsymbol{L}_{bb}^T \begin{bmatrix} \boldsymbol{M}_{\text{int}}^{(1)} & \boldsymbol{0} \\ \boldsymbol{0} & \boldsymbol{M}_{\text{int}}^{(2)} \end{bmatrix} \boldsymbol{L}_{bb} \\ \boldsymbol{K}_{int}^{(1+2)} &= \boldsymbol{L}_{bb}^T \begin{bmatrix} \boldsymbol{K}_{\text{int}}^{(1)} & \boldsymbol{0} \\ \boldsymbol{0} & \boldsymbol{K}_{\text{int}}^{(2)} \end{bmatrix} \boldsymbol{L}_{bb} \end{aligned}$$

Here, $\boldsymbol{L}_{bb}$ is the part of the total Boolean matrix $\boldsymbol{L}$ that operates on the interface DoF. By solving the eigenproblem of the interface equations above, the *interface modes* and *interface eigenfrequencies* are obtained, i.e.:

$$\left(\boldsymbol{K}_{\text{int}}^{(1+2)} - \omega_i^2\boldsymbol{M}_{\text{int}}^{(1+2)}\right)\boldsymbol{\phi}_{P,i} = \boldsymbol{0}$$

The obtained interface modes ($\boldsymbol{\Phi}_P$) are actually mode shapes, meaning the principle of modal superposition can be applied using these interface modes. The response of the boundary DoF can thus be expressed as a summation of the interface modes times their modal amplitudes:

$$\boldsymbol{u}_b = \sum_{j=1}^{n_b} \boldsymbol{\phi}_{P,j}\eta_{P_j} \tag{38}$$

The interface reduction is performed by only including the first k ($k < n_b$) interface modes in (38). The new set of generalized coordinates can now be rewritten according to:

$$\begin{bmatrix} \boldsymbol{q}_i \\ \boldsymbol{u}_b \end{bmatrix} = \begin{bmatrix} \boldsymbol{I} & \boldsymbol{0} \\ \boldsymbol{0} & \boldsymbol{\Phi}_P \end{bmatrix} \begin{bmatrix} \boldsymbol{q}_i \\ \boldsymbol{\eta}_b \end{bmatrix} = \boldsymbol{R}_b \begin{bmatrix} \boldsymbol{q}_i \\ \boldsymbol{\eta}_b \end{bmatrix} \tag{39}$$

Here $\boldsymbol{q}_i$ is the set of DoF that is inactive in this reduction step and could either be the original set of (internal) DoF ($\boldsymbol{u}_i$) or modal amplitudes due to a reduction of the internal DoF. The most computationally intensive step in this interface reduction method is computing the static modes Ψ_C. However, in some cases this step is "free", for instance when the substructure models have already been reduced using the Craig-Bampton method (section 3.2.1), since the reduction basis already contains these constraint modes. Computing (37) then simply requires assembling the interface part of the reduced matrices and solving the (relatively small) eigenproblem of the interface DoF as described above.

This interface reduction method described in this section can be an effective way of further reducing the number of DoF for substructures with a large number of interface DoF.

Interface Reduction of Dual Assembled Systems

The approach discussed in the previous section is suited for the reduction of substructures of which the interface is described in terms of displacements. However, when substructures are assembled in a dual way, the interface DoF are no longer interface displacements but interface forces (or interface force intensities). In that case a different approach has to be taken.

Since the interface of a dual assembled system is described in terms of forces, interface modes corresponding to force distributions are required to perform the reduction. To this end, an approach inspired by the work described in [13] or [20] could be taken. Currently such a methodology is under development, but the results are too premature to be described here.

4 CASE STUDY ON A SIEMENS WIND TURBINE

To illustrate the potential of the dynamic substructuring approach in wind turbine engineering, a DS analysis has been performed on the yaw system of a 2.3 MW Siemens wind turbine (SWT-2.3-93). This system is an important part of every modern wind turbine and is an interesting test case for the DS methodology, since it comprises many components and complex interfaces. Furthermore, the yaw system is generally not taken into account in a detailed way in aero-elastic codes, but is in some cases thought to influence the overall turbine dynamics. The case study will be discussed in more detail in this section, starting in the next subsection with a brief description of the system at hand.

4.1 System Description

Yawing denotes the rotation of the nacelle and the rotor about the vertical tower axis. By yawing the wind turbine, the rotor can be positioned such that rotor plane is orthogonal to the wind direction. Basically, the yawing of the wind turbine can be performed passively or actively. In passive yawing, the wind force itself is utilized to keep the turbine aligned with the wind. One way to achieve passive yawing is to construct the turbine in a "downwind" fashion, with the rotor plane behind the tower. For "upwind" turbines, passive yawing can be achieved by using a tail vane and a cone-shaped rotor. However, passive yawing can generate high yawing rates, leading to excessive gyroscopic moments on the wind turbine tower. Twisting of the cable that runs from the generator in the nacelle to the transformer in the tower base is also an issue. To overcome these problems, the vast majority of the modern multi-MW wind turbines is constructed according to the "Danish concept", that is, with a three bladed upwind rotor and equipped with an active yaw system. One such modern turbine is the SWT-2.3-93 from Siemens Wind Power. As the name suggest, this is a 2.3MW turbine with a rotor diameter of 93 meters. Its yaw system is depicted in figure 4. In the yaw system of this wind turbine we can identify a number of components:

- *Bedplate*: The bedplate can be seen as the "chassis" of the nacelle; it serves as a platform for mounting the main turbine components, such as the main gearbox, main bearing, main shaft, etcetera. Furthermore, the bedplate houses the interface between the tower and the rest of the turbine.
- *Tower top*: The tower top is the upper section of the tower, with an integrated top flange for assembly with the yaw ring.
- *Yaw ring*: The yaw ring is a big sprocket wheel which is driven by the yaw gearbox – motor assemblies. The yaw ring is attached to the tower top and enclosed by the yaw pads.
- *Yaw pads*: The yaw pads are attached to the bedplate and serve as a friction-type bearing for the yaw ring. The yaw pads are made of polyamide material and thereby pose an additional challenge in their modeling.
- *Yaw gearboxes and motors*: The yaw motors are electric motors controlled by the yaw controller. Via the yaw gearboxes their rotational speeds are greatly reduced, while their torque is increased. This is needed in order to overcome the inertia of the nacelle and the friction of the yaw pads so that the nacelle can be rotated.
- *Yaw controller*: The yaw controller is a central controller for the yaw system and is instructed by the global turbine controller. The yaw controller regulates the rotational speed and torque of the yaw motors.

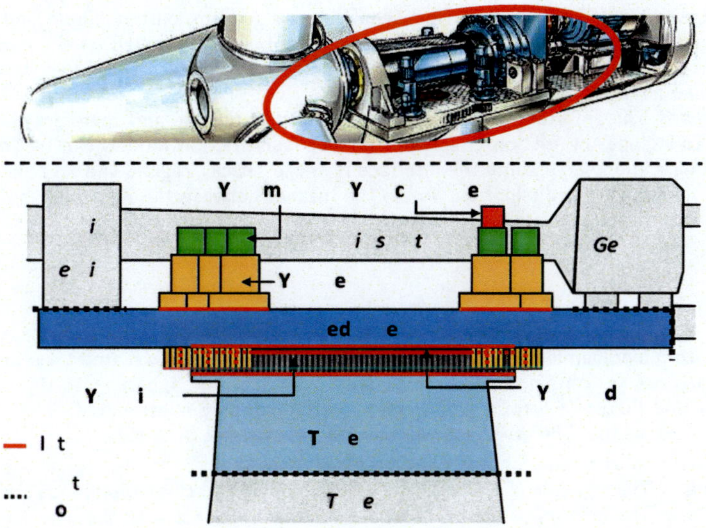

Figure 4: Yaw system of a 2.3 MW Siemens wind turbine

Each of the above mentioned components is modeled as a separate substructure, although the yaw motors and yaw controller are not yet included in the current DS analysis.

As outlined before, a successful DS analysis requires both accurate substructure models and proper interface descriptions are paramount. As we will see for the analysis of the yaw system some interfaces will allow an "exact" coupling of the substructures, while others could show additional effects. In the following subsection the component and interface modeling for each substructure will be discussed in more detail.

4.2 Component and Interface Modeling

Due to the fact that extensive stress analysis is performed on the structural components of a wind turbine, finite element models of most components are often already available. Furthermore, most components are made from steel and are hence very well suited for FE modeling. Therefore, existing FE models can be used in a DS analysis with only some minor changes, which benefits the practical usability of the DS approach.

4.2.1 BEDPLATE

The bedplate is made of steel. It is meshed using 10-node tetrahedral elements and irrelevant geometric features are removed in order to avoid a too fine mesh. The connections of the bedplate to the main bearing housing and the generator are outside of the system boundaries and will therefore not be discussed. From figure 4, one can see that two interfaces remain. These are:

- *Bedplate* ↔ *yaw gearbox* This interface is assumed to behave like a rigid section, the interface is therefore "rigidified" (as outlined in section 3.3.1) and coupling is done using a single "master" node with six DoF.
- *Bedplate* ↔ *yaw pad* This interface is assumed to behave fully flexible and therefore the original set of interface DoF is retained. "Rigidification" of the interface would significantly stiffen the bedplate model, since the yaw pads cover a large part of the bedplate. Hence the interface is modeled fully flexible, although the meshes are incompatible. To overcome this, the technique from section 3.3.3 has been applied.

4.2.2 TOWER TOP AND YAW RING

The tower top and yaw ring are integrated into one substructure. The compatibility between the two steel structures is enforced by the bolts and can be assumed to be "exact", thereby allowing them to be combined to a single substructure. The tower top and yaw ring are also meshed using 10-node tetrahedral elements. From figure 4 we see two types of interfaces for this component:

- *Yaw ring ↔ yaw gearboxes* The interaction between the yaw gearbox output pinion and the yaw ring is through the gear tooth contact. An equivalent gear tooth stiffness has been determined for the connection between the yaw ring and yaw gearbox output pinion using ISO 6336 [12]. The assembly of these two structures with the interface stiffness is performed as outlined in section 3.3.2.
- *Yaw ring ↔ yaw pads* This interface is assumed to behave fully flexible and therefore the original set of interface DoF is retained. To include the friction effects on the interface, friction models can be added in the coupling. For the purpose of simplification, we assume the interface is in the "stick" regime and an equivalent viscous damping is added to account for the energy dissipation. Again the meshes between the pads and the ring are non-conforming, this is again solved using the node collocation method.

4.2.3 YAW PADS

The yaw pads are made from a polyamide with a high wear resistance and a low (dynamic) friction coefficient. The full set of mechanical properties is not yet obtained and/or measured. We therefore approximate their mechanical properties by the modulus of elasticity and Poisson's ratio at 20°C, the material damping is estimated. The yaw pads are also meshed using 10-node tetrahedral elements. The yaw pads will have an interface at both sides:

- *Yaw pads ↔ bedplate* The top side of the yaw pad will be coupled to the bedplate, as described before.
- *Yaw pads ↔ yaw ring* The bottom of the yaw pad will have an interface with the yaw ring, as described before.

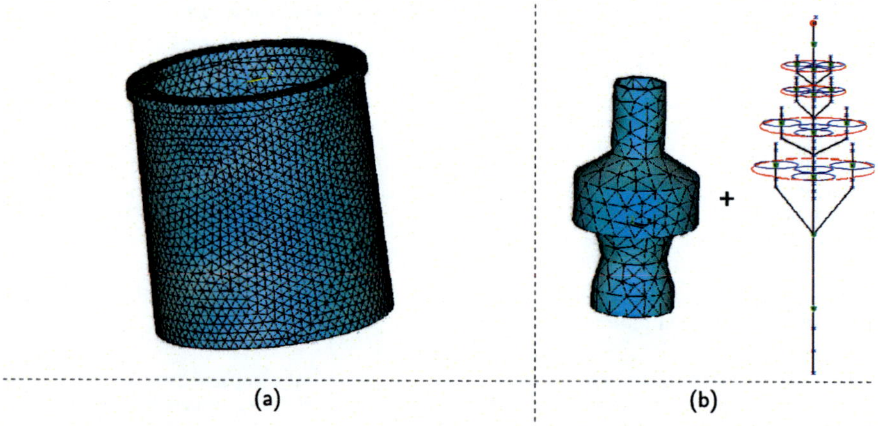

Figure 5: (a) Tower top and yaw ring model and (b) model of the yaw gearbox.

4.2.4 YAW GEARBOX

The yaw gearbox consists of a set of planetary gears that reduce the input speed of the yaw motor. The gearbox can be divided into two parts; the housing (external) and the running gears (internal). A finite element model of the internal part of the gearbox has been created in MatLab, according to the methodology outlined in [14, 19], whereas for the gearbox housing an FE model is created using 10-node tetrahedral elements. The parts are assembled through the bearing stiffness and the ring gears of the planetary stages. Both interfaces of the yaw gearboxes with other components of the yaw system have already been described.

- *Yaw gearbox ↔ bedplate* As described in modeling of the bedplate.
- *Yaw gearbox ↔ yaw ring* As described in modeling of the tower top and yaw ring.

The yaw gearbox model is shown in figure 5 (b).

4.3 Component Model Validation

In order to gain confidence in the substructure modeling it is important to validate their numerical models using measurements. In a DS analysis one can identify the two types of modeling errors;

- Errors in the substructure models

- Errors in the interface models

By validating the substructure models, the first source of errors is minimized. Furthermore, in order to validate the interface models one needs to validate the substructure models first, since interface model validation requires measurements on the assembled system and thus validated substructure models. Since the research described in this paper is still ongoing, only the bedplate's dynamic and interface models have been validated. The validation of the remaining substructures is planned for the near future.

4.3.1 BEDPLATE MODEL VALIDATION

Measurements have been performed to validate the FE model of the bedplate. The bedplate was suspended using low stiffness air springs and accelerations were measured at 33 locations using triaxial ICP accelerometers. Excitation of the bedplate was done by a shaker using a random noise signal. The identified modes were expanded using the SEREP technique [17] and a MAC analysis was performed to visualize correlation between the identified modes and the finite element modes. The low cross-correlation at FE mode 8 and mode 9 (which are missing in the set of measured modes) is

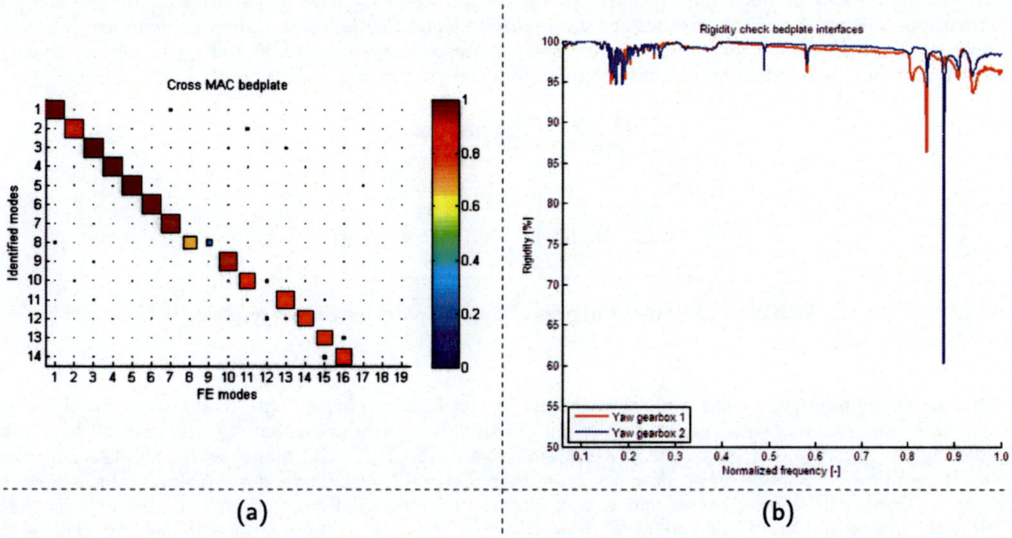

Figure 6: MAC of the bedplate FE modes and measured modes (a), Rigidity plot of the yaw gearbox interfaces (b).

due to the fact that both seem to be in-plane modes, whereas the excitation was out-of-plane. FE mode 10 shows a good correlation to the 9th identified mode. The difference between the measured eigenfrequencies and the FE eigenfrequencies was less than 2%. See figure 6 (a).

As described in the modeling section, we have assumed that the interfaces to the gearboxes behave as local rigid sections. In order to validate this assumption, two yaw gearbox interfaces have each been equipped with 4 triaxial accelerometers during the bedplate measurements. By projecting the measured FRFs onto the rigid motions and dividing their norm by the norm of the FRFs, a measure for the rigidity is obtained (see [9]). It can be seen from figure 6 (b) that the interfaces indeed behave rigidly up to a normalized frequency of approximately 0.85, while the frequency range of interest is up to a normalized frequency of 0.5.

From these measurements on the bedplate one can thus conclude that both the bedplate FE model itself and the assumptions made for its interface model can be considered valid.

4.4 Assembled System Analysis & Results

The substructure models described in section 4.1 have all been reduced using the method of Craig-Bampton and the Dual Craig-Bampton method (as described in section 3.2). By assembling the reduced substructures and the full FE component models, a dynamic model of the total system as shown in section 4 is obtained. In order to show that the reduced system can be used to accurately describe the global dynamic behavior of the yaw system, a modal analysis is performed. The results are compared to those of the full FE model.

The following cases have been considered:

- Full – Assembly of the full FE models of all components (this is the reference solution);
- CB – Assembly of Craig-Bampton reduced component models, where every component has been reduced using 30 fixed interface eigenmodes;
- DCB – Assembly of Dual Craig-Bampton reduced component models, using 30 eigenmodes (including rigid body modes);
- CB-IR 100 – Assembly of Craig-Bampton reduced component models (with 30 fixed interface eigenmodes each) and interface reduction using 100 interface modes for the total system;
- CB-IR 200 – Assembly of Craig-Bampton reduced component models (with 30 fixed interface eigenmodes each) and interface reduction using 200 interface modes for the total system;

The non-conforming meshes between the bedplate, yaw pads and yaw ring were in all cases assembled using the node collocation method. The results of the analyses are summarized below. Firstly, table 1 gives an overview of the size of the different assembled systems. One can see that the full system contains almost 250.000 DoF, whereas the Craig-Bampton reduced system with interface reduction has only close to 500 DoF, which is a reduction of almost a factor 500.

Method	Number of DoF
Full	230.895
CB	7.716
DCB	8.232
CB-IR 100	490
CB-IR 200	590

TABLE 1: Number of degrees of freedom of the assembled systems.

The performance of the different methods is shown in the figures below. Figure 7 (a) shows accuracy of the computed eigenfrequencies of the methods with respect to the full solution. It can be seen that the reduced models perform very well in predicting the eigenfrequencies of the yaw system. Even the CB-IR 100 model with only 490 DoF can predict the frequencies up to mode 50 with an accuracy of more than 99% with respect to the reference solution. As expected, the CB-IR 200 model performs even better and is very close to the original Craig-Bampton model at most frequencies. However, the accuracy of the Dual Craig-Bampton reduced system seems to degrade after mode no. 30. This is due to the fact that of the 30 modes used in the reduction basis of each component, 6 were rigid body modes. Hence, less "flexible" information is taken into account which affects the accuracy of the higher modes. Nonetheless, the Dual Craig-Bampton method performs very well at the low frequencies. The accuracy of the corresponding mode shapes is shown in figure 7

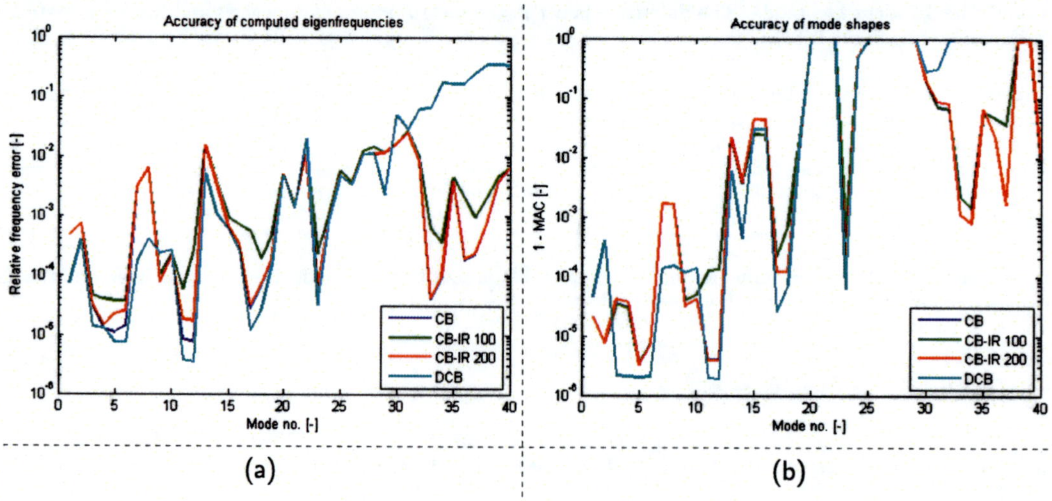

Figure 7: Accuracy of the computed eigenfrequencies (a) and the accuracy of the corresponding mode shapes (b).

(b). First, a modal assurance criterion (MAC) analysis was performed for the assemblies of reduced systems with respect to the full system. The resulting MAC values for corresponding mode numbers (i.e. the entries on the diagonal of the MAC matrix) have been subtracted from 1 (to indicate their deviation from "perfect" correlation) and are plotted in figure 7 (b). As one can see, the reduced models perform very good at the low frequencies. However, at modes 20 to 30 the correlation jumps to zero. This is due to the fact that all reduced models seem to miss one mode shape. This is indeed confirmed by the full MAC plots shown in figure 8.

Again, the Dual Craig-Bampton method performs very well up to approximately mode 30, at higher frequencies the model suffers from the fact that less flexible modes were used for the reduction in comparison to the Craig-Bampton system.

Finally it should be remarked that due to the current implementation (many different solvers are used) it was not possible to compare computation times of the different methods. As a general remark one can say that the computation cost of building the reduced matrices and subsequently performing a modal analysis is of roughly the same order as performing a modal analysis on the full system. The computational gain is obtained when multiple simulations need to be performed (e.g. different load cases) or when certain reduced components need to be interchanged. In such situations the initial cost of building reduced models is easily recovered by much shorter analysis times.

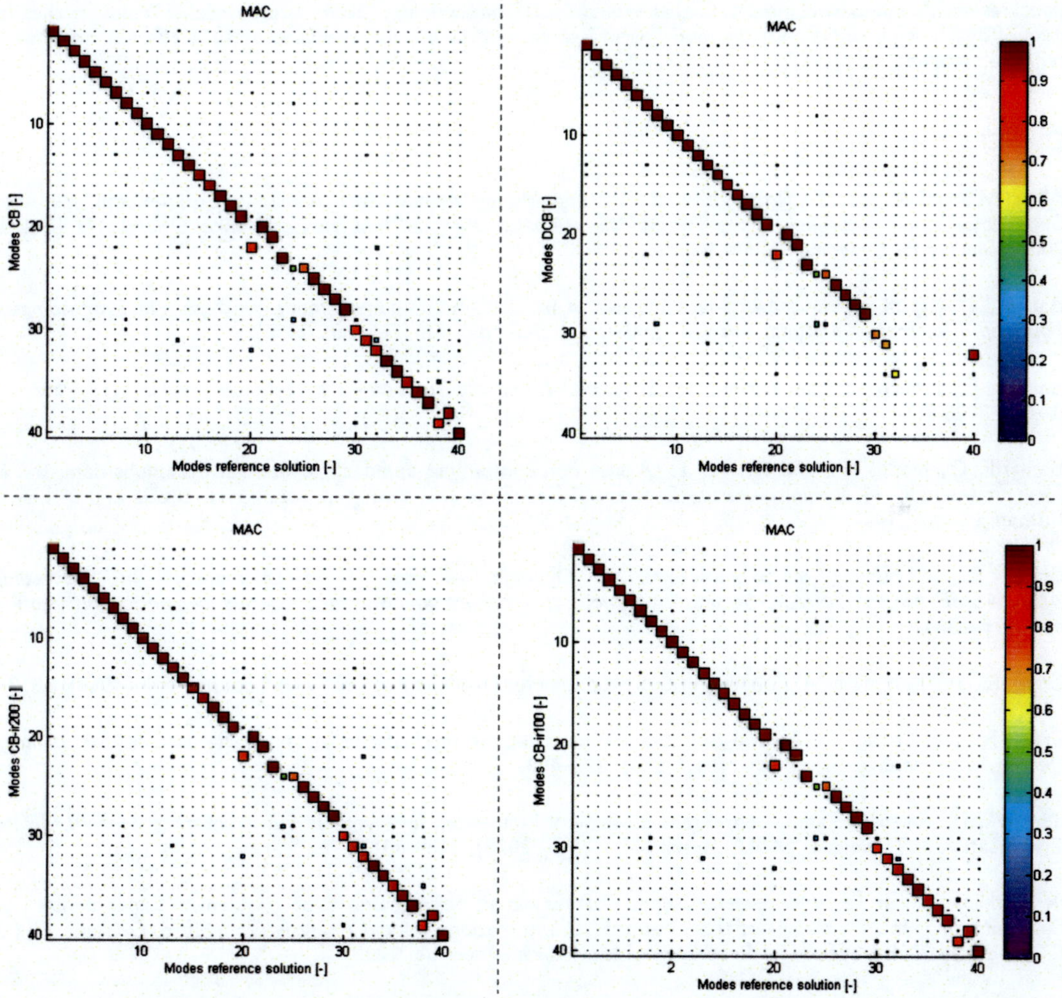

Figure 8: MAC plots of the computed eigenmodes with respect to the eigenmodes of the full system.

5 CONCLUSIONS & RECOMMENDATIONS

In this paper the use of dynamic substructuring (DS) has been proposed as a structural dynamic analysis tool in wind turbine engineering. It is felt that there is a need for such techniques in addition to the commonly used aero-elastic dynamic simulation codes, as wind turbine designs become optimized and local dynamic behavior and its interaction with the global dynamics can no longer be neglected.

This paper outlined a number of practical issues that are encountered when applying a DS analysis to practical industrial sized problems. The most challenging aspect of such a DS analysis is the modeling of the interfaces between substructures. To this end, a number of interface modeling techniques have been outlined, from simple rigid section models to reduction of full interface models with non-conforming meshes. We believe that with the mix of methods and techniques introduced here, tackling real DS problems has become one step closer. To illustrate this, the yaw system of a Siemens 2.3MW wind turbine has been modeled and analysed. Good results were obtained; the accuracy of the reduced models was satisfactory while these models at the same time allow for short computation times when multiple load cases are considered. Moreover, using such reduced models, changing component models and performing a reanalysis is very easy and computationally efficient.

Additional research is however required to further extend the DS methodology. Some future research topics are the interface reduction of dual assembled systems, the modeling of systems with non-linear interfaces and the efficient time integration of assembled systems.

REFERENCES

[1] Allen, M., and Mayes, R. Comparison of FRF and Modal Methods for Combining Experimental and Analytical Substructures. In *Proceedings of the Twentyfifth International Modal Analysis Conference, Orlando, FL* (Bethel, CT, 2007), Society for Experimental Mechanics.

[2] Balmès, E. Use of generalized interface degrees of freedom in component mode synthesis. In *Proceedings of the Fourteenth International Modal Analysis Conference, Dearborn, MI* (February 1996).

[3] Balmès, E., Bianchi, J., and Leclère, J. *Structural Dynamics Toolbox - User's Guide Version 6.1.* Paris, France, August 2008.

[4] Bernardi, C., Maday, Y., and Patera, T. A new non conforming approach to domain decomposition: the Mortar Element Method. In *Nonlinear Partial Differential Equations and their Applications*, H. Brezis and J. Lions, Eds. Pitman, London, 1994.

[5] Craig, R. Coupling of Substructures for Dynamic Analyses – An Overview. In *Proceedings of AIAA/ASME/ASCE/AHS/ASC Structures, Structural Dynamics, and Materials Conference and Exhibit* (April 2000), pp. 1573–1584.

[6] Craig, R., and Bampton, M. Coupling of Substructures for Dynamic Analysis. *AIAA Journal 6*, 7 (1968), 1313–1319.

[7] Craig, R., and Chang, C. A review of substructure coupling methods for dynamic analysis. *NASA. Langley Res. Center Advan. in Eng. Sci. 2*, CR-2781 (1976), 393–408.

[8] de Klerk, D., Rixen, D., and Voormeeren, S. General Framework for Dynamic Substructuring: History, Review and Classification of Techniques. *AIAA Journal 46*, 5 (May 2008), 1169–1181.

[9] de Klerk, D., Rixen, D., Voormeeren, S., and Pasteuning, F. Solving the RDoF Problem in Experimental Dynamic Substructuring. In *Proceedings of the Twenty Sixth International Modal Analysis Conference, Orlando, FL* (Bethel, CT, February 2008), Society for Experimental Mechanics. Paper no. 129.

[10] Duncan, W. The admittance method for obtaining the natural frequencies of systems. *Philosophical Magazine 32* (1941), 401–409.

[11] Guyan, R. Reduction of Stiffness and Mass Matrices. *AIAA Journal 3* (February 1965), 380.

[12] ISO. *ISO 6336-1: Calculation of load capacity of spur and helical gears – Part 1: Basic principles, introduction and general influence factors*, second ed. Geneva, Switzerland, 2006.

[13] Junge, M., Brunner, D., Becker, J., and Gaul, L. Interface-reduction for the CraigBampton and Rubin method applied to FEBE coupling with a large fluidstructure interface. *International Journal for Numerical Methods in Engineering 77* (2009), 1731–1752.

[14] Kubur, M., Kahraman, A., Zini, D. M., and Kienzle, K. Dynamic Analysis of a Multi-Shaft Helical Gear Transmission by Finite Elements: Model and Experiment. *Journal of Vibrations and Acoustics 126* (July 2004), 398–406.

[15] MacNeal, R. A Hybrid Method of Component Mode Synthesis. *Computers & Structures 1*, 4 (1971), 581–601.

[16] Mayes, R., and Stasiunas, E. Combining Lightly Damped Experimental Substructures with Analytical Substructures. In *Proceedings of the Twentyfifth International Modal Analysis Conference, Orlando, FL* (Bethel, CT, 2007), Society for Experimental Mechanics.

[17] O'Callahan, J., Avitabile, P., and Riemer, R. System Equivelant Reduction Expansion Process (SEREP). In *Proceedings of the Seven International Modal Analysis Conference, Las Vegas, NV* (Bethel, CT, February 1989), Society for Experimental Mechanics, pp. 29–37.

[18] Passon, P., Kühn, M., Butterfield, S., Jonkman, J., Camp, T., and Larsen, T. OC3-Benchmark Exercise of Aero-elastic Offshore Wind Turbine Codes. *Journal of Physics: Conference Series 75* (2007), 1–12.

[19] Peeters, J., Vandepitte, D., and Sas, P. Analysis of internal drive train dynamics in a wind turbine. *Wind Energy 9* (2006), 141–161.

[20] Rixen, D. J. *Substructuring and Dual Methods in Structural Analysis.* PhD thesis, Université de Liège, 1997.

[21] Rixen, D. J. A dual Craig-Bampton method for dynamic substructuring. *Journal of Computational and Applied Mathematics 168* (2004), 383–391.

[22] Rubin, S. Improved Component-Mode Representation for Structural Dynamic Analysis. *AIAA Journal 13* (1975), 995–1006.

[23] Sofrin, T. The combination of dynamical systems. *Journal of the Aeronautical Sciences 13*, 6 (1946), 281–288.

[24] Tran, D.-M. Component Mode Synthesis Methods Using Interface Modes: Application to Structures with Cyclic Symmetry. *Computers & Structures 79* (2001), 209–222.

[25] Witteveen, W., and Irschik, H. Efficient modal formulation for vibration analysis of solid structures with bolted joints. In *Proceedings of the Twenty Fifth International Modal Analysis Conference, Orlando, FL* (Bethel, CT, February 2007), Society for Experimental Mechanics.

Appendix: Construction of Boolean Matrices

This appendix illustrates the construction of the Boolean matrices $\boldsymbol{B}$ and $\boldsymbol{L}$. To this end, the general system shown in figure 9 is considered: this figure schematically shows the coupling of two general substructures. Both substructures consist of 3 nodes; substructure A has 4 degrees of freedom while substructure B holds 5 DOF.

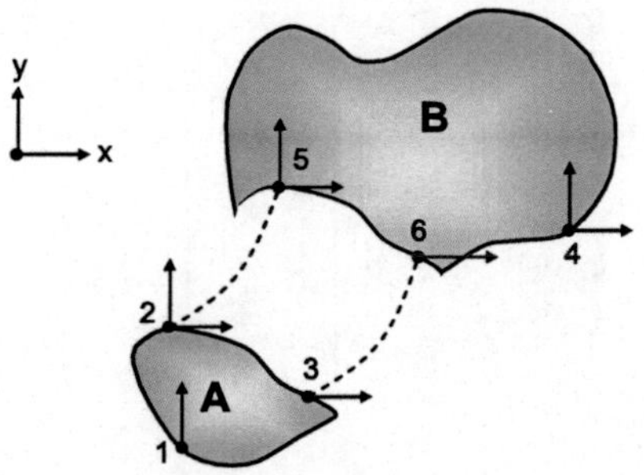

Figure 9: *Coupling of two general substructures*

In this example, nodes 2 and 3 of substructure A are coupled to nodes 5 and 6 of substructure B, respectively. So, three compatibility conditions should be satisfied:

$$\begin{cases} u_{2x} = u_{5x} \\ u_{2y} = u_{5y} \\ u_{3x} = u_{6x} \end{cases} \tag{40}$$

To express this condition as in equation 3, i.e. $\boldsymbol{Bu} = \boldsymbol{0}$, the signed Boolean matrix $\boldsymbol{B}$ must be constructed. The total vector of degrees of freedom $\boldsymbol{u}$ is:

$$\boldsymbol{u} = \begin{bmatrix} u_{1y} & u_{2x} & u_{2y} & u_{3x} & u_{4x} & u_{4y} & u_{5x} & u_{5y} & u_{6x} \end{bmatrix}^T$$

The signed Boolean matrix $\boldsymbol{B}$ is now found as:

$$\begin{array}{cc} & \begin{array}{ccccccccc} u_{1y} & u_{2x} & u_{2y} & u_{3x} & u_{4x} & u_{4y} & u_{5x} & u_{5y} & u_{6x} \end{array} \\ \boldsymbol{B} = & \begin{bmatrix} 0 & 1 & 0 & 0 & 0 & 0 & -1 & 0 & 0 \\ 0 & 0 & 1 & 0 & 0 & 0 & 0 & -1 & 0 \\ 0 & 0 & 0 & 1 & 0 & 0 & 0 & 0 & -1 \end{bmatrix} \end{array}$$

Every coupling term, or equivalently, every compatibility condition, corresponds to a line in the Boolean matrix $\boldsymbol{B}$. Therefore, in the general case where the coupled substructures comprise n degrees of freedom of which m are coupled interface DOF, the matrix $\boldsymbol{B}$ has size m-by-n. In this example, $n = 9$ and $m = 3$; the size of $\boldsymbol{B}$ is 3-by-9. It can easily be seen that the condition $\boldsymbol{Bu} = \boldsymbol{0}$ is equivalent to the three compatibility equations in equation (40).

From this signed Boolean matrix, the Boolean localization matrix $\boldsymbol{L}$ is found by computing the nullspace. In this example, this gives:

$$\boldsymbol{L} = \begin{bmatrix} 1 & 0 & 0 & 0 & 0 & 0 \\ 0 & 0 & 0 & 1 & 0 & 0 \\ 0 & 0 & 0 & 0 & 1 & 0 \\ 0 & 0 & 0 & 0 & 0 & 1 \\ 0 & 1 & 0 & 0 & 0 & 0 \\ 0 & 0 & 1 & 0 & 0 & 0 \\ 0 & 0 & 0 & 1 & 0 & 0 \\ 0 & 0 & 0 & 0 & 1 & 0 \\ 0 & 0 & 0 & 0 & 0 & 1 \end{bmatrix}$$

The set of unique interface DOF that is chosen for this example is found as[3]:

$$\boldsymbol{q} = \begin{bmatrix} u_{1y} & u_{4x} & u_{4y} & u_{5x} & u_{5y} & u_{6x} \end{bmatrix}^T$$

Indeed, the Boolean matrix $\boldsymbol{L}$ transforms this unique set of degrees of freedom to the total set of DOF:

$$\boldsymbol{u} = \boldsymbol{Lq} = \begin{bmatrix} u_{1y} \\ u_{5x} = u_{2x} \\ u_{5y} = u_{2y} \\ u_{6x} = u_{3x} \\ u_{4x} \\ u_{4y} \\ u_{5x} \\ u_{5y} \\ u_{6x} \end{bmatrix} = \begin{bmatrix} 1 & 0 & 0 & 0 & 0 & 0 \\ 0 & 0 & 0 & 1 & 0 & 0 \\ 0 & 0 & 0 & 0 & 1 & 0 \\ 0 & 0 & 0 & 0 & 0 & 1 \\ 0 & 1 & 0 & 0 & 0 & 0 \\ 0 & 0 & 1 & 0 & 0 & 0 \\ 0 & 0 & 0 & 1 & 0 & 0 \\ 0 & 0 & 0 & 0 & 1 & 0 \\ 0 & 0 & 0 & 0 & 0 & 1 \end{bmatrix} \begin{bmatrix} u_{1y} \\ u_{4x} \\ u_{4y} \\ u_{5x} \\ u_{5y} \\ u_{6x} \end{bmatrix}$$

[3] The interface DOF of substructure B are retained.

In addition, the Boolean localization matrix $\boldsymbol{L}$ describes the force equilibrium naturally as well:

$$\boldsymbol{L}^T\boldsymbol{g} = \begin{bmatrix} 1 & 0 & 0 & 0 & 0 & 0 & 0 & 0 & 0 \\ 0 & 0 & 0 & 0 & 1 & 0 & 0 & 0 & 0 \\ 0 & 0 & 0 & 0 & 0 & 1 & 0 & 0 & 0 \\ 0 & 1 & 0 & 0 & 0 & 0 & 1 & 0 & 0 \\ 0 & 0 & 1 & 0 & 0 & 0 & 0 & 1 & 0 \\ 0 & 0 & 0 & 1 & 0 & 0 & 0 & 0 & 1 \end{bmatrix} \begin{bmatrix} 0 \\ g_{2x} \\ g_{2y} \\ g_{3x} \\ 0 \\ 0 \\ g_{5x} \\ g_{5y} \\ g_{6x} \end{bmatrix} = \begin{bmatrix} 0 \\ 0 \\ 0 \\ g_{2x} + g_{5x} \\ g_{2y} + g_{5y} \\ g_{3x} + g_{6x} \end{bmatrix} = \boldsymbol{0}$$

In order to satisfy the equilibrium condition, the connection forces on dual degrees of freedom must thus sum to zero.

Proceedings of the IMAC-XXVIII
February 1–4, 2010, Jacksonville, Florida USA
©2010 Society for Experimental Mechanics Inc.

Developments in Large Wind Turbine Modal Analysis Using Point Tracking Videogrammetry

Uwe Schmidt Paulsen, National Laboratory for Sustainable Energy, Risoe DTU, Roskilde, Denmark
Tim Schmidt, Trilion Quality Systems, 500 Davis Drive, Suite 200, Plymouth Meeting, PA 19462
Oliver Erne, GOM GmbH, Mittelweg, Braunschweig, Germany

ABSTRACT

Wind turbine installations are increasing significantly in numbers as well as capacity as part of the global demand for sustainable energy sources. This leads to stronger need for comprehensive dynamic test data to support reliable cost reductions while meeting safety requirements. The complete wind turbine structural mechanical response at different load conditions is of particular interest. Detailed analysis of the component operational resonances during loading, such as the interaction of the blade - and tower deformations is of basic engineering importance. A testing of the structure with traditional equipment would be quite extensive in sensor instrumentation, calibration and analysis effort.

Dynamic field measurements were performed on a 500 kW Nordtank wind turbine at the Risoe DTU campus in Denmark using a customized Pontos point tracking photogrammetry system. Data was acquired at 100 Hz for 7 seconds from more than 50 targets within a 50 meter wide field of view. Practical aspects of the measurement setup including illumination requirements, blade and tower targeting, the ability to automatically track numerous targets through multiple rotations, rigid body correction (de-rotation), and the use of projected angles and trajectory analysis are discussed.

Results include movies with animated vectors and associated time history plots for 3D directional and resultant displacements of all blades and the support tower, as well as trajectory plots.

INTRODUCTION

Wind turbines are increasing significantly in numbers as well as capacity as part of the global demand for sustainable energy sources. Many commercial wind turbines have a rotor diameter of 80 meters and more, with a five-year goal to achieve 120 meter blade lengths (with future generation blades-FGB), and therefore 240 meter diameter rotors for a 20 MW wind turbine. This leads to stronger need for comprehensive dynamic test data to support reliable cost reductions while meeting safety requirements. The complete wind turbine structure (mainly rotor and tower) structural mechanical response is of interest for prototype testing and certification.

Large wind turbines present significant challenges for measurements. Their height precludes a stable vantage point normal to the rotor; measurements must be conducted off-axis, and preferably from ground level. Any rotating component, regardless of size, is more difficult to measure than a non-rotating one. Loads and stresses on rotor and tower are measured with strain gauges, and wind turbine electrical/mechanical power output and load inputs like wind speed is measured as recommended for certification [1]. Typical load validations rely on about 60 strain gauges and other signals, totally about 100-150 channels are not uncommon for documentation. Strain gauges with wireless data transfer, typically mounted as shown in Figure 1, allow some degree of condition monitoring, but give and will not without intelligent data processing provide operating deflection shapes. The installation and preparatory work for such calibrated signals, data transfer and analysis are laborious and time consuming, and hence prototype measurements are expensive. Intensive measurements on a component such as the rotor blade prior to full-scale testing are made as in Figure 2, but dependent on the clamping conditions.

Scanning laser vibrometers are often used for modal analysis, and manufacturers such as Polytec have introduced de-rotator systems for use on tires, consumer and automotive fans, and small turbines. However, this setup requires an axial view of the test object, and the time required for sequential data acquisition precludes use on certain transient events.

T. Proulx (ed.), *Structural Dynamics and Renewable Energy, Volume 1*, Conference Proceedings of the Society for Experimental Mechanics Series 10, DOI 10.1007/978-1-4419-9716-6_17, © The Society for Experimental Mechanics, Inc. 2011

The state of the art of experimental modal testing of structures is that analysis has advanced from signal- or system analysis techniques towards operational modal analysis. Analysis approaches such as physical modeling (mainly FEM) and system identification techniques such as non-parametric, e.g. frequency peak picking, frequency domain decomposition, and parametric, e.g. ARMA, correlative function estimation, state space identification) are elements of modal testing [2-5]. In principal, a component such as a 19 m wind turbine rotor blade can be modally analyzed by using acoustic or force excitation [6-7], and later visualized with software tools such as ME'scope, accounting for noise input and robust estimation of modal frequencies, damping and mode shapes.

However, there is still the entire multi-body wind turbine to consider under operational conditions, which will include both symmetric and nonsymmetrical mode shapes. The structure is assembled from numerous flexible components, which makes the complete vibration picture complex and a testing of the structure quite extensive in sensor instrumentation and analysis effort. System responses in operation may be quite different than what is expected. A system analysis approach can provide answers in particular low frequency modes of the vibration components. The technique of using the stroboscopic light effect on camshafts triggered the idea to use this on wind turbines. Operating deflection shapes on rotating shafts have been analyzed [8], and ways were identified to carry out similar measurements on the complete wind turbine structure when under normal operation. Laser vibrometry [9] has been used on rotating shafts for torsional deflection mode shape analysis.

In the absence of measured damping-, mass- and stiffness matrices, we apply the mathematical notation that the structure's vibration response Z is a result of a mechanical load with associated displacement X: $Z=H \cdot X$ [10]. With measured 3D coordinate **y** on a wind turbine component, displacements of points relative to a reference point **x** could be used for applying modal analysis techniques [11-12], including transmissibility functions, operating deflection shapes, parameter estimation of damping and resonances, and mode shapes. For the instrumented blade shown in Figure 1, the origin is regarded as a reference position with respect to other sensor positions on the blade. The blade is interpreted as a lumped system with degree of freedom (DOF) elements, each represented as a SDOF system. With the reference position moving in space, the response functions for the vector displacement **z=y-x** typically resembles those in Figure 2. For the wind turbine under normal (non-transient) conditions, transmissibility functions contain responses excited by Gaussian stochastic wind.

There are also guidelines for sensor positioning in terms of uncertainty for the mode shape calculation, again referring to Figure 1 on the relative variation of DOF_{i+2} and DOF_{i+1} [7]. This was also kept in mind for targeting. The distance between two targets measuring the flapwise accelerations should be as large as (practical) possible, and the angle between their measuring axes should be as close as possible to zero.

The wind turbine is instrumented with a variety of sensors on the rotor, nacelle and tower and measurement equipment for 35Hz to 10 kHz sampling. Measurement records are stored for post analysis in a data base [13-14]. Analysis of the data is carried out on the basis of 35 Hz sampled records.

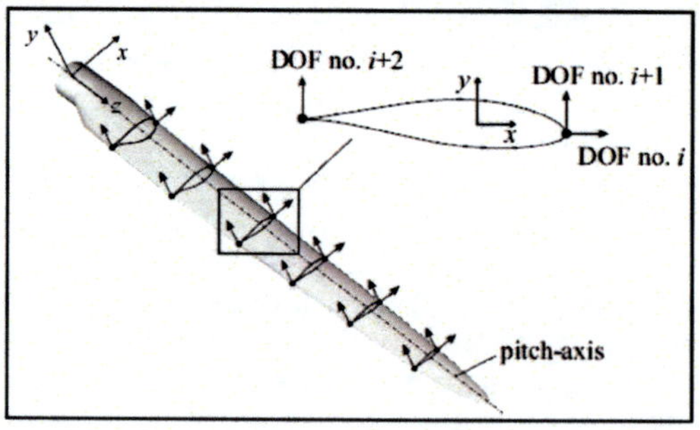

Figure 1: Typical strain gauge locations and degrees of freedom for a wind turbine blade [7].

Figure 2: Strip wise excitation by hammer on a 19 m blade, and response output by accelerometers [7]

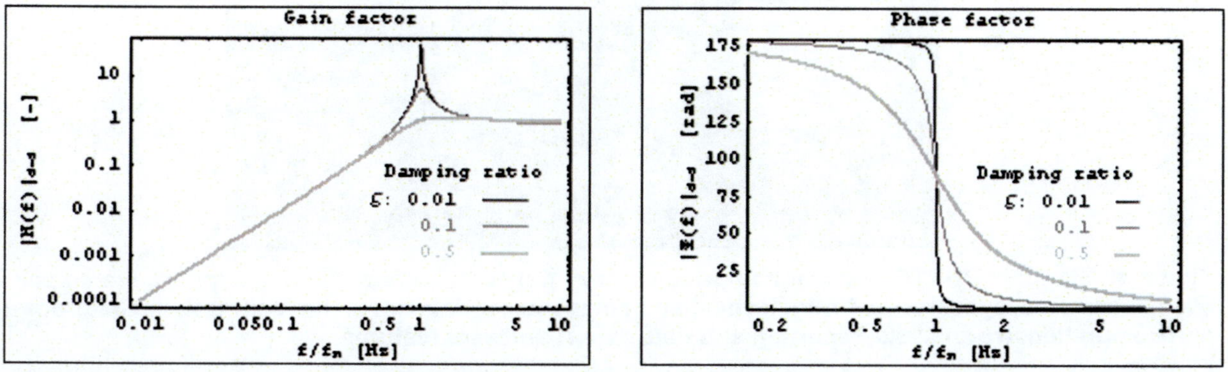

Figure 3: Frequency response functions, SDOF of mechanical system.

Full-field optical techniques, particularly stereo photogrammetry and videogrammetry systems, have some intrinsic features and capability that are extremely advantageous for the present challenge of measuring the operational deflection shapes of huge rotating objects. PONTOS is a 3D point target tracking system developed by GOM. It uses a calibrated pair of cameras to triangulate the x,y,z coordinates of the centers of circular dot targets, which typically have a 10-15 pixel diameter. It is desirable to have an angle between the cameras of about 25 degrees to facilitate accurate triangulation of the z coordinate, although camera angles from 15 to 45 degrees can be used. The practical implication of this is that the distance between the two cameras needs to be between 1/3 - 1/2 of the working distance from the cameras to the test object. A robust ellipse finder algorithm makes a best-fit outline and then the centers of the dot targets can be determined with an accuracy of approximately 1/50 pixel. Therefore, the nominal accuracy of the measurement is 30 microns (0.001") per meter of field of view. The coordinates of each target at a reference condition are subtracted from the current coordinates in each test image to compute the 3D displacements. A large number of experimental data points can be obtained with negligible mass loading.

The commercially available fully integrated PONTOS system comprises a heavy duty carbon fiber camera support, a pair of cameras that can take 500 images per second at full resolution of 1280 x 1024 pixels, synchronized ring flashes for each camera, and a computer for camera and flash control, image processing to determine the target coordinates in each image, and comprehensive deformation analysis, reporting and export software tools. The standard sensor, shown in Figure 3, is normally used for fields of view of 1-3 meters. Common applications include car door slam, engine startup, wind tunnel testing, etc. [15]. Modified calibration methods have been developed to meet the desire for larger field of view capability, including civil structure measurements. Fields of view up to about 15 meters have been addressed, for example to study the deflections of beams, and to track orientation and deceleration of aerospace vehicles during swing and ground impact tests [16]. Using the same hardware with different surface preparation and full-field pattern matching (ARAMIS 3D image correlation), thousands of data points are obtained, comparable to the number of nodes in an FEA model, allowing for extensive correlation between experimental and analytic data. ARAMIS has been applied for modal analysis of small rotating devices such as fans, and complex shapes such as dryer panel stampings and a model helicopter [17-18]. Using both high speed cameras and slower cameras with a phase stepping method for harmonic excitation, detailed comparisons have been made to impact hammer and laser vibrometry results, including assessment of modal assurance criteria (MAC) for flexible modes. Because of the potential advantages, the use of stereo optical techniques is being contemplated for modal analysis of passenger aircraft during ground vibration tests (GVT) [20].

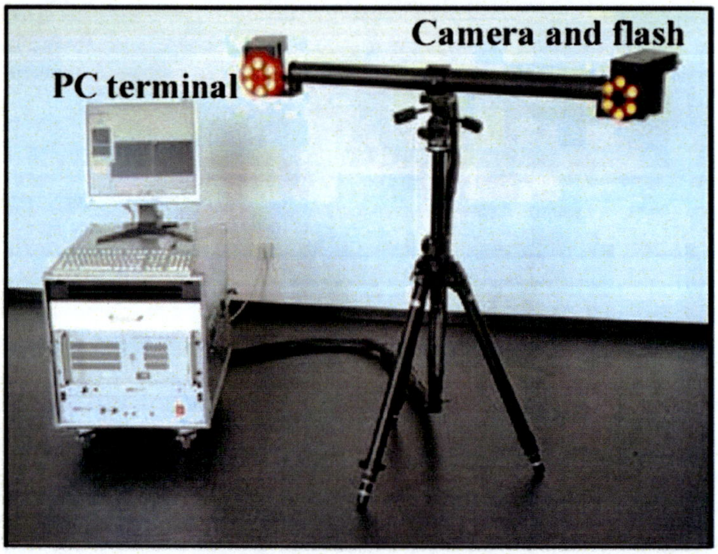

Figure 4: Standard PONTOS sensor comprising heavy-duty carbon fiber camera support, stereo pair of 500 fps cameras, synchronized ring flashes, and computer for data acquisition control, image processing, and comprehensive analysis, reporting and data export software features.

EXPERIMENTAL CONSIDERATIONS AND SETUP

The goal of this project was to apply the PONTOS system on an operating full-scale wind turbine for the first time ever, which involved some significant experimental challenges. The PONTOS field setup is shown in Figure 5. The cameras were placed on tripods approximately 55 meters away from each other, and approximately 110 meters away from the tower of the wind turbine, resulting in a camera angle of 20-25 degrees. High power strobes synchronized to the cameras were used to illuminate retro reflective targets on the blades and tower. The targets had a diameter of 0.3 meters in order to meet the 10 pixel diameter requirement for accurate determination of the center point coordinates. Testing indicated that targets wetted by rain could still be successfully identified.

The targets, shown in Figure 5, were positioned with certain criteria in mind. Lines consisting of at least three markers were established to define pitch deflection axes for each blade. The minimum spacing between targets was two to three times the diameter, in order to ensure that target identification would be consistent from frame to frame despite multiple rotations. The brightness of the retro reflective targets enhanced the contrast to the blade backgrounds, so a high contrast surround was not required. A high density of targets was applied to ensure that no spatial aliasing of modes would occur. Based on prior knowledge and modeling of blade deflections, spacing between targets were kept to a minimum near the tips, and increased somewhat towards the roots.

Relative camera positions and lens distortions were determined using the targeted rotors, similar to the use of standard calibration crosses for smaller fields of view.

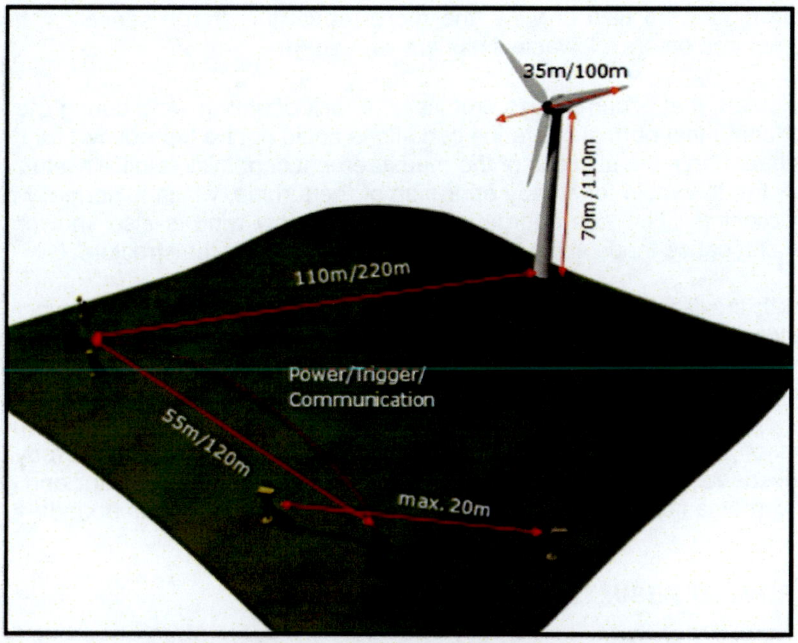

Figure 5: Schematic of the PONTOS field setup for measurement of a large wind turbine.

Figure 6: View of the fully targeted rotor and tower during flash testing. It can be seen that the high power flash has successfully illuminated all of the retro reflective targets.

The tower targets are momentarily occluded each time a blade passes in front of it, but these dropouts are of very short duration compared to the complete rotation time, and are therefore acceptable.

The camera tripods and also the flashes and computers had to be placed on ploughed land, which was far from the ideal of stable concrete pads. There were changes in soil softness between setup and testing due to rain and frost. The primary concern was that relative shifts between the two cameras could cause de-calibration and reduction in accuracy, particularly for the coordinates representing out-of-plane motion. The PONTOS software monitors the intersection deviation for every target in every image, and it was found that de-calibration did not occur. In general, the bigger the field of view, the more robust the stereo sensor is, because a much larger relative movement is required before allowable limits are exceeded.

The major practical problem that occurred was, ironically, a lack of strong wind during the available test periods. It was extremely unfortunate that normal operating conditions could not be reproduced for detailed study, although this certainly did not detract from the success of the measurement capability feasibility study. The low and shifting wind speed was below the threshold for steady operation of the turbine, which is normally required for application of modal analysis processing. The less energetic loading from low wind is also known to affect the operating deflection shapes and can cause mode shifts due to the light damping of the structure.

In order to proceed with the feasibility study and to acquire some useful data, the asynchronous generator was forced to act as a motor, thereby increasing the rotor speed. The motor was then idled at various speeds, while the available wind acted on the blades. In addition, an emergency stop simulating grid loss such as from a lightning strike or downed transmission cable was performed starting from a condition near normal operating speed. The 500 kW Nordtank turbine used for this test is equipped with a fail-safe brake mechanism whereby the outermost 1.7 meters of the blades will be released and automatically act as aerodynamic brakes, quickly reducing the rotational speed. In addition, the dynamic equilibrium between the rotor and gearbox and generator is lost, so the rotor becomes a free-rotating mass, which is a significant change in boundary conditions.

TRADITIONAL INSTRUMENT RESULTS

Figure 7 shows a comparison of measured spectral load responses (PSD) from strain gauges indicating the edgewise blade bending, rotor shaft torsion and tower bending near the base.

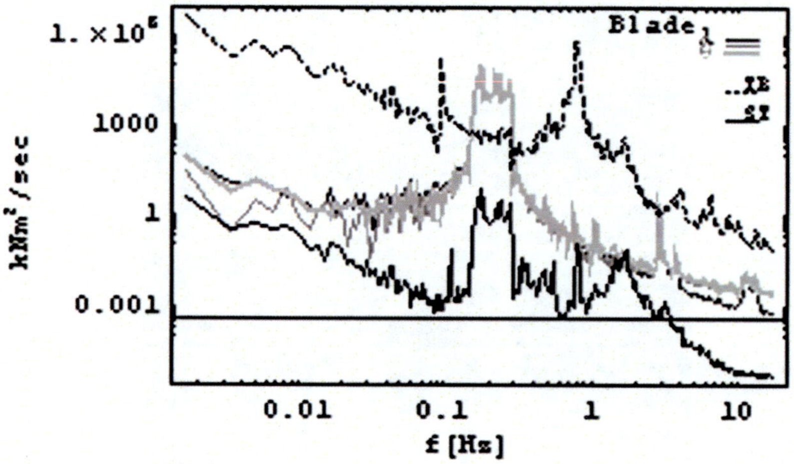

Figure 7: Power spectral density (PSD) of edgewise bending moments, tower base bending moment (TB) and rotor shaft torsion (ST).

Here at this atypical operating condition, the turbine is rotating slowly at a speed of 0.17-0.29 Hz, demonstrating the edgewise blade vibration and in the rotor shaft. Furthermore the blade masses rotating with the rotor plane excite the tower natural frequency at about 0.8 Hz, and the first symmetrical mode of the edgewise blade frequency at around 3.0 Hz.

The wind power spectrum is a straight sloping line following the main trend of the signals in the Figure 7. A similar explanation on the influence of this ambient excitation is applicable for Figure 8, which also reveals a similar result as in Figure 7 on the coupled vibrations in the structure, with a major flapwise motion correlated to the tower bending response and rotor shaft bending.

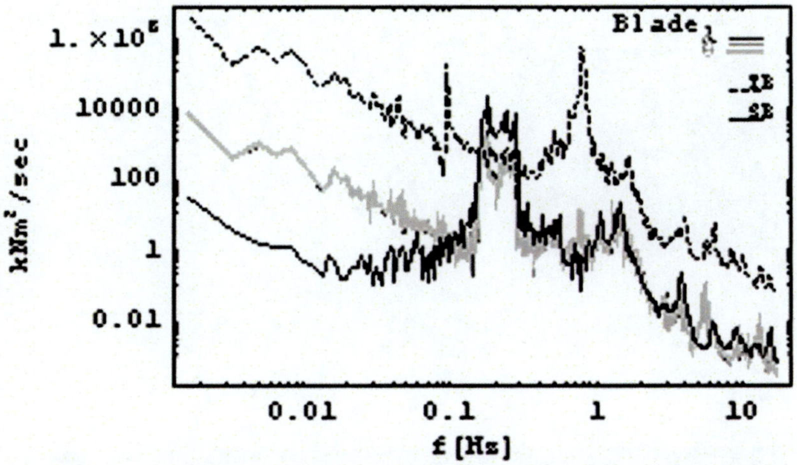

Figure 8: PSD of flapwise bending moments, tower base bending moment (TB) and rotor shaft bending(SB).

The flapwise deflection of a cantilevered 19 m blade has been measured and computed [7] and consists of a relative measure of the deflection shape along the blade for the different modes. The operational deflection measured here is by magnitude superimposed with the wind loading and the mode shapes at this operating condition. To decouple these effects, detailed analysis of the test data are planned for the next measurement, which will involve stronger wind forces.

PONTOS RESULTS

The initial PONTOS data shows all motion relative to the cameras, with a default coordinate system with z normal to the cameras. A coordinate system transformation is applied to set the global coordinate system relative to the turbine. In this test, the x-axis was set normal to the rotor, such that the positive x-axis direction is opposite to the wind direction. The positive z direction represents motion upwards, and the positive y direction is movement sideways towards the right. The data can now be used to quantify the rigid body motion, for example to measure tip velocities. For modal analysis, it is also necessary to perform a rigid body motion correction, to subtract out the rotational motion of the rotor as a whole, leaving only relative deformations. This is done using points on each blade closest to the rotor. A best-fit plane among the chosen points is created and mapped back to itself using all six degrees of freedom, separately in each measurement stage. The 3D coordinates of all points are updated with the transformation results, leaving only the blade deformations. The tower point deformations are unaffected by this local transformation.

The first symmetrical flapwise blade frequency mode is found at 1.5 Hz from the traditional instrument PSD. In comparison, the optical displacement offers the ability to investigate the vibrations which occur in the structure, as seen in Figure 9. It can be seen that the response is primarily 1 Hz sinusoidal cycles from out-of-plane deflections correlated and synchronized with the first bending mode of the main shaft are also observed from the spectral load response results. Relative phase information can also be observed through the presented comparison of deflections at the indicated radial position on the three different blades. The three blades are nominally in phase, though there are some notable differences in peak to peak amplitude, and several harmonics are present. The much higher frequency content is noise from the finite accuracy of the 3D coordinates; note that its magnitude is acceptably low compared to the actual structural response.

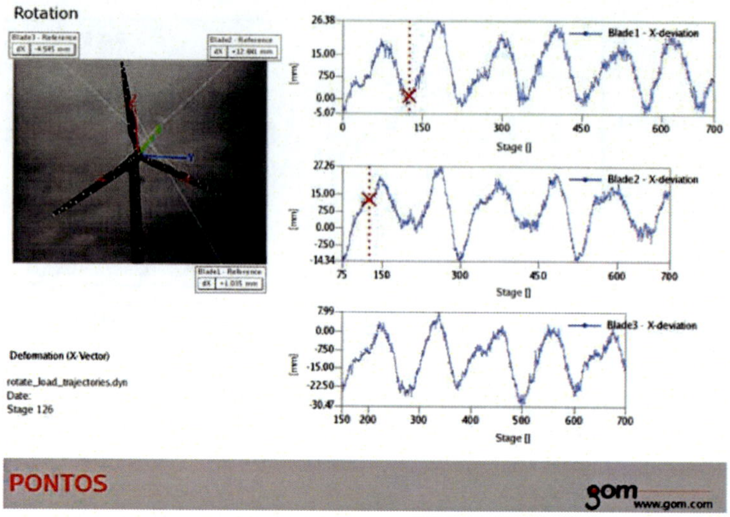

Figure 9: Blade deflections in the windward direction.

Figure 10 is an example of the usefulness of the initial PONTOS data, before movement correction is applied, quantifying the beginning of the emergency stop, in which the turbine is slowed down from 27 rpm to standstill. The decrease in rotational frequency is evident in the upper graph. The middle graph is tip velocity, which can be seen to decrease rapidly, leading to significant lateral tower deformations that are visible in the lower graph. The blade passing frequency is clearly dominating the tower response.

Figure 11 uses a combination of the original and rigid-body corrected data to further explore the effects of the beginning of the emergency stop. Tip velocity is again shown for reference. Note the higher order harmonics appearing and increasing in amplitude.

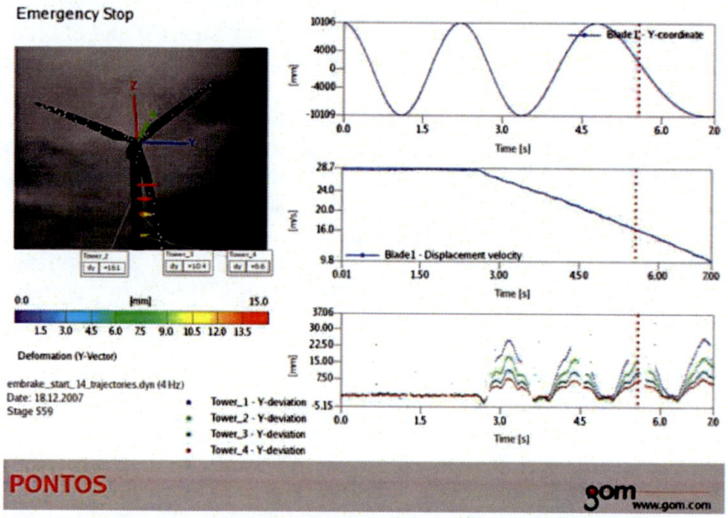

Figure 10: PONTOS data without movement correction quantifies the beginning of the emergency stop. The middle graph is tip velocity, which can be seen to decrease rapidly, leading to significant tower deformations. The decrease in rotational frequency is evident in the upper graph.

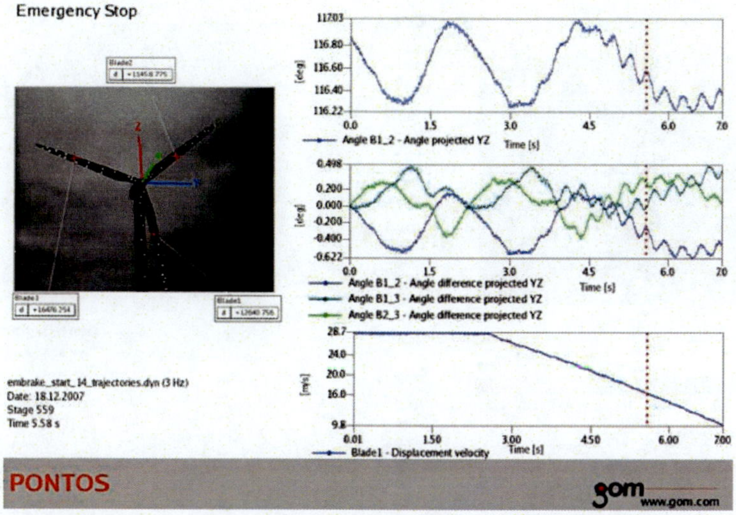

Figure 11: Phase angle analysis provides further insight into the beginning of the emergency stop. Tip velocity is again shown for reference. Note the higher frequency response appearing and increasing in amplitude.

Figure 12 is a comparison between the tower and blade deformations, showing the relative amplitudes and phases. When the two blades are at the upside position, as indicated on the image, they bend downwards, causing an elongation of the distance between the markers. At the opposite position the situation has reversed to a shortening of the distance between the markers. Knowledge of these magnitudes (~25 mm in this case) and the bending shape can be used for stress analysis. At the end of the stop, the rotor is pitching back and forth due to backlash in the gearbox. This motion was analyzed by monitoring the distance between the markers on two blades, thus providing insight into the coupling of the multi-body system. Further post-processing is planned to better understand these interactions.

Figure 13 shows an example of 3D resultant displacement vector plots for similar points on all three blades. These can be animated during review of the captured data, and are part of the standard PONTOS analysis tools. Here it can be seen that the phases are very orderly, with some amplitude variations.

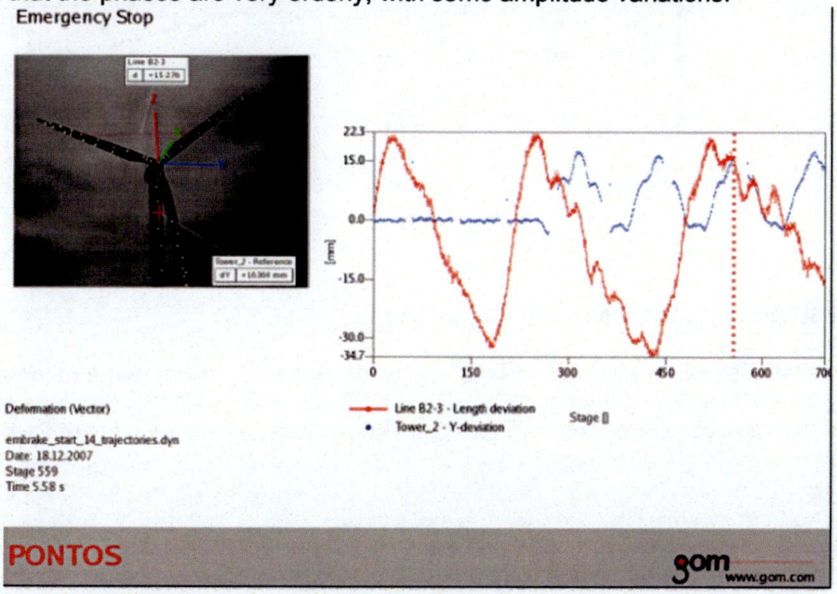

Figure 12: Tower versus blade deflection during an emergency stop.

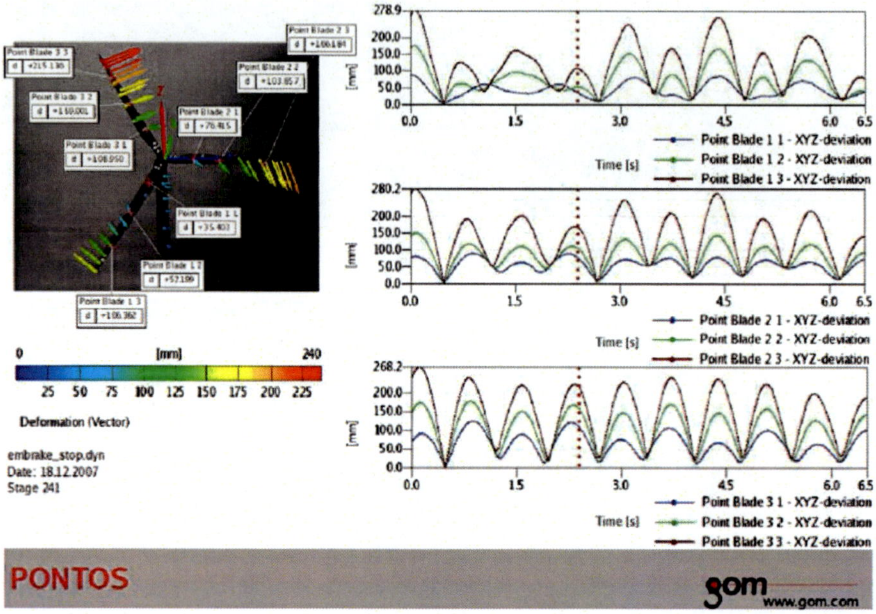

Figure 13: Vector plots that can be animated during review of captured data show 3D resultant displacements for three similar points on each blade. The phases are very orderly, with some amplitude variations.

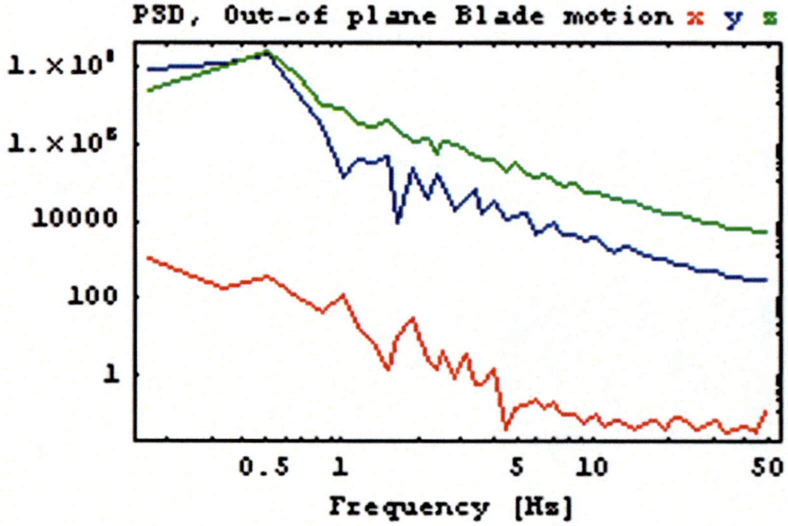

Figure 14: PSD of 3D motion analysis of the blade tips

Figure 14 shows an example of 3D motion analysis of the blade tips in the out of plane (flapwise) mode. A comparison with Figure 8 shows this motion is real in the 0.8 - 2 Hz range, and that the low spectral resolution is plausible because of the high sampling. This can be improved by performing with a lower frame rate.

CONCLUSIONS

The operating deflection shapes of a 41 meter diameter wind turbine were successfully measured during motoring representing normal operation, and also during an emergency stop. This had never been done before at this size scale, and the calibration, measurement and analysis techniques that were developed for the PONTOS full-field optical system are applicable for even larger turbines, up to at least 100 meters. PONTOS provides unique capability for measurements on extremely large objects and rotating components, with sufficient bandwidth and accuracy to capture and quantify 3D displacements, bending shapes, blade torsion, tower deformation, etc. The PONTOS data compared well with traditional data acquisition in terms of frequency response, but provided dynamic deformation data at numerous locations along the length and chord of all blades as well as the tower. The success of these measurements is extremely encouraging regarding future modal analysis efforts on even bigger turbines as well as other large civil structures.

ACKNOWLEDGMENTS

Jeppe Herbsgaard Laursen, Zebicon A/S is greatly appreciated for bringing people from GOM and RISØ DTU together, and for his valuable, practical and encouraging support and input into success.
GOM is gratefully acknowledged for support, to participate in the experiment and assisting with analysis, in particular Dirk Behring, Oliver Erne, Markus Klein, Gunther Sanow and Theodor Moeller.

The technical staff at RISØ DTU campus (BAS), especially Anders B. Møller and Oluf Høst is acknowledged for their technical assistance before, during and after the tests. Thanks to Kirsten A. Frydensberg for valuable help with necessary approvals concerning air traffic safety and car traffic issues.

Finally, the wind energy division, test and measurement program at RISO DTU is gratefully acknowledged for financial support.

REFERENCES

1) IEC 61400-13 Wind Turbine Generator Systems - Part 13: Measurement of Mechanical Loads
2) Andersen, P., Brincker, R., The Stochastic Subspace Identification Techniques.www.svibs.com.
3) Overschee, E. and de Moor, P. Subspace Identification for Linear Systems: Theory, Implementation, Applications. Kluwer Academic Publishers, 1996.
4) Ventura C. E. and Gade, S., IOMAC Pre-Course. IOMAC 2005. Copenhagen, Denmark, April 2005.
5) Andersen, P. Identification of Civil Engineering Structures Using ARMA Models, Ph.D. Thesis, Aalborg University, Denmark, 1997.
6) Gade, S., Modal Parameters from a Wind Turbine Wing by Operational Modal Analysis. Internoise, Korea, 2003.
7) Larsen, G.C. et al. Modal Analysis of Wind Turbine Blades. Risø-R-1181(EN), 2002.
8) Døssing, Ole, Structural Stroboscopy -Measurement of Operational Deflection Shapes, Brüel & Kjær Application Note (BO 0212).
9) Gatzwiller, K. B&K Application Note, Measuring Torsional Operational Deflection Shapes of Rotating Shafts. BO 0402. www.BKSV.com.
10) Bendat, J. et al. Random Data: Analysis and Measurement Procedures. Wiley – Interscience, 1971.
11) McHargue, P. L., Richardson, M.H. Operating Deflection Shapes from Time versus Frequency Domain Measurements. 11th IMAC Conference Florida USA, 1993.
12) Richardson, M.H. Is It a Mode Shape or an Operating Deflection Shape? Sound & Vibration, 1997.
13) Helgesen, K.O.et al, Wind Turbine Measurement Technique-an Open Laboratory for Educational Purposes, Wind Energy 11 pp.281-295 Wiley Interscience, 2006.
14) Paulsen. S., Preliminary Results with A Novel Drive Train Measurement System, EWEC2008, Brussels, Belgium, 2007.
15) Erne, O, Friebe, H., and Galanulis, K., Is it possible to replace conventional displacement and acceleration sensor technology? Solution methods using optical 3D measuring technology, GOM white paper, 2007.
16) Schmidt, T., Tyson, J., Some Common and Not So Common Applications of 3D Image Correlation, Proceedings of the GOM International User Meeting, Braunschweig, Germany, Sept 25, 2007.
17) Helfrick, M., Niezrecki, C., and Avitabile, P., "Optical Non-contacting Vibration Measurement of Rotating Turbine Blades," Proceedings of IMAC-XXVII, Orlando, FL, February 2009.
18) Helfrick, M., Niezrecki, C., Avitabile, P., and Schmidt, T., "3D Digital Image Correlation Methods for Full-Field Vibration Measurement," Proceedings of IMAC-XXVI, Orlando, FL, February 2008.

19) Niezrecki and Avitable, Digital Image Correlation Applied to Structural Dynamics, Proceedings of IMAC-XXVII, Orlando, FL, February 2009.
20) Pickrel, C.R., A Possible Hybrid Approach for Modal Testing of Aircraft, Proceedings of IMAC-XXVII, Orlando, FL, February 2009.

Proceedings of the IMAC-XXVIII
February 1–4, 2010, Jacksonville, Florida USA

Optimal Design of Magnetostrictive Transducers for Power Harvesting From Vibrations

Viktor Berbyuk

Department of Applied Mechanics

Chalmers University of Technology, SE-412 96, Gothenburg, Sweden

e-mail: viktor.berbyuk@chalmers.se

ABSTRACT

Methodology is proposed for designing of magnetostrictive electric generator having maximal mean power output for a given amount of active material in a transducer and a prescribed vibration excitation. The methodology is based on dimensional analysis of constitutive linear equations of magnetostriction and numerical solution of constrained optimization problem in transducer's dimensionless design parameters space by using Sequential Quadratic Programming algorithm. The methodology has been used to design optimal Terfenol-D based transducer for power harvesting from vibrations. It was shown that for steady state operations there exists possibility to choose only 4 new design parameters being the functions of dimensionless parameters of the transducer. Magnetostrictive strain derivative, Young's modulus and magnetic permeability were determined as functions of magnetic bias and prestress by using experimental data of Terfenol-D. Contour plots and numerical analysis of design parameters show that within the considered concept of magnetostrictive electric generator there exists a set of structural parameters of the transducer that lead to its optimal performance with given amount of active material and prescribed vibration excitation. Examples of solution of optimal design problem demonstrate that for harmonic kinematic excitation with amplitude 0,0002m and frequency 100Hz it is possible to design a magnetostrictive electric generator with 3,2W mean power output having mass of active material 0,01kg.

NOMENCLATURE

ε, σ, H, B – strain, stress, applied magnetic field strength and magnetic flux density in active material,

$\mathbf{y} = (y_1, y_2, ..., y_n)$ -vector of generalized coordinates of transducer's hosting system,

E^H - Young's modulus at constant applied magnetic field strength,

d_{33} - magnetostrictive strain derivative (linear coupling coefficient),

d_{33}^* - magnetomechanical effect,

μ^σ - magnetic permeability at a constant stress,

R - electric load resistance,

R_{coil} , L_{coil} - coil resistance and coil inductance,

C - capacitance of transducer electrical circuit,

σ_0 – mechanical presstress,

H_0- magnetic bias,

a_k ,ω_k , φ - amplitude, frequency and phase angle of harmonic external excitation,

$j = \sqrt{-1}$.

T. Proulx (ed.), *Structural Dynamics and Renewable Energy, Volume 1*, Conference Proceedings of the Society for Experimental Mechanics Series 10, DOI 10.1007/978-1-4419-9716-6_18, © The Society for Experimental Mechanics, Inc. 2011

INTRODUCTION

Recent developments in miniaturized sensors, digital processors, self-powered electronics and wireless communication systems have many desirable applications. The realization of these applications however, is still limited by the lack of a similarly sized power sources. Powering the above mentioned systems can be a significant engineering problem, as traditional solutions such as batteries are not always appropriate. Many proposed power harvesting systems employ a piezoelectric component to convert the mechanical energy to electrical energy. While piezoelectric transducers are now widely used for harvesting energy [1-5] other methods do exist. For example, methods which are based on Villari effect of magnetostrictive materials [6-11]. In developing of power harvesting transducers one of the most important issues is efficiency of energy transduction processes. Several studies were devoted to optimization problems of harvesting devices in different formulations [12-19].

In this paper the methodology is proposed for designing of magnetostrictive electric generator (MEG) having maximal mean power output for a given amount of active material in a transducer and a prescribed vibration excitation. The methodology is based on dimensional analysis of constitutive linear equations of magnetostriction and numerical solution of constrained optimization problem in transducer's dimensionless parameters space. The methodology has been implemented in Matlab environment by using Optimization Toolbox and developed graphical user interface (GUI). Results of exemplary solution of the problem of design of mean power output optimal MEG are presented.

MATHEMATICAL MODEL OF MAGNETOSTRICTIVE ELECTRIC GENERATOR

A characteristic property of magnetostrictive materials is that the mechanical strain will occur if they are subjected to a magnetic field in addition to strain originated from pure applied stresses. Also, their magnetization changes due to changes in applied mechanical stresses in addition to the changes caused by changes of the applied magnetic field. These dependencies can be described by mathematical functions [7]:

$$\varepsilon = \varepsilon[\sigma(y,\dot{y}),H], \qquad B = B[\sigma(y,\dot{y}),H] \tag{1}$$

The most important mode of operating magnetostrictive materials is the longitudinal. Therefore, stresses, strains, and magnetic field quantities are directed parallel to the longitudinal axis. In this case, from (1) the linearized constitutive equations that describe the magnetostriction in active element of magnetostrictive transducer (MT) appear as follows:

$$\varepsilon(\mathbf{y},\dot{\mathbf{y}},\sigma,H) = S^H\sigma(\mathbf{y},\dot{\mathbf{y}}) + d_{33}H, \qquad B(\mathbf{y},\dot{\mathbf{y}},\sigma,H) = d_{33}^{*}\sigma(\mathbf{y},\dot{\mathbf{y}}) + \mu^{\sigma}H \tag{2}$$

where $S^H = \frac{\partial\varepsilon}{\partial\sigma}_{|H=const} = \frac{1}{E^H}$, $d_{33} = \frac{\partial\varepsilon}{\partial H}_{|\sigma=const}$, $d_{33}^{*} = \frac{\partial B}{\partial\sigma}_{|H=const}$, $\mu^{\sigma} = \frac{\partial B}{\partial H}_{|\sigma=const}$.

In equations (2) the stress $\sigma(\mathbf{y},\dot{\mathbf{y}})$, the strain $\varepsilon(\mathbf{y},\dot{\mathbf{y}},\sigma,H)$, and the magnetic flux $B(\mathbf{y},\dot{\mathbf{y}},\sigma,H)$ characterize the interaction between dynamics of hosting system and dynamics of transducer.

The main components of any type of MT are an active material and a solenoid with a coil wrapped around a central object including active material. Solenoid produces a magnetic field when an electric current is passed through it. The magnetic field will change the shape of active material and it gives possibility to use MT as an actuator (*Joule effect*). Opposite, if the active material of a transducer is under the acting of external excitation, then magnetization of the active material varies due to the magnetostrictive effect. The flux variation obtained in the material induces an electromotive force in a coil surrounding the material. This process in magnetostrictive materials is called *Villari effect* and is used in magnetostrictive sensors [20] and power harvesting from vibration [6, 8, 9, 15, 21]. Often, the design of MT comprises permanent magnet creating bias which is helpful for optimising performance characteristics of transducer.

One possible general sketch of MEG looks like the sketch depicted in Fig. 1. Several other possible sketches of MT can be found in [6],[7],[9].

Let the solenoid of MT has the number of the coil turns N_{coil}, the coil length l_{coil} and the coil area A_{coil}. The change of magnetic flux in active material induces a voltage V in a coil surrounding the active material. According to Faraday's – Lent's law the voltage can be expressed as follows:

$$V = -N_{coil} A_{coil} \frac{dB}{dt} \tag{3}$$

This voltage will generate a current, I, in the coil, that according to the Ampèré's law will give rise to an opposing field $H_r = N_{coil} I / l_{coil}$. Using the Ohm's law in the form of $I = V / Z_{eq}$, the opposing magnetic field H_r can be written in the following way:

$$H_r = \frac{N_{coil}}{l_{coil}} \frac{V}{Z_{eq}} \tag{4}$$

In equation (4) Z_{eq} is the transducer electrical circuit total impedance $Z_{eq} = R + R_{coil} + j(\omega L_{coil} - \frac{1}{\omega C})$.

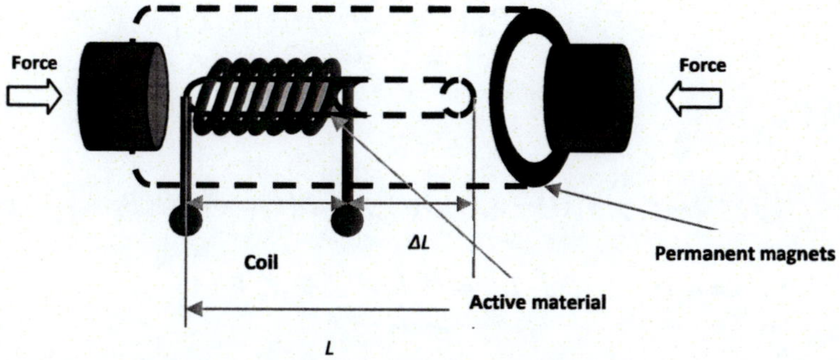

Fig. 1 Sketch of a magnetostrictive transducer

If the MT design includes permanent magnets (see Fig. 1), then the total magnetic field intensity, H, can be written as sum of a constant quantity, namely the magnetic bias, H_0, and a varying quantity, H_r, that depends on the magnetostriction process in the generator, i.e. $H = H_0 + H_r$.

Using the equations (3) and (4), the total magnetic field intensity is determined by:

$$H = H_0 - \frac{1}{b} \frac{dB}{dt}, \qquad b = \frac{Z_{eq} l_{coil}}{N_{coil}^2 A_{coil}} \tag{5}$$

Equations (2)-(5) constitute the mathematical model of MEG and describe internal electro-magneto-elastic transduction processes in the transducer.

POWER OUTPUT OF DISPLACEMENT DRIVEN MEG

Consider the case when interaction between the hosting system and the MEG is resulting in strain $\varepsilon(t)$ in active material of transducer which can be calculated or measured for a given motion of hosting system. Below it's assumed that

$$\varepsilon = \varepsilon_u(\mathbf{y}, \dot{\mathbf{y}}, \mathbf{C}, t), \qquad t \in [t_0, t_1] \tag{6}$$

where the function $\varepsilon_u(\mathbf{y}, \dot{\mathbf{y}}, \mathbf{C}, t)$ is determined by given electro-magnetic design of MEG, the design of adaptive structure that needed to integrate the transducer into hosting system, and by the motion of hosting system, $\mathbf{C}$ is the vector of design parameters of MEG.

In the paper it will be assumed that the parameters $E^H, d_{33}, d_{33}^*, \mu^\sigma$ of active material of the MEG are constant during the transducer operation.

Problem 1. Let the strain in active material of transducer, i.e. the function (6), is known and the structural parameters of the design of MEG, the transducer electrical circuit total impedance, Z_{eq}, and the magnetic field intensity at initial instance of time, $H_0 = H(0)$, are given. It is required to determine the electrical power output, $P_{out}(t)$, of the transducer.

It is possible to show [10] that by using (2)-(5) and (6), the solution to the *Problem 1* is determined by the following formulae:

$$P_{out}(t) = \frac{R l_{coil}^2}{N_{coil}^2}[H(t) - H_0]^2 \tag{7}$$

Here in equation (7):

$$\begin{aligned} &H(t) = H_0 - \frac{1}{b}[f_\varepsilon(t) - a_\varepsilon(t)B(t)], \qquad B(t) = e^{-A_\varepsilon(t)}\left(B_0 + \int_0^t f_\varepsilon(\tau)e^{A_\varepsilon(\tau)}d\tau\right), \\ &f_\varepsilon(t) = a_\varepsilon(t)\left[\varepsilon_u(t)d_{33}E^H + H_0\left(\mu^\sigma - d_{33}d_{33}^*E^H\right)\right], \\ &a_\varepsilon(t) = \frac{b}{\mu^\sigma - d_{33}d_{33}^*E^H}, \qquad A_\varepsilon(t) = \int_0^t a_\varepsilon(\tau)d\tau, \qquad t \in [0, T] \end{aligned} \tag{8}$$

Then the mean power output is calculated by the following expression:

$$\overline{P}_{out} = \frac{1}{T}\int_0^T P_{out}(t)\,dt \tag{9}$$

As soon as controlled motion $\mathbf{y} = (y_1, y_2, \ldots, y_n)$ of the hosting system is predefined or calculated and the generated strain in active material of the transducer, i.e. the function (6) is known, the formulae (7)- (9) constitute algorithm for determining the mean power output of MEG incorporated into hosting system.

Some results of numerical simulation of performance of displacement driven MEG as well as numerical analysis of mean power output can be found in [10], [22]. Physical prototype of MEG (see Fig. 2) was built at the Vibrations and Smart Structures Lab of the Department of Applied Mechanics, Chalmers University of Technology.

In Fig. 3. the transducer is shown being incorporated into the test rig (hosting system) comprising amplifier, oscilloscope (HP, 4 channel 100 MHz), data acquisition unit and frequency converter. The test rig comes to the MEG with high frequency vibration excitation (up to 1000 Hz) via a cam coupled to an electric motor. A video showing the MEG in its real-time operation as power harvesting device can be found via web link (~15MB)

http://www.mvs.chalmers.se/~berbyuk/chalmers_magnetostrictive_generator.MOV

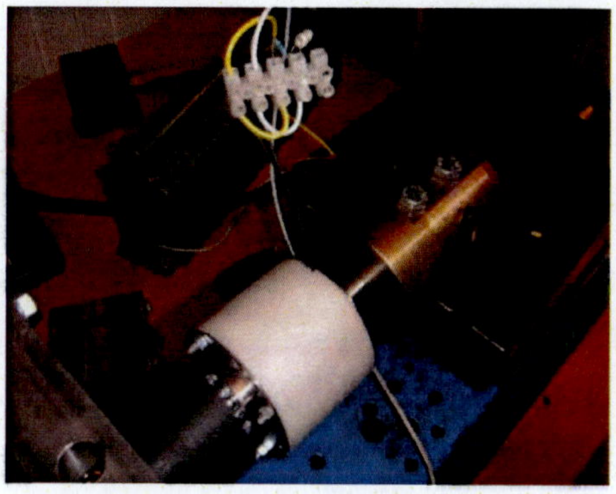

Fig. 2 Chalmers MEG

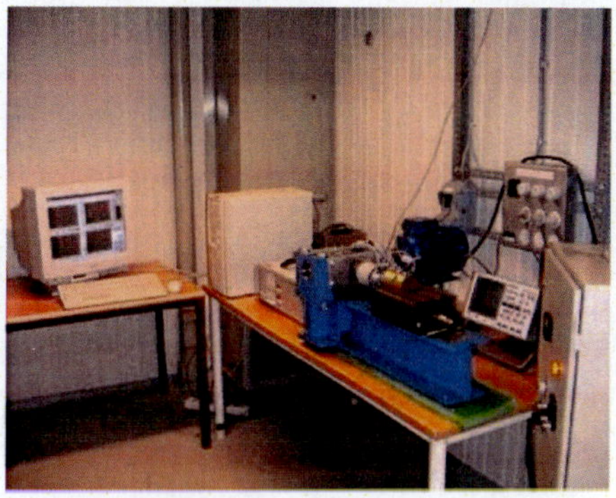

Fig. 3 Test rig with MEG

OPTIMAL DESIGN OF MEG

In the paper it is assumed that the Terfenol-D rod with the length l_{rod} and the mass $m_{Terfenol}$ is used as active material of the transducer. The MEG is under the action of external harmonic kinematic excitation given by:

$$\varepsilon_u = \frac{a_\kappa}{l_{rod}} \sin(\omega_\kappa t + \phi) + \varepsilon_0, \qquad \varepsilon_0 = -\frac{\sigma_0}{E^H} + d_{33} H_0 \tag{10}$$

Problem 2. Let the concept of the MEG is given, i.e. it is determined by the sketch of the transducer depicted in Fig. 1. It's required to find the vector of design parameters, $\mathbf{C} = \mathbf{C}^*$, which satisfies the variational equation

$$\bar{P}_{out}(\mathbf{C}^*) = \max_{\mathbf{C} \in \Omega_{\mathbf{C}}} [\bar{P}_{out}(\mathbf{C})] \tag{11}$$

subject to the equations (7)-(9), kinematic constraints (10) and given mass of the active material $m_{Terfenol} = m_0$.

So, the optimal design problem for MEG is formulated as nonlinear programming problem. Solution of this problem is very hard due to large number of design parameters as well as high complexity of constraints and uncertainties which define the set of admissible values of design parameters $\Omega_{\mathbf{C}}$.

To solve the Problem 2 the following methodology is proposed. Firstly, by using dimensional analysis introduce a set of dimensionless parameters of MEG and write the function $\bar{P}_{out}(\mathbf{C})$ in dimensionless form. Secondly, perform qualitative and quantitative analysis of the set $\Omega_{\mathbf{C}}$, and finally solve the obtained nonlinear programming problem by using Matlab's optimization toolbox.

By eliminating details for the brevity the nondimensional expression for mean power output for steady state operation of the MEG can be written as follows:

$$\bar{P}_{out,d} = \frac{\zeta_1^2 \zeta_2^2 \zeta_3}{2} - \frac{\zeta_1^4 \zeta_2^2 \zeta_3 \zeta_4^2}{\zeta_1^2 \zeta_4^2 + 1} + \frac{1}{2} \frac{\zeta_1^4 \zeta_2^2 \zeta_3 \zeta_4^2}{\left(\zeta_1^2 \zeta_4^2 + 1\right)^2} + \frac{1}{2} \frac{\zeta_1^6 \zeta_2^2 \zeta_3 \zeta_4^4}{\left(\zeta_1^2 \zeta_4^2 + 1\right)^2} \tag{12}$$

where $\zeta_1 = \left(\mu^\sigma - d_{33} d_{33}^{(*)} E^H\right)^{-1}, \quad \zeta_2 = \frac{a_\kappa}{l_{rod}} d_{33} E^H, \quad \zeta_3 = \frac{R_{load} l_{coil}^2}{N_{coil}^2}, \quad \zeta_4 = \frac{Z_{eq} l_{coil}}{A_{coil} N_{coil}^2}.$

The parameter ς_2 can be divided into an "external" part and a "material" part. For the displacement driven transducer it becomes $\zeta_2 = \zeta_{2,ext} \cdot \zeta_{2,mat}$, where $\zeta_{2,ext} = \frac{a_K}{l_{rod}}$, $\zeta_{2,mat} = d_{33} E^H$.

NUMERICAL ANALYSIS AND SOLUTION

In numerical implementation of the proposed methodology the determination of active material parameters E^H, d_{33}, μ^σ as a function of magnetic bias H_0 and prestress σ_0 is needed. The parameters were calculated as follows:

$$[E^H, d_{33}, \mu^\sigma]^T = [H_0^4, H_0^3, H_0^2, H_0, 1]\mathbf{A}_{pol}(*)[\sigma_0^4, \sigma_0^3, \sigma_0^2, \sigma_0, 1]^T \tag{13}$$

Here $\mathbf{A}_{pol}(*)$ are 4x4 matrices of polynomial coefficients which were determined by using experimental data for Terfenol-D (data on magnetostriction over magnetic field strength for different level of prestress [7] and data for the changing Young's modulus [23]). The obtained matrices $\mathbf{A}_{pol}(*)$ for the material parameters are the following:

Parameters	Dimension of Output	Matrix of Polynomial Coefficients $\mathbf{A}_{pol}(*)$
E^H	$[E^H] = GPa$	$\begin{pmatrix} 0 & 0 & 0 & 0.0001 \\ 0 & -0.0001 & 0.0043 & -0.0324 \\ 0 & 0.0048 & -0.2811 & 2.5768 \\ -0.0005 & 0.0200 & 2.4359 & -7.1961 \end{pmatrix}$
d_{33}	$[d_{33}] = 10^{-6} \frac{m}{kA}$	$\begin{pmatrix} 0 & 0 & 0 & 0.0001 \\ 0 & -0.0002 & 0.0043 & -0.0246 \\ -0.0001 & 0.0112 & -0.2499 & 1.4313 \\ 0.0017 & -0.1267 & 2.7046 & -14.4499 \end{pmatrix}$
μ^σ	$[\mu^\sigma] = 10^{-3} \frac{kgm}{A^2 s^2}$	$10^{-3} \cdot \begin{pmatrix} 0 & 0 & 0 & 0.0001 \\ 0 & -0.0001 & 0.0027 & -0.0162 \\ -0.0001 & 0.0071 & -0.1687 & 1.0202 \\ 0.0012 & -0.0947 & 2.3501 & -10.9910 \end{pmatrix}$
Dimensions of Input: Prestress $[\sigma_0] = MPa$, Magnetic Bias $[H_0] = kA/m$		

Table 1. Matrices of polynomial coefficients

As example, the Young's modulus and the magnetic permeability which were calculated by using (13) are presented in the Fig. 4-5.

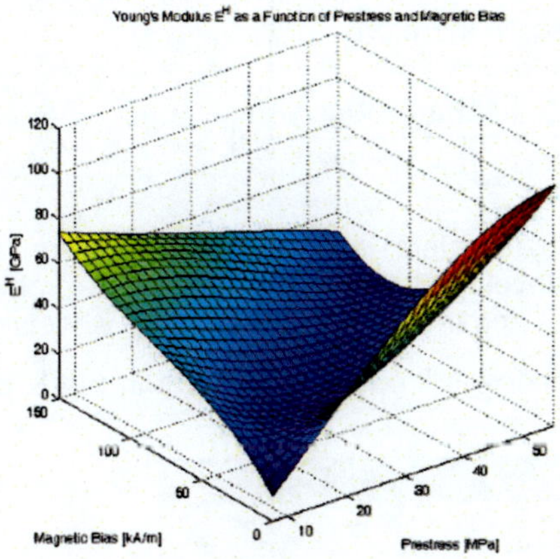

Fig. 4 E^H as a function of H_0 and σ_0

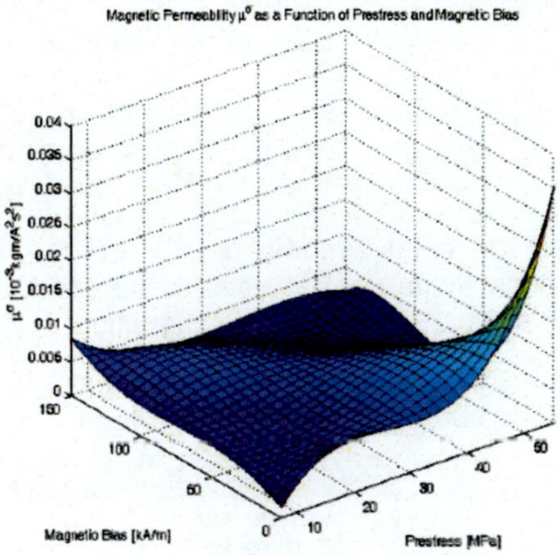

Fig. 5 μ^{σ} as a function of H_0 and σ_0

Both surfaces demonstrate high nonlinear properties of material parameters.

In implementation of the optimization algorithm, only the expedient function values can be used. As follows from analysis of $\zeta_1 = \left(\mu^{\sigma} - d_{33} d_{33}^{(*)} E^H\right)^{-1}$ the feasible region of selectable parameters of prestress and magnetic bias diminishes to a region shown in Fig. 6. The feasible region follows nearly the areas of a low Young's modulus shown in Fig. 4. To handle the difficulties described above, optimization algorithm includes auxiliary *m-function* where all values that are below zero and above a globally defined threshold, are set to zero. A contour plot of the resulting function is shown in Fig 7.

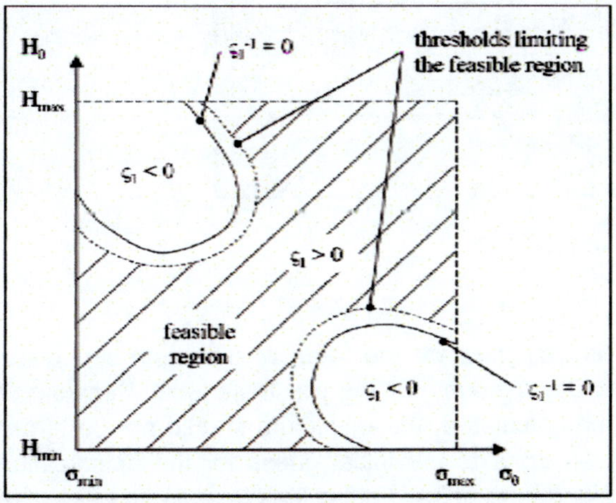

Fig. 6 Feasible region of H_0 and σ_0

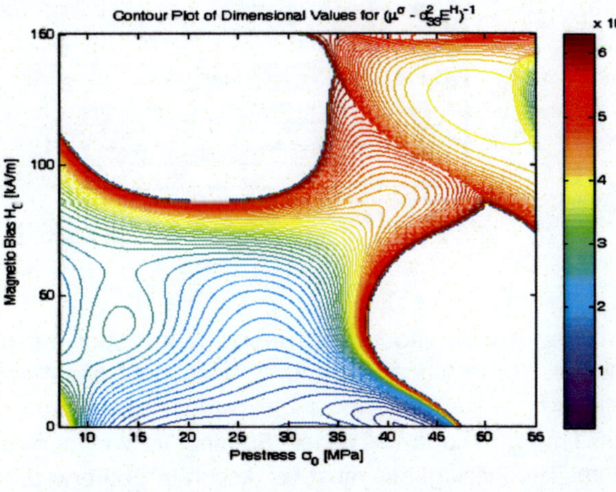

Fig. 7 Contour plot of dimensional values for $(\mu^{\sigma}-d_{33}{}^2E^H)^{-1}$

As seen at the color bar, all values are in the same dimensional range. Further one can recognize two "plateaus", one at around 14 MPa and 40 kA/m and the second larger one at 48 MPa and 130 kA/M. The two main "valleys" are separated by a "pass" following a nearly straight line from (35 MPa, 135 kA/m) to (50 MPa, 90 kA/m).

The lower and upper boundaries for prestress σ_0 and magnetic bias H_0 are determined by the range of the respective characteristic diagrams. The upper and lower bounds become therefore

$$b_{l,\sigma,H} = [\frac{7.2\,\text{MPa}}{\sigma_{char}}, \frac{0\,\text{kA/m}}{H_{char}}]^T, \quad b_{u,\sigma,H} = [\frac{55.1\,\text{MPa}}{\sigma_{char}}, \frac{150\,\text{kA/m}}{H_{char}}]^T \tag{14}$$

Because the setting of the boundaries for the nondimensional parameters is a difficult procedure, several optimization loops usually have to be conducted. The GUI was developed to solve optimization problems in interactive way. Below a short explanation on how the developed GUI and the associated program work is given.

On running the script run_gui.m in the main folder, the main GUI window will open. It is shown in Fig 8. Here all input has to be made.

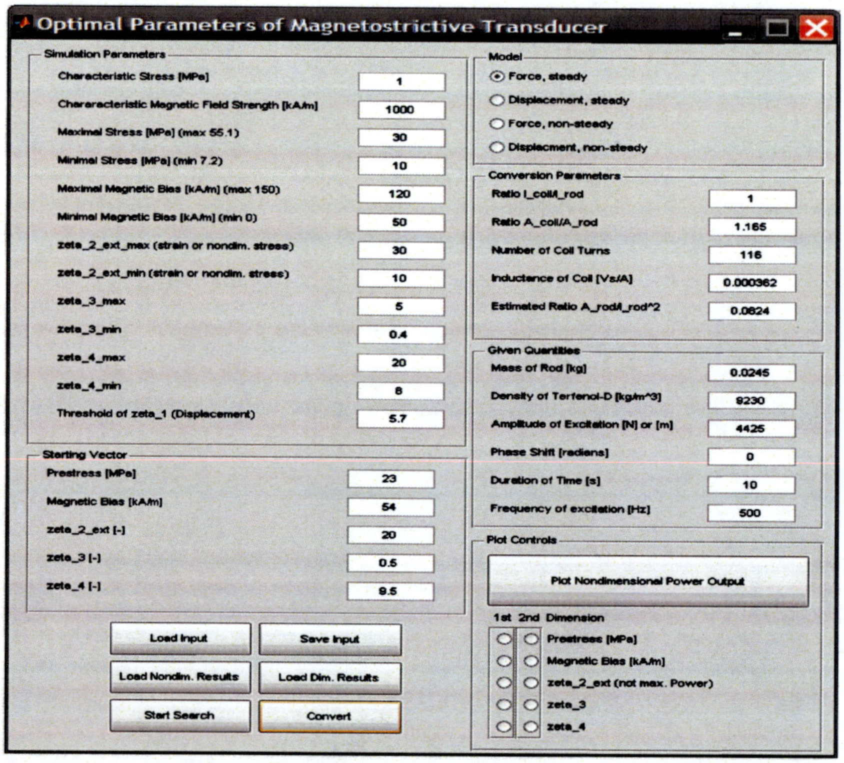

Fig. 8 Main GUI window

The window is split into several areas. The upper left part is assigned for the input of the main simulation parameters. For instance, the data in "Maximal Stress" defines upper bound for the prestress level. Because of restrictions from the range in which material parameters can be calculated, the maximum should not be larger than 55.1 MPa. The lower panel "Starting Vector" serves as an input to the initial guess for the optimization algorithm. The initial guess must be "accurate", otherwise the algorithm will have difficulties to come to a feasible solution. In the upper right panel "Model", the user can choose the function of power calculation. It is possible to choose the displacement or force driven model of MEG in steady state and non steady state of operations.

Below this panel one can find the part to enter the "Conversion Parameters". Since the algorithm computes a set of optimal nondimensional parameters, the program needs additional information on which set of dimensional parameters to choose. For instance, in "Inductance of Coil" the inductance of the coil has to be known to calculate the capacitance of the transducer. It is assumed that the right capacitance will reduce the amount of idle power produced to zero. The data to be entered in the panel "Given Quantities" consists of the quantities that are already given, like mass of transducer, density of Terfenol-D, etc.

On the lower left part of the Fig. 8, one can see a group of six buttons. They will save the entered data to a mat.file or load it from it. If there are already results in nondimensional or dimensional form, they will be displayed as it is shown in Fig. 9 and Fig. 10.

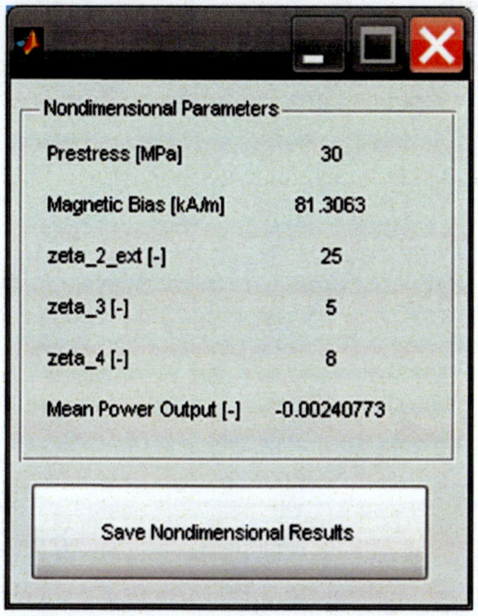

Fig. 9 Nondimensional results GUI window

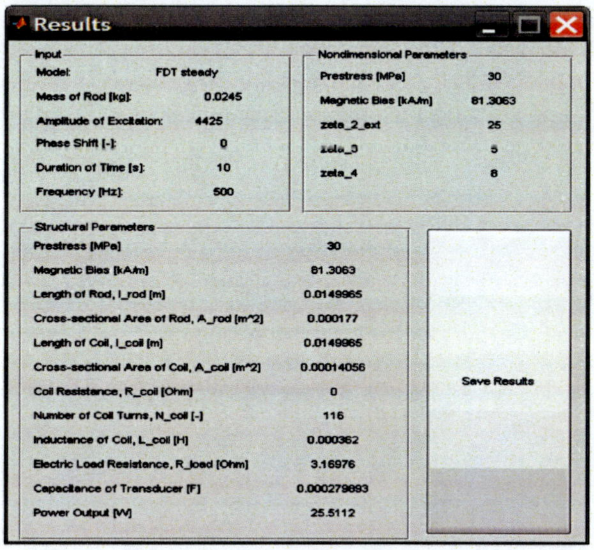

Fig. 10 Results GUI-window

The button "Start Search" will call the optimization routine. At the end of it, the results will be saved to a temporary mat-file and will be displayed as in Fig. 9. The button "Convert" will call the conversion routine. As above, it will save the results to a temporary file and display the data as shown in Fig. 10.

In the lower right part of the main GUI window one can see a panel "Plot Controls". Here the user can choose to plot 2D plots or surface plot of the nondimensional power output. The first and second dimension can be freely chosen. If a parameter is chosen both as first and second dimension, only a two dimensional plot will be generated, otherwise it will be a surface plot. The bounds of the independent parameters are taken from the respective fields of the panel "Simulation Parameters". The values of the rest of the parameters are taken from the entered initial guess.

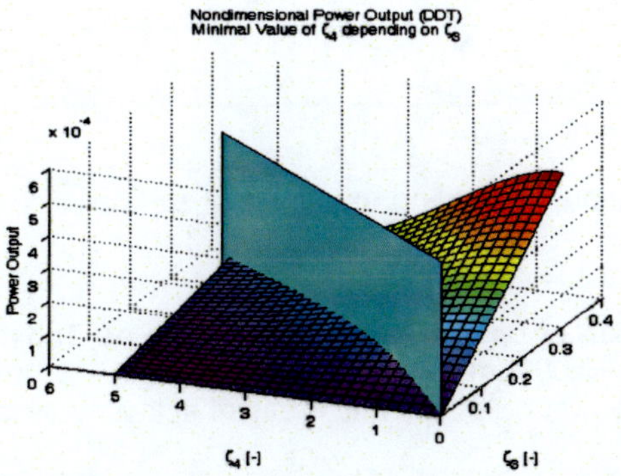

Fig. 11 Constraint and power output over the two factors of electrical circuit

When choosing zeta_3 and zeta_4 as independent parameters, not only a surface plot is generated, but also a plane that indicates the relation between the two parameters. This relation can be written as $\zeta_3 - u_l u_A \frac{A_{rod}}{l_{rod}^2}\zeta_4 \leq 0$, where u_l and u_A are the entered ratios and A_{rod}/l_{rod}^2 is the entered estimated ratio. Because this constraint cannot be observed by the solving algorithm, it must be handled manually. Fig. 11 shows this constraint. A point of maximal power output can be recognized clearly.

In the Table 2 the exemplary solution of the Problem 2 obtained by developed algorithm and GUI is presented. The values for ς_3 and ς_4 are defined manually by using the plot function displaying graphs similar to the one in Fig. 11 and adapting the boundaries iteratively.

Parameters	Values
mass of rod [kg]	0.01
amplitude of excitation [m]	2e-5
frequency [Hz]	100
prestress [MPa]	32.86
magnetic bias [kA/m]	74.11
$\varsigma_{2,ext}$ [-]	7e-7
ς_3 [-]	0.15
ς_4 [-]	2.78
length of rod [m]	0.0286
cross sectional area of rod [m^2]	3.79e-5
length of coil [m]	0.0286
cross sectional area of coil [m^2]	4.4e-5
number of coil turns [-]	30
el. load resistance [Ω]	2.42e-3
power output [W]	3.20

Table 2. Input and important results of exemplary calculations

Analysis of the results of the solution of optimal design problem (See Table 2) demonstrates that for harmonic kinematic excitation with amplitude 0,0002m and frequency 100Hz it is possible to design a magnetostrictive electric generator with 3,2W mean power output having mass of active material 0,01kg.

CONCLUSIONS AND OUTLOOK

The paper has discussed the issues related to optimal design of magnetostrictive transducers for power harvesting from vibrations. The main idea of the methodology proposed for optimal design of MEG is to perform dimensional analysis of mathematical model and cost function to be used in optimization, and then to consider the transducer performance in the space of dimensionless design parameters. In this way one can decrease a number of independent design variables and get insight into the characteristic properties of optimal design. Last but not least, this approach makes it possible to perform all design process in interactive form involving designer actively into the process via convenient GUI.

The optimal design problem is formulated as a nonlinear programming problem and Sequential Quadratic Programming algorithm is used to solve the respective multidimensional constrained optimization problem. The methodology has been implemented in MATLAB environment for the designing of mean power output optimal MEG with given mass of active material and prescribed external kinematic harmonic excitations. The methodology also includes utilization of experimental data to be able to estimate variation of active material parameters over magnetic bias and prestress (like Young's modulus, magnetic permeability, other).

To get insight into the behaviour of the optimized harvesting device, the developed tools were applied and exemplary calculations were performed for the transducer within the concept which was realized early in a physical prototype of MEG. The results of these calculations show that for harmonic kinematic excitation with amplitude 0,0002m and frequency 100Hz it is possible to design a magnetostrictive electric generator with 3,2W mean power output having mass of active material 0,01kg.

This work has focused on the power output of the transducer in steady state operation. An extension could be to change the active material properties, which are stationary and currently only influenced by prestress and magnetic bias, to a time dependent functions.

It is also important further development of the methodology and algorithms to be able to solve optimal design problem for MEG not only within the frame of specified transducer concept (like e.g. concept depicted in Fig. 1) but to find optimal solution on a set of physically admissible transducer concepts.

ACKNOWLEDGEMENTS

The research was partially supported by the Swedish Agency for Innovation Systems, VINNOVA, via the project P35195-1, Dnr: 2008-04106.

The author is grateful to Holger Fuchs who was working with him on optimization problems of magnetostrictive transducers at the department of Applied Mechanics Chalmers University of Technology within the Erasmus Exchange Program.

Thanks also go to Thomas Nygårds and Jan Möller for fruitful collaborative work on experimental study and modelling of magnetostrictive electric generators.

REFERENCES

1. Sodano, H.A., D.J. Inman, and G. Park, *A review of power harvesting from vibration using piezoelectric materials. The Shock and Vibration Digest*, 2004. **36**(3): p. 197-205.
2. Steven, R.A. and H.A. Sodano, *A review of power harvesting using piezoelectric material (2003-2006)*. Smart Materials and Structures, 2007. **16**: p. R1-R21.
3. Erturk, A., Inman, D., *On Mechanical Modeling of Cantilevered Piezoelectric Vibration Energy Harvesters*. Journal of Intelligent Material Systems and Structures, 2008. **OnlineFirst, published on April 21, 2008**
4. Daqaq, M.F., Stabler, C., Qaroush, Y., Seuaciuc-Oso'rio, T., *Investigation of Power Harvesting via Parametric Excitations*. Journal of Intelligent Material Systems and Structures, 2009. **20**: p. 545-557.
5. Priya, S., Inman, D.J., ed. *Energy Harvesting Technologies*. 2009, Springer.
6. Lundgren, A., et al., *A magnetostrictive electric generator*. IEEE Trans. Magn., 1993. **29**(6): p. 3150-3152.
7. Engdahl, G., ed. *Handbook of Giant Magnetostrictive Materials*. 2000, Academic Press: San Diego, USA.

8. Berbyuk, V., J. Sodhani, and J. Möller, *Experimental Study of Power Harvesting from Vibration using Giant Magnetostrictive Materials.* Proc. of 1st International Conference on Experiments, Process, System Modelling, Simulation and Optimization, Athens, 6-9 July, 2005, Ed. Demos T. Tsahalis, Patras University Press, 2005: p. 1-8.
9. Berbyuk, V., *Controlled Multibody Systems with Magnetostrictive Electric Generators*, in *Proc. The ECCOMAS Thematic Conference Multibody Dynamics 2005 on Advances in Computational Multibody Dynamics, Madrid, June 21-24, 2005, Eds. J.M. Goicolea, J. Cuardrado and J.C. Garcia Orden*. 2005.
10. Berbyuk, V., *Towards Dynamics of Controlled Multibody Systems with Magnetostrictive Transducers.* Multibody System Dynamics, 2007. **18**: p. 203-216.
11. Zhao, X. and D.G. Lord, *Application of the Villari effect to electric power harvesting.* Journal of Applied Physiscs, 2006. **99**(08M703): p. 08M701-08M703.
12. Engdahl, G. *Design procedure for optimal use of giant magnetostrictive materials in magnetostrictive applications.* in *ACTUATOR2002, 8th International Conference on New Actuators*. 2002. Bremen, Germany.
13. Oscarsson, M. and G. Engdahl, *Key numbers in design of magnetostrictive actuators and generators*, in *ACTUATOR 2006, 10th International Conference on new Actuators*. 2006: Bremen, Germany. p. 774-777.
14. Roundy, S.e.a., *Improving power output for vibration-based energy scavengers.* IEEE Cs and IEEE ComSoc, Pervasive Computing, 2005: p. 28-36.
15. Berbyuk, V., *Terfenol-D based transducer for power harvesting from vibration*, in *Proceedings of IDETC/CIE 2007 ASME 2007 International Conference*. 2007: Las Vegas, Nevada, USA. p. paper DETC2007-34788.
16. McInnes, C.R., D.G. Gorman, and M.P. Cartmell, *Enhanced vibration energy harvesting using non-linear stochastic resonance.* Journal of Sound and Vibration, 2008. **318**(4-5): p. 655-662.
17. Scruggs, J.T., *An optimal stochastic control theory for distributed energy harvesting network.* Journal of Sound and Vibration, 2009. **320**: p. 707-725.
18. Shahruz, S.M., *Increasing the Efficiency of Energy Scavengers by Magnets.* ASME, Journal of Computational and Nonlinear Dynamics, 2008. **3**: p. 041001-1 041001-12
19. Nakano, K., Elliot, S.J.,Rustighi, E., *A unified approach to optimal conditions of power harvesting using electromagnetic and piezoelectric transducers.* Smart Materials and Structures, 2007. **16**: p. 948-958.
20. Yamamoto, Y., et al., *Three-dimensional magnetostrictive vibration sensor:development, analysis and applications.* Alloys and Componds, 1997. **258**: p. 107-113.
21. Berbyuk, V., Sodhani, J., *Towards modeling and design of magnetostrictive electric generators. Computers and Structures*, 2008. **86**: p. 307-313.
22. Berbyuk, V. and T. Nygårds. *Power Harvesting from Vibration Using Magnetostrictive Materials.* in *Joint Baltic-Nordic Acoustics Meeting*. 2006. Gothenburg.
23. Kellogg, R. and A. Flatau, *Experimental investigation of Terfenol-D elastic modulus.* Intelligent Material Systems and Structures, 2008. **19**(5): p. 583-595.

Proceedings of the IMAC-XXVIII
February 1–4, 2010, Jacksonville, Florida USA

Experimental Investigations of a Bistable Energy Harvester

A.J. Sneller, B.A. Owens & B.P. Mann

Department of Mechanical Engineering & Materials Science
Duke University
144 Hudson Hall
Durham, NC 27708

ABSTRACT

This paper investigates the design and analysis of a nonlinear energy harvester that uses magnet interactions to create an inertial generator with a bistable potential well. The motivating hypothesis for this work was that nonlinear behavior could be used to improve the performance of an energy harvester by broadening its frequency response. Both numerical and experimental investigations study the harvester's response when directly powering an electrical load. Both sets of results show that the potential well escape phenomenon can be used to broaden the frequency response of an energy harvester.

NOMENCLATURE

$\mathbf{B}$ Magnetic Field
I Electrical Current
L Inductance
R Resistance
U Potential Energy
d Half length Between End Magnets
m Mass
$\mathbf{p}$ Magnetic Moment
r Radius of Outer-Ring of Magnets
$\mathbf{s}$ Position Vector
v Volume
x Global Coordinate of Center Magnet
y Local Coordinate of Center Magnet
y_e Equilibrium Position of y
z Base Excitation
β Residual Magnetic Flux Density
γ Electromechanical Coupling Constant
ζ Linear Viscous Damping Ratio
μ_0 Magnetic Constant
ω Linear Natural Frequency

T. Proulx (ed.), *Structural Dynamics and Renewable Energy, Volume 1*, Conference Proceedings of the Society for Experimental Mechanics Series 10, DOI 10.1007/978-1-4419-9716-6_19, © The Society for Experimental Mechanics, Inc. 2011

1 INTRODUCTION

Due to recent reductions in the power consumption of MEMS devices [1,2], scavenging environmental energy in order to power these devices has become a viable area of investigation [3]. In addition to solar, chemical, and thermal energy sources, the ubiquity of environmental vibrations has recently brought mechanical energy harvesting to the forefront [1,4–10].

Vibrational energy harvesting has been primarily concerned with linear inertial generators [10]. Due to the relatively poor response of such devices to all but a narrow band of excitation frequencies, linear generators have limited utility. Matching the resonance of a system to a known frequency is a possible, albeit difficult endeavor [7]. However, natural vibration sources have the added complication of being difficult to predict and/or varying over time. This suggests that an ideal generator would have a relatively large response over a broad range of frequencies [11].

Recent efforts have attempted to overcome the narrow frequency response of linear devices through the use of tuning the device's resonance and/or the use of multiple oscillators to broaden the device's response [7,12]. In contrast to these works, the present study attempts to overcome the inherent shortcomings of linear generators through the purposeful implementation of nonlinearities into the energy-harvesting device. Specifically, we investigate a bistable electromagnetic-inductive generator, and the utility of potential well escapes as a means to promote energy harvesting. The analytical and experimental results of this work show the device to feature a broad frequency response, making it an ideal energy harvester.

2 EXPERIMENTAL APPARATUS

The experimental system is shown in Fig. 1; it consists of one magnet free to move horizontally within a set of fixed magnets, which are oriented such that the free magnet features two stable equilibria. The left and right magnets both repel the center magnet, thus suspending it within the frame of the device [4,5,11]. Four additional magnets located in the device's mid-plane are oriented such as to repel the center magnet away from the midpoint, thus making the system bistable. The motion of the center magnet will be used to power an electrical circuit via electromagnetic induction.

3 ENERGY GENERATOR MODEL

In this section, a mathematical model describing the behavior of the energy harvester will be developed. The device along with its geometry are shown in Fig. 1, and the derivation of the equations of motion will use the definitions presented therein.

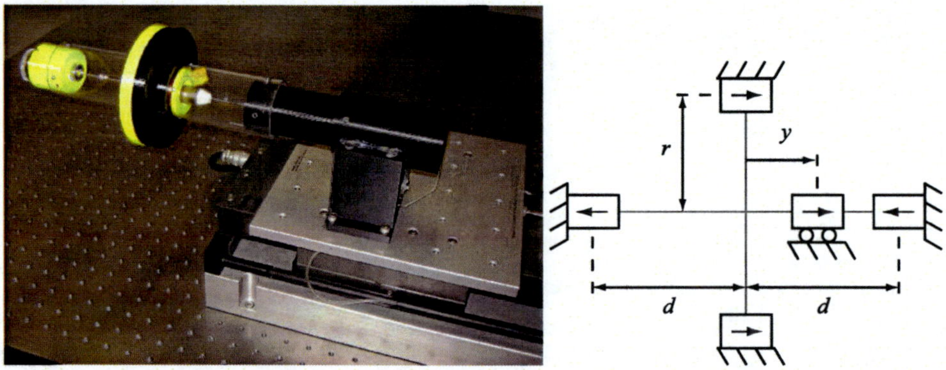

(a) (b)

Figure 1 - (a) Image, and (b) schematic of the experimental apparatus, showing the local coordinate y as well as the half-length d and the radius r. Each magnet is marked by an arrow that points from its south pole to its north pole. Note that the actual system features 4, rather than 2, radial magnets.

The potential energy of the center magnet due to the presence of the fixed magnets is defined as

$$U_m = -\mathbf{p}_c \cdot \sum_{i=1}^{6} \mathbf{B}_i(\mathbf{s}_c) \,, \tag{1}$$

where $\mathbf{p}_c$ is the magnetic moment of the center magnet, and $\mathbf{B}_i(\mathbf{s}_c)$ is the magnetic field generated by the ith fixed magnet evaluated at the position $\mathbf{s}_c$ of the center magnet. Using the dipole model to approximate the magnetic fields [13–16], Eq. (1) can be rewritten in terms of the local coordinate y as

$$U_m = \frac{\beta_c v_c}{2\pi\mu_0}\left[\beta_e v_e\left(\frac{1}{(d+y)^3}+\frac{1}{(d-y)^3}\right)+\tfrac{1}{2}\beta_o v_o N\left(\frac{r^2}{(y^2+r^2)^{5/2}}\right)\right], \tag{2}$$

where $N = 4$ is the number of outer-ring magnets, v and β are a magnet's volume and residual flux density, respectively, which are both indexed by c, o, and e to indicate the center, outer-ring, and left/right end magnets, respectively. Plots of the potential energy curve for representative values of the outer-ring radius r are presented in Fig. 2. The stable equilibria y_e of the system occur at the minima of U_m, and for the parameter values given in Table 1 it can be found that $y_e = \pm 41.1$ mm.

The equations of motion for the system can be found using Kirchoff's Voltage Law and Newton's Second Law:

$$L\dot{I} + (R_L + R_i)\,I + \gamma\dot{y} = 0\,, \tag{3a}$$

$$\ddot{y} + 2\zeta\omega\dot{y} + \tfrac{1}{m}F_m(y) - \tfrac{1}{m}\gamma I = -\ddot{z}\,, \tag{3b}$$

where an overdot indicates a time-derivative, ω is the natural frequency about either equilibrium position, ζ is a linear viscous damping coefficient, $\ddot{z}$ is the base acceleration, I is the electrical current, L is the inductance of the coil, R_i and R_L are the internal and load resistances, m is the center magnet's mass, $F_m = \partial U_m/\partial y$ is the magnetic restoring force, and γ is a transducer constant - derived from Faraday's Law of Induction - that couples the mechanical and electrical systems.

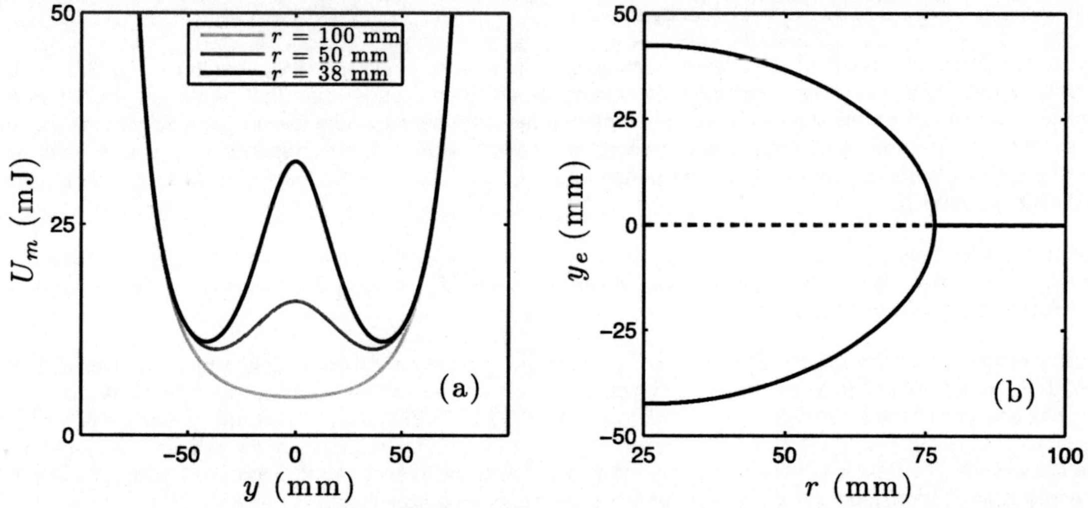

Figure 2 - (a) Potential energy curves for several magnet outer-ring radii r, and (b) a bifurcation diagram showing the stability and existence of equilibria over a range of r values. Holding all other parameters constant (see Table 1), the system is bistable for r <76.5mm.

TABLE 1 - Parameter values that were used to model the system.

System Parameters			
m	35.6 g	v_c	4826 mm^3
ω	32.19 rad/s	v_o	679 mm^3
ζ	0.02	v_e	1609 mm^3
L	76 mH	β_c	1.32 T
R_i	127 Ω	β_o	1.48 T
R_L	150 Ω	β_e	1.32 T
γ	5 T · m	d	106 mm
y_e	±41.1 mm	r	38 mm

4 FREQUENCY RESPONSE INVESTIGATIONS

Of particular interest to the design of an energy-harvesting device lies with its ability to operate over a broad range of frequencies. In this section, numerical and experimental frequency sweeps will be conducted in order to investigate the frequency response of the system. Single-frequency excitations will be considered, with the form $\ddot{z} = A\cos\Omega t$, where A is the acceleration amplitude and Ω is the excitation frequency in rad/s.

4.1 Numerical response

To determine the ability of the present system to achieve this goal, the equations of motion - Eqs. (3a) and (3b) - were simulated numerically for both linearly increasing and decreasing excitation frequencies. The model parameters were assigned in order to closely match the analytical and experimental systems, and they are summarized in Table 1.

The graphs of Fig. 3 show the numerical results for the displacement y of the center magnet for both up and down sweeps, at several levels of base excitation. Overlaid onto each time-series plot is a set of stroboscopic samples, taken once per forcing period. The stroboscopic points indicate the periodicity of these responses with respect to the excitation frequency, which is an important aspect of the system's frequency-dependent behavior.

For low amplitude excitation (1.5 m/s^2), the system responded in a nearly linear fashion (Fig. 3(a) and (d)). The system responds in a periodic manner with large amplitude responses only in the vicinity of resonance. In addition, there is no perceivable difference between the forward and reverse frequency sweeps.

For the middle excitation tested (4 m/s^2), the system can be seen to feature weakly nonlinear behavior (Fig. 3(b) and (e)). The response, while largest near the linear resonance, can be observed to be broader than that of the low amplitude excitation. Visible nonlinear effects include the slight broadening of the frequency response, and a loss of periodicity in the region of peak response, as can be observed by inspecting the stroboscopic points. However, the presence of a visible peak near linear resonance and a lack of difference between the response of the forward and reverse frequency sweeps indicate this to be a weakly nonlinear response.

The largest acceleration amplitude (10 m/s^2) elicited a qualitatively different response than did the smaller excitations (Fig. 3(c) and (f)). In addition to an overall broader peak response, it can be observed that the system exhibits hysteresis between multiple solutions, including potentially chaotic attractors.

The frequency responses of the system can be used to ascertain the frequency dependence of the power delivered to an electrical load. Fig. 4 (a), (b), and (c) show the power spectrum for each of the excitation amplitudes investigated in the numerical study. Each plot was generated by overlaying the power dissipation $R_L I^2$ for both the up and down sweeps. For the lowest level of base excitation, the power graph shows a rather narrow peak near the natural frequency of the system. This type of response, which is reminiscent of classical linear behavior, no longer dominates the responses for increased levels of base excitation; instead, the frequency response begins to broaden for higher levels of base excitation.

4.2 Experimental response

Fig. 1 (a) shows an image of the fabricated nonlinear generator attached to a shaker table. The generator was held in a horizontal orientation to avoid the influence of gravity. The generator's coil was wired to directly power a resistive load; the resistive load was set to $R_L = 150\Omega$ and the voltage drop across the resistive load was measured with a data acquisition system. Harmonic base

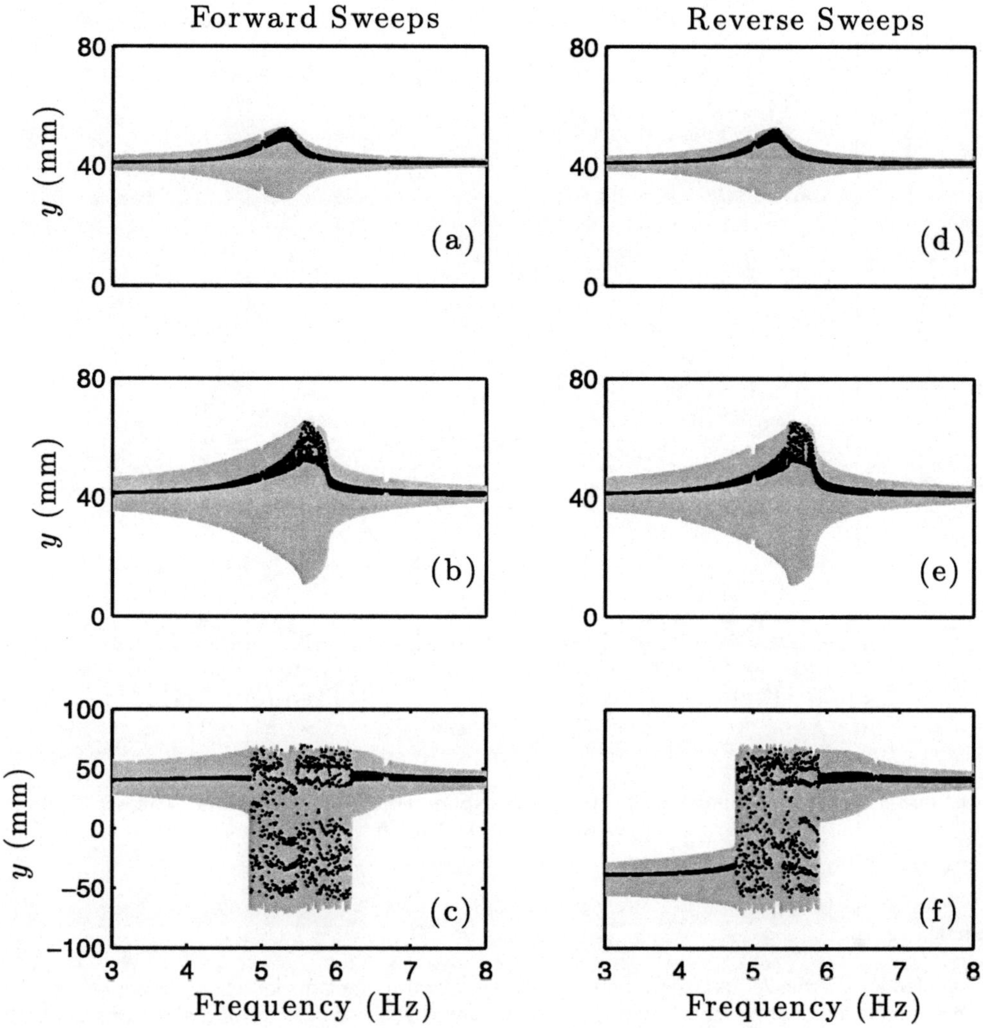

Figure 3 - Graphs showing the response of the center magnet to up and down frequency sweeps at 1.5, 4, and 10 m/s^2.

excitation was applied by mounting the inertial generator to the air bearing shake table of an APS Dynamics model 129 shaker. To alleviate the possibility of magnetic field interference from the shaker, the attachment support was designed to distance the energy harvester away from the shaker base. During the experimental tests, the base acceleration was measured by mounting a PCB Model 3713D1FD20G accelerometer to the shake table. The voltage drop across the resistive load was recorded and was used to analyze the system's behavior.

As with the numerical studies, forward and reverse frequency sweeps between 3 and 8 Hz were conducted at three separate acceleration amplitudes (1.5, 4, and 10 m/s^2). The power spectra were developed by overlaying plots of V^2/R_L for the forward and reverse frequency sweeps, where V is the voltage drop across the load. Fig. 4 shows a comparison between the numerical and experimental power responses for each of the excitation amplitudes tested. It can be seen that larger amplitudes of excitation yield increasingly nonlinear behavior, coupled with a broadening of the peak response.

It can be observed that the system responds in a nearly linear manner at relatively low excitation levels. Specifically, a narrow peak response near resonance, periodic behavior, and no hysteresis is obvious when comparing the forward and reverse sweeps. As anticipated from the numerical investigations, the experimental frequency response broadens at relatively higher levels of excitation.

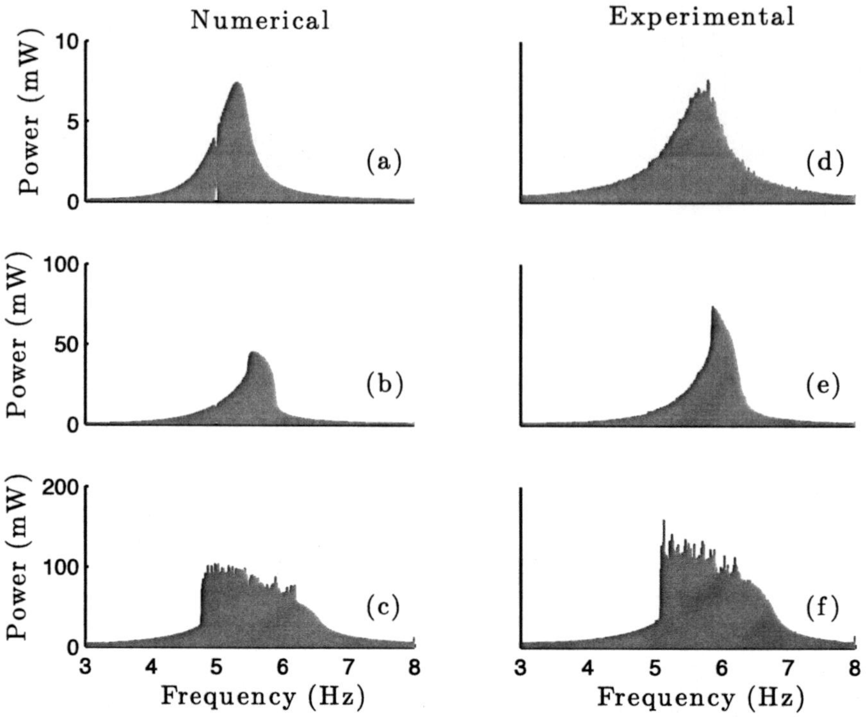

Figure 4 - Power responses for both numerical (a, b, c) and experimental (d, e, f) results. The excitation amplitudes used were 1.5, 4, and 10 m/s^2.

5 CONCLUSIONS

This study investigated an alternative paradigm to harvesting energy from ambient vibrations. To date, much of the work in this field has been restricted to linear generators and frequency matching. These concepts were abandoned in favor of a generator designed to feature purposeful and inventive nonlinearities. The bistable energy harvester that was used in this study was designed to utilize the potential well escape phenomenon, with the intention of broadening the frequency response of the device. The paper introduced a mathematical model for the energy-harvesting system, which allowed for the system parameters to be tailored to induce the desired system behavior. Numerical and experimental investigations reveal regions of coexisting solutions (high- and low-energy responses) and hysteresis in the frequency sweeps. The present investigations also reveal that a nonlinear generator, particularly one with bistable potential well, can be used to broaden the frequency response.

The presence of multiple attractors provides an additional complication that is not observed in linear harvesters. Specifically, both high- and a low-energy responses can coexist for the same parameter combinations. However, we note that the coupling between the electrical circuit and mechanical oscillations could be used to trigger a jump to the more desirable attractor; alternatively, mechanical and electrical perturbations have been observed to induce jumps between different attractors in the laboratory.

In summary, the present studies show that a nonlinear phenomenon - the escape from a potential well - can be used to broaden the frequency spectrum of an energy harvester. However, optimizing the performance of the nonlinear design was not attempted, as this is likely to present a formidable challenge. In addition, it is likely that many other nonlinear phenomena could also be used to achieve enhanced performance from energy harvesters.

ACKNOWLEDGMENTS

The authors would like to acknowledge support from an ONR Young Investigator Award through program manager Ronald Joslin.

REFERENCES

[1] **Glynne-Jones, P., Tudor, M. J., Beeby, S. P.** and **White, N. M.**, *An electromagnetic, vibration-powered generator for intelligent sensor systems*, Sensors and Actuators A, Vol. 110, pp. 344–349, 2004.

[2] **Roundy, S., Wright, P. K.** and **Rabaey, J. M.**, Energy Scavenging for Wireless Sensor Networks, Springer-Verlag, New York, 2003.

[3] **Beeby, S. P., Tudor, M. J.** and **White, N. M.**, *Energy harvesting vibration sources for microsystems applications*, Measurement Science and Technology, Vol. 17, pp. 175–195, 2006.

[4] **Saha, C. R., O'Donnell, T., Wang, N.** and **McCloskey, P.**, *Electromagnetic generator for harvesting energy from human motion*, Sensors and Actuators A, Vol. 147, No. 1, pp. 248–253, 2008.

[5] **von Buren, T.** and **Troster, G.**, *Design and optimization of a linear vibration-driven electromagnetic micro-power generator*, Sensors and Actuators A, Vol. 135, pp. 765–775, 2007.

[6] **Kulkarni, S., Koukharenko, E., Torah, R.** and **Tudor, J.**, *Design, fabrication and test of integrated macro-scale vibration-based electromagnetic generator*, Sensors and Actuators A, Vol. 145–146, pp. 336–342, 2008.

[7] **Sari, I., Balkan, T.** and **Kulah, H.**, *An electromagnetic micro power generator for wideband environmental vibrations*, Sensors and Actuators A, Vol. 145–146, pp. 405–413, 2008.

[8] **Shahruz, S. M.**, *Design of mechanical band-pass filters for energy scavenging*, Journal of Sound and Vibration, Vol. 292, No. 3-5, pp. 987–998, 2006.

[9] **Shahruz, S. M.**, *Limits of performance of mechanical band-pass filters used in energy scavenging*, Journal of Sound and Vibration, Vol. 293, No. 1-2, pp. 449–461, 2006.

[10] **Williams, C. B.** and **Yates, R. B.**, *Analysis of a microgenerator for microsystems, Proceedings of the 8th International Conference on Solid-State Sensors and Actuators,* Stockholm, Sweden, pp. 87–B4, Eurosensors IX, 1995.

[11] **Mann, B. P.** and **Sims, N. D.**, *Energy harvesting from the nonlinear oscillations of magnetic levitation*, Journal of Sound and Vibration, Vol. 319, pp. 515–530, 2009.

[12] **Leland, E. S.** and **Wright, P. K.**, *Resonance tuning of piezoelectric vibration energy scavenging generators using compressive axial load*, Smart Material and Structures, Vol. 15, pp. 1413–1420, 2006.

[13] **Yung, K. W., Landecker, P. B.** and **Villani, D. D.**, *An analytical solution for the force between two magnetic dipoles*, Magnetic and Electrical Separation, Vol. 9, pp. 39–52, 1998.

[14] **Landecker, P. B., Villani, D. D.** and **Yung, K. W.**, *An analytical solution for the torque between two magnetic dipoles*, Magnetic and Electrical Separation, Vol. 10, pp. 29–33, 1999.

[15] **Furlani, E. P.**, Permanent magnet and electromechanical devices, Academic Press, New York, 1st edn., 2001.

[16] **Vanderlinde, J.**, Classical Electromagnetic Theory, Springer, Dordrecht, The Netherlands, 2nd edn., 2004.

Proceedings of the IMAC-XXVIII
February 1–4, 2010, Jacksonville, Florida USA
©2010 Society for Experimental Mechanics Inc.

Response of Uni-Modal Duffing-Type Harvesters to Random Excitations

Mohammed F. Daqaq

Nonlinear Vibrations and Energy Harvesting Lab. (NoVEHL)
Department of Mechanical Engineering
Clemson University, Clemson, SC 29634
mdaqaq@clemson.edu

Abstract

The common linear energy harvesters have an inherent shortcoming in their operation concept because they operate efficiently only when the excitation frequency is very close to the fundamental frequency of the harvester. To extend the harvester's bandwidth, some recent solutions call for utilizing energy harvesters with stiffness-type nonlinearities. From a steady-state perspective, this hardening nonlinearity can extend the coupling between the excitation and the harvester to a wider range of frequencies. In this effort, we investigate the response of such harvesters to Gaussian White and Colored excitations. For Gaussian White excitations, we solve the Fokker-Plank-Kolmogorov equation for the exact joint probability density function of the response. We show that the expected value of the output power is not even a function of the non-linearity. As such, under White Gaussian excitations, nonlinearities in the stiffness do not provide any enhancement over the typical linear harvesters. Furthermore, we demonstrate that nonlinearities in the damping or inertia should be sought to enhance the output power. We also use the Van Kampen expansion to analyze the response to Colored excitations of different bandwidths and center frequencies. Again, we show that, regardless of the bandwidth or the center frequency, the expected value of the power is always less than that associated with a linear harvester.

1 Introduction

Today, many critical electronic devices, such as health-monitoring sensors,[1,2] pace makers,[3] spinal stimulators,[4] electric pain relievers,[5] wireless sensors,[6–8] etc., require minimal amounts of power to function. Such devices have, for long time, relied on batteries that have not kept pace with the devices' demands, especially in terms of energy density.[9] In addition, batteries have a finite life span, and require regular replacement or recharging, which, in many of the previously mentioned examples, is a very cumbersome process. In light of such challenges, vibration-based energy harvesting has flourished as a major thrust area of micro power generation. Various devices have been developed to transform mechanical motions directly into electricity by exploiting the ability of active materials and some mechanisms to generate an electric potential in response to mechanical stimuli and external vibrations.[10–12]

However, there are still two major issues limiting the efficiency of energy harvesters. First, traditional *linear* energy harvesters have a very narrow frequency bandwidth and, hence, operate efficiently *only* when the excitation frequency is very close to the fundamental frequency of the harvester. Small variations in the excitation frequency around the harvester's fundamental frequency drops its small energy output even further making the energy harvesting process inefficient. Second, most environmental excitations have a broad band or time dependent characteristics in which the energy is distributed over a wide spectrum of frequencies or the dominant frequencies vary with time. Together, these two factors have a negative influence that reduces the output power and hinders the efficiency of the harvester.

To resolve these two issues, a large portion of the energy harvesting research is currently directed toward designing harvesters capable of scavenging energy from non-stationary and random excitations.[13–21] One proposed solution is based on purposefully introducing nonlinearities into the harvester's dynamics. A class of such harvesters utilizes a nonlinear compliance to extend the coupling between the environmental excitation and the harvester to a wider range of frequencies as shown in Fig. 1. The nonlinearities can be introduced using nonlinear magnetic levitation,[18,20] Fig. 2(b), or by other means,[22] Fig. 2(a). These new designs provide a *steady-state* power response that extends over a wider range of frequencies.

While such solutions have been proposed to resolve the issue of excitations' non-stationarities and randomness, the associated analysis and predicted power enhancements were usually based on the steady-state response which assumes a harmonic fixed-frequency excitation. As of today, we still do not understand how the nature of excitation influences the output power, or what role nonlinearities play in the transduction of energy harvesters under realistic excitations. Furthermore, it is not even known if the steady-state fixed-frequency analysis currently adopted in the literature is a valid performance indicator.

T. Proulx (ed.), *Structural Dynamics and Renewable Energy, Volume 1*, Conference Proceedings of the Society for Experimental Mechanics Series 10, DOI 10.1007/978-1-4419-9716-6_20, © The Society for Experimental Mechanics, Inc. 2011

To fill this gap in the current knowledge, we shed some light into this issue by studying the response of uni-modal duffing-type harvesters, similar to those proposed in[18,20,22] to random excitations. Specifically, we aim to understand how the statistical characteristics of the output power are affected by the statistics, (e.g. spectral density, bandwidth, and center frequency) of the excitation. We also seek to delineate the influence of stiffness nonlinearities on the performance of these harvesters under random excitations. To achieve this goal the rest of the paper is organized as follows: In Section 2, we obtain an analytical expression for the expected value of the power under Gaussian White excitations by solving the Fokker-Plank-Kolmagarov (FPK) equation for the exact probability density function (PDF) of the response. In Section 3, we use the Van Kampen expansion to generate a set of linearly-coupled ordinary differential equations governing the evolution of the response statistics under Colored excitations. In Section 4, we solve the equations and study the influence of the excitation bandwidth and center frequency on the mean power. Finally, in Section 5, we present our conclusions and recommendations for future work.

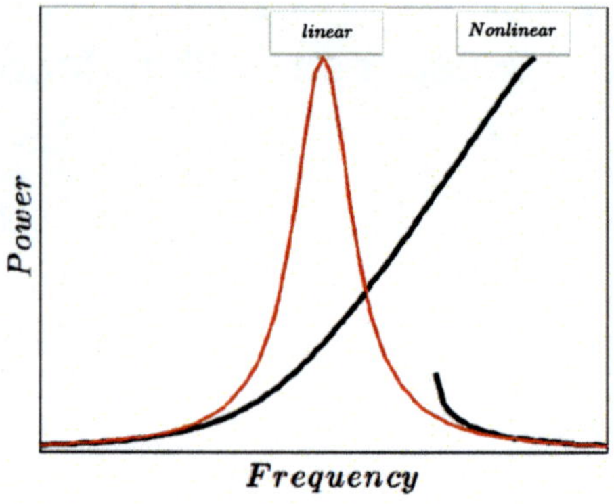

Figure 1: Power-frequency curves for a linear and a nonlinear harvester with stiffness type nonlinearities.

2 Response to White Gaussian Excitations

We consider an electormagnetic duffing-type harvester, similar to the one shown in Fig. 2 (b). The equations of motion can be written as

$$\begin{aligned} &\ddot{u} + 2\zeta\omega_n\dot{u} + \omega_n^2 u + \theta i + \beta u^3 = F(t) \\ &\theta\dot{u} = iR \end{aligned} \tag{1}$$

where u represents the position of the mass, ζ is the mechanical damping ratio, ω_n is the natural frequency, θ is electromechanical coupling coefficient, i is the current, β is a cubic nonlinearity coefficient, and R is the load resistance. The force $F(t)$ is assumed to be a stationary Gaussian White noise process with zero mean and autocorrelation function

$$\langle F(t)F(t+\tau)\rangle = 2\pi S_0\delta(\tau) \tag{2}$$

where $\langle\rangle$ denotes the expected value, S_0 is the spectral density of the excitation, and δ is the dirac-delta function. To obtain an analytical expression for the mean output power, we cast the problem in the Itô stochastic differential form where Equation (1) can be rewritten as[23]

$$\begin{aligned} &dx_1 = x_2 dt \\ &dx_2 = -\{c_{eff}x_2 - \omega_n^2 x_1 - \beta x_1^3\}dt + SdB \end{aligned} \tag{3}$$

where $x_1 = u$, $x_2 = \dot{u}$, $c_{eff} = 2\zeta\omega_n + \frac{\theta^2}{R}$, $S = \pi S_0$, and B is a Brownian motion process such that $\frac{dF}{dt} = B(t)$. The joint PDF, $P(x_1, x_2, t)$ of the response can be obtained by solving the linear diffusion FPK equation which can be expressed for system (3) as[24]

$$\frac{\partial P}{\partial t} = -\frac{\partial}{\partial x_1}(x_2 P) + \frac{\partial}{\partial x_2}\left(\{c_{eff}x_2 + \omega_n^2 x_1 + \beta x_1^3\}P\right) + S\frac{\partial^2 P}{\partial x^2} \tag{4}$$

subjected to the boundary conditions $P(-\infty, t) = P(\infty, t) = 0$. While an analytical solution of Equation (4) is not attainable in the general sense, a stationary solution can be obtained by setting $\frac{\partial P}{\partial t}$ equal to zero. In this case, Equation (4) admits

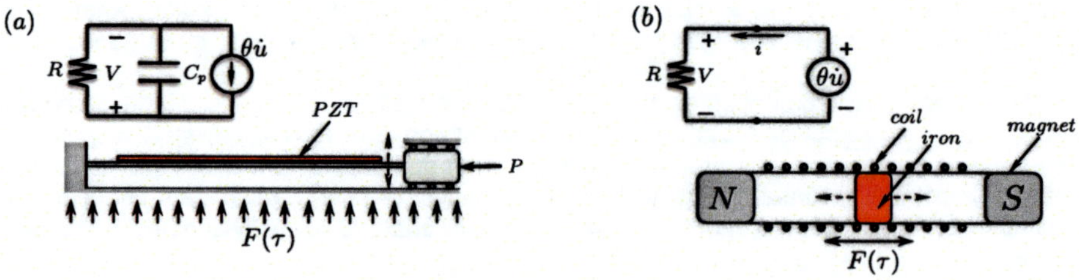

Figure 2: Schematics of two uni-modal duffing-type harvesters.

a solution of the form:

$$P(x_1, x_2, t) = P_1(x_1, t) \times P_2(x_2, t) = A \exp\left\{\frac{c_{eff}}{2S}(\omega_n^2 x_1^2 + \frac{\beta}{2}x_1^4)\right\} \times \exp\left\{\frac{c_{eff}}{2S}x_2^2\right\} \tag{5}$$

where A is a normalization constant given by

$$A^{-1} = \int_{-\infty}^{\infty}\int_{-\infty}^{\infty} \exp\left\{\frac{c_{eff}}{2S}x_2^2\right\} \times \exp\left\{\frac{c_{eff}}{2S}(\omega_n^2 x_1^2 + \frac{\beta}{2}x_1^4)\right\} dx_1 dx_2 \tag{6}$$

Note that the stationary PDF of the response is factored into a function of the displacement, x_1, and a function of the velocity, x_2. This implies that x_1 and x_2 are independent random variables with the PDF of x_2 being independent of the nonlinearity coefficient, β. With that, the expected value of the mean square velocity can be obtained using

$$\langle \dot{u}^2 \rangle = \langle x_2^2 \rangle = \int_{-\infty}^{\infty} x_2^2 P_2(x_2, t) dx_2 = \frac{S}{c_{eff}} \tag{7}$$

From Equations (1) and (7), the expected valued of the mean square output current passing through the load can be written as

$$\langle i^2 \rangle = \frac{\theta}{R}\langle x_2^2 \rangle = \frac{\theta S}{R c_{eff}} \tag{8}$$

and the mean power is

$$\langle \mathcal{P} \rangle = \frac{\theta S}{c_{eff}} \tag{9}$$

Equation (9) clearly indicates that the expected value of the power (mean power) is not a function of the nonlinearity and is equal to that obtained for a linear harvester with $\beta = 0$. Hence, under White Gaussian excitations both a linear and a uni-modal duffing-type harvester provide the same mean output power.

This stationary PDF of the response given by Equation (5) also suggests that in order to alter the value of the mean square velocity, and hence power, from the linear case, one should seek other types of nonlinearities. Damping type nonlinearities that are function of the velocity, x_2, alone for example can alter the velocity part of the joint PDF, which, in turn, can improve the output power. Inertia nonlinearities as well as other nonlinearities which are a combined function of the velocity and displacement will make the PDF inseparable and hence are also expected to alter the mean power.

3 Response to Colored Excitations

While many environmental excitations exhibit the characteristics of white excitations, many others have most of their energy trapped within a narrow bandwidth possessing the characteristics of a narrow-band (colored) excitation. To analyze the influence of the center frequency, bandwidth, and variance of such excitations on the output power for the harvester considered, we couple Equations (1) with a second-order by pass filter through which we pass a white Gaussian excitation according to

$$\ddot{F} + \gamma\dot{F} + \omega_c^2 F = \gamma^{1/2}\omega_c W \tag{10}$$

where ω_c is the center frequency of the filter and hence the excitation, γ is its bandwidth, W is a White Gaussian excitation of a spectral density, S_0. Again, to find the an expression for the mean power, we cast the problem in the Itô stochastic form as

$$\begin{aligned} dx_1 &= x_2 dt; \\ dx_2 &= (-c_{eff}x_2 - \omega_n^2 x_1 - \beta x_1^3 + x_3)dt; \\ dx_3 &= x_4 dt; \\ dx_4 &= (-\gamma x_4 - \omega_c^2 x_3)dt + \gamma^{1/2}\omega_c S dB; \end{aligned} \tag{11}$$

where $[x_1, x_2, x_3, x_4] = [u, \dot{u}, F, \dot{F}]$. With that, the joint PDF, $P(x_1, x_2, x_3, x_4, t)$, of the response can be obtained by solving the linear diffusion FPK equation which can be expressed for system (11) as

$$\frac{\partial P}{\partial t} = -\frac{\partial}{\partial x_1}\left\{x_2 P\right\} + \frac{\partial}{\partial x_2}\left\{P(c_{eff}x_2 + \omega_n^2 x_1 + \beta x_1^3 - x_3)\right\} - \frac{\partial}{\partial x_3}\left\{x_4 P\right\} + \frac{\partial}{\partial x_4}\left\{P(\gamma x_4 + \omega_c^2 x_3)\right\} + \alpha\frac{\partial^2 P}{\partial^2 x_4} \tag{12}$$

subjected to the boundary conditions $P(-\infty, t) = P(\infty, t) = 0$. Here $\alpha = \gamma^{1/2}\omega_c S$. Even in the steady-state case, an exact solution of Equation (12) is not attainable. To obtain an approximate solution, we use the Van Kampen expansion.[25]

In the Van Kampen expansion, which was introduced in the context of some statistical physics problem, the variables are expanded in a successive powers of the excitation's spectral density α. That is,

$$\begin{aligned} x_1 &= \alpha^{1/2}\eta_1 + \mathcal{O}(\alpha^{3/2}) \\ x_2 &= \alpha^{1/2}\eta_2 + \mathcal{O}(\alpha^{3/2}) \\ x_3 &= \alpha^{1/2}\eta_3 + \mathcal{O}(\alpha^{3/2}) \\ x_4 &= \alpha^{1/2}\eta_4 + \mathcal{O}(\alpha^{3/2}) \end{aligned} \tag{13}$$

The reason for expanding x_i in orders of $\alpha^{1/2}$ stems from our previous knowledge that the variance of x_i or $\langle x_i^2 \rangle$ which is a measure of the response amplitude will turn out to be proportional to α. With this expansion, the PDF becomes a function of the new variables η_i as

$$P(x_1, x_2, x_3, x_4, t) = P(\alpha^{1/2}\eta_1, \alpha^{1/2}\eta_2, \alpha^{1/2}\eta_3, \alpha^{1/2}\eta_4, t) = G(\eta_1, \eta_2, \eta_3, \eta_4, t) \tag{14}$$

In terms of the new PDF, the FPK equation becomes

$$\frac{\partial G}{\partial t} = -\eta_2 \frac{\partial G}{\partial \eta_1} + \frac{\partial G}{\partial \eta_2}\left\{\omega_n^2 \eta_1 + \beta\alpha\eta_1^3 - \eta_3\right\} + c_{eff}\frac{\partial}{\partial \eta_2}\{G\eta_2\} - \eta_4 \frac{\partial G}{\partial \eta_3} + \omega_c^2 \eta_3 \frac{\partial G}{\partial \eta_4} + \frac{\partial^2 G}{\partial^2 \eta_4} + \mathcal{O}(\alpha^{3/2}) \tag{15}$$

subjected to the boundary conditions $G(-\infty, t) = G(\infty, t) = 0$. Next, we generate the equations governing the response statistics (statistical moments). For a general function $\Phi(\eta_1, \eta_2, \eta_3, \eta_4)$, the response statistics, i.e., $\langle \Phi \rangle$, can be obtained by multiplying both sides of Equation (15) by Φ and integrating by parts over the entire space. That is

$$\begin{aligned} \int_{-\infty}^{\infty}\int_{-\infty}^{\infty}\int_{-\infty}^{\infty}\int_{-\infty}^{\infty} \Phi \frac{\partial G}{\partial t} d\eta_1 d\eta_2 d\eta_3 d\eta_4 = \int_{-\infty}^{\infty}\int_{-\infty}^{\infty}\int_{-\infty}^{\infty}\int_{-\infty}^{\infty} \Phi \Big\{ -\eta_2 \frac{\partial G}{\partial \eta_1} + \frac{\partial G}{\partial \eta_2}\left\{\omega_n^2 \eta_1 + \beta\alpha\eta_1^3 - \eta_3\right\} \\ + c_{eff}\frac{\partial}{\partial \eta_2}\{G\eta_2\} - \eta_4 \frac{\partial G}{\partial \eta_3} + \omega_c^2 \eta_3 \frac{\partial G}{\partial \eta_4} + \frac{\partial^2 G}{\partial^2 \eta_4} \Big\} d\eta_1 d\eta_2 d\eta_3 d\eta_4 \end{aligned} \tag{16}$$

For the sake of illustration, we find the expected mean square square value of the variable η_1, i.e., $\Phi = \eta_1^2$. In this case, we can write

$$\begin{aligned} \int_{-\infty}^{\infty}\int_{-\infty}^{\infty}\int_{-\infty}^{\infty}\int_{-\infty}^{\infty} \eta_1^2 \frac{\partial G}{\partial t} d\eta_1 d\eta_2 d\eta_3 d\eta_4 = \int_{-\infty}^{\infty}\int_{-\infty}^{\infty}\int_{-\infty}^{\infty}\int_{-\infty}^{\infty} \eta_1^2 \Big\{ -\eta_2 \frac{\partial G}{\partial \eta_1} + \frac{\partial G}{\partial \eta_2}\left\{\omega_n^2 \eta_1 + \beta\alpha\eta_1^3 - \eta_3\right\} \\ + c_{eff}\frac{\partial}{\partial \eta_2}\{G\eta_2\} - \eta_4 \frac{\partial G}{\partial \eta_3} + \omega_c^2 \eta_3 \frac{\partial G}{\partial \eta_4} + \frac{\partial^2 G}{\partial^2 \eta_4} \Big\} d\eta_1 d\eta_2 d\eta_3 d\eta_4 \end{aligned} \tag{17}$$

The left-hand side of the equation can be rewritten as

$$\int_{-\infty}^{\infty}\int_{-\infty}^{\infty}\int_{-\infty}^{\infty}\int_{-\infty}^{\infty} \eta_1^2 \frac{\partial G}{\partial t} d\eta_1 d\eta_2 d\eta_3 d\eta_4 = \frac{\partial}{\partial t}\int_{-\infty}^{\infty}\int_{-\infty}^{\infty}\int_{-\infty}^{\infty}\int_{-\infty}^{\infty} \eta_1^2 G d\eta_1 d\eta_2 d\eta_3 d\eta_4 = \frac{d}{dt}\langle \eta_1^2 \rangle \tag{18}$$

The right-hand side can be integrated by parts. Since the boundary conditions on the FPK equation are identically zero, all the terms, except the first one vanish upon integration by parts. This leads to

$$\begin{aligned} \frac{d}{dt}\langle \eta_1^2 \rangle &= -\int_{-\infty}^{\infty}\int_{-\infty}^{\infty}\int_{-\infty}^{\infty}\int_{-\infty}^{\infty} \eta_1^2 \eta_2 \frac{\partial G}{\partial \eta_1} d\eta_1 d\eta_2 d\eta_3 d\eta_4 \\ &= -\eta_1^2 \eta_2 G\Big|_{-\infty}^{\infty} + 2\int_{-\infty}^{\infty}\int_{-\infty}^{\infty}\int_{-\infty}^{\infty}\int_{-\infty}^{\infty} \eta_1 \eta_2 G d\eta_1 d\eta_2 d\eta_3 d\eta_4 \\ &= 2\langle \eta_1 \eta_2 \rangle \end{aligned} \tag{19}$$

In a similar manner, we generate the equations governing the other response statistics. Keeping only moments up to fourth order, we obtain 45 linearly-coupled equations that need to be solved together for the response statistics. These equations are listed in Appendix A.

4 Results and Discussion

To analyze the effect of the excitation bandwidth and center frequency on the expected stationary value of the mean power, $\langle \mathcal{P} \rangle$, we study variation of the stationary mean square velocity, $\langle \eta_2^2 \rangle$, (proportional to the power), with the aforementioned parameters. To that end, we set the time derivatives in the equations governing the response statitics to zero and solve the resulting 45 linearly-coupled algebraic equations for the stationary response statistics. Figure 3 depicts variation of $\theta\gamma^2\langle \eta^2 \rangle$ with the nonlinearity coefficient β for an excitation centered at the natural frequency of the harvester, i.e. $\omega_c = \omega_n$.

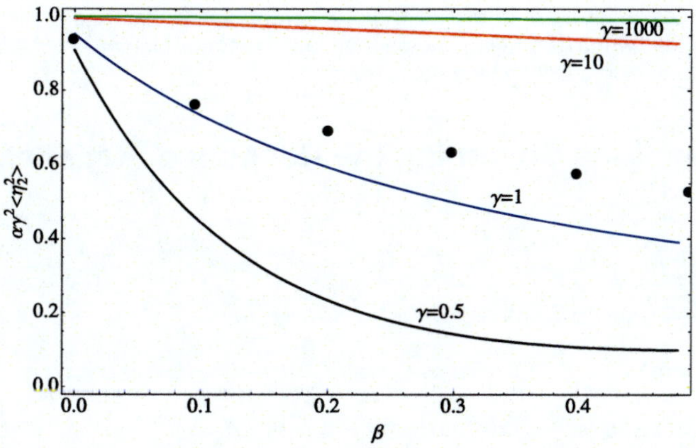

Figure 3: Variation of the mean square velocity with the nonlinearity coefficient β for different values of γ. Results are obtained for $c_{eff} = 0.05$, $S = 0.05$, $\omega_n = 1$, and $\omega_c = 1$. Circles represent solutions obtained via long time integration ($100000sec$) of the equations of motion at $\gamma = 1$. The integration was carried using the Stochastic Communication Toolbox in Matlab.

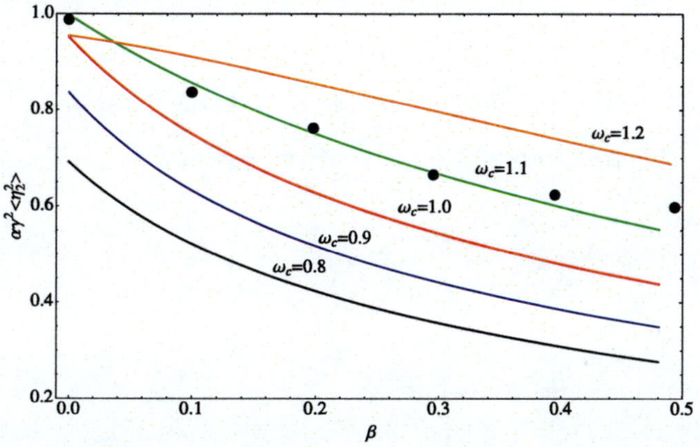

Figure 4: Variation of the mean square velocity with the nonlinearity coefficient β for different values of ω_c. Results are obtained for $c_{eff} = 0.05$, $S = 0.05$, $\omega_n = 1$, and $\gamma = 1$. Circles represent solutions obtained via long time integration ($100000sec$) of the equations of motion at $\omega_c = 1.1$. The integration was carried using the Stochastic Communication Toolbox in Matlab.

We note that, for large bandwidths approaching the White noise limit, the mean square velocity and hence power are insensitive to variations in the nonlinearity. This corroborates the results of Equation (7). Indeed for the parameters used in the simulations, the expected value of the velocity as calculated from Equation (7) is $\langle x_2^2 \rangle = 1$, which represents the same limit the curves of Fig. 3 are approaching when γ approaches infinity.

As the bandwidth of the excitation is decreased, the mean power becomes sensitive to variations in the nonlinearity coefficient. It is shown that the mean power decreases as the nonlinearity increases for small excitation bandwidths. This indicates that the nonlinearity does not improve the output power even when the excitations are colored. In other words, the best performance is always attained when $\beta = 0$.

Figure 4 depicts variation of the mean square velocity with the nonlinearity coefficient for different center frequencies. It is shown that, in general, the trend of the mean power decreasing with the nonlinearity continues even for excitations that are not centered at the natural frequency of the harvester. Furthermore, since the nonlinearity is of the hardening type, and large amplitude steady-state responses under deterministic excitations occur for excitations that are slightly larger than ω_n, it can be beneficial to tune the harvester at a center frequency that is slightly larger then the natural frequency of the harvester. This can be recommended for energy harvesters that posses small inherent stiffness-type nonlinearities.

5 Conclusions and Future Work

In this manuscript, we studied the response of uni-modal duffing harvesters to Gaussian White and Colored excitations. We showed that the mean output power of the harvester under White Gaussian excitations is not influenced by stiffness-type nonlinearities. We also demonstrated that other type of nonlinearities such as damping and inertia nonlinearities can be beneficial to the harvester's operation. Furthermore, our results show that stiffness-type nonlinearities hinder the efficiency of the harvester under Colored excitations of different bandwidths and center frequencies. Hence, such nonlinearities should be avoided when harvesting energy from environmental Colored excitations. Current work is focused on extending this analysis to bi-modal duffing type harvesters.

A Appendix: Equations Governing the Response Statistics

$$\frac{d}{dt}\langle\eta_1^2\rangle = 2\langle\eta_1\eta_2\rangle, \tag{A.1}$$

$$\frac{d}{dt}\langle\eta_2^2\rangle = -2\omega_n^2\langle\eta_1\eta_2\rangle - 2\beta\alpha\langle\eta_1^3\eta_2\rangle + 2\langle\eta_2\eta_3\rangle - 2c_{eff}\langle\eta_2^2\rangle \tag{A.2}$$

$$\frac{d}{dt}\langle\eta_3^2\rangle = 2\langle\eta_3\eta_4\rangle, \tag{A.3}$$

$$\frac{d}{dt}\langle\eta_1\eta_2\rangle = \langle\eta_2^2\rangle - \omega_n^2\langle\eta_1^2\rangle - \beta\alpha\langle\eta_1^4\rangle + \langle\eta_1\eta_3\rangle - c_{eff}\langle\eta_1\eta_2\rangle \tag{A.4}$$

$$\frac{d}{dt}\langle\eta_1\eta_3\rangle = \langle\eta_2\eta_3\rangle + \langle\eta_1\eta_4\rangle, \tag{A.5}$$

$$\frac{d}{dt}\langle\eta_1\eta_4\rangle = \langle\eta_2\eta_4\rangle - \gamma\langle\eta_1\eta_4\rangle - \omega_c^2\langle\eta_1\eta_3\rangle \tag{A.6}$$

$$\frac{d}{dt}\langle\eta_4^2\rangle = -2\gamma\langle\eta_4^2\rangle - 2\omega_c^2\langle\eta_3\eta_4\rangle + 2 \tag{A.7}$$

$$\frac{d}{dt}\langle\eta_2\eta_4\rangle = -\omega_n^2\langle\eta_1\eta_4\rangle - \alpha\beta\langle\eta_1^3\eta_4\rangle + \langle\eta_3\eta_4\rangle - (c_{eff}+\gamma)\langle\eta_2\eta_4\rangle - \omega_c^2\langle\eta_2\eta_3\rangle \tag{A.8}$$

$$\frac{d}{dt}\langle\eta_2\eta_3\rangle = -\omega_n^2\langle\eta_1\eta_3\rangle - \beta\alpha\langle\eta_1^3\eta_3\rangle + \langle\eta_3^2\rangle - c_{eff}\langle\eta_2\eta_3\rangle + \langle\eta_2\eta_4\rangle; \tag{A.9}$$

$$\frac{d}{dt}\langle\eta_3\eta_4\rangle = \langle\eta_4^2\rangle - \gamma\langle\eta_3\eta_4\rangle - \omega_c^2\langle\eta_3^2\rangle \tag{A.10}$$

$$\frac{d}{dt}\langle\eta_2\eta_1^3\rangle = 3\langle\eta_1^2\eta_2^2\rangle - \omega_n^2\langle\eta_1^4\rangle + \langle\eta_1^3\eta_3\rangle - c_{eff}\langle\eta_1^3\eta_2\rangle, \tag{A.11}$$

$$\frac{d}{dt}\langle\eta_1^4\rangle = 4\langle\eta_1^3\eta_2\rangle \tag{A.12}$$

$$\frac{d}{dt}\langle\eta_1^3\eta_4\rangle = 3\langle\eta_1^2\eta_2\eta_4\rangle - \gamma\langle\eta_1^3\eta_4\rangle - \omega_c^2\langle\eta_1^3\eta_3\rangle, \tag{A.13}$$

$$\frac{d}{dt}\langle\eta_1^3\eta_3\rangle = 3\langle\eta_1^2\eta_2\eta_3\rangle + \langle\eta_1^3\eta_4\rangle \tag{A.14}$$

$$\frac{d}{dt}\langle\eta_1^2\eta_2^2\rangle = 2\langle\eta_1\eta_2^3\rangle - 2\omega_n^2\langle\eta_1^3\eta_2\rangle + 2\langle\eta_1^2\eta_2\eta_3\rangle - 2c_{eff}\langle\eta_1^2\eta_2^2\rangle \tag{A.15}$$

$$\frac{d}{dt}\langle\eta_1^2\eta_2\eta_4\rangle = 2\langle\eta_1\eta_2^2\eta_4\rangle + \langle\eta_1^2\eta_3\eta_4\rangle - (\gamma+c_{eff})\langle\eta_1^2\eta_2\eta_4\rangle - \omega_c^2\langle\eta_1^2\eta_2\eta_3\rangle - \omega_n^2\langle\eta_1^3\eta_4\rangle \tag{A.16}$$

$$\frac{d}{dt}\langle\eta_1^2\eta_2\eta_3\rangle = 2\langle\eta_1\eta_2^2\eta_3\rangle + \langle\eta_1^2\eta_3^2\rangle + \langle\eta_1^2\eta_2\eta_4\rangle - c_{eff}\langle\eta_1^2\eta_2\eta_3\rangle - \omega_n^2\langle\eta_1^3\eta_3\rangle \tag{A.17}$$

$$\frac{d}{dt}\langle\eta_1\eta_2^3\rangle = \langle\eta_2^4\rangle + 3\langle\eta_1\eta_2^2\eta_3\rangle - 3c_{eff}\langle\eta_1\eta_2^3\rangle - 3\omega_n^2\langle\eta_1^2\eta_2^2\rangle \tag{A.18}$$

$$\frac{d}{dt}\langle\eta_1\eta_2^2\eta_4\rangle = \langle\eta_2^3\eta_4\rangle + 2\langle\eta_1\eta_2\eta_3\eta_4\rangle - (\gamma+2c_{eff})\langle\eta_1\eta_2^2\eta_4\rangle - \omega_c^2\langle\eta_1\eta_2^2\eta_3\rangle - 2\omega_n^2\langle\eta_1^2\eta_2\eta_4\rangle \tag{A.19}$$

$$\frac{d}{dt}\langle\eta_1^2\eta_3\eta_4\rangle = 2\langle\eta_1\eta_2\eta_3\eta_4\rangle + \langle\eta_1^2\eta_4^2\rangle - \gamma\langle\eta_1\eta_2\eta_3\eta_4\rangle - \omega_c^2\langle\eta_1^2\eta_3^2\rangle \tag{A.20}$$

$$\frac{d}{dt}\langle\eta_1\eta_2^2\eta_3\rangle = \langle\eta_2^3\eta_3\rangle + 2\langle\eta_1\eta_2\eta_3^2\rangle + \langle\eta_1\eta_2^2\eta_4\rangle - 2c_{eff}\langle\eta_1\eta_2^2\eta_3\rangle - 2\omega_n^2\langle\eta_1^2\eta_2\eta_3\rangle \tag{A.21}$$

$$\frac{d}{dt}\langle\eta_1^2\eta_3^2\rangle = 2\langle\eta_1\eta_2\eta_3^2\rangle + 2\langle\eta_1^2\eta_3\eta_4\rangle \tag{A.22}$$

$$\frac{d}{dt}\langle\eta_2^4\rangle = 4\langle\eta_2^3\eta_3\rangle - 4c_{eff}\langle\eta_2^4\rangle - 4\omega_n^2\langle\eta_1\eta_2^3\rangle \tag{A.23}$$

$$\frac{d}{dt}\langle\eta_2^3\eta_4\rangle = 3\langle\eta_2^2\eta_3\eta_4\rangle - (\gamma+3c_{eff})\langle\eta_2^3\eta_4\rangle - \omega_c^2\langle\eta_2^3\eta_3\rangle - 3\omega_n^2\langle\eta_1\eta_2^2\eta_4\rangle \tag{A.24}$$

$$\frac{d}{dt}\langle\eta_1\eta_2\eta_3\eta_4\rangle = \langle\eta_2^2\eta_3\eta_4\rangle + \langle\eta_1\eta_3^2\eta_4\rangle + \langle\eta_1\eta_2\eta_4^2\rangle - (\gamma+c_{eff})\langle\eta_1\eta_2\eta_3\eta_4\rangle - \omega_c^2\langle\eta_1\eta_2\eta_3^2\rangle - \omega_n^2\langle\eta_1^2\eta_3\eta_4\rangle \tag{A.25}$$

$$\frac{d}{dt}\langle\eta_1^2\eta_4^2\rangle = 2\langle\eta_1^2\rangle + 2\langle\eta_1\eta_2\eta_4^2\rangle - 2\gamma\langle\eta_1^2\eta_4^2\rangle - 2\omega_c^2\langle\eta_1^2\eta_3\eta_4\rangle \tag{A.26}$$

$$\frac{d}{dt}\langle\eta_2^3\eta_3\rangle = 3\langle\eta_2^2\eta_3^2\rangle + \langle\eta_2^3\eta_4\rangle - 3c_{eff}\langle\eta_2^3\eta_3\rangle - 3\omega_n^2\langle\eta_1\eta_2^2\eta_3\rangle \tag{A.27}$$

$$\frac{d}{dt}\langle\eta_1\eta_2\eta_3^2\rangle = \langle\eta_2^2\eta_3^2\rangle + \langle\eta_1\eta_3^3\rangle + 2\langle\eta_1\eta_2\eta_3\eta_4\rangle - c_{eff}\langle\eta_1\eta_2\eta_3^2\rangle - \omega_n^2\langle\eta_1^2\eta_3^2\rangle \tag{A.28}$$

$$\frac{d}{dt}\langle\eta_2^2\eta_3\eta_4\rangle = 2\langle\eta_2\eta_3^2\eta_4\rangle + \langle\eta_2^2\eta_4^2\rangle - (\gamma + 2c_{eff})\langle\eta_2^2\eta_3\eta_4\rangle - \omega_c^2\langle\eta_2^2\eta_3^2\rangle - 2\omega_n^2\langle\eta_1\eta_2\eta_3\eta_4\rangle \tag{A.29}$$

$$\frac{d}{dt}\langle\eta_1\eta_3^2\eta_4\rangle = \langle\eta_2\eta_3^2\eta_4\rangle + 2\langle\eta_1\eta_3\eta_4^2\rangle - \gamma\langle\eta_1\eta_3^2\eta_4\rangle - \omega_c^2\langle\eta_1\eta_3^3\rangle \tag{A.30}$$

$$\frac{d}{dt}\langle\eta_1\eta_2\eta_4^2\rangle = \langle\eta_2^2\eta_4^2\rangle + \langle\eta_1\eta_3\eta_4^2\rangle - (2\gamma + c_{eff})\langle\eta_1\eta_2\eta_4^2\rangle - 2\omega_c^2\langle\eta_1\eta_2\eta_3\eta_4\rangle - \omega_n^2\langle\eta_1^2\eta_4^2\rangle + 2\langle\eta_1\eta_2\rangle \tag{A.31}$$

$$\frac{d}{dt}\langle\eta_2^2\eta_3^2\rangle = 2\langle\eta_2\eta_3^3\rangle + 2\langle\eta_2^2\eta_3\eta_4\rangle - 2c_{eff}\langle\eta_2^2\eta_3^2\rangle - 2\omega_n^2\langle\eta_1\eta_2\eta_3^2\rangle \tag{A.32}$$

$$\frac{d}{dt}\langle\eta_1\eta_3^3\rangle = \langle\eta_2\eta_3^3\rangle + 3\langle\eta_1\eta_3^2\eta_4\rangle \tag{A.33}$$

$$\frac{d}{dt}\langle\eta_2\eta_3^2\eta_4\rangle = \langle\eta_3^3\eta_4\rangle + 2\langle\eta_2\eta_3\eta_4^2\rangle - (\gamma + c_{eff})\langle\eta_2\eta_3^2\eta_4\rangle - \omega_c^2\langle\eta_2\eta_3^3\rangle - \omega_n^2\langle\eta_1\eta_3^2\eta_4\rangle \tag{A.34}$$

$$\frac{d}{dt}\langle\eta_2^2\eta_4^2\rangle = 2\langle\eta_1\eta_3\eta_4^2\rangle - 2(\gamma + c_{eff})\langle\eta_2^2\eta_4^2\rangle - 2\omega_c^2\langle\eta_2^2\eta_3\eta_4\rangle - 2\omega_n^2\langle\eta_1\eta_2\eta_4^2\rangle + 2\langle\eta_2^2\rangle \tag{A.35}$$

$$\frac{d}{dt}\langle\eta_1\eta_3\eta_4^2\rangle = \langle\eta_2\eta_3\eta_4^2\rangle + \langle\eta_1\eta_4^3\rangle - 2\gamma\langle\eta_1\eta_3\eta_4^2\rangle - 2\omega_c^2\langle\eta_1\eta_3^2\eta_4\rangle + 2\langle\eta_1\eta_3\rangle \tag{A.36}$$

$$\frac{d}{dt}\langle\eta_2\eta_3^3\rangle = \langle\eta_3^4\rangle + 3\langle\eta_2\eta_3^2\eta_4\rangle - c_{eff}\langle\eta_2\eta_3^3\rangle - \omega_n^2\langle\eta_1\eta_3^3\rangle \tag{A.37}$$

$$\frac{d}{dt}\langle\eta_3^3\eta_4\rangle = 3\langle\eta_3^2\eta_4^2\rangle - \gamma\langle\eta_3^3\eta_4\rangle - \omega_c^2\langle\eta_3^4\rangle \tag{A.38}$$

$$\frac{d}{dt}\langle\eta_2\eta_3\eta_4^2\rangle = \langle\eta_3^2\eta_4^2\rangle + \langle\eta_2\eta_4^3\rangle - (2\gamma + c_{eff})\langle\eta_2\eta_3\eta_4^2\rangle - 2\omega_c^2\langle\eta_2\eta_3^2\eta_4\rangle - \omega_n^2\langle\eta_1\eta_3\eta_4^2\rangle + 2\langle\eta_2\eta_3\rangle \tag{A.39}$$

$$\frac{d}{dt}\langle\eta_1\eta_4^3\rangle = \langle\eta_2\eta_4^3\rangle - 3\gamma\langle\eta_1\eta_4^3\rangle - 3\omega_c^2\langle\eta_1\eta_3\eta_4^2\rangle + 6\langle\eta_1\eta_4\rangle \tag{A.40}$$

$$\frac{d}{dt}\langle\eta_3^4\rangle = 4\langle\eta_3^3\eta_4\rangle \tag{A.41}$$

$$\frac{d}{dt}\langle\eta_3^2\eta_4^2\rangle = 2\langle\eta_3\eta_4^3\rangle - 2\gamma\langle\eta_3^2\eta_4^2\rangle - 2\omega_c^2\langle\eta_3\eta_4^3\rangle + 2\langle\eta_3^2\rangle \tag{A.42}$$

$$\frac{d}{dt}\langle\eta_2\eta_4^3\rangle = \langle\eta_3\eta_4^3\rangle - (3\gamma + c_{eff})\langle\eta_2\eta_4^3\rangle - 3\omega_c^2\langle\eta_2\eta_3\eta_4^2\rangle - \omega_n^2\langle\eta_1\eta_4^3\rangle + 6\langle\eta_2\eta_4\rangle \tag{A.43}$$

$$\frac{d}{dt}\langle\eta_3\eta_4^3\rangle = \langle\eta_4^4\rangle - 3\gamma\langle\eta_3\eta_4^3\rangle - 3\omega_c^2\langle\eta_3^2\eta_4^2\rangle + 6\langle\eta_3\eta_4\rangle \tag{A.44}$$

$$\frac{d}{dt}\langle\eta_4^4\rangle = -4\gamma\langle\eta_4^4\rangle - 4\omega_c^2\langle\eta_3\eta_4^3\rangle + 12\langle\eta_4^2\rangle \tag{A.45}$$

References

[1] A. J. duPlessis, M. J. Huigsloot, and F. D. Discenzo. Resonant Packaged Piezoelectric Power Harvester for Machinery Health Monitoring. In *Proceedings of Smart Structures and Materials Conference, SPIE*, page 5762, San Diego, CA, 2005.

[2] D. J. Inman and B. L. Grisso. Towards Autonomous Sensing. In *Proceedings of Smart Structures and Materials Conference, SPIE*, page 61740T, San Diego, CA, 2006.

[3] R. S. Sanders and M. T. Lee. Implantable Pacemakers. *Proceedings of the IEEE*, 84(3), 1995.

[4] I. D. Capel, H. M. Dorrell, E. P. Spencer, and M. W. Davis. The Amelioration of the Suffering Associated with Spinal Cord Injury with Subperception Transcranial Electrical Stimulation. *Spinal Cord*, 41:109–117, 2003.

[5] G. Renzenbrink and M. J Jzerman. Percutaneous Neuromuscular Electrical Stimulation for Treating Shoulder Pain in Chronic Hemiplegia. Effects on Shoulder Pain and Quality of Life. *Clinical Rehabilitation*, 18:359–365, 2004.

[6] S. Roundy, P. K. Wright, and J. Rabaey. A Study of Low Level Vibrations as a Power Source for Wireless Sensor Nodes. *Computer Communications*, 26:1131–1144, 2003.

[7] S. Roundy and P. K. Wright. A Piezoelectric Vibration-Based Generator for Wireless Electronics. *Journal of Intelligent Materials and Structures*, 16:809–823, 2005.

[8] S. W. Arms, C. P. Townsend, D. L. Churchill, G. H. Galbreath, and S. W. Mundell. Power Management for Energy Harvesting Wireless Sensors. In *Proceedings of the Smart Structures and Materials Conference, SPIE*, pages 5763 267–75, San Diego, CA, 2005.

[9] J. A. Paradiso and Starner T. Energy Scavenging for Mobile and Wireless Electronics. *IEEE Pervasive Computing*, 4:18–27, 2005.

[10] H. Sodano, D. J. Inman, and G.Park. A Review of Power Harvesting from Vibration Using Piezoelectric Materials. *The Shock and Vibration Digest*, 36:197–205, 2004.

[11] H. Sodano, D. J. Inman, and G.Park. Generation and Storage of Electricity from Power Harvesting Devices. *Journal of Intelligent Material Systems and Structures*, 16:67–75, 2005.

[12] S. Roundy. On the Effectiveness of Vibration-based Energy Harvesting. *Journal of Intelligent Materials and Structures*, 16:809–823, 2005.

[13] S. Roundy and Y. Zhang. Toward Self-Tuning Adaptive Vibration-Based Micro-Generators. In *Smart Materials, Nano- and Micro-Smart Systems*, Sydney, Australia, 2005.

[14] W. Wu, Y. Chen, B. Lee, J. He, and Y. Peng. Tunable Resonant Frequency Power Harvesting Devices. In *Proceedings of Smart Structures and Materials Conference, SPIE*, page 61690A, San Diego, CA, 2006.

[15] V. Challa, M. Prasad, Y. Shi, and F. Fisher. A Vibration Energy Harvesting Device with Bidirectional Resonance Frequency Tunability. *Smart Materials and Structures*, 75:1–10, 2008.

[16] S. M. Shahruz. Design of Mechanical Band-Pass Filters for Energy Scavenging. *Journal of Sound and Vibrations*, 292:987–998, 2006.

[17] S. M. Shahruz. Limits of Performance of Mechanical Band-Pass Filters Used in Energy Harvesting. *Journal of Sound and Vibrations*, 294:449–461, 2006.

[18] B. Mann and N. Sims. Energy Harvesting from the Nonlinear Oscillations of Magnetic Levitation. *Journal of Sound and Vibrations*, 319:515–530, 2008.

[19] A. Erturk, J. Renno, and D. Inman. Modeling of Energy Harvesting form an L-shaped Beam-mass Structure with an Application to UAVs. *Journal of Intelligent Material Systems and Structures*, 20:1–16, 2008.

[20] D. Barton, S. Burrow, and L. Clare. Energy Harvesting from Vibrations with a Nonlinear Oscillator. In *Proceedings of the ASME 2009 International Design Engineering Technical Conference and Computers and Information in Engineering Conference*, San Diego, CA, 2009.

[21] D. Quinn, D. Vakakis, and L. Bergman. Vibration Energy Harvesting with Essential Nonlinearities. In *Proceedings of the ASME 2007 International Design Engineering Technical Conference and Computers and Information in Engineering Conference*, Las Vegas, NV, 2007.

[22] M. F. Daqaq. New Insight into Energy Harvesting via Axially-Loaded Beams. In *Proceedings of the ASME 2009 International Design Engineering Technical Conference and Computers and Information in Engineering Conference*, San Diego, CA, 2009.

[23] K. Ito. Stochastic Integral. *Proceedings of the Imperial Academy, Tokyo*, 20:519–524, 1944.

[24] A. H. Jazwinski. *Stochastic Processes and Filtering Theory*. Academic Press, NY, 1970.

[25] R. F. Rodriguez and N. G. Van Kampen. Systematic Treatment of Fluctuations in a Nonlinear Oscillator. *Physica A*, 85:347–362, 1976.

Proceedings of the IMAC-XXVIII
February 1–4, 2010, Jacksonville, Florida USA

Self-powered Active Control of Structures with TMDs

Xiudong Tang and Lei Zuo

Department of Mechanical Engineering, State University of New York at Stony Brook

Stony Brook, New York 11794

Email: lei.zuo@stonybrook.edu

NOMENCLATURE

M, K, C	Mass, stiffness and damping matrices
k_t,k_e	Thrust constant and back electromotive force coefficient of electromagnetic motor
R, L, C	Electrical resistance, inductance and capacitance
C_d	The object's drag coefficient,
f, μ	Turning ratio, mass ratio of TMD
ξ	Damping ratio
F, u	Excitation and control forces
x_s, x_1	Displacements of structure and TMD

ABSTRACT

This paper studies the feasibility of self-powered active vibration control of structures with Tuned Mass Dampers (TMDs). Without consuming external energy, the proposed self-powered vibration control strategy can provide better performance than the passive TMD in mitigating the vibration of buildings induced by wind load. The vibration-dissipative element of TMD is replaced with an electromagnetic machine, which serves as actuator and harvester at the same time. First, the desired force is obtained by adopting Linear Quadratic Regulator (LQG) optimal control. Then, the self-powered active control strategy is designed based on the essence of the desired force, which can be passive or active types. The energy of the vibration can be harvested as electricity in *energy harvesting mode*, and pumped back into the mechanical system when instant active force is required. Switch based circuits with capability of bi-directional power flow are presented for the implementation of self-powered active TMDs. With taking the efficiency of the harvesting and driving circuits into account, numerical simulations are carried out based on a tall building with a TMD under random and harmonic excitations. The results indicate the feasibility and effectiveness of self-powered active TMDs.

1. INTRODUCTION

Vibration has been a serious concern since the early days when the tall buildings or long-span bridges were built. These structures are subjected to huge dynamic loadings from the winds, earthquakes, water waves, traffics, and human motions. The large vibration amplitude can damage the structures or the secondary components, or cause discomfort to its human occupants [1]. Extensive research has been conducted to mitigate this harmful vibration, by, for example, structure design [2], vibration isolation systems [3], and auxiliary damping systems. Among these methods, the TMD has been proved to be a very simple and effective vibration suppression device, with many

T. Proulx (ed.), *Structural Dynamics and Renewable Energy, Volume 1*, Conference Proceedings of the Society for Experimental Mechanics Series 10, DOI 10.1007/978-1-4419-9716-6_21, © The Society for Experimental Mechanics, Inc. 2011

practical implementations on tall buildings, such as Taipei 101 in Taipei, Citi Group in New York, and many others. The original TMD, which was invented by Frahm in 1911 [4], only consists of an auxiliary mass and a spring connected to the primary system to mitigate vibration in a very narrow frequency range near the resonance. Ormondroyd and Den Hartog [5] increased the working frequency range by introducing an additional damper.

Although TMD composes the greatest percentage of the supplemental damping systems currently in use, it has its inherent drawbacks. When the parameters of the primary systems change a small amount, the performance of TMDs will be greatly defected. It is so called off-tuning problem, which has been investigated by researchers, for example, Bergman [6] and Setareh [7]. Hence, active TMD is developed to handle this problem. It has been proved that the active TMD can provide better vibration mitigation performance than the passive one, at the cost of large amount energy [8]. However, the active TMD system is more complex and costly, which limits its practical implementations. Particularly active TMDs suffer from the inherent disadvantage that it relies on external energy, which is not countable in the hazardous situations like earthquakes or hurricanes. Hybrid active TMD, which is composed of passive TMD and one additional actuator, can reduce the power consumption and increase the robustness in the event of power failure [9, 10].

Realizing both the vibration mitigation performance as well as the limitations of active TMDs, researchers proposed semi-active TMDs to provide better vibration mitigation performance than the passive one without the drawbacks of active TMDs. Semi-active control using Magneto Rheological (MR) or variable orifice dampers has been investigated by many researchers [11-15], especially in vibration control of buildings with TMDs [14] and vehicle suspension systems [15].

In this paper, we proposed a self-powered active vibration control of TMD, which can provide better vibration mitigation performance than the passive TMD, without consuming external energy. The auxiliary damper of classic TMD is replaced with an electromagnetic motor, which acts both actuator and harvester in this self-powered active TMD. The kinetic energy of vibration is harvested and stored, and then used when instant active force is required. At first, the feasibility of self-powered active vibration control is studied. We classify the desired force obtained by LQG optimal control design into two categories: passive force and active force, resulting two working modes of self-powered active TMD, named as energy harvesting mode and driving mode. A switch based circuit is proposed to realize the bi-directional power flow control. With taking the efficiency and parasitic power of the circuit into account, numerical simulations are carried out based on a tall building with TMD. The results under random and harmonic excitation show that the self-powered active TMD more effective than the passive one in mitigating the vibration.

This paper is organized as follows: In Section 2, the feasibility of self-powered active vibration control of TMD will be conducted, using LQG optimal control methods. The realization of self-powered active TMD is introduced in Section 3, where the circuit capable of bi-direction power flow is also discussed. In Section 4, the extensive numerical study will be carried out, where the vibration performances of building structures with passive and active TMDs are compared and the energy balance is also analyzed with taking the efficiency and parasitic power into account. Then conclusions follow as Section 5.

2. FEASIBILITY STUDY OF SELF-POWERED ACTIVE TMD

Figure 2.1 shows the modeling of building with TMD, where the building is considered as single degree-of-freedom (SDOF) system and the damper of the TMD is replaced with an electromagnetic motor. The primary structure is subjected to the wind load disturbance F. The force u is exerted by the electromagnetic motor. Then, the dynamics equations of this building with TMD can be written as:

$$\begin{cases} m_s\ddot{x}_s + k_s x_s + c_s\dot{x}_s = k_1(x_1 - x_s) + F - u \\ m_1\ddot{x}_1 + k_1(x_1 - x_s) = u \end{cases} \tag{1}$$

which can be further expressed in the following form:

$$\mathbf{M}\ddot{\mathbf{x}}(t) + \mathbf{K}\mathbf{x}(t) + \mathbf{C}\dot{\mathbf{x}}(t) = \mathbf{G}F(t) + \mathbf{H}u(t) \tag{2}$$

where $\mathbf{M} = \begin{bmatrix} m_s & 0 \\ 0 & m_1 \end{bmatrix}$, $\mathbf{K} = \begin{bmatrix} k_s + k_1 & -k_1 \\ -k_1 & k_1 \end{bmatrix}$, $\mathbf{C} = \begin{bmatrix} c_s & 0 \\ 0 & 0 \end{bmatrix}$, $\mathbf{G} = \begin{bmatrix} 1 \\ 0 \end{bmatrix}$, $\mathbf{H} = \begin{bmatrix} -1 \\ 1 \end{bmatrix}$, $\mathbf{x}(t) = [x_s \ x_1]^T$

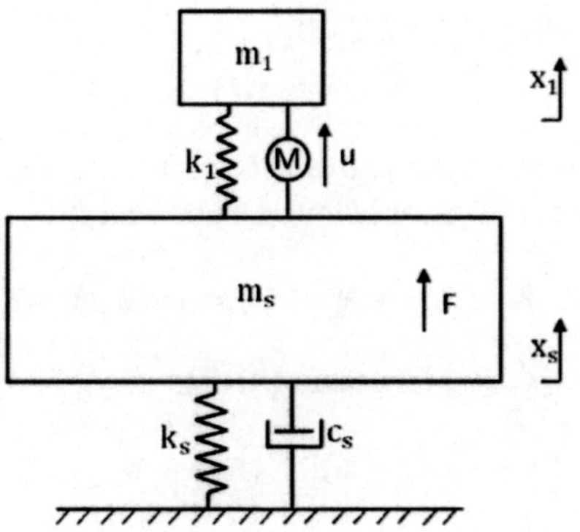

Figure 2.1 - Modeling of Building with TMD

Defining the state space as $\mathbf{q} = [q_1, q_2, q_3, q_4]^T = [x_s, x_1, \dot{x}_s, \dot{x}_1]^T$, (2) can be further written in the state space form:

$$\dot{\mathbf{q}}(t) = \mathbf{A}\mathbf{q}(t) + \mathbf{B}_f F(t) + \mathbf{B}_u u(t) \tag{3}$$

where $\mathbf{A} = \begin{bmatrix} \mathbf{0} & \mathbf{I} \\ -\mathbf{M}^{-1}\mathbf{K} & -\mathbf{M}^{-1}\mathbf{C} \end{bmatrix}$, $\mathbf{B}_f = \begin{bmatrix} \mathbf{0} \\ \mathbf{M}^{-1}\mathbf{G} \end{bmatrix}$, $\mathbf{B}_u = \begin{bmatrix} \mathbf{0} \\ \mathbf{M}^{-1}\mathbf{H} \end{bmatrix}$.

Tall buildings are usually subjected to tremendous wind load disturbance, which accounts as the main reason for the overall vibration of the building, since the local load is governed by the seismic stress and the overall motion is governed by the wind load [16]. The amplitude of dynamic wind load on the structures can be calculated as [17]:

$$F = \frac{1}{2} \rho A V^2 C_d \tag{4}$$

where ρ is the air density, V is the wind speed, and A is the projected area of the building perpendicular to the wind velocity, C_d is the object's drag coefficient, which depends on the shape of the object (about 1 for a cylinder). The wind force is usually very large, for example, using (4), the dynamic load of breeze at the speed of 12 MPH on a high-rise building of 1440ft (Taipei101) height and slenderness (height to width) ratio 9 is estimated to be 400kN, or 40 tons. Besides, the frequencies of wind load vary in a range, which can be modeled as Gaussian white noise with certain power density. Hence, it is more suitable to use LQG control as the optimal control strategy. The feedback measurements we choose are the accelerations of the main structure and auxiliary mass.

First, the desired active force u_{des} is obtained by LQG active controller design. Then u_{des} is clipped into two categories: passive force and active force. Then the self-powered active control strategy is design to make the actual force follow the desired force as much as possible. To get the desired force, the root-mean-square (RMS) value is used as the performance index to evaluate the output performance, where the output is the acceleration of the primary system, expressed as:

$$z(t) = \mathbf{C}_2\mathbf{q}(t) + \mathbf{D}_{22}F \tag{5}$$

where $\mathbf{C}_2 = \left[-\left(\frac{k_s}{m_s} + \frac{k_1}{m_s}\right),\ \frac{k_1}{m_s},\ -\frac{c_s}{m_s},\ 0\right]$, $\mathbf{D}_{22} = [1/m_s]$.

Hence the performance index in chosen to minimize the RMS acceleration of the building under limited control effort:

$$J = \int_0^\infty (z^2 + ru^2)\, dt = \int_0^\infty \begin{bmatrix} \mathbf{q} \\ \mathbf{u} \end{bmatrix}' \mathbf{T} \begin{bmatrix} \mathbf{q} \\ \mathbf{u} \end{bmatrix} dt \tag{6}$$

where $\mathbf{T}$ is obtained as:

$$\mathbf{T} = \begin{bmatrix} \mathbf{C_2}'\mathbf{C_2} & \mathbf{C_2}'\mathbf{D_{22}} \\ \mathbf{D_{22}}'\mathbf{C_2} & \mathbf{D_{22}}'\mathbf{D_{22}} + \mathrm{r} \end{bmatrix} \tag{7}$$

By minimizing the quadratic cost, we can minimize the response to Gaussian white noise input. Denoting $\mathbf{Q} = \mathbf{C_2}'\mathbf{C_2}, \mathbf{N} = \mathbf{C_2}'\mathbf{C_2}, \mathrm{R} = \mathbf{D_{22}}'\mathbf{D_{22}} + \mathrm{r}$, the optimal control law is expressed:

$$\mathrm{u_{des}} = -\mathrm{R}^{-1}(\mathbf{B_u^T S} + \mathbf{N^T})\mathbf{q}(t) = -\mathbf{Kq}(t) \tag{8}$$

where **S** is the solution of the algebraic Riccati Equation (ARE):

$$\mathbf{A^T S} + \mathbf{SA} - (\mathbf{SB_u} + \mathbf{N})\mathrm{R}^{-1}(\mathbf{B_u^T S} + \mathbf{N^T}) + \mathbf{Q} = 0 \tag{9}$$

Since we use acceleration feedback, not all the states are available as feedback. LQG state-estimator is employed to estimate the state vector q(*t*) in (3). The feedback measurement $\mathbf{y}(t)$ can be expressed in the form of

$$\mathbf{y}(t) = \mathbf{C_1 q}(t) + \mathbf{D_{12} F} + \mathbf{D_{11} u} \tag{10}$$

This LQG state-estimator is so called Kalman filter, the equation of which can be expressed as:

$$\dot{\tilde{\mathbf{q}}}_{(t)} = \mathbf{A}\tilde{\mathbf{q}}_{(t)} + \mathbf{B_u}\mathrm{u_{des}} + \mathbf{L}\left(\mathbf{y}_{(t)} - \mathbf{C_1}\tilde{\mathbf{q}}_{(t)} - \mathbf{D_{12}}\mathrm{u_{des}}\right) \tag{11}$$

where $\tilde{\mathbf{q}}_{(t)}$ is the estimated state vector. The observer gain matrix **L** is obtained as:

$$\mathbf{L} = \mathbf{PC_1^T V^{-1}} \tag{12}$$

where **V** is the covariance matrix of the measurement noise, and **P** is the solution of ARE:

$$\mathbf{AP} + \mathbf{PA^T} - \mathbf{PC_1^T V^{-1} C_1 P} = 0 \tag{13}$$

In order to verify the feasibility of self-powered active vibration control, the numerical simulation using the active controller design described above is carried out based on Taipei101, one of the tallest buildings in the world. It is installed with TMD of 730 tons (mass ratio μ=0.78%), where μ is ratio of total auxiliary masses over the modal mass of the primary system, which is typically 1/3 of the building mass [16]. The parameters of the stiffness and damping coefficient of the passive TMD are optimized using the method in the paper by Warburton [18], where the tuning ratio f_1 and damping ratio ζ_1 are given by:

$$f_1 = \frac{1}{(1+\mu)^{1/2}} \tag{14}$$

$$\zeta_1 = \sqrt{\frac{3\mu}{8(1+\mu/2)^3}} \tag{15}$$

For the active TMD, the tuning ratio is the same as the passive one. However the damper is replaced with an electromagnetic motor. The active TMD is obtained using the LQG control design described above, when subjected to a wind load disturbance with the power spectral density of $S_0=4\cdot10^{14}N^2s$. Without considering practical implementation, the power is estimated by multiplying the desired force and relative velocity between the auxiliary mass and primary structure.

$$P = u_{des} \cdot v_r \tag{16}$$

Figure 2.2 shows the transient response of energy and power using active control, when the building is subjected to random wind load disturbance. It can be seen from the figure that the power on the transducer may be positive or negative, which means the power flow is bi-directional. From the total energy goes through the transducer, we can see that it is negative after 40 seconds, which means the system is still dissipating energy. Ideally, more energy can be harvested than that used for active force. It should be noted that Nakano and Suda [19] investigated self-powered active control of vehicle suspension using sky-hook control and Scruggs [20] also investigated the feasibility based on PID control law. From figure 2.2, we can also conclude that there is minimum value of the capacity of the external reservoir. At the beginning, active force is required more frequently. As time goes by, the external reservoir is charged and the energy can be further used for active control. Without the initial energy reservoir, the system will be uncontrolled at the beginning, however, after a while the reservoir will be charged and the system can work continuously. And the excessive harvested energy can be used for other purposes, for example, to power the utilities or feed the electricity grid.

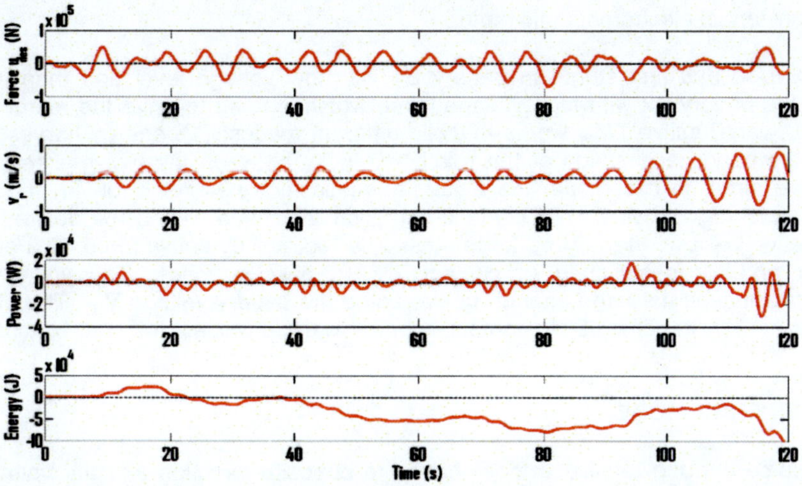

Figure 2.2 - Transient response of force, velocity, energy and power active TMD under random excitation

3. DESIGN OF THE SELF-POWERED ACTIVE TMD

In section 2, we discussed the feasibility of self-powered active vibration control of TMD, and found that even the pure active controlled TMD is essentially dissipative, which means system is still dissipating energy. In this section, we are discussing the realization of self-powered active TMD, including the control strategy of the switches based circuit, which is capable of bi-directional power flow.

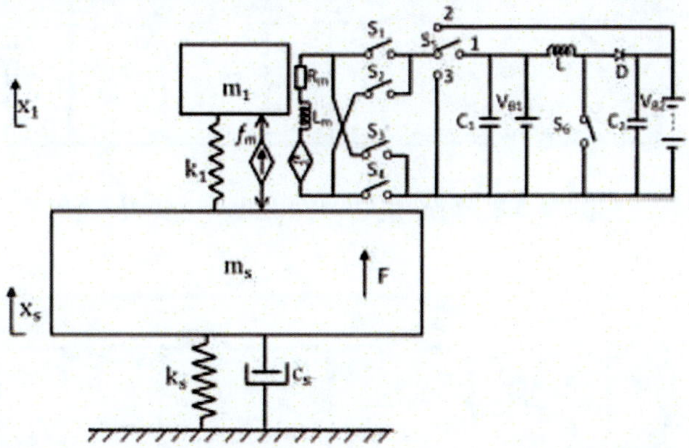

Figure 3.1 - Schematic of the self-powered active TMD system

Figure 3.1 shows the schematic of the self-powered active TMD, with the mechanical system coupled with electric circuit, which is controlled by the switches S_1-S_6. The electromagnetic motor is modeling as a voltage source with the inherent conductor L_m and resistor R_m connected in series. The relative motion between the two masses can induce a voltage e_m in the coils, which is proportional to the relative velocity of stator and rotor $v_r = \dot{x}_1 - \dot{x}_s$. Magnification mechanisms may exist, therefore, the relative speed between the stator and mover can be *M* times the relative motion between the TMD and the structure.

$$e_m = Mk_e v_r \tag{17}$$

where k_e is the back electromotive force coefficient (EMF) of the electromagnetic motor. Meanwhile, the current flows inside of the motor coil will induce a back electromotive force proportional to the current:

$$f_b = Mk_t i \tag{18}$$

where k_t is thrust constant of electromagnetic motor.

The energy is harvested and first stored in battery B1 with low voltage level and then the voltage is further boosted to charge the battery B2 with high voltage level, which is used to drive the motor when active force is required. The self-powered active TMD works in three different modes: (1). energy harvesting mode. When the desired force has the opposite direction of the relative velocity between the two masses, the desired force is essentially a passive force. In this mode, the electromagnetic motor works in driven mode, acting as energy harvester and the switch S_5 is switched to contact 1. (2). driving mode. When the desired force has the same direction as the relative velocity, the desired force should be realized by active force. The electromagnetic motor works as actuator and S_5 is switched to contact 2. (3). passive mode. The voltage generated by the electromagnetic motor maybe not large enough to overcome the battery voltage V_{B1}. The electromagnetic motor is set to be closed-circuit by switching S_5 to contact 3. In these three modes, the control schemes of the switches S_1-S_6 are different.

3.1. ENERGY HARVESTING MODE

When the desired force and the relative velocity have the opposite direction and the voltage generated by the electromagnetic mode is larger than the voltage of the first battery V_{B1}, the system will work in energy harvesting mode. In this mode, S_5 is switched to contact 1. Energy is harvested and stored in battery B_1 and booster DC-DC converter is used to step-up the voltage to further charge the battery B_2 which has high voltage.

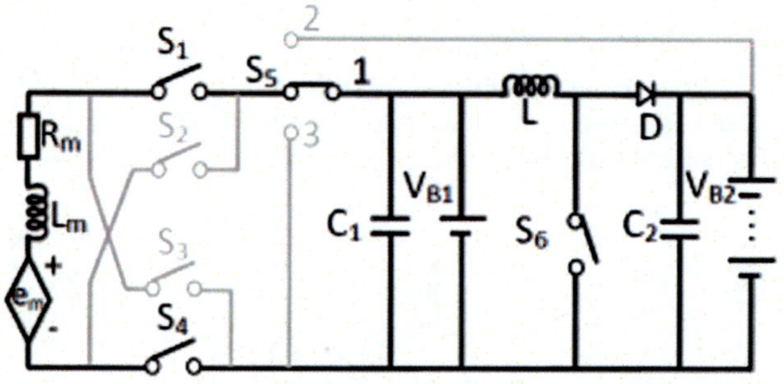

Figure 3.2 - Energy harvesting mode (e_m>0)

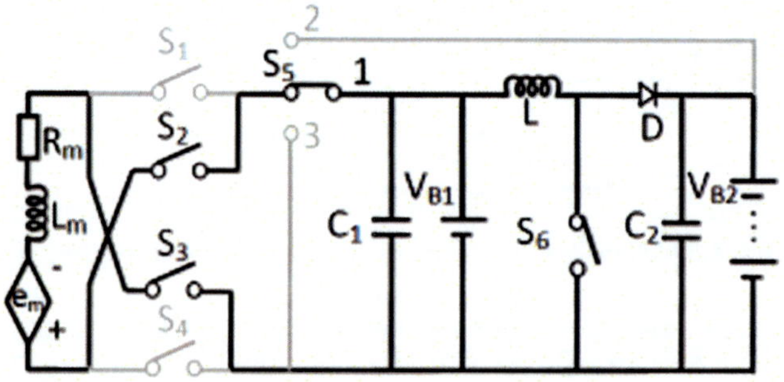

Figure 3.3 - Energy harvesting mode (e_m<0)

Figure 3.2 and figure 3.3 show the circuit in this mode. Either switch pair S_1 and S_4, or S_2 and S_3 is involved, since the direction of the relative velocity varies. In this way, the two switch pairs serve as synchronous rectifier. The corresponding passive force can be expressed as follows, when the inductance L_m is negligible:

$$f_m = -sign(v_r)\frac{Mk_t(Mk_e|v_r|-V_{B1})}{R_m} \tag{19}$$

where sign() is the signum function. In energy harvesting mode the passive force is composed of a viscous damping component $-M^2k_tk_ev_r/R_m$ and a force $sign(v_r)V_{B1}/R_m$ due to the battery B_1. Vibration mitigation performance can be further improved, if the actual force can be controlled to follow the desired force, by applying Pulse-Width Modulation (PWM) to the switch pairs. Hence, the force provided by the electromagnetic motor can be controlled by adjusting the duty cycle of the switch pair S_1 and S_4, or S_2 and S_3. The work principle is similar to the switching amplifier. Hence the force can be expressed as:

$$f_m = -sign(v_r)\frac{Mk_t(DMk_e|v_r|-V_{B1})}{R_m} \tag{20}$$

The duty cycle corresponding to the desired force u_{des} can be expressed as:

$$D = \frac{|u_{des}|R_m}{M^2k_ek_t|v_r|} + \frac{V_{B1}}{Mk_t} \tag{21}$$

The passive damping force can be provided in this mode is limited due to the fact that D should be smaller than one. The damping force is set to be maximum (D=1) when it is out of the limitation:

$$f_{mmax} = -sign(v_r)\frac{Mk_t(Mk_e|v_r|-V_{B1})}{R_m} \tag{22}$$

3.2. DRIVING MODE

When the desired force and the relative velocity have the same direction, the self-powered active TMD will work in driving mode, where the electromagnetic motor acts as actuator. The circuit involved in this mode is actually classic D amplifier [21], where the direction and amplitude of the voltage applied on the electromagnetic motor can be controlled.

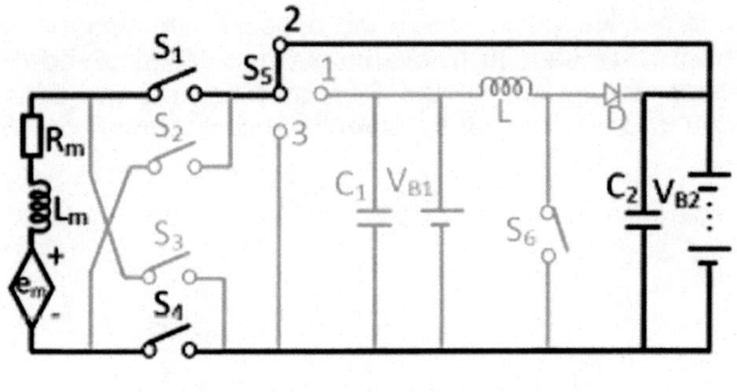

Figure 3.4 – Driving mode (f_m>0)

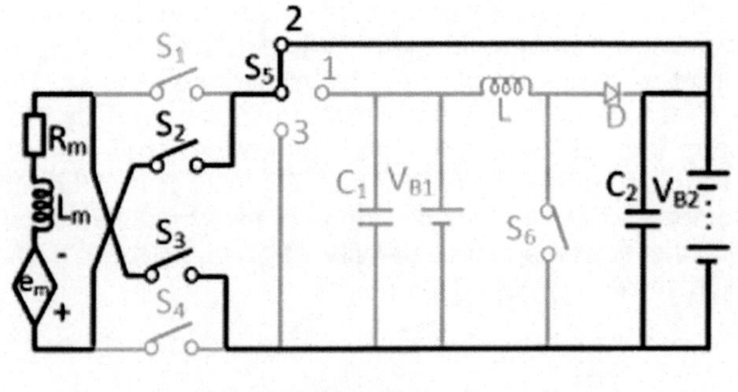

Figure 3.5 – Driving mode (f_m<0)

As shown in figure 3.4 and figure 3.5, the direction of the voltage is controlled by switching the two pairs of switches. The voltage amplitude is controlled by the duty cycle of the PWM applied on the switch pair. The circuit is classic D amplifier, the output voltage of which is proportional to the duty cycle. Hence, the output active force of the electromagnetic motor is:

$$f_m = \frac{Mk_t(DV_{B2} - Mk_e|v_r|)}{R_m} \tag{23}$$

The duty cycle provided by the controller in order to follow the corresponding desired force u_{des} can be expressed as:

$$D = \frac{|u_{des}|R_m}{Mk_t V_{B2}} + \frac{Mk_e|v_r|}{V_{B2}} \tag{24}$$

3.3. PASSIVE MODE

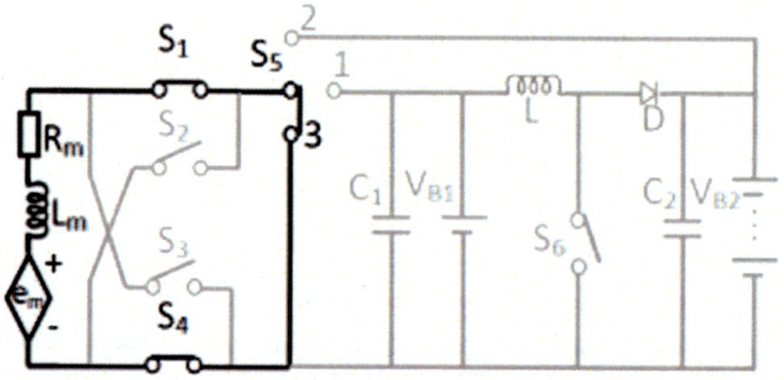

Figure 3.6 - Passive mode

When the desired force and the relative velocity have the opposite direction, however, the voltage generated by the electromagnetic motor is not large enough to overcome the voltage of the first battery V_{B1}, the system will work in this mode. In this mode, the switch S_1 and S_4 is switched on and S_5 is switched to contact 3. In the passive mode, the electromotive force will appear to be an ideal viscous damping force:

$$f_m = \frac{M^2 k_t k_e v_r}{R_m} \tag{25}$$

4. NUMERICAL SIMULATIONS

The simulations are carried out based on the building Taipei 101. The parameters of Taipei 101 have been described in section 2 (m_1=730 tons, mass ratio μ=0.78%). And the damping ratio of the primary system is assumed to be 1% [22]. For the electromagnetic motor, the thrust constant k_t, Back-EMF constant k_e and the resistance R_m are the inherent parameters of the electromagnetic motor. From the specification data of commercialized *IC44 series* linear electromagnetic motors manufactured by *Kollmorgen* [23], k_t ranges from 72.7N/A to 1210 N/A, k_e ranges from 59.3V·s/m to 988 V·s/m, R_m ranges from 0.37Ω to 38.6Ω. Hence, in this simulation k_t and k_e are chosen to be 700N/A and 700V·s/m. R_m is 20 Ω and the motion magnification mechanism *M* is set to be 2. Figure 4.1 compares the transient response of the structure's acceleration when subjected to the random wind disturbance with the power spectral density of S_0=4.0x10^{14} N^{2}s. And the RMS of the output for structures with different TMDs is shown in Table 4.1.

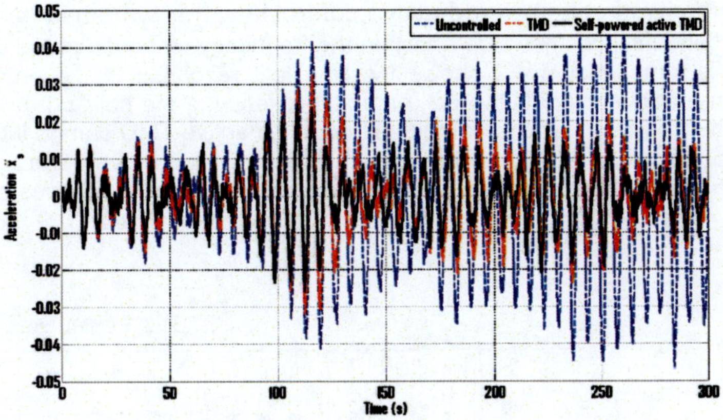

Figure 4.1- Transient acceleration response under random excitation

Table 4.1 – RMS acceleration of the building under random wind load excitation with the power spectral density of S_0=4.0x10^{14} N^{2}s

TMD type	RMS of the acceleration (m/s^2)
Uncontrolled	0.018903
TMD	0.008267
Active TMD	0.006805
Self-powered active TMD	0.007345

Figure 4.2 shows the transient force response of self-powered TMD and the active TMD. As we can seen from the figure, there are three types of forces expressed by (20), (23), (25), corresponding to the threes working modes: harvesting, driving, and passive.

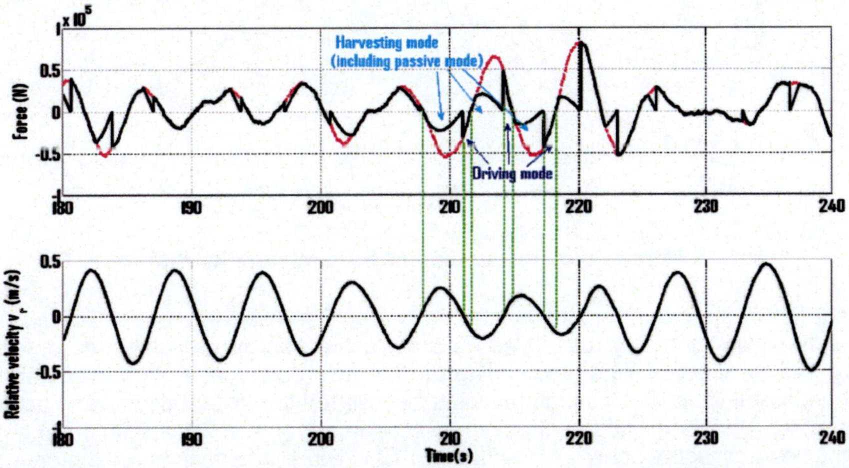

Figure 4.2- Transient force response under random excitation. (Passive mode is a small segment marked in the harvesting mode in this figure when the relative velocity $|v_r|$ is less than Mk_eV_{B1}/R_m.)

Because of the nonlinearity of self-powered active control law, the system is no longer linear. However, we find that the excitation amplitude has little effect on the frequency response of transmissibility ratio, when the wind load disturbance is large. Figure 4.3 shows the transient response of the primary system subjected to harmonic excitation with a frequency of 0.146 Hz (natural frequency of the building) and amplitude of 250kN. As seen from Figure 4.3, the transient response of the self-powered active TMD can go into steady state after a few periods. This figure also indicates that at this frequency the self-powered active TMDs can reduce the vibration to 56.6% over the passive TMD. For this reason, we compare the frequency responses of transmissibility ratios of different TMDs. Figure 4.4 shows the frequency response of the self-powered TMD, compared with the passive and active TMD.

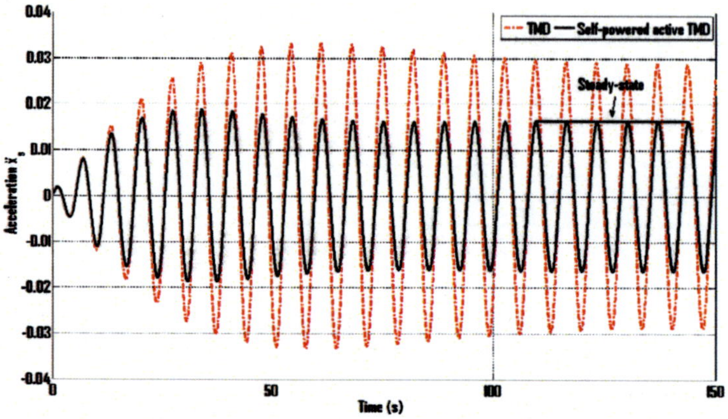

Figure 4.3 – Transient response under harmonic excitation

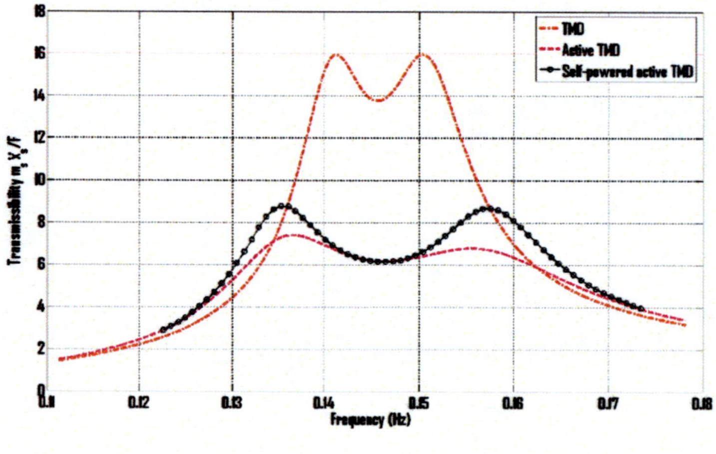

Figure 4.4 Frequency responses of transmissibility ratio $m_s\ddot{x}_s$ /F.

As discussed in section 3, the circuit is actually classic D amplifier in driving mode. The classic D amplifier has very high power conversion efficiency, usually>90% [24]. In the energy harvesting mode, the booster DC-DC converter has a typical efficiency of 78%, accounting for the main power loss. With taking both the self-powered active control law and parasitic power loss into account, we plotted the instant power and accumulated energy of the system in figure 4.5. shows that the self-powered active control is still feasible when subjected to random disturbance with the power spectral density of S_0=4.0x$10^{14}N^2s$. The system can harvest more energy than needed for the active control part.

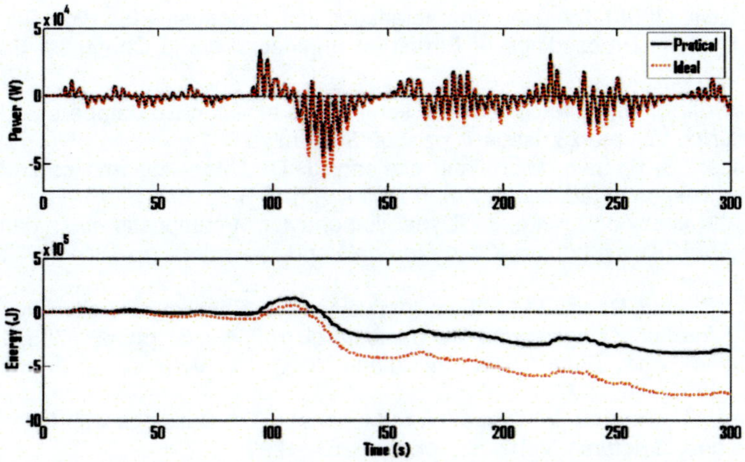

Figure 4.5 Power and energy transient response. The ideal case is the one doesn't consider the any power loss, while the practical take the efficiency of both harvesting and driving circuit into account.

5. CONCULSION

This paper investigated the feasibility and realization of self-powered active control of TMD. Without consuming external energy, the proposed self-powered active TMD can provide better vibration mitigation performance compared with the passive TMD. Switch based circuits with capability of bi-directional power flow are presented for the implementation of self-powered active TMD. The power balance is analyzed with taking the efficiency of circuit and parasitic power loss into account. It also should be noted that a pre-charged energy reservoir may be used to jump-start the system before the system maintains the performance itself, but this pre-charged energy reservoir is not necessary.

6. REFERENCE

[1] Kareem, A., Kijewski, T. and Tamura, Y., "Mitigation of Motions of Tall Buildings with Specific Examples of Recent Applications," *Wind and Structures*, Vol 2, n3, pp 201-251, 1999.

[2] Ali, M.M. and Moon K.S., "Structural Developments in Tall Buildings: Current Trends and Future Prospects", *Architectural Science Review*, Volu 50 n3, pp 205-223, 2007

[3] Morgan,T.A and Mahin, S.A., "Development of a Design Methodology for Seismic Isolated Buildings Considering a Range of Performance Objectives", *Proceedings, 4th International Conference on Earthquake Engineering, NCREE*, Taipei, Taiwan, 2006

[4] Frahm, H., "Device for Damping Vibrations of Bodies," *U.S. Patent No.989, 958*, 1911.

[5] Den Hartog, J. P., *Mechanical Vibration*, McGraw-Hill, New York, 1947.

[6] Kerschen, G., Kowtko, J. J., McFarland, D. M., Bergman, L. A. and Vakakis, A. F., "Theoretical and experimental study of multimodal targeted energy transfer in a system of coupled oscillators", *Nonlinear Dynamics*, vol. 47, pp. 285.309, 2007.

[7] Setareh, M., Ritchey, J.K., Baxter, A.J. and Murray, T.M., "Pendulum tuned mass dampers for floor vibration control". *J Perf Constr Fac* 20(1): 64–73, 2006

[8] Koshika, N., Sakamoto, M., Sasaki, K., Ikeda, Y. and Kobori, T., "Control effect of active mass driver system during earthquakes and winds", *Proc 1st MOVIC Conf, Yokohama*, Japan, 1992, pp 261-266

[9] Fujino, Y., Soong, T.T. and Spencer, B. F., "Structural Control: Basic Concepts and Applications", *the Proceedings of the 1996 ASCE Structures Congress,* 1996

[10] Tomoo, S. and S. Keiji. (1998), Dynamic Characteristics of a Triangular-Plan High-Rise Building and Its Active Vibration Control System, Proceedings of Structural Engineers World Congress, San Francisco, CD-ROM: T198-5.

[11] Pinkaew, T. and Fujino, Y., "Effectiveness of semi-active tuned mass dampers under harmonic excitation", *Engineering Structures*, Volume 23, Issue 7, pp. 850-856, July 2001.

[12] Dyke, S. J., Spencer, B. F, Sain, M. K. and Carlson, J. D., "Modeling and control of magnetorheological dampers for seismic response reduction", *Smart Mater. Struct.* 5 pp. 565-575,1996.

[13] Lewandowski, R. and Grazymislawska, J., "Dynamic behavior of composite mass damper for control of wind-excited vibration", *AMAS Workshop on Smart Materials and Structures*. pp.131–140 Jadwisin, September 2-5, 2003.

[14] Hrovat, D., Barak, P. and Rabins, M., "Semi-Active versus Passive or Active Tuned Mass Dampers for Structural Control", *Journal of Engineering Mechanics*, Vol. 109, No. 3, pp. 691-705, 1983.

[15] Karnopp, D., "Active and Semi-Active Vibration Isolation", *Journal of Mechanical Design*, Volume 117, pp. 177-186, 1995.

[16] Kareem, A., Kijewski, T., Tamura, Y., Mitigation of Motions of Tall Buildings with Specific Examples of Recent Applications, *Wind and Structures*, Vol 2, n3, pp 201-251, 1999

[17] Rao, S., *Mechanical Vibrations*, Prentice Hall, 2003

[18] Warburton, G., "Optimum absorber parameters for various combinations of response and excitation parameters", *Earthquake Eng. Struct. Dyn.,* 10, pp.381-401, 1982.

[19] Nakanno, K., Suda, Y. and Nakadai, S., "Self-powered active vibration control using a single electric actuator", *Journal of Sound and Vibration*, 260 21-235,2003

[20] Scruggs, J. and Lindner, D., "Active energy control in civil structures", *SPIE Symposium on Smart Structures and Materials*, Newport Beach, CA, 1999.

[21] McCorkle, D., *US Patent*, 5,160,896, 1992

[22] Joseph, L., Poon, D. and Sbieb, S., "Ingredients of high-rise design", technical report, 2006

[23] *Kollmorgen Linear Motors Aim to Cut Cost of Semiconductors and Electronics Manufacture,* Kollmorgen, Radford, VA, U.S.A.,1997.

[24] Honda, J. and Admas, J., "Class D Audio Amplifier Basics," technical paper on International rectifier website, http://www.irf.com/technicalinfo/appnotes.htm.

Proceedings of the IMAC-XXVIII
February 1–4, 2010, Jacksonville, Florida USA

ENERGY HARVESTING UNDER INDUCED BEST CONDITIONS

Sebastian Pobering, CiS GmbH Erfurt, Steffi Ebermayer, Norbert Schwesinger,
Technische Universitaet Muenchen

Nomenclature

A	cross section of a generator	V	voltage
c_B	Beetz-Number	v_f	velocity of flowing media
D	characteristic fluidic dimension	v_s	velocity behind the harvester
d_{31}	piezoelectric coefficient	w(x)	bending line
f	frequency	W_{el}	electrical energy
f_v	detachment frequency of vortices	W_m	mechanical energy
P_{el}	electrical power	Y	Young's Module
P_{max}	maximum power	Δp	pressure difference
PZT	Lead-Zirconate-Titanate	λ_{KVS}	Period length of vortex street
Re	Reynolds-Number	ϱ	density of a flowing medium
Sr	Strouhal-Number	$\varepsilon_0\varepsilon_r$	permittivity number
T_{PZT}	thickness of PZT cantilever	η	dynamic viscosity
U_{pp}	peak-to-peak-voltage	σ_{max}	break strength

ABSTRACT

Resonant energy harvesters require for optimum energy conversion an oscillation with resonance frequency. A piezoelectric resonant power converter has been developed. It operates under all circumstances with resonance frequency. The harvester takes the energy out of flowing medium surrounding him. Flowing media can generate forces behind bluff bodies in shape of localized low pressure regions with a high periodicity. These low pressure regions are vortices released from the bluff body in flowing direction. Piezoelectric cantilevers fixed with one end onto a bluff body are used as energy harvesters. If a vortex meets the free end of the cantilever it will be forced to oscillate. This oscillation always occurs with the Eigenfrequency of the cantilever. Since best energy conversion can be achieved in case of resonance with cantilever based harvesters, best conversion conditions are induced by the harvester itself. Former simulations have shown the geometry of the bluff body affects strongly the pressure and the velocity distribution in the flow. Based on those results basic design rules for the dimensions of the cantilever and the bluff body have been developed. Measurements have been carried out in gaseous media.

1. Introduction

Distributed sensor systems are getting more and more attractive in technical and natural environments. These sensors detect selected physical/chemical/biological parameters, convert the information into electrical signals and transmit the signals to data receivers for further processing. Since data transmission occurs wireless wiring is only necessary for the energy supply. Batteries are often used to omit completely wiring. Unfortunately, batteries lead to an increase of avoidable operating expenses. The best option in these circumstances is to scavenge or harvest energy of the environment surrounding the sensors. Especially for microscopic sensors with low energy requirements energy converters could be fabricated in microscopic dimensions. This leads to smart system integration possibilities, as the sensor, the energy supply, the power management and communication circuit could be placed in just one small module. Therefore, it is necessary to develop energy converters with well adapted dimensions and high conversion efficiencies. Typical energy sources are radiation, heat or mechanical movement. These sources are used in commercial available converters. Previously wasted types of energy like

T. Proulx (ed.), *Structural Dynamics and Renewable Energy, Volume 1*, Conference Proceedings of the Society for Experimental Mechanics Series 10, DOI 10.1007/978-1-4419-9716-6_22, © The Society for Experimental Mechanics, Inc. 2011

vibration, heat development due to friction forces or kinetic energy of flowing media are presently in focus of increasing research activities.

Aim of this work was the conversion of kinetic energy of a flowing media like air or water. This kind of energy offers a relatively high energy density in combination with a high availability. Basic principle of common converters using this type of energy is the rotation of electrical coils in a magnetic field. Kinetic energy will be transferred into rotational mechanical energy first. Second, electrical energy will be generated in the coil due to the time dependent changes of the magnetic field strength. Macroscopic devices, like generators, are using this principle for an optimized energy conversion. Scaling down this principle into the microscopic area two important effects can be observed. The efficiency of the energy conversion decreases and friction forces increase the smaller the dimensions become. Thus, rotational energy conversion will fail completely in the microscopic area since energy conversion is not dominant in whished manner. Therefore, it is necessary to operate with converters free of mechanical rotation.

Drug of choice in this work is the oscillation of piezoelectric cantilevers. This principle is research topic of many groups around the world. Most of them try to convert vibration energy into electrical energy due to induced oscillation of piezoelectric cantilevers. Several small scale energy converters for scavenging energy of vibrating sources are published recently [1] and first commercially products are available [2]. But, vibrations can be observed often in a very wide frequency range. Vibrations with selected frequencies are the exception. Conversion of vibration energy into electrical energy is optimal if the Eigenfrequency of the converter is similar with the frequency of the vibration. Unfortunately, this condition is not given in most cases. So most research groups try to adjust the resonance frequency of their cantilever based harvesters to the frequency of the corresponding source [3,4,5]. This leads to more complicated systems with more elements. Finally, the systems increase in size and in complexity controlling the best power conversion conditions. An optimal solution is a harvester that operates permanently with its own Eigenfrequency independent from the frequency of the source.

2. Development of the conversion principle

a) Basic idea

Since piezoelectric energy conversion is a time dependent operation it is necessary that the piezoelectric cantilevers oscillate. A simple bending of the cantilever without any other time dependent mechanical movements would lead to a short electric pulse and nothing more. The cantilever has to be forced to oscillate. Forces perpendicular to the streaming direction are not available in regular flows with Reynolds-Numbers below the critical values. Therefore, it is necessary to disturb the laminar flow and to generate vortices with flow directions perpendicular to the regular flow. A permanent generation of vortices can be obtained in a flowing medium in "von Kármán's vortex streets". Such a vortex street is characterized by periodic vortices with alternating directions of rotation. It can be induced by a bluff body similar to a flagpole which forces the flag to flutter in the wind. In the same way can piezoelectric cantilevers (as a flag) be forced to flutter (oscillate). They react to the flow disturbance (see Figure 1). The cantilever bends into the direction of the vortex since the vortex induce low pressure regions across the piezoelectric cantilever.

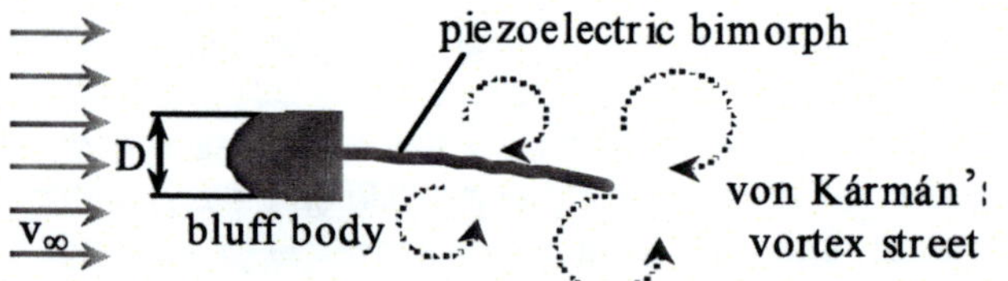

Figure 1: Piezoelectric cantilever in a von Kármán's vortex street

For generation purposes it is necessary to work with cantilevers that oscillate just in the first oscillation mode. Under these conditions it is guaranteed, that one polarity of charges will be collected on one surface while the opposite polarity of charges will be gathered on the opposite side of the cantilever surface. Thus, a charge separation appears between both sides of the cantilever. In case of metalized surfaces, a voltage can be measured between both sides due to this charge separation. This voltage oscillates in the same way as the cantilever will be forced to oscillate up- and downwards. Hence, a sinusoidal voltage is generated. The result is comparable with harvesters driven by vibrations, while the source is completely different.

We have developed an energy converter that allows the energy conversion without any rotational movements and that is independent of the driving frequency. Kinetic energy will be transferred into electricity utilizing piezoelectric cantilevers.

b) Modeling and Dimensioning

A perspective generator should contain multiple, three dimensionally arranged piezoelectric cantilevers. Each single cantilever consists of two active piezoelectric layers. PZT(lead zirconate titanate) is used as piezoelectric material. Sheet like surface metallizations act on top, on bottom and in between the layers as electrodes as shown in Figure 2. Downward bending of the cantilever leads to tensile stress in the top layer and simultaneously to compressive stress in the bottom layer. So polarization in top and bottom layer show opposite orientation respectively and causes a collection of charges of different polarities at the electrodes on top and bottom.

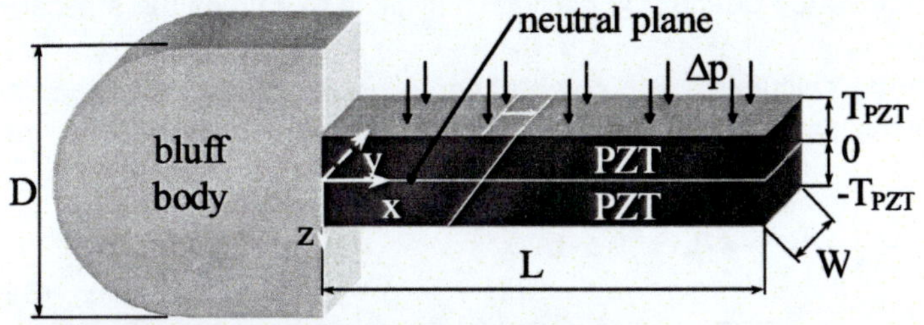

Figure 2: Schematic of the generator setup

The bluff body serves as flow disturbance and as mechanical bearing for the cantilever. It is fixed with its ends to an external carrier (not shown in Figure 2). Behind the bluff body a von Kármán's Vortex Street appears. Stable von Kármán's Vortex Streets are developing at Reynolds-Numbers Re between 50 and 10 000. Above theses values vortices become more stochastically until the critical Reynolds number is reached and the flow turns to fully turbulent.

In turbulent flows as "von Kármán's Vortex Streets" most energy will be transferred inside microscopic vortices. Therefore, the usable energy increases with reduced dimensions of the cantilever. To gain usable power ratings numerous small cantilevers need to be combined. Based on these thoughts it is possible to develop a model that allows to estimate the dimensions of a system consisting of a bluff body and a cantilever.

First for fluidic systems it is important to estimate the characteristic dimension D. D stands for the overall dimension of the bluff body as shown in figure 2. The range of D can be calculated using the Reynolds-Number.

$$D = \frac{Re \cdot \eta}{v_f \cdot \rho} \tag{1}$$

One can achieve a value-range for D = 25 μm…5 mm at a flow velocity of 2 m/s in water and D = 67 μm…13 mm at a velocity of 10 m/s in air, respectively.

The detachment frequency f_V of vortices depends on the flow velocity v_f, the characteristic dimension D and the Strouhal-Number Sr. Sr has a dimensionless value of about 0.2 in the Reynolds-Number range of possible applications in water and air.

$$f_V = \frac{Sr \cdot v_f}{D} \tag{2}$$

Vortices behind the bluff body have a flow velocity which can be calculated to $v_S = 0.85 v_f$. Therefore, the pressure oscillates across the piezoelectric cantilever with a respectively reduced frequency [3]. Thus, the period length λ_{KWS} of such a "von Kármán's vortex street" can be calculated as follows:

$$\lambda_{KWS} = \frac{0.85 \cdot v_f}{Sr \cdot \frac{v_f}{D}} = 4.25 \cdot D \tag{3}$$

Cantilevers must oscillate in their first harmonic mode to prevent charge compensation across one electrode. Logically, the length of the cantilever L has to be adapted to a half of the period length of the "von Kármán's vortex street". Under these conditions one can find a basic ratio of L to D written in equation (6)

$$\frac{L}{D} = 2.125 \tag{4}$$

The pressure difference Δp can be calculated under the assumption that the rotational flow velocity of the vortices is one third of the flow velocity of the undisturbed fluid v_f. Assuming furthermore a gap between two vortices traveling on opposite sides of the cantilever one can assume:

$$\Delta p = \frac{\rho}{2} \cdot \left(\left(v_f - \frac{v_f}{3} \right)^2 - v_f^2 \right) \tag{5}$$

In the following calculations the difference Δp is assumed to be uniformly distributed across the whole cantilever. Furthermore, the semi statically description covers only the deflection to one side from the neutral position to the maximum value. Therefore, the amount of energy converted during one oscillation cycle is four times the statically calculated energy.

Hence, the bending line w(x) of the cantilever can be calculated with given dimensions (see Figure 2), material properties (Y – Young's Module) of the piezoelectric cantilever and the calculated pressure difference Δp as follows.

$$w(x) = \frac{\Delta p}{16 \cdot Y \cdot T_{PZT}^3} \left(x^4 - 4 \cdot L \cdot x^3 + 6 \cdot L^2 \cdot x^2 \right) \tag{6}$$

Important is furthermore the minimum thickness T_{PZT} of the PZT-cantilever. This value can be calculated applying the maximum deflection w_{max} and the break strength of the PZT-Material σ_{max}. The following expression shows these dependencies.

$$T_{PZT} = \frac{1}{2} \cdot \sqrt{3 \cdot \Delta p \frac{L^2}{\sigma_{max}}} \tag{7}$$

c) Energy conversion

Mechanical energy W_m will be converted partially during the deformation into electrical energy W_{el}. The ratio of the electrical energy W_{el} to mechanical energy W_m depends finally on material properties only as shown in equation (10). This expression gives an idea about the conversion efficiency of the piezoelectric harvesters.

$$\frac{W_{el}}{W_m} = \frac{5}{13} \cdot \frac{d_{31}^2 \cdot Y}{\varepsilon_0 \cdot \varepsilon_r} \tag{8}$$

A peak to peak voltage U_{PP} at the electrodes during the oscillation can be calculated as:

$$U_{PP} = \frac{1}{2} \cdot \Delta p \cdot \frac{d_{31}}{\varepsilon_0 \cdot \varepsilon_r} \cdot \frac{L^2}{T_{PZT}} \tag{9}$$

The overall electrical power P_{el} of the whole cantilever is given as:

$$P_{el} = 8 \cdot f \cdot W_{el} \tag{10}$$

Hereby, it is taken into consideration that the cantilever contains two active piezoelectric layers with two electrodes each.

d) Power in flowing media

The power available from flowing media can be estimated using the equation for wind power of Albert Betz developed in 1920 [6]. Because of the universal approach it can be used for flowing water, too.

$$P_{max} = \frac{\rho}{2} \cdot A \cdot v_f^3 \cdot c_B \tag{12}$$

The Betz-Number c_B takes into account the reduction of the flow velocity due to the generator. Hereby, the velocity ratio before v_f and after the generator v_S will be considered especially.

$$c_B = \frac{1}{2} \cdot \left(1 - \frac{v_S^2}{v_f^2}\right) \cdot \left(1 + \frac{v_S}{v_f}\right) \tag{13}$$

Hence, flow velocity slows down during energy extraction. Therefore, a maximum Betz Number of 0.59 can be obtained at a velocity ratio of $v_S/v_f = 0.33$. Assuming a Betz-Number of $c_B = 0.4$, it is possible to calculate the power that can be extracted in principle out of a flowing media. One can find the following scale.

Flowing water at a velocity of 2m/s → P_{max}= 1.6kW/m²

Flowing air at a velocity of 10m/s → P_{max}= 256W/m²

3. Experimental setup

Different types of harvesters have been used in the experiments. Three cantilevers were arranged in a row and fixed on a mechanical carrier in a first series. Hereby, the carrier acts also as bluff body (Figure 3). Electrical connections inside were realized before the bluff body was encapsulated with PDMS (Polydimethylsiloxan). PDMS enables as synthetic rubber examinations in fluidic solutions, like water. Piezoelectric substrates used have not fulfilled the requirements for harvesting purpose. Figure 3 shows the internal structure of the piezoceramic substrates. Each substrate consists of several ceramic and metal layers. Piezoelectric active material can be found only in two small layers of the substrate. All other layers with exception of the electrodes are electrical inactive and serve as dielectric layers. The ratio of active material to passive material is about 1:3. Those assemblies shown in Figure 3 (left) were fixed in a frame allowing arrangements behind or above each other. A change of the distance between the assemblies was also possible. Piezoelectric substrates of a second series show a much better internal design (Figure 4). Each substrate consists of an active piezoelectric layer of 640µm in thickness. The ratio active to passive material is equal to 4:1. However, in all cases bluff bodies have been designed in agreement with equation (4). All measurements were made in artificial water channels or wind channels. A typical measurement arrangement for air is shown in Figure 5. A Flow sensor is arranged above the cantilevers. It allows the monitoring of the of the air flow.

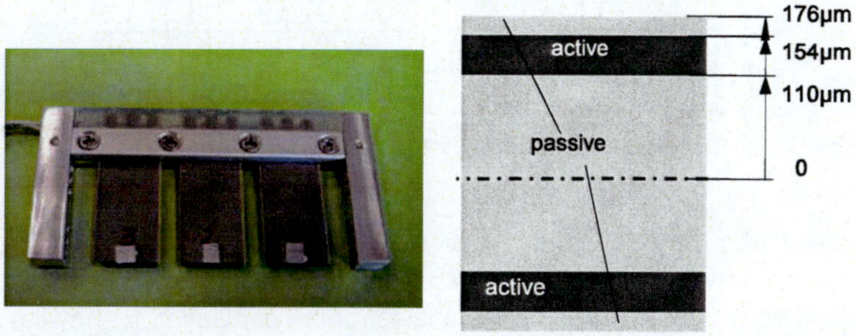

Fig. 3: Plates of the first series; active thickness: 88µm; arrangement in a row

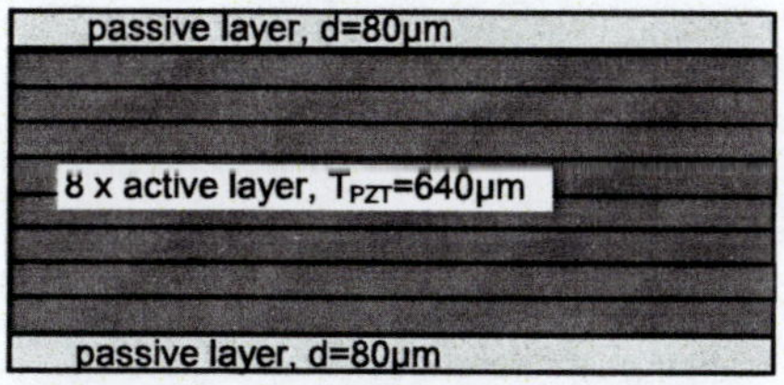

Figure 4: Piezoelectric cantilevers with a piezoelectric multilayer structure

Figure 5: Measurement set up in air

Voltage and oscillation frequency generated by each cantilever were measured simultaneously. Therefore, each cantilever was connected to an external control unit. Furthermore, the frequency spectra have been controlled for the undisturbed and the disturbed flow.

4. Results and discussion

First, we have considered the behavior of the cantilevers in view of their electrical output. It was found that all cantilevers were forced to oscillate. Obviously the experiment has fulfilled the basic thoughts about the principle. Considering Figure 6 one can see a typical snapshoot of the voltage output generated by 3 cantilevers arranged behind each other. Each cantilever shows a generation of voltage. The voltage characteristic is sinusoidal as predicted and it seems to verify the periodicity of the vortices crossing the cantilever. Differences in the amplitude are expected since they should depend on the strength of the related vortex. It appears as would each cantilever oscillates with a fixed frequency. Therefore, a phase shift is found comparing the cantilevers 2 and 3. Furthermore, the frequencies of all cantilevers appear to be in a similar range. Since we could observe this behavior in each experiment we have measured the frequency of the undisturbed flow, of the flow behind the bluff body and the oscillation of the cantilever. The result was very surprising since we have not expected a characteristic which was found. Figure 7a shows an example that was detected. It shows the frequency spectra which are expected for the flow with and without the bluff body. Unusual seems the answer of the cantilever. It shows a very sharp resonant peak at about 320Hz and a saddle at 120Hz. This behavior explains possibly the curves shown in figure 6 – cantilevers oscillate with their own resonance frequency. This assumption will be fortified by experiments done at higher flow rates.

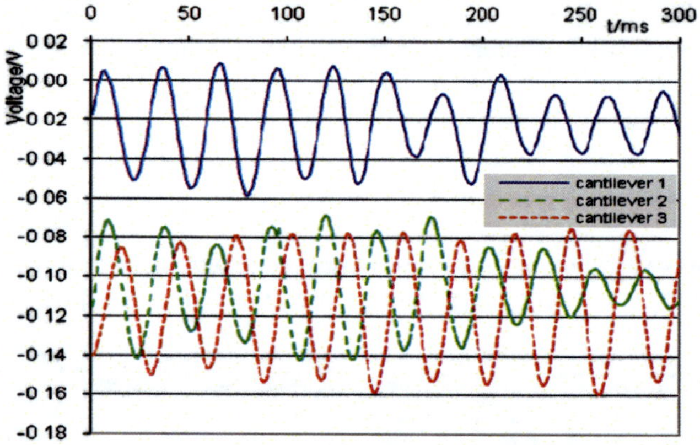

Figure 6: Oscillation of 3 cantilevers in a row @30m/s in air

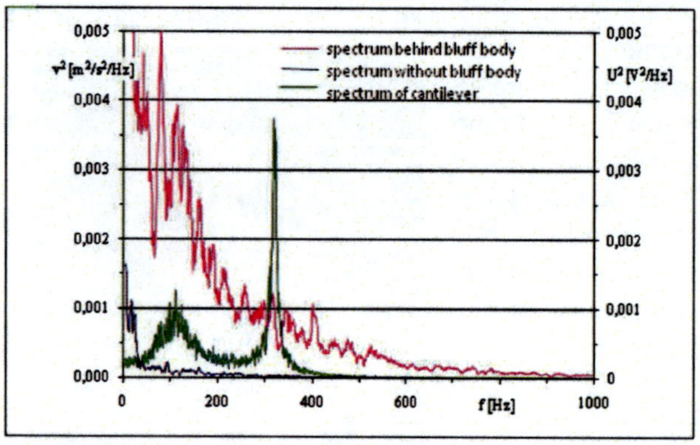

Figure 7a: Frequency spectra measured at 10m/s flow velocity

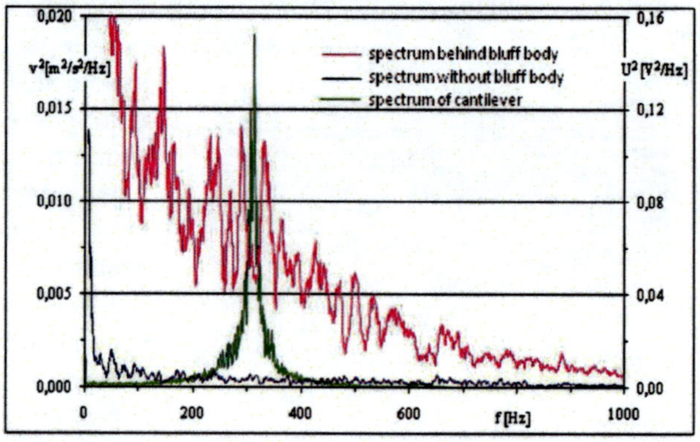

Figure 7b: Frequency spectra measured at 30m/s flow velocity

In Figure 7b the same measurement is done at a flow velocity of 30m/s. Again, a sharp resonance peak can be observed at a frequency of 320Hz. The resonant frequency of the cantilever has shown no dependency on the flow velocity. A cantilever length of 22 mm was used in these cases.

Other measurements have been carried out with various lengths of the cantilevers. Hereby, the optimal relation L/D was used with shortened cantilevers. Lengths of 15.6mm and 9.2mm were used and examined. For these observations we found resonance values of 620Hz and 1330Hz respectively. Again, no dependency on the flow velocity was found. Under these circumstances it is necessary to questioning the basic assumption described in chapter 2. Obviously, vortices have not the same meaning to force the oscillation as described above. Bending of the cantilever will be induced by one ore more vortices travelling across the surface. Spring forces inside the cantilever lead to a mechanical restoring movement – the cantilever starts to oscillate. The impact of following vortices is comparable low. Only when the oscillation amplitude is relatively low following vortices could lead to pulses renewing the oscillation. An example of this behavior is shown in Figure 8. Velocity differences can be found behind the bluff body. These differences seem to be stochastically distributed. Amplitudes of velocity seem not to agree with the oscillation of the cantilever. This oscillation shows a sinusoidal shape with increasing and decreasing amplitudes. The frequency appears to be stable.

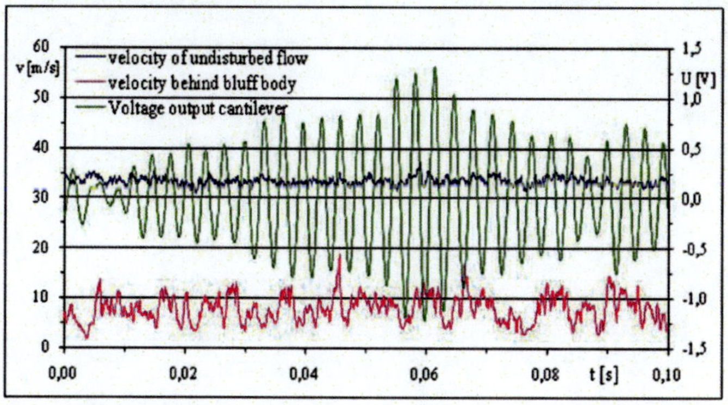

Figure 8: Velocity and Voltage measurements behind the bluff body @34m/s in air

This all over behavior of the cantilevers in flowing media is very promising. It is not required to adapt the frequency of the cantilevers to any external parameter. Cantilevers find independent of the flow velocity their own resonance frequency and oscillate only with this frequency. If cantilevers with the same resonance frequency could be successfully produced they could be forced to work in a synchronized mode. This would simplify the electrical read out. Instead of two diodes for each cantilever one could work with just two diodes for the whole synchronized system.

Voltage measurements of the two series described in chapter 3 have delivered in air at a flow velocity of 8 m/s the following values:

Series 1 (S1)	U_{pp}= 130mV
Series 2 (S2)	U_{pp}= 1320mV

Comparing the piezoelectric layer thickness relation $T_{PZT}(S2)/T_{PZT}(S1) = 640/88 = 7.27$ and the voltage output relation $V(S2)/V(S1) = 1320/130 = 10.15$ a much better power conversion with cantilevers of series 2 can be observed. This difference is sensible since more material is involved in the conversion process. A determination of optimized PZT layer thicknesses is still under progress.

5. References

[1] Paul D. Mitcheson, Tim C. Green, Eric M. Yeatman, Andrew S. Holmes, "Architectures for Vibration-Driven Micropower Generators", in Journal of Microelectromechanical Systems, vol. 13, no. 3, June 2004, pp. 429-440.

[2] Website: Mide Technology Corporation, http://www.mide.com/products/volture/volture_catalog.php

[3] Charnegie, David (2007) *Frequency Tuning Concepts For Piezoelectric Cantilever Beams And Plates For Energy Harvesting.* Master's Thesis, University of Pittsburgh. MSME

[4] V. Challa, et al. *A vibration energy harvesting device with bidirectional resonance frequency tenability*, Smart Mater. Struct. **17** (2008) 015035

[5] Ch. Eichhorn et al; *Frequency tunable energy converters for energy scavenging in environments with variable vibrational frequencies*, Mikrosystemtechnik Kongress 2009, Berlin, Paper 59

[6] Albert Betz; *Das Maximum der theoretisch möglichen Ausnutzung des Windes durch Windmotoren*, Zeitschrift für das gesamte Turbinenwesen, 1920.

Proceedings of the IMAC-XXVIII
February 1–4, 2010, Jacksonville, Florida USA

Frequency Domain Solution of a Piezo-aero-elastic Wing for Energy Harvesting

Wander Gustavo Rocha Vieira[1], Carlos De Marqui Junior[2]
Department of Aeronautical Engineering, Engineering School of Sao Carlos, University of Sao Paulo. Av. Trabalhador Sancarlense 400, 13566-590, Sao Carlos, SP, Brazil. [1]wander.vieira@usp.br, [2]demarqui@sc.usp.br

Alper Erturk
Center for Intelligent Material Systems and Structures, Department of Engineering Science and Mechanics, Virginia Tech. 310 Durham Hall, MC 0261. erturk@vt.edu

Daniel J. Inman
Center for Intelligent Material Systems and Structures, Department of Mechanical Engineering, Virginia Tech. 310 Durham Hall, MC 0261. dinman@vt.edu

Abstract

Multifunctional structures are pointed out as a future breakthrough technology for Unmanned Air Vehicle (UAV) design. These structures can perform tasks additional to their primary functions. Based on the concept of vibration-based energy harvesting, the structure (lifting surfaces) of a UAV can perform the additional function of providing electrical energy by converting aeroelastic vibrations to electricity. In this paper, frequency-domain piezo-aero-elastic modeling of a cantilevered plate-like wing with embedded piezoceramics is presented for energy harvesting. The electromechanical finite-element plate model is based on the Kirchhoff assumptions and a resistive load is considered in the external circuit. The subsonic unsteady aerodynamic model is accomplished with the doublet lattice method. The electromechanical and the aerodynamic models are combined to obtain the piezo-aero-elastic equations, which are solved using a modified P-K scheme. This way the evolution of the aerodynamic damping and the frequency of each mode is obtained with changing airflow speeds for a given load resistance. Piezo-aero-elastically coupled frequency response functions (voltage, current and electrical power as well relative tip motion) are defined for a given airflow speed and load resistance. Hence the piezo-aero-elastic evolution can be investigated under several different conditions. A procedure is also presented in order to obtain the optimum load resistance (for maximum power and for maximum damping) for a given airflow speed.

Nomenclature

a_B = base acceleration
$\bar{\mathbf{C}}_p$ = diagonal global capacitance matrix
$\bar{\mathbf{C}}$ = global damping matrix

T. Proulx (ed.), *Structural Dynamics and Renewable Energy, Volume 1*, Conference Proceedings of the Society for Experimental Mechanics Series 10, DOI 10.1007/978-1-4419-9716-6_23, © The Society for Experimental Mechanics, Inc. 2011

$\tilde{\mathbf{\Theta}}$	=	global electromechanical coupling matrix
$\bar{\mathbf{M}}$	=	global mass matrix
$\bar{\mathbf{K}}$	=	global stiffness matrix
$\mathbf{\Psi}$	=	global vector of mechanical coordinates
$\mathbf{F}$	=	global vector of mechanical forces
R_l	=	load resistance
$\mathbf{m}^*$	=	mass per unit area of the finite element
v_p	=	voltage across the electrodes
$\mathbf{A}$	=	transformation matrix
w	=	downwash
V	=	airflow speed
ρ_o	=	density
p	=	pressure
K	=	Kernel function
M	=	Mach number
ω	=	circular frequency
C_p	=	coefficient of pressure
$\mathbf{A}$	=	matrix of influence coefficients
$\mathbf{\Phi}$	=	modal matrix
q	=	dynamic pressure

1 Introduction

Nowadays there is an interesting trade-off between the increasing military demand to reduce the size and weight of Unmanned Air Vehicles (UAVs) and the total flight duration of these aerial vehicles. A major limitation for practical small UAVs is the energy required for long endurance missions [1]. The limited available space and limited energy sources reduce the endurance and range. Harvesting solar and vibration energy has been investigated to provide an additional source of energy for UAVs [2-5]. Generating usable e electrical energy during their mission has the important practical value of relieving auxiliary power drains for aircraft by providing an additional energy source.

The concept of energy harvesting has been pointed out as a future breakthrough technology for UAV design [6]. The resulting structure can perform not only the primary load carrying function but also it can generate electricity. Researchers have added solar panels over the wing skin of UAVs to power small electronic devices or to charge batteries to support night flight missions [2,3]. Another possible source of energy for UAVs is the mechanical vibration energy due to unsteady aerodynamic loads during the flight [2] or due to ground excitation in perching [3,5]. Although other transduction mechanisms exist, piezoelectric transduction has received the most attention for vibration-based energy harvesting due to large power density it provides and several review articles have appeared in the last four years [7-10]. Recently a new multifunctional concept of self-charging structures has been presented [11]. The proposed multilayer, composite structure consists of piezoceramic layers for vibration to electrical energy conversion, thin-film battery layers for storing the generated energy and a metallic substructure layer as the primary load-bearing layer.

The time-domain piezo-aero-elastic modeling of a piezoelectric generator wing with embedded piezoceramics (two identical layers, on the top and bottom surfaces) has been presented in the literature [4,12]. The model is obtained from the combination of an electromechanically coupled FE model [13] with an unsteady vortex lattice model. The conversion of aeroelastic vibrations into electrical energy is investigated at several airflow speeds for a set of electrical load resistances. The aeroelastic behavior, and consequently the power generation, is dependent on aerodynamic damping which is modified with increasing airflow speed. At the flutter speed, the aerodynamic damping vanishes and the oscillations are persistent. Although this condition is avoided in a real aircraft, this is the best condition as a concept demonstration for the generator wing using the linear piezo-aero-elastic model. The response history with the largest power output at the flutter speed shows a decaying behavior which is due to the strong shunt damping effect of power generation. The effect of segmented electrodes on the piezo-aero-elastic response of the same generator wing and same set of load resistances has also been investigated [4]. The electrodes are segmented on the center line (mid-chord position) and properly combined to the electrical load to avoid the cancelation of the potential electrical output of the torsion-dominated modes (which is strongly cancelled when continuous electrodes are used). As a consequence of the improved electromechanical coupling, better power generation and shunt damping effects are obtained for the aeroelastic behavior since the piezoelectric reaction of the torsional modes in the coupled aeroelastic motions of flutter are taken into account with the segmented-electrode configuration. Although the time-domain linear piezo-aero-elastic model used for a concept demonstration for the generator wing (aeroelastic vibrations to electricity) has yielded interesting results, it is not practical (in terms of computational cost) to determine the optimum elements in the energy harvesting circuit of the generator wing.

In this paper, the frequency domain piezo-aero-elastic analysis of a generator wing with continuous electrodes is presented for energy harvesting. The piezo-aero-elastic model is obtained by combining an unsteady aerodynamic model and an electromechanically coupled FE model [13]. The subsonic unsteady aerodynamic model is accomplished here with the doublet lattice method [14]. The piezo-aero-elastic equations are solved using the P-K scheme modified to account the electrical domain of the problem. The evolution of damping for each mode with increasing airflow speed for a given load resistance is obtained. A procedure is also presented in order to obtain the optimum load resistance (for maximum power and for maximum damping) for a given airflow speed. Piezo-aero-elastically coupled frequency response functions (FRFs) are defined here for a desired airflow speed by combining the base excitation condition and the unsteady aerodynamic influence in the piezo-aero-elastic equations. The piezo-aero-elastic FRFs (relative tip motion, voltage, current and electrical power) are presented here for several different airflow speeds and load resistances. The effect of aerodynamic damping and mode coupling with increasing airflow speed over the aeroelastic evolution of the generator wing as well as its energy harvesting performance are discussed.

2 Piezo-Aero-Elastic Model

The piezo-aero-elastic model is obtained by combining an electromechanically coupled FE model and an unsteady doublet lattice aerodynamic model. The electromechanically coupled FE model is based on the Kirchhoff assumptions to model the thin cantilevered wing with embedded piezoceramic layers shown in Fig. 1. The substructure and the piezoceramic layers are assumed to be perfectly bonded to each other. The piezoceramic layers (which are poled in the thickness direction) are covered by continuous electrodes (which are assumed to be perfectly conductive) with negligible thickness. A resistive load is considered in the electrical domain and the purpose is to estimate the power generated in the electrical domain due to the aeroelastic vibrations of the energy harvester wing. A rectangular finite element with four nodes and three mechanical degrees of freedom per node is used to model the substructure. An electrical degree of freedom is added to the finite element to model the piezoceramic layers (13 degrees of freedom in total). A transformation is imposed in order to account for the presence of continuous and conductive electrodes bracketing each piezoceramic layer. This way a single electrical output is obtained from each piezoceramic layer. Refer to [13] for the detailed derivation and validation of the electromechanically coupled FE model against analytical and experimental results.

Since the electrical boundary condition due to a resistive load (R_l) is $\dot{Q}=v_p/R_l$, the governing piezo-aero-elastic equations for the generator wing are

$$\mathbf{M}\ddot{\mathbf{\Psi}}+\mathbf{C}\dot{\mathbf{\Psi}}+\mathbf{K}\mathbf{\Psi}-\tilde{\mathbf{\Theta}}v_p=\mathbf{F} \tag{1a}$$

$$\bar{C}_p\dot{v}_p+\frac{v_p}{R_l}+\tilde{\mathbf{\Theta}}'\dot{\mathbf{\Psi}}=0 \tag{1b}$$

where the global mass, stiffness and damping matrices as well as the equivalent electromechanical coupling vector and equivalent capacitance of the piezoceramic in the left-hand-side of the equations are obtained from the electromechanically coupled FE model. It is known from the literature [15,16] that the electrode pairs covering each piezoceramic layer can be connected in series or in parallel to the external electrical load (for better voltage or current). In general, the piezoceramic layers are poled in the same direction for parallel connection whereas they are poled in the opposite direction for series connection. For the parallel connection case, the effective electromechanical coupling vector is the sum of the individual contribution of each layer and the effective capacitance is the sum of each individual capacitances. For the series connection case, the effective electromechanical coupling vector is equal to that of one piezoceramic layer and the effective capacitance is one half of the capacitance of one piezoceramic layer. In the case study presented here, the continuous electrodes covering the piezoceramic layers (poled in the opposite directions) are connected in series to a resistive electrical load (Fig. 1). The right-hand-side of the mechanical equation (Eq. 1a) is the unsteady aerodynamic loads obtained with the unsteady doublet lattice solution.

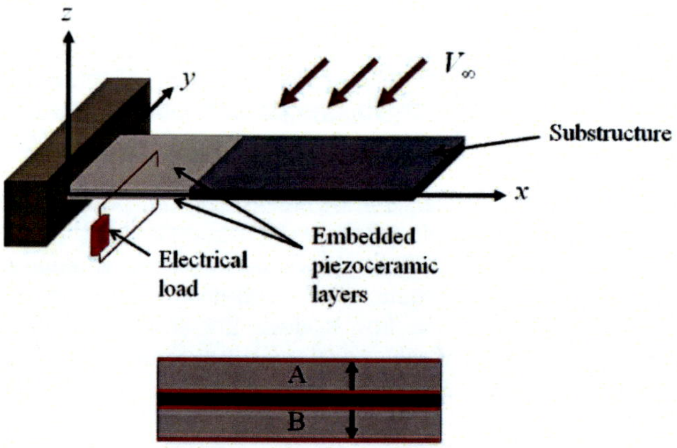

Figure 1 - Thin cantilevered wing with embedded piezoceramics layers and cross section.

2.1 The Doublet Lattice Model

The linearized formulation for the oscillatory, inviscid, subsonic lifting surface theory relates the normal velocity at the surface of a body (for example, a wing) with the aerodynamic loads by the pressure distribution [14]. The formulation is derived using the unsteady Euler equations of fluid. The doublet singularity (or a sheet of doublets) is a solution of the aerodynamic potential equation. The unsteady behavior as well as the resultant differential pressure across the surface of a wing can be represented with this solution.

The relation between the pressure difference across the surfaces and the velocity normal to the surface of a wing is given by a kernel function. The kernel function is a closed-form solution of the integro-differential equation with the assumption of harmonic motion. This way the distribution of velocity normal to the surface of a wing is given by the equation,

$$w(x,y,z)=\frac{-1}{4V\pi\rho_o}\iint_S \Delta p(x,y,z)K(x-\xi,y-\eta,z)\,d\xi d\eta \tag{2}$$

where $\Delta p(x,y,z)$ is the differential pressure, V is the free-stream velocity, ρ_o is the density of the air, ξ and η are dummy variables of integration over the area *S* of the wing in *x* (chord-wise) and *y* (span-wise) direction, *z* is the transverse direction and *K* is the kernel function. The kernel function is given as

$$K(x-\xi,y-\eta,z)=\exp\left(\frac{-j\omega(x-\xi)}{V}\right)\frac{\partial^2}{\partial z^2}\left[\frac{1}{R}\exp\left[\frac{j\omega}{V\beta^2}(\lambda-M\bar{R})\right]d\lambda\right] \tag{3}$$

where $\beta^2=1-M^2$ and $R=\sqrt{(x-\xi)^2+(y-\eta)^2+z^2}$, ω is the frequency of excitation, *M* is the mach number and λ is a dummy variable.

The Doublet Lattice Method (DLM) provides a numerical approximation for the solution of the kernel function. The wing is represented by a thin lifting surface and it is divided into a number of elements (panels or boxes represented by the blue and red lines in Fig. 1) with doublet singularities of constant strength in chord-wise and parabolic strength in span-wise direction. A line of doublets (distribution of acceleration potential) is assumed at the ¼ chord line (green lines in Fig. 1) of each panel, which is equivalent to a pressure jump across the surface. A control point (red circles in Fig. 1) is defined in the half span of each box at the ¾ chord line (the point where the boundary condition is verified). The strength of the oscillating potential placed the ¼ chord lines are the unknowns of the problem.

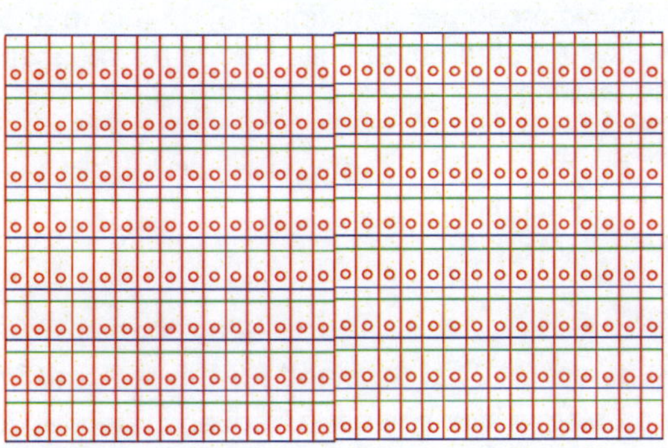

Figure 2 – Example of a rectangular lifting surface divided into boxes (blue and red lines); doublets lines represented by green lines and control points (red circles).

The prescribed downwash (as the solution is assumed to be harmonic) introduced by the lifting lines is checked at each control point. The solution of the resulting matrix equation is

$$\frac{\bar{\mathbf{w}}}{V}=\mathbf{A}\Delta\mathbf{C}_p \tag{4}$$

which gives the strength of the lifting line at each panel and consequently the pressure distribution across the surface. Here, $\mathbf{A}$ is the matrix of influence (which is related to the kernel function) between the normal velocity and non-dimensional pressure distribution $\Delta\mathbf{C}_p$. Integration over the surface gives the local and total aerodynamic force coefficients.

2.2 Piezo-aero-elastically Coupled Equations of Motion

The aerodynamic loads can be included in the piezo-aero-elastic equations as an aerodynamic matrix of influence coefficients. The aerodynamic loads and the structural motion are obtained from distinct numerical methods with distinct meshes. Transformation matrices are determined using a surface splines scheme in order to interpolate the forces obtained in the doublet lattice mesh to the nodes of the FE mesh [17]. The resulting displacements from the structural mesh are also interpolated to the corners of the aerodynamic mesh.

As harmonic response is assumed for the unsteady aerodynamic solution, the piezo-aero-elastic equations (Eqs. 1a and 1b) in modal domain can be presented as,

$$\left(-\omega^2\bar{\mathbf{M}}+j\omega\bar{\mathbf{C}}-q\mathbf{Q}\right)\boldsymbol{\eta}-\boldsymbol{\Phi}^t\tilde{\boldsymbol{\Theta}}v_p=0 \tag{5a}$$

$$C_p j\omega v_p+\frac{v_p}{R_l}+j\omega\tilde{\boldsymbol{\Theta}}^t\boldsymbol{\Phi}\boldsymbol{\eta}=0 \tag{5b}$$

where $\bar{\mathbf{M}}$ is the modal mass matrix, $\bar{\mathbf{K}}$ is the modal stiffness matrix, $\bar{\mathbf{C}}$ is the modal damping matrix assumed here as proportional to the mass and stiffness, $\boldsymbol{\eta}$ is the vector of modal displacement, $\boldsymbol{\Phi}$ is the modal matrix, q is the dynamic pressure, $\mathbf{Q}$ is the aerodynamic influence matrix, j is the unit imaginary number and ω the circular frequency.

The conventional P-K scheme is one of the available ways to address the flutter equations for unsteady aerodynamic theories with the harmonic motion assumption [18]. In this method, the evolution of the frequencies and damping is iteratively investigated for different airflow speeds (or reduced frequencies) solving the following eigenvalue problem for a conventional wing,

$$p\begin{Bmatrix}\mathbf{h}_1\\ \mathbf{h}_2\end{Bmatrix}=\begin{bmatrix}0 & \mathbf{I}\\ -\bar{\mathbf{M}}^{-1}\left(\bar{\mathbf{K}}-q\mathbf{Q}^R\right) & -\bar{\mathbf{M}}^{-1}\left(\bar{\mathbf{C}}-q\mathbf{Q}^I\right)\end{bmatrix}\begin{Bmatrix}\mathbf{h}_1\\ \mathbf{h}_2\end{Bmatrix} \tag{6}$$

where $\mathbf{h}_1=\boldsymbol{\eta}$ and $\mathbf{h}_2=p\boldsymbol{\eta}$, the superscripts R and I stand for the real and imaginary parts of the aerodynamic matrix, p is the eigenvalue of the problem which gives the frequency (related to the imaginary part) and damping (related to the real part). However, in this work, Eqs. (5a) and (5b) differ from the conventional flutter equation due to the presence of piezoelectric layers and an external generator circuit (electromechanical coupling in the mechanical equation and the electrical equation, Eq. 5b). Therefore the conventional P-K scheme is modified to solve the piezo-aero-elastic problem. This way an augmented system is solved to examine the piezo-aero-elastic behavior with increasing airflow speed and one specific load resistance,

$$p\begin{Bmatrix}\mathbf{h}_1\\ \mathbf{h}_2\\ h_3\end{Bmatrix}=\begin{bmatrix}0 & \mathbf{I} & 0\\ -\bar{\mathbf{M}}^{-1}\left(\bar{\mathbf{K}}-q\mathbf{Q}^R\right) & -\bar{\mathbf{M}}^{-1}\left(\bar{\mathbf{C}}-q\mathbf{Q}^I\right) & -\bar{\mathbf{M}}^{-1}\left(\boldsymbol{\Phi}\tilde{\boldsymbol{\Theta}}\right)\\ 0 & \dfrac{\tilde{\boldsymbol{\Theta}}\boldsymbol{\Phi}}{C_p} & -\dfrac{1}{C_pR_l}\end{bmatrix}\begin{Bmatrix}\mathbf{h}_1\\ \mathbf{h}_2\\ h_3\end{Bmatrix} \tag{7}$$

where $\mathbf{h}_1=\boldsymbol{\eta}$, $\mathbf{h}_2=p\boldsymbol{\eta}$ and $h_3=v_p$ (voltage output). The modified solution still gives the evolution of frequency and damping of the modes with increasing airflow speed as in the conventional P-K solution, but here the effect of a load resistance connected to the piezoceramic layer is considered. Although the main motivation here is electrical power generation this solution can be used to investigate the influence of the electrical domain (a load resistance here, or a resistive inductive circuit, or a more complex circuit for example) over the aeroelastic behavior of a generator wing.

2.3 Piezo-aero-elastically coupled FRFs

In addition the modified PK scheme (which gives the neutral stability limit) the piezo-aero-elastic behavior can be investigated in terms of piezo-aero-elastically coupled FRFs. The FRFs are defined using the Eqs. (5a) and (5b) assuming the base excitation condition in the piezo-aero-elastic problem. Therefore, the forcing term in Eq. (1a) is modified as

$$\mathbf{F} = \mathbf{F}_{aero} + \mathbf{F}_{b} \tag{8}$$

where $\mathbf{F}_{aero}$ is the unsteady aerodynamic loads determined by the doublet lattice method ($\mathbf{F}_{aero} = q\mathbf{Q}$). As discussed in the literature [19], if the base is vibrating in the transverse direction (z-direction), the effective force on the structure is due to the inertia of the structure in the same direction. Therefore the forcing term $\mathbf{F}_b$ is represented as

$$\mathbf{F}_b = \mathbf{m}^* a_b \tag{9}$$

where $\mathbf{m}^*$ is a vector of mass per unit area obtained from the FE solution (including both the piezoceramic and/or the substructure layers) and a_b is the base acceleration. Assuming the harmonic motion of the cantilevered base of the generator wing with the influence of the unsteady aerodynamics the piezo-aero-elastically coupled FRFs are defined by the matrix equation,

$$\mathbf{FRF}s = \begin{bmatrix} -\omega^2\bar{\mathbf{M}} + j\omega\bar{\mathbf{B}} + \bar{\mathbf{K}} - q\mathbf{Q} & -\tilde{\mathbf{\Theta}} \\ j\tilde{\mathbf{\Theta}}\omega & jC_p\omega + \dfrac{1}{R_l} \end{bmatrix} \mathbf{m}^* \tag{10}$$

where **FRFs** is a $(n+1)$ x 1 vector containing the *n* modal displacement vectors per base acceleration (*n* is the number of modes considered in the solution) for a desired airflow speed. The line $n+1$ gives the steady state voltage FRF defined here as the voltage per base acceleration for a desired airflow speed. In addition to the voltage FRF we can define the electrical current and electrical power FRFs. The electric current FRF is obtained by dividing the voltage FRF to the load resistance R_l of the energy harvesting circuit. The electrical power FRF is the product of voltage and current FRFs and it is defined as the ratio of electrical power output to the square of the base acceleration.

The variation of power output with load resistance at the short circuit resonance frequency of a specific mode at a desired airflow speed can be investigated with the formulation presented here. This way the optimum load resistance for maximum power or maximum shunt damping can be determined for a desired airflow speed and mode of interest. Although the flutter condition is avoided in a real aircraft, this is a useful practice to obtain an idea of how much energy can be harvested from such a cantilevered wing. The typical aeroelastic behavior at this speed results in continuous power generation, i.e., modes are coupled at the flutter frequency and self-sustained oscillations are obtained (zero aerodynamic damping). The optimum load resistance can be determined at the flutter speed (or at airflow speeds slightly smaller than the flutter speed) exciting the generator wing at the short circuit flutter frequency and investigating the power output. The optimum load resistance for maximum resistive shunt damping effect can also be determined by exciting the generator wing at the short circuit flutter frequency and investigating the relative tip motion.

3 Results

This section presents the piezo-aero-elastic behavior of a cantilevered plate-like wing with embedded piezoceramics. The electromechanical behavior of the generator wing is first presented here by piezo-aero-elastically coupled FRFs close to the short-circuit condition (assuming $R_l = 100\Omega$) and several airflow speeds (from low airflow speeds to the flutter speed). After that the optimum load resistance for maximum power generation and maximum shunt damping effect at the flutter speed are determined. Using the optimal load resistance for power the piezo-aero-elastically coupled FRFs are compared with the short circuit FRFs.

Two identical layers of PZT-5A are embedded into the top and on the bottom of the plate at the root. Conductive electrodes covering the upper and lower faces of the piezoceramic layers and connected in series to a resistive electrical load as depicted in Fig. 1. The dimensions of the plate-like wing used in this work are 1200 × 240 × 3 mm^3. The identical piezoceramic layers have the same width as the wing chord. The embedded piezoceramics layers cover 30% of the wing span (from the root to the tip) and each one has a thickness of 0.5 mm. The geometric and material properties of the wing (aircraft aluminum alloy Al 2024-T3) are presented in Table 1. Note that the length - to - thickness ratio of the wing is large enough to neglect the shear deformation and rotary inertia effects for the practical vibration modes.

Table 1. Geometric and material properties of the aluminum wing with embedded piezoceramics

Length of the wing (mm)	1200
Width of the wing (mm)	240
Thickness of the wing (mm)	3
Young's modulus of the wing (GPa)	70.0
Mass density of the substructure (kg/m^3)	2750
Proportional constant – α (rad/s)	0.1635
Proportional constant – β (s/rad)	4.1711×10^{-4}

The material and electromechanical properties of PZT-5A are given in Table 2. It is worth mentioning that manufacturers typically provide limited number of properties for piezoelectric ceramics. For instance, in the predictions of their analytical model, Erturk and Inman [15] used the data provided by Piezo Systems Inc. [20] as the data provided by the manufacturer was sufficient for a beam-type formulation. However, the plate-type formulation given here requires more than what is provided in the manufacturer's data sheet (see, for instance, the properties required for the calculation of the plane-stress elastic, piezoelectric and dielectric components in [13]). Therefore, the 3-D properties of PZT-5A [21] displayed in Table 2 are used here.

Table 2. Material and electromechanical properties of PZT-5A

Mass density (kg/m^3)	7800
Permittivity (nF/m)	$1800 \times \varepsilon_0$
c_{11}^E, c_{22}^E (GPa)	120.3
c_{12}^E (GPa)	75.2
c_{13}^E, c_{23}^E (GPa)	75.1
c_{33}^E (GPa)	110.9
c_{66}^E (GPa)	22.7
e_{31}, e_{32} (C/m^2)	-5.2
e_{33} (C/m^2)	15.9

The mode sequence and the undamped natural frequencies for the plate-like wing obtained from the FE model for close to short-circuit conditions (very low load resistance) are presented in Table 3. The first five modes are listed where B and T stand for the bending modes and torsion modes, respectively. It is important to note that the span-wise elastic axis and the center of gravity are coincident at 50% of chord.

Table 3. Natural frequencies and mode shapes

MODE	MODE SHAPE	ω_{sc} [Hz]
1	1B	1.68
2	2B	10.46
3	1T	16.66
4	3B	27.74
5	2T	48.65

The aeroelastic behavior of the generator wing close to the short circuit conditions ($R_l = 100\Omega$) is presented in terms of damping and frequency with increasing airflow speed in Figs. 3. The evolution of damping with airflow speed (Fig. 3a) shows the flutter instability for the short circuit condition at 40 m/s. The frequency evolution with increasing airflow speed is observed in Fig. 3b. Clearly the coalescence of the second and third modes is observed around the flutter condition.

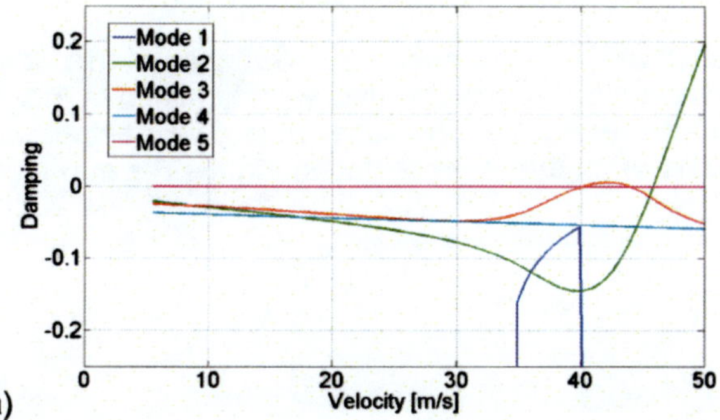

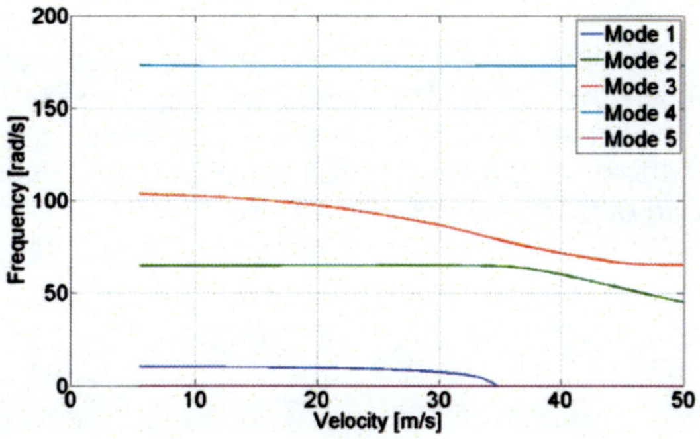

Figure 3 – (a) Damping evolution with increasing airflow speed and (b)frequency evolution with increasing airflow speed.

The relative tip motion FRF and the electrical power output FRF are presented for several airflow speed [from the no flow condition (V=0m/s) to the flutter speed] in Figs. 4. The peaks relative to the first bending and second bending modes are observed for the no flow condition. As discussed in this work and in the literature [19] the forcing term in the base excitation problem is related to the inertia of the structure in the direction of base motion (z-direction in this work). For the symmetric structures (as the generator wing with symmetric mass distribution used here), one cannot observe the peaks related to pure torsional modes in the electromechanical FRFs for the base excitation condition without unsteady aerodynamic influence (V=0 m/s). For instance, the resonance frequency for the first torsional mode of the wing is 16.6 Hz and no peak is observed for this frequency in Figs. 4 (a) and (b) when V=0 m/s. In typical aeroelastic response, modes are coupled with increasing airflow speed. Therefore a peak is observed around 16 Hz for the airflow speed of 20 m/s in Fig 4 (a). However, this peak is not observed in the power FRF (Fig. 4b). At this airflow speed this is a bending-torsion coupled mode dominated by torsional motion. The electrical output from torsional vibrations is canceled when continuous electrodes are covering the piezoceramic layers of the generator wing [4]. At the airflow speed of 35 m/s this peak is shifted for 13 Hz and still represents a bending torsion mode. However, at this airflow speed it is dominated by bending motion. As a result, one can observe a peak at this frequency in the power FRF of Fig. 4b(no cancellation). At the flutter speed (40 m/s), aerodynamic damping is zero and modes are coupled at the flutter frequency (11.5 Hz) and maximum tip displacement and power output are achieved. Power could be optimized if segmented electrodes were used to avoid the cancelation of electrical outputs from torsional motions at the coupled bending torsion motions of flutter. The effect of increasing airflow speed over damping can also be observe in Figs. 4 (a) and (b). Aerodynamic damping increases from 5 m/s 35 m/s due to unsteady aerodynamic effects. The maximum damping is observed at an airflow speed of 35 m/s and this is not a favorable condition for power harvesting.

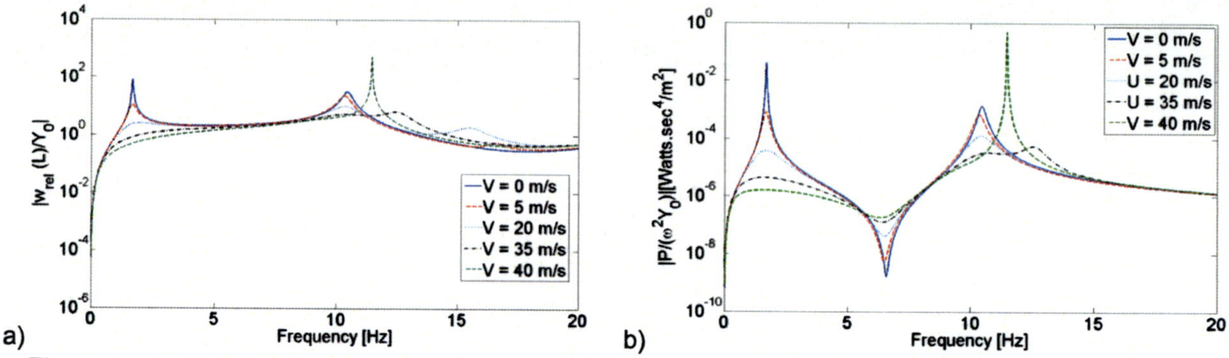

Figure 4 – (a) Relative tip motion FRF for several airflow speed and $R_l = 100\Omega$ and (b) Power FRF for several airflow speed and $R_l = 100\Omega$.

The optimum load resistance for maximum power output at the flutter speed is determined in this work. The cantilevered end of the generator wing is excited at the short circuit flutter frequency (determined in Fig. 4) and the maximum power output can be obtained for a certain load resistance. The variation of power output with load resistance at the short circuit flutter frequency (11.5 Hz) and V=40 m/s is presented in Fig. 5. The maximum power output is observed for $R_l = 15.8k\Omega$.

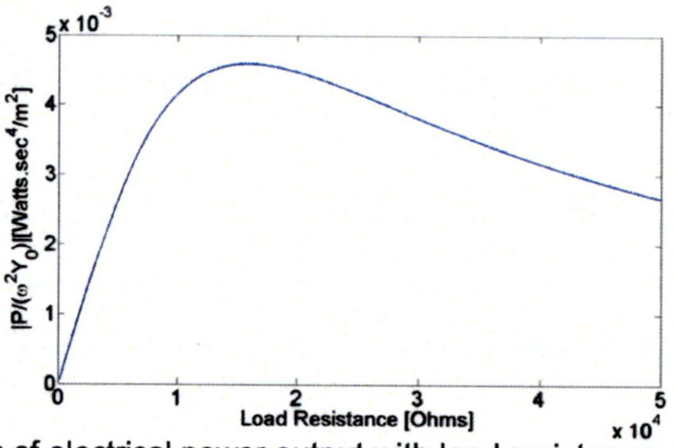

Figure 5 – Variation of electrical power output with load resistance at flutter conditions.

The electrical power output and relative tip motion FRF at the flutter speed are presented in Figs. 6 (a) and (b) for two values of load resistance. The first load resistance is $R_l = 100\Omega$ (short circuit condition) and the second one is the optimum load resistance for maximum power. Power amplitude is larger for the optimum load resistance over the entire range of frequencies considered. The system is vibrating at slightly different frequencies at the short circuit condition and the optimum load condition, a typical behavior due to the electromechanical coupling [13,19]. The strong shunt damping effect of resistive power dissipation is observed in the relative tip motion FRF. Although damping is introduced into the system and the amplitude of motion at the wing tip is reduced the flutter speed is not significantly increased due to the resistive power harvesting. The flutter speed for the optimum load resistance is 40.5 m/s. A resistive inductive generator circuit can be easily included in the present frequency domain formulation. This way increased power output and increased flutter speed are expected by adjusting the inductor to the target frequency (flutter frequency) and searching for the optimum load resistance for maximum damping.

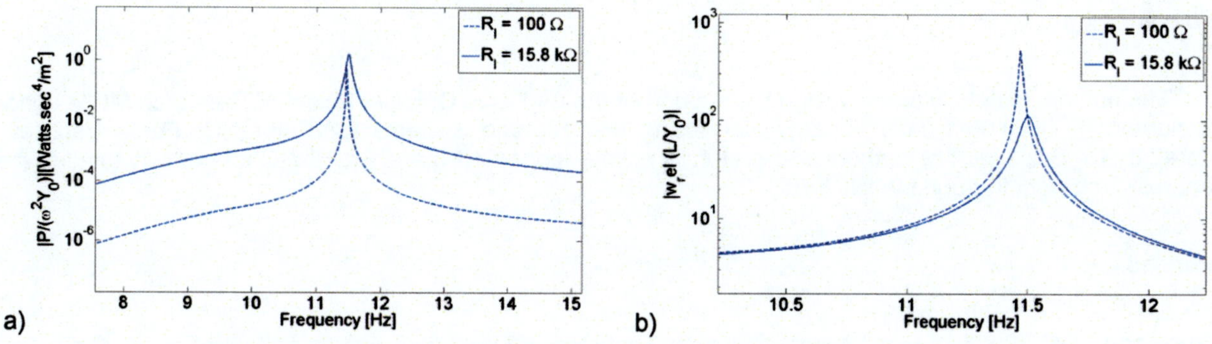

Figure 6 – (a) Power FRF and (b) relative tip motion FRF at the flutter speed for the short circuit condition and using the optimum load resistance for maximum power output.

4 Conclusions

In this paper, piezo-aero-elastic modeling in frequency domain of a cantilevered plate-like wing with embedded piezoceramics is presented for energy harvesting. The electromechanical finite element plate model is based on the Kirchhoff assumptions. A resistive load is considered in the external circuit. The subsonic unsteady aerodynamic model is accomplished using the doublet lattice method. The electromechanical and the aerodynamic models are combined in order to obtain the piezo-aero-elastic equations. A modified P-K scheme is presented in order to solve the piezo-aero-elastic equations. The piezo-aero-elastic evolution with airflow speed is investigated. The flutter speed can be obtained for any value of load resistance used in the electrical domain.

An imposed base excitation is also considered in the piezo-aero-elastic model. This way piezo-aero-elastically coupled FRFs (base excitation with unsteady aerodynamic effects) are obtained for an airflow speed and/or load resistance. The effect of the aerodynamic damping over the resonance frequencies and the modes coupling with airflow speed are clearly observed in the FRFs. The cancelation of electrical output from torsion-dominated modes for the continuous electrodes case is observed by comparing the mechanical FRF (relative tip motion) with the power output FRF. The peak relative to a torsion dominated coupled mode is observed around 16 Hz in the mechanical FRF after a certain airflow speed (20 m/s). However, this peak is not observed in the electrical FRF for the same speed since continuous electrodes are used in this work. The coupling is modified with increasing airflow speed. At 35 m/s the coupled mode oscillates around 13 Hz. It is a bending torsion mode dominated by bending motion. Therefore, with this airflow speed, an electrical output is observed for base excitation at this frequency. As a consequence, at the flutter speed and flutter frequency (where bending torsion motion is observed) the cancelation of the out of phase electrical output occurs. The use of segmented electrodes could avoid the cancelation and improve the piezo-aero-elastic behavior.

The frequency domain piezo-aero-elastic model presented here has the great advantage of reduced computational cost when compared to the time domain piezo-aero-elastic solution previously presented by the authors. Therefore the optimum load resistance for maximum power output (or for maximum shunt damping) can be obtained using the unsteady aerodynamics along with an imposed base excitation for a desired airflow speed. In addition to the maximum power output at flutter speed, the optimum load gives a strong shunt damping effect reducing the amplitude of motion at the wing tip.

Acknowledgments

The authors gratefully acknowledge the support of the Air Force Office of Scientific Research MURI under Grant number F 9550-06-1-0326 "Energy Harvesting and Storage Systems for Future Air Force Vehicles" monitored by Dr. B.L. Lee. The authors also gratefully acknowledge CNPq and FAPEMIG for partially funding the present research work through the INCT-EIE.

References

[1] Langelaan, J.W., 2007, "Long Distance/Duration Trajectory Optimization for Small UAVs" AIAA Guidance, Navigation and Control Conference and Exhibit.

[2] Anton, S.R. and Inman D.J., "Vibration energy harvesting for unmanned air vehicles," Proceedings of SPIE 6928, March 10-13, San Diego, CA (2008).

[3] Magoteaux, K.C., Sanders, B. and Sodano, H.A., "Investigation of energy harvesting small unmanned air vehicle" Proceedings of SPIE 6928, March 10-13, San Diego, CA (2008).

[4] De Marqui, Jr., C., Erturk, A., and Inman, D.J. 2009 Effect of Segmented Electrodes on Piezo-Elastic and Piezo-Aero-Elastic Responses of Generator Plates, Proceedings of the ASME Conference on Smart Materials, Adaptive Structures and Intelligent Systems, Oxnard, CA, 20-24 September 2009.

[5] Erturk, A., Renno, J.M. and Inman, D.J., "Modeling of piezoelectric energy harvesting from an L-shaped beammass structure with an application to UAVs," Journal of Intelligent Material Systems and Structures 20, 529-544 (2009).

[6] Pines, D.J. and Bohorquez, F., 2006, "Challenges Facing Future Micro-Air-Vehicle Development," Journal of Aircraft, 43, pp. 290-305.

[7] Sodano, H.A., Inman, D.J. and Park, G., 2004, "A review of power harvesting from vibration using piezoelectric materials," The Shock and Vibration Digest, 36, pp. 197-205.

[8] Priya, S., 2007, "Advances in energy harvesting using low profile piezoelectric transducers," Journal of Electroceramics, 19, pp. 167–184.

[9] Anton, S.R. and Sodano, H.A., 2007, "A review of power harvesting using piezoelectric materials 2003-2006," Smart Materials and Structures, 16, R1-R21.

[10] Cook-Chennault, K.A., Thambi, N. and Sastry, A.M., 2008, "Powering MEMS portable devices – a review of non-regenerative and regenerative power supply systems with emphasis on piezoelectric energy harvesting systems," Smart Materials and Structures, 17, 043001.

[11] Anton, S.R., Erturk, A., Kong, N., Ha, D.S. and Inman, D.J. 2009, Self-charging structures using piezoceramics and thin-filmbatteries, Proceedings of the ASME Conference on Smart Materials, Adaptive Structures and Intelligent Systems, Oxnard, CA, 20-24 September 2009.

[12] De Marqui, Jr., C., Erturk, A., and Inman, D.J. 2009 Piezo-Aero-Elastically Coupled Modeling and Analysis of Electrical Power Generation and Shunt Damping for a Cantilever Plate, Proceedings of the 17th International Conference on Composite Materials, Edinburgh, UK, 27-31 July 2009.

[13] De Marqui Junior, C., Erturk, A. and Inman, D.J., 2009, "An electromechanical finite element model for piezoelectric energy harvester plates," Journal of Sound and Vibration, V. 327, pp 9-25.

[14] Albano, E. and Rodden, W.P., 1969, A doublet-lattice method for calculating lift distributions on oscillating surfaces in subsonic flow. AIAA Journal 7, pp. 279–285

[15] Erturk, A. and Inman, D.J., 2009, "An experimentally validated bimorph cantilever model for piezoelectric energy harvesting from base excitations," Smart Materials and Structures, 18, 025009.

[16] Wang, Q.M. and Cross, L.E., 1999, "Constitutive equations of symmetrical triple layer piezoelectric benders," IEEE Transactions on Ultrasonics, Ferroelectrics, and Frequency Control, 46, pp. 1343–51.

[17] Harder, R. L.; Desmarais, R. N., 1972, Interpolation using surface splines, Journal of Aircraft, V. 9, pp. 189-191.

[18] Hassig, H. J.,"An approximate true damping solution of the flutter equation by determinant iteration" Journal of Aircraft, Vol.8, no.11 (885-889), November, 1971, doi: 10.2514/3.44311

[19] Erturk A and Inman D J 2009 Issues in Mathematical Modeling of Piezoelectric Energy Harvesters Smart Materials and Structures 17 065016.

[20] /http://www.piezo.com/prodsheet1sq5A.htmlS, Accessed December 2008.

[21] /http://www.efunda.com/materials/piezo/material_dataS, Accessed December 2008.

Proceedings of the IMAC-XXVIII
February 1–4, 2010, Jacksonville, Florida USA
©2010 Society for Experimental Mechanics Inc.

Mechanical effect of combined piezoelectric and electromagnetic energy harvesting

Mickaël LALLART* and Daniel J. INMAN
Department of Mechanical Engineering, Center for Intelligent Material Systems and Structures
Virginia Tech, Blacksburg, Virginia, 24061, USA

ABSTRACT
The recent progress in microelectronics, combined with the increasing demand from numerous industrial fields in terms of autonomous sensor networks, has placed a particular attention on ways to power up such devices. In order to bypass the drawbacks of batteries, harvesting power from ambient sources has been proposed. Among all the available sources, particular attention has been placed on scavenging energy from vibrations, which are commonly available in many environments. In this case, converting mechanical energy into electrical energy is typically done using either piezoelectric or electromagnetic transduction. Recently, the combination of these two conversion mechanisms has been proposed, allowing benifits from the advantages of the two techniques and leading to the concept of hybrid energy harvesting. This paper proposes an investigation of the combination of the two conversion effects on the mechanical behavior of the host structure, both in terms of damping and stiffness changes. Particularly, for highly coupled, weakly damped systems, it is shown that combining the piezoelectric and electromagnetic effects does not lead to a power increase, but allows enhancing both the bandwidth and the load independency of the harvester.

1 INTRODUCTION

The increasing demand in terms of consumer electronics has raised the issue of powering up devices using ambient harvested energy in order to replace batteries that raise maintenance and environmental issues. Hence, the future of sensing should be self-powered, allowing devices to operate using the surrounding energy sources for supplying electrical circuits ([1, 2, 3]). Such a trend is encouraged by progresses in micropower electronics, as well as an increasing demand from various industrial fields (for instance aeronautic, civil and biomedical engineering, and home automation) in terms of "place and forget" sensors and sensor networks.

Among all the available energy for small-scale devices (solar, magnetic or thermal), a particular attention has been placed on vibration energy harvesting, as mechanical energy from vibrating parts is one of the most commonly available sources in many environments ([4]). When dealing with electromechanical conversion, several physical effects can be considered, but the two most common ones rely on piezoelectricity ([5, 6, 7, 8, 9, 10, 11]) and electromagnetism ([12, 13, 14]). While a significant number of studies on energy harvesting has been devoted to only one or the other conversion effect, a recent trend consisted in combining piezoelectric and magnetic energy harvesting into a single device, leading to the concept of hybrid energy harvesting ([15]).

The purpose of this paper is to investigate the mechanical effects that arise when using such a harvester, demonstrating both the frequency shift and damping effects generated by the harvesting processes. The paper is organized as follows. Sections 2 and 3 review the basic principles of energy harvesting using piezoelectric and magnetic transducers, and expose the derivation of the expected output power and the effect of harvesting on mechanical vibrations when using both transduction mechanisms. Section 4 aims at validating the proposed model through experimental measurements, and Section 5 finally concludes the paper.

2 HYBRID HARVESTING PRINCIPLES

When designing an energy harvester that aims at supplying an electrical circuit from vibrations, the basic operation consists of rectifying the voltage of the transducer. This is typically done using AC/DC converters. However, from

*mickael.lallart@insa-lyon.fr

the transducer's point of view and assuming that only the first harmonic of the vibrations has a significant effect, the converter can be seen as a single linear load represented by its internal impedance (Figure 1)[1].

Under these assumptions, the energy harvesting processes can be considered separately in an electrical point of view (but not in terms of mechanical effect). As well, the dual nature of the piezoelectric and electromagnetic effects leads to the following differences ([16]):

- Mechanical aspects and positioning:
  - Piezomaterials are sensitive to the absolute strain within the structure, and have to be placed near a clamped edge for simple structures such as cantilever beams.
  - Electromagnetic elements are sensitive to the relative speed of the structure with respect to the second part of the transducer (*e.g.*, coil), and have to be placed near a free end for simple structures such as cantilever beams (large displacement).
- Electrical aspects:
  - Piezomaterials act as voltage sources, these latter being proportional to the strain or stress (or, in a macroscopic point of view, to the displacement).
  - Magnetic elements act as current sources, these latter being proportional to the relative speed.
- Representation:
  - Piezomaterials are equivalent in the electrical point of view to a current source in parallel with a capacitor, and in a mechanical point of view to a force in parallel with a spring.
  - Magnetic elements are equivalent in the electrical point of view to a voltage source in series with an inductor, and in a mechanical point of view to a force in series with a mass.

3 THEORETICAL DEVELOPMENT

Based on the previous statements, this section proposes to derive the power expression obtained when using a hybrid device, as well as to investigate the mechanical effect of harvesting, in terms of resonance frequency shift and vibration damping, induced by the combination of piezoelectric and magnetic energy harvesting.

3.1 Modeling

In the followings, a simple single degree of freedom (SDOF) model of the mechanical structure that relates quite well the mechanical behavior of the device near one of its resonance frequencies ([17, 18]) is considered. It can be shown that the governing electromechanical equations of the entire device are given by ([16]):

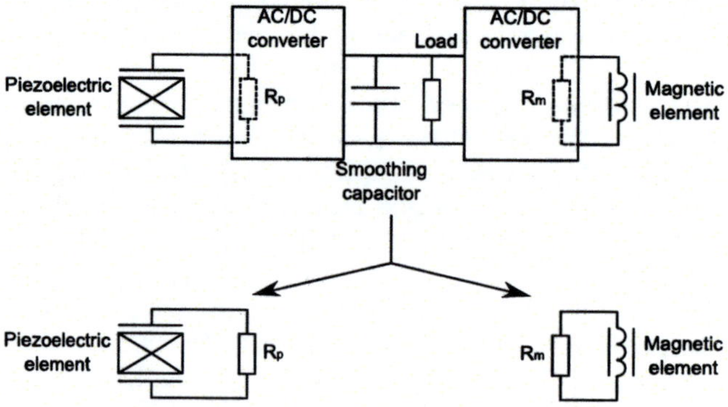

Figure 1: Hybrid energy harvesting schematic

[1]which can nevertheless depends on the load and frequency

$$\begin{cases} M\ddot{u} + C\dot{u} + K_E u = -\mu_1 Ma - \alpha V_p - AI_m \\ I_p = \alpha\dot{u} - C_0\dot{V}_p \\ V_m = A\dot{u} - L_0\dot{I}_m - r_L I_m \end{cases}, \tag{1}$$

where the parameters are those defined in Table 1. The duality of the transducers can also be demonstrated from these equations (*e.g.*, piezovoltage *vs.* magnetic current, piezocapacitance *vs.* magnetic inductance...).

3.2 Power derivation

When connecting each transducer to its corresponding converter that presents a given input impedance, the electrical equation of Eq. (1) becomes:

$$\begin{cases} \dfrac{V_p}{R_p} = \alpha\dot{u} - C_0\dot{V}_p \\ R_m I_m = A\dot{u} - L_0\dot{I}_m - r_L I_m \end{cases}, \tag{2}$$

where R_p and R_m represent the piezoelectric and electromagnetic equivalent loads, respectively. Hence, expressing these relationships in the frequency domain yields:

$$\begin{cases} V_p(\omega) = \dfrac{j\omega R_p \alpha}{1 + jR_p C_0 \omega} u(\omega) \\ I_m(\omega) = \dfrac{j\omega A}{(R_m + r_L) + jL\omega} u(\omega) \end{cases}, \tag{3}$$

and therefore the transfer function giving the displacement as a function of the applied acceleration is given by:

$$\frac{u(\omega)}{a(\omega)} = \frac{-\mu_1 M}{-M\omega^2 + jC\omega + K_E + \dfrac{j\omega R_p \alpha^2}{1 + jR_p C_0 \omega} + \dfrac{j\omega A^2}{(R_m + r_L) + jL\omega}}. \tag{4}$$

Hence, for weakly damped systems, the resonance frequency ω_0 can be found by cancelling the real part of the denominator, leading to:

$$\left[(R_m + r_L)^2 + (L\omega_0)^2\right] + C_0 (R_P \alpha \omega_0)^2 \left[(R_m + r_L)^2 + (L\omega_0)^2\right] + L (A\omega_0)^2 \left[1 + (R_p C_0 \omega_0)^2\right] = 0, \tag{5}$$

and the corresponding vibration magnitude is given by:

Parameter	Definition
Mechanical	
M	Dynamic mass
C	Structural damping coefficient
K_E	Natural stiffness
μ_1	Correction factor ([18])
a	Base acceleration
Electrical	
V_p	Piezoelectric voltage
I_p	Piezoelectric current
C_0	Piezoelectric transducer clamped capacitance
V_m	Electromagnetic voltage
I_m	Electromagnetic current
L_0	Electromagnetic transducer inductance
r_L	Electromagnetic transducer leakage resistance
Electromechanical	
α	Piezoelectric force factor
A	Magnetic force factor

Table 1: Model parameter definition

$$u_m\big|_{res} = \frac{\mu_1 M a_M}{C\omega_0 + \dfrac{\omega_0 R_p \alpha^2}{1+(R_p C_0 \omega_0)^2} + \dfrac{\omega_0 A^2 (R_m + r_L)}{(R_m + r_L)^2 + (L\omega_0)^2}}. \tag{6}$$

Consequenly, these last two equations demonstrate that the harvesting process induces both a resonance frequency shift of the structure (Eq. (5)) and a damping of the vibrations at the resonance (Eq. (6)), depending on both the piezoelectric and magnetic energy harvesting processes.

The power harvested only by the piezoelectric element can be expressed as:

$$P_p = \frac{V_p {V_p}^*}{2R_p} = \frac{1}{2}\frac{R_p {\omega_0}^2 \alpha^2}{1+(R_p C_0 \omega_0)^2}{u_M}^2? \tag{7}$$

with u_M the vibration magnitude at the considered frequency (obtained using Eq. (4)), while the power harvested only by the magnetic element is given by:

$$P_m = \frac{1}{2} R_m I_m {I_m}^* = \frac{1}{2}\frac{R_m {\omega_0}^2 A^2}{(R_m + r_L)^2 + (L\omega_0)^2}{u_M}^2. \tag{8}$$

Hence, when combining both transduction mechanisms, the total output power yields:

$$P_{\text{total}} = \left[\frac{R_p \alpha^2}{1+(R_p C_0 \omega_0)^2} + \frac{R_m A^2}{(R_m + r_L)^2 + (L)^2}\right] \times \frac{(\omega_0 u_M)^2}{2}. \tag{9}$$

3.3 Theoretical comparison

Here it is proposed to compare and evaluate the performance and mechanical effects of harvesting energy by combining both the piezoelectric and electromagnetic conversion principles, with regard to systems using only one of them.

When considering that the system is excited at its resonance frequency (Eq. (5)), the corresponding expected output power and displacementt magnitude are given in Figure 2 when each transducer has the same coupling coefficient defined as:

$$\begin{aligned} k_p &= \sqrt{\alpha^2/(C_0 K_E + \alpha^2)} \quad \text{for the piezo device} \\ k_m &= \sqrt{A^2/(L_0 K_E + A^2)} \quad \text{for the magnetic device} \end{aligned}, \tag{10}$$

and the resistive losses of the magnetic transducer are null, Figure 3 when each transducer has the same coupling coefficient but the magnetic transducer presents resistive losses, and Figure 4 when the coupling coefficients of the two transducers are different and taking into account resistive losses of the magnetic device. These charts are normalized along the x and y-axis according to the optimal loads given by:

$$\begin{aligned} (R_p)_{opt} &= \frac{1}{C_0 \omega_0} \quad \text{for the piezo device} \\ (R_m)_{opt} &= \sqrt{(L_0 \omega_0)^2 + {r_L}^2} \quad \text{for the magnetic device} \end{aligned}, \tag{11}$$

where the magnetic transducer is considered as perfect (*i.e.* $r_L = 0$), and along the z-axis according to the maximal harvested power ([7]):

$$P_{max} = \frac{(\mu_1 M a_M)^2}{8C}. \tag{12}$$

These figures are indexed with repsect to the figure of merits given by the product of the mechanical quality factor Q_M of the structure (giving the amount of energy that is available) with the squared coupling coefficient k^2 (reflecting the part of the available energy that can actually be harvested).

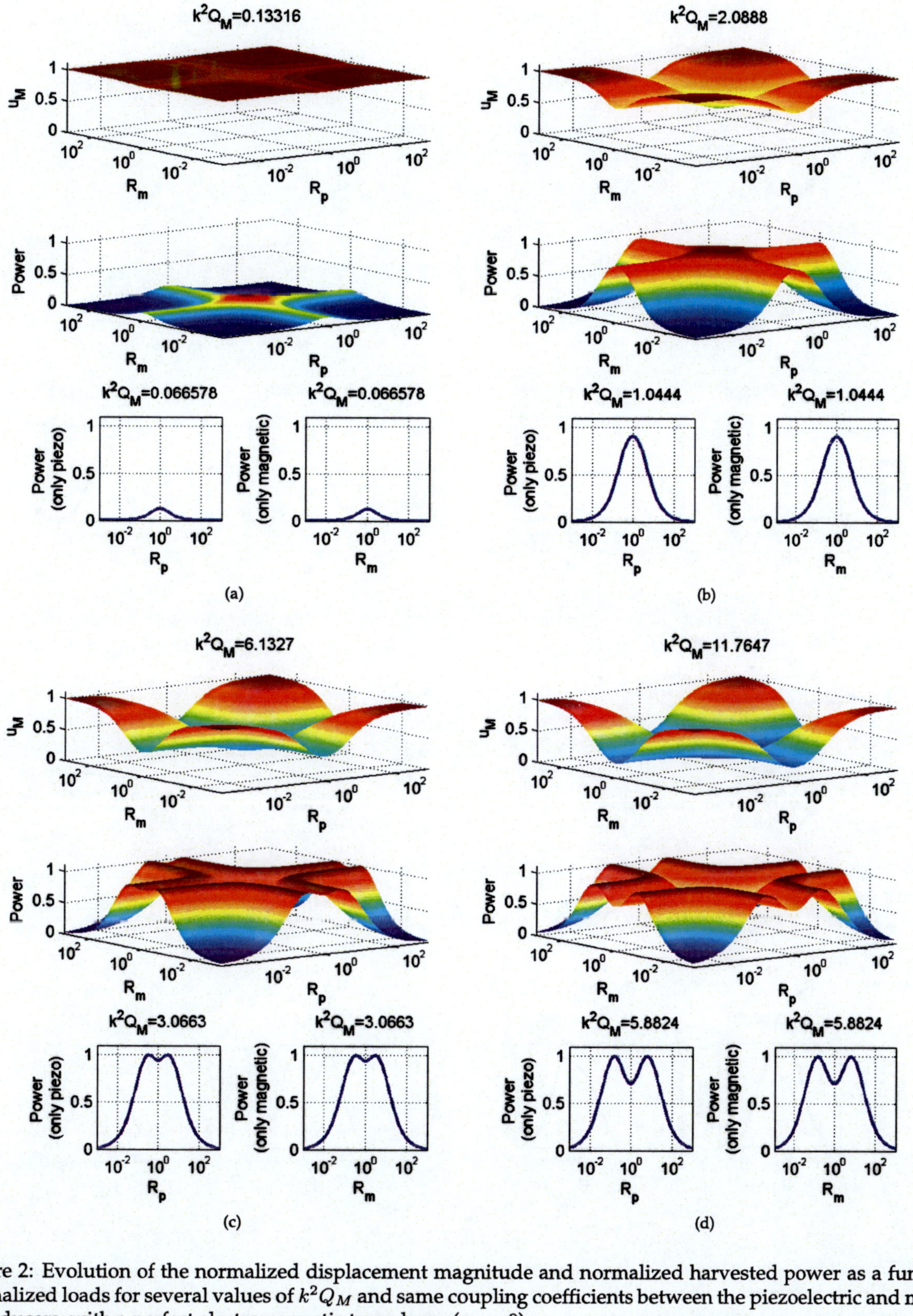

Figure 2: Evolution of the normalized displacement magnitude and normalized harvested power as a function of normalized loads for several values of $k^2 Q_M$ and same coupling coefficients between the piezoelectric and magnetic transducers, with a perfect electromagnetic transducer ($r_L = 0$)

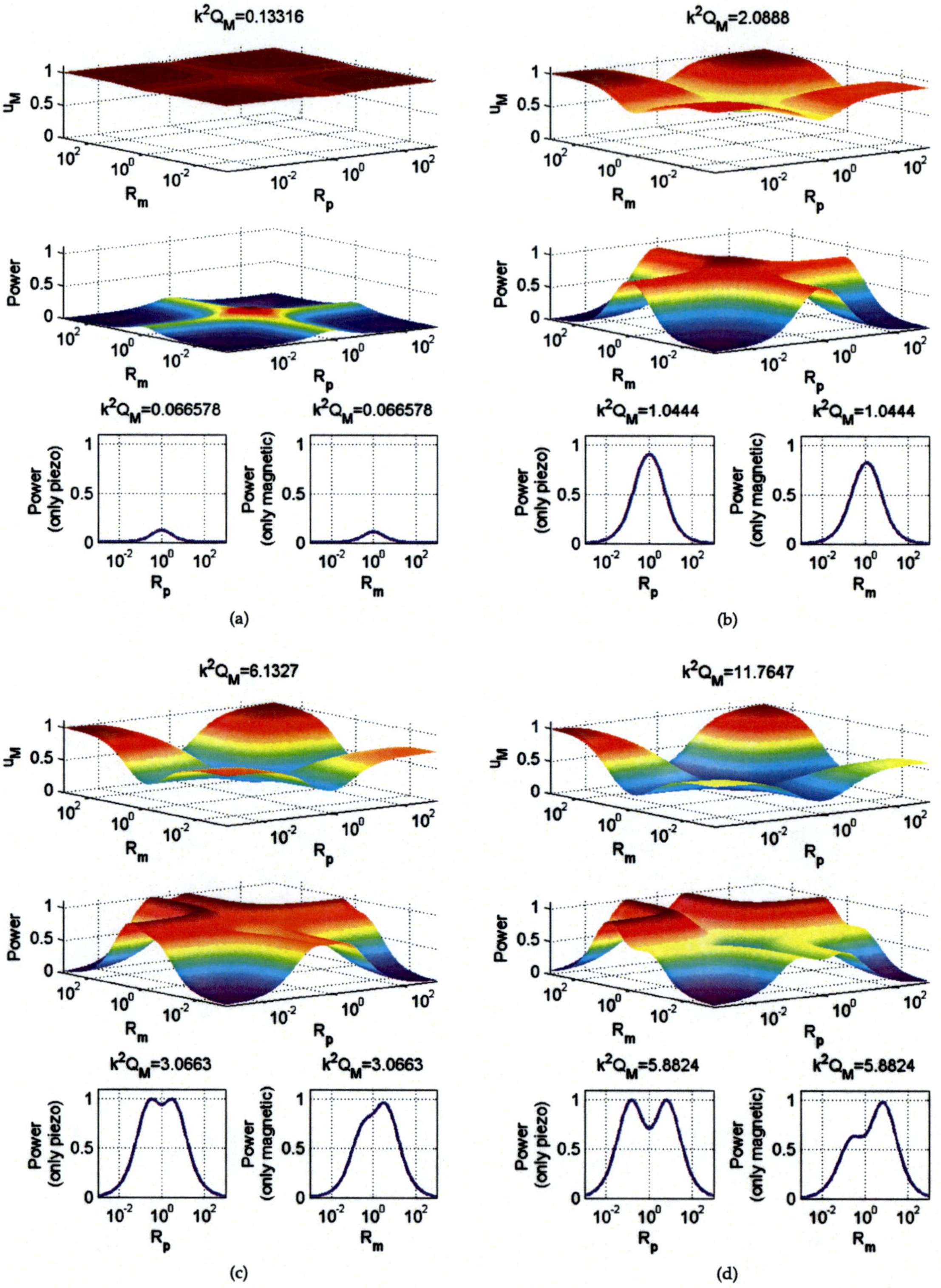

Figure 3: Evolution of the normalized displacement magnitude and normalized harvested power as a function of normalized loads for several values of k^2Q_M and same coupling coefficients between the piezoelectric and magnetic transducers, with an imperfect electromagnetic transducer (angle of losses: $\varphi_L = 84^\circ$)

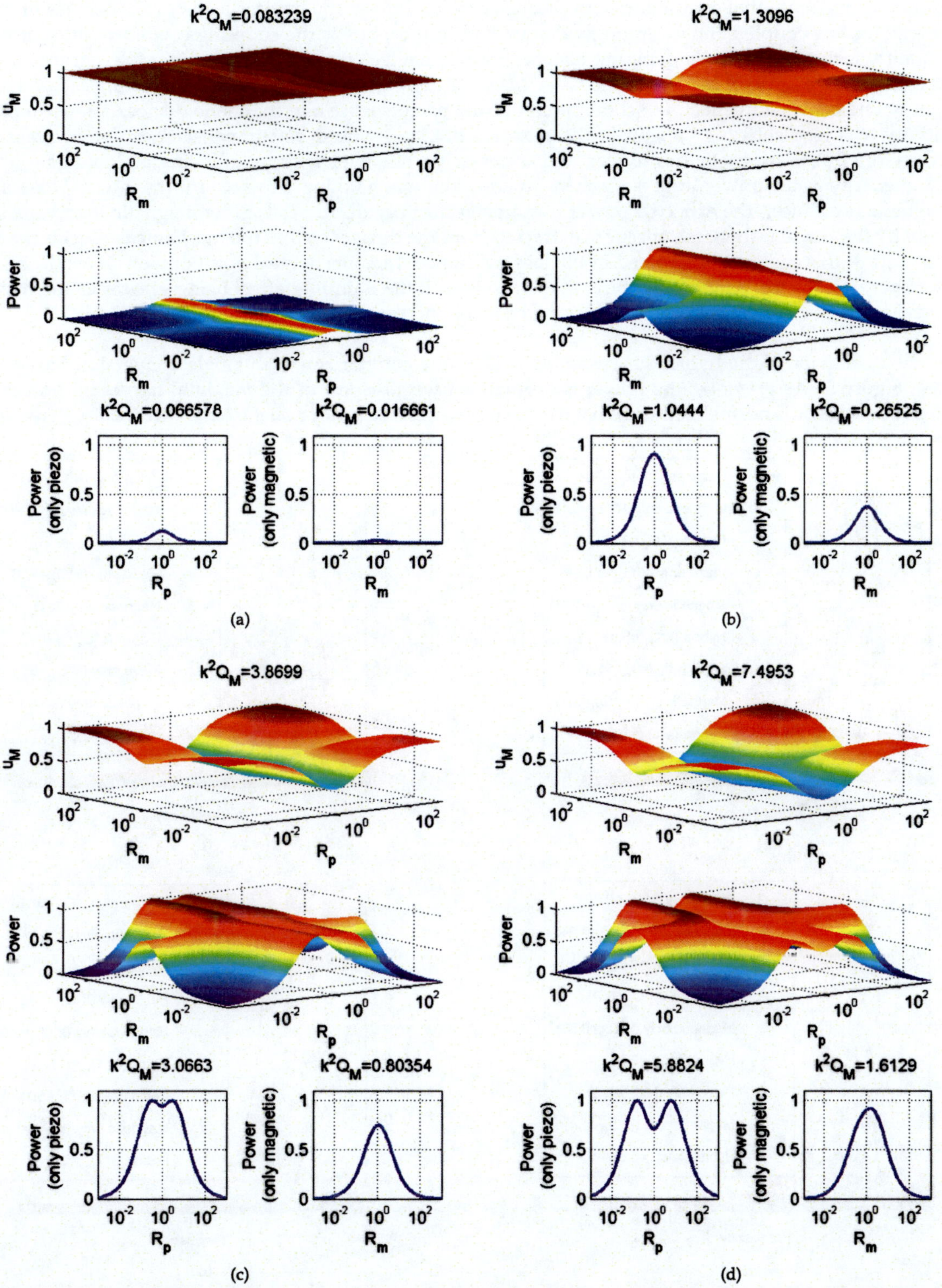

Figure 4: Evolution of the normalized displacement magnitude and normalized harvested power as a function of normalized loads for several values of k^2Q_M and different coupling coefficients between the piezoelectric and magnetic transducers, with an imperfect electromagnetic transducer (angle of losses: $\varphi_L = 84^\circ$)

These figures demonstrate that combining both piezoelectric and magnetic devices leads to an increase of the harvested energy for low coupled and/or strongly damped structures, but as the coupling coefficient and/or the mechanical quality factor increase, the damping effect cannot be neglected any longer, leading to a harvested power limit and a split in the optimal loads. In this latter case, compared to classical energy harvesting techniques that consider only one conversion effect, the combination of the piezoelectric and magnetic devices allows decreasing the sensitivity to a load shift (*e.g.* a shift in the piezoelectric load would be compensated by an increase of the harvested energy by the magnetic transducer, and conversely), however without allowing a global gain in terms of harvested energy as the power limit is reached. In addition, when taking into account the internal losses in the electromagnetic transducer, the expected power output significantly decreases for low magnetic load values. This is explained by the fact that the same energy is extracted from the device (so that the mechanical effect is the same), but a great part of this energy is dissipated in the loss resistance. In a mechanical point of view, it can be seen that harvesting energy leads to vibration damping at the resonance, the damping effect being greater as the harvested energy increases, whether by a better load matching or by an increase of the figure of merit k^2Q_M.

The frequency behavior of the hybrid harvester, as well as the comparison with single conversion harvesters, is depicted in Figure 5. As previous, the power is normalized with respect of the maximal harvested power P_{max}, while the frequency axis is normalized such that 0 is the resonance frequency and ± 0.5 the -3 dB cut-off frequencies.

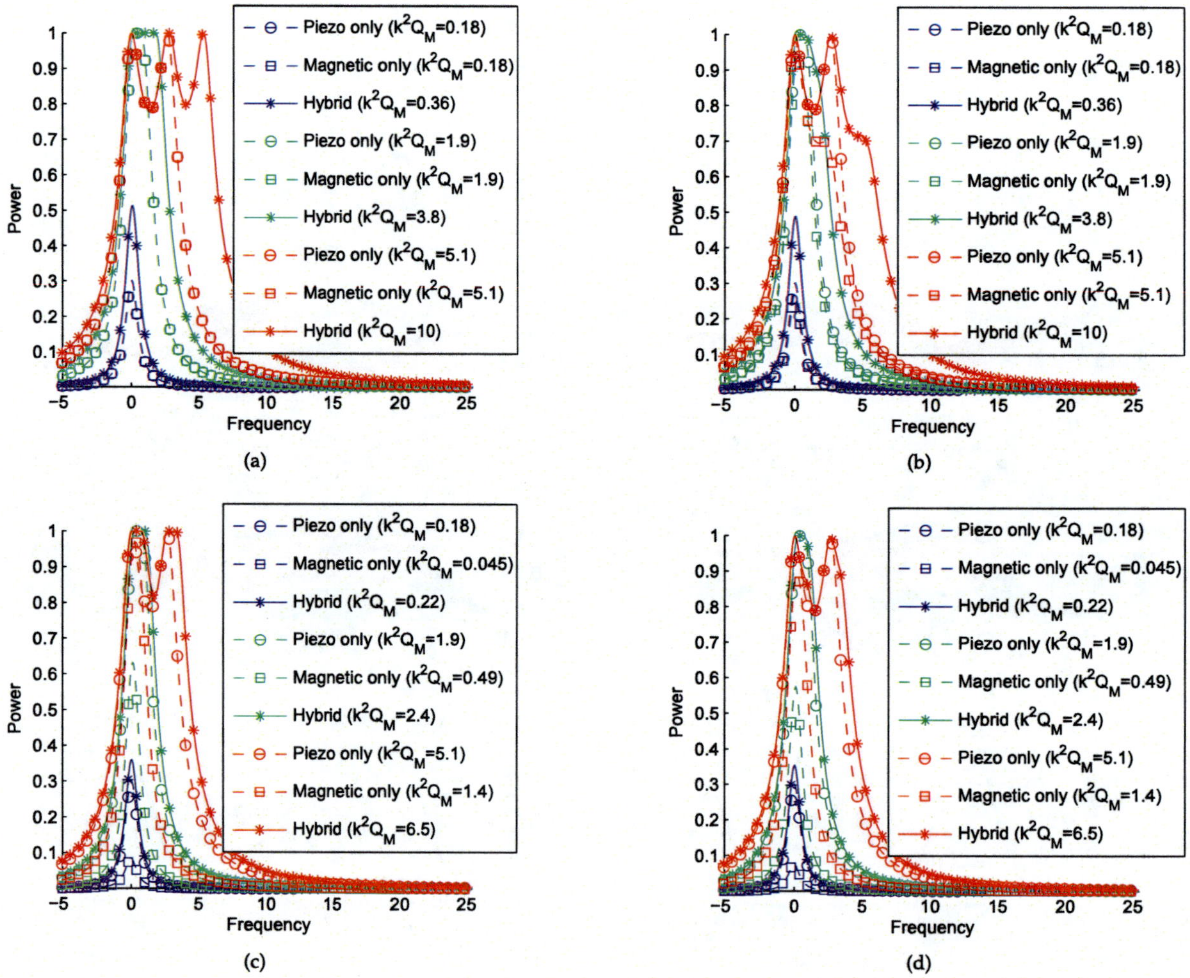

Figure 5: Evolution of the normalized maximal harvested power as a function of the normalized frequency for several values of k^2Q_M, same ((a);(b)) and different ((c);(d)) coupling coefficients between the piezoelectric and magnetic transducers, with a perfect ((a);(c)) and imperfect ((b);(d)) electromagnetic transducer (angle of losses: $\varphi_L = 84^\circ$)

Parameter	Value
Mechanical	
Dynamic mass M	40 g
Structural damping coefficient C	0.14 N.s.m^{-1}
Natural stiffness K_E	450 N.m^{-1}
Correction factor μ_1	1^2
Electrical	
Piezoelectric clamped capacitance C_0	54 nF
Magnetic transducer inductance L_0	3 mH
Magnetic transducer leakage resistance r_L	4.3 Ω
Electromechanical	
Piezoelectric force factor α	0.82 mN.V^{-1}
Electromagnetic force factor A	1.2 N.A^{-1}

Table 2: Experimental model parameter identification

This chart clearly shows that the use of the hybrid configuration allows a great enhancement of the harvester, as the resonance frequency range is extended, starting from the case where the piezoelectric element is short-circuited and the electromagnetic element left in open circuit, with the corresponding resonance frequency given by:

$$(\omega_0)_{low} = \sqrt{K/M}, \tag{13}$$

to the configuration where the piezoelectric element is in open circuit condition and the electromagnetic element short-circuited, where the resonance frequency is equal to:

$$(\omega_0)_{high} = \sqrt{\left(K + \alpha^2/C_0 + A^2/L_0\right)/M}. \tag{14}$$

Hence the combined effect of piezoelectric and magnetic actuation permits to siginificantly extend the effective energy harvesting bandwidth.

4 EXPERIMENTAL VALIDATION

4.1 Experimental set-up

This section aims at validating the previously exposed results giving the theoretical output power and displacement magnitude. The experimental setup, depicted in Figure 6, consists of a cantilever beam bonded with piezoelectric inserts on each side, hence shapping a bimorph piezoelectric transducer. One of the bimorphs is used for vibration sensing ; hence only one piezoelectric material is connected to the harvesting circuit. The beam also features a tip mass made of rare earth permanent magnets that go into a coil, therefore constituing the magnetic transducer. The system is vibrated at approximately at 0.1 g peak using a shaker, driven by a function generator through a power amplifier. The piezoelectric and magnetic transducers are each connected to a pure resistor, simulating the input impedance of the AC/DC converter, in a independent fashion, and the signal waveforms are monitored using a digital oscilloscope. Preliminary measurements have also been performed in order to identify the model parameters, which are listed in Table 2.

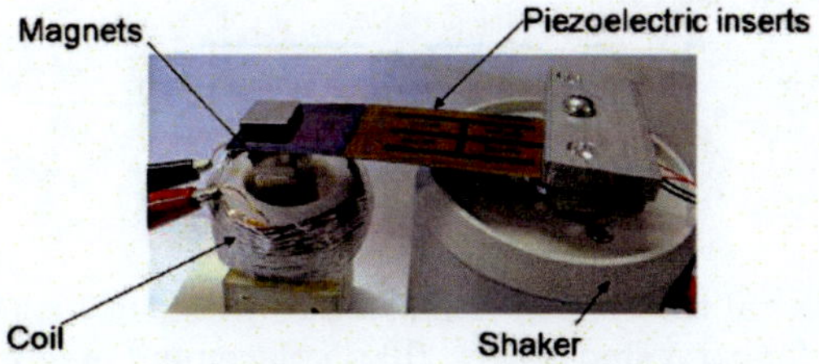

Figure 6: Hybrid energy harvesting experimental setup

4.2 Experimental results and discussion

The first set of measurements consisted of exciting the structure at its resonance frequency[3] for several piezoelectric and electromagnetic load values. Results showing the harvested power and displacement magnitude as a functioin of the connected load are depicted in Figure 7(a) and in Figure 7(b), respectively. Theoretical predictions are also shown in these chart for comparison.

These results show a good agreement between the previously proposed model and the reality. Hence, for low magnetic load values, the parasitic resistance of the electromagnetic transducer absorbs the major part of the extracted energy, leading to a low harvested energy and a significant damping effect. As the electromagetic load increases, both displacement and harvested power increase, almost independently to the value of the piezoelectric load, until the electromagnetic load reaches its optimal value. Around this point, the converse piezoelectric effect also affects the mechanical behavior of the structure, and the displacement is strongly related to the value of the piezoelectric load. Once the magnetic load becomes far greater than its optimal value, only the piezoelectric effect is active in terms of energy harvesting (the magnetic transducer only influencing the resonance frequency of the system, not the damping ratio).

It can also be noted that there is no split in the optimal loads, as the figure of merit k^2Q_M is below its critical value and the leakage resistance of the magnetic transducer is relatively large. Another interesting point, also shown by the theoretical predictions, is that the optimal load pair giving the maximum harvested power is not exactly equal to the couple given by the optimal load of the piezoelectric element alone and the optimal load of the electromagnetic element alone.

The behavior of the system in terms of harvested power as a function of the frequency is also depicted in Figure 8, along with theoretical predictions (considering either hybrid, piezoelectric or eletromagnetic harvesting) for comparison. Again, the experimental results closely match the theoretical predictions. As the figure of merit k^2Q_M of each transducer is below its critical value and as the leakage resistor r_L is important, the combination of the piezoelectric and the electromagnetic transducers allows a gain in terms of harvested energy. This gain is equal to 10% compared to the piezoelectric element alone and 30% compared to the electromagnetic element alone. The bandwidth is also slightly enhanced, going from 0.9 Hz and 0.83 Hz in the purely electromagnetic and purely piezolectric

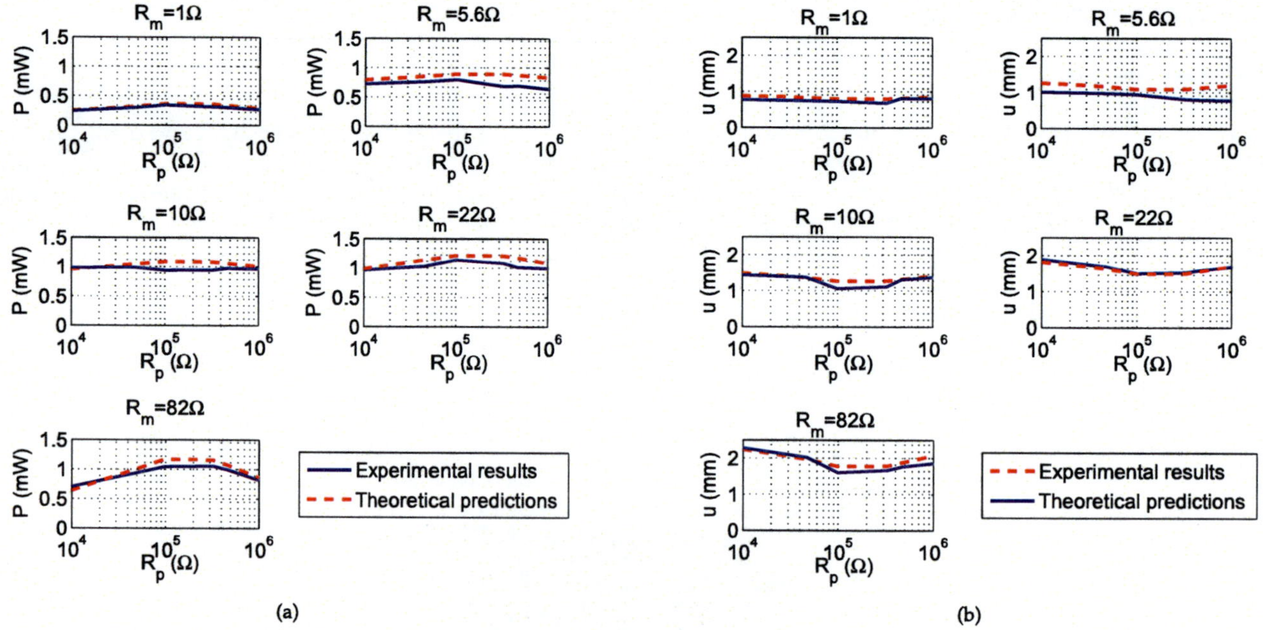

Figure 7: Experimental and theoretical results at the resonance frequency, as a function a the piezoelectric and electromagnetic transducers' load: (a) harvested power; (b) displacement magnitude

[3] which varies with the connected load according to Eq. (5)

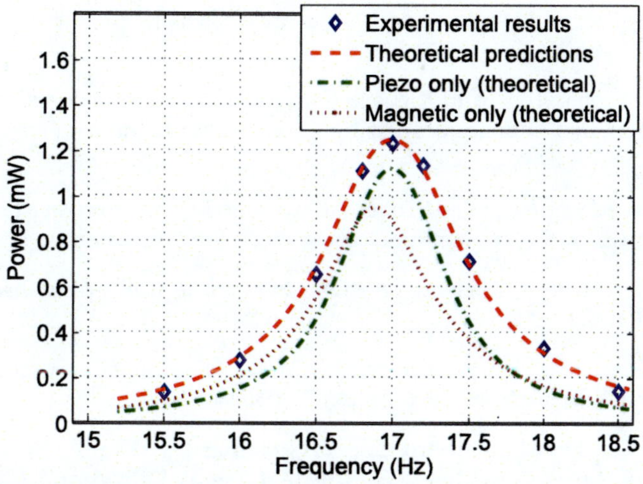

Figure 8: Theoretical and experimental maximum harvested power at a function of the frequency

case respectively, to 1.1 Hz in the hybrid case. This bandwitdh magnification would be even more important in the case of strongly coupled systems, as shown in Section 3.3.

5 CONCLUSION

This paper investigated the effect of combined piezoelectric and electromagnetic energy harvesting from vibrations. Based on a simple SDOF model that approximates quite well the behavior of such a harvester near one of its resonance frequencies, it has been demonstrated that both conversion effects lead to damping and a shift in the resonance frequency. The study also showed that, although the use of a hybrid energy harvesting can increase the energy level for low coupled, highly damped structures, the harvested energy has the same limit as devices using only one conversion principle due to damping effect when the electromechanical coupling is high and the structure weakly damped. However, in this case, it has been also demonstrated that the use of both piezoelectric and magnetic energy harvesting leads to a greater robustness when facing load drifts, as well as an enhanced bandwitdh in terms of harvested power and vibration magnitude.

Acknowledgements

The authors would gratefully acknowledge the support of the U.S. Department of Commerce, National Institute of Standards and Technology, Technology Innovation Program, Cooperative Agreement Number 70NANB9H9007, and the Air Force Office of Scientific Research MURI Grant No. F955-06-1-0326.

References

[1] Paradiso J. A. and Starner T., "Energy scavenging for mobile and wireless electronics", *IEEE Pervasive Computing*, 4, pp. 1827, 2005.

[2] Guyomar D., Jayet Y., Petit L., Lefeuvre E., Monnier T., Richard C. and Lallart M., "Synchronized Switch Harvesting applied to Self-Powered Smart Systems : Piezoactive Microgenerators for Autonomous Wireless Transmitters", *Sens. Actuators A: Phys.*, 138(1), pp. 151-160, 2007.

[3] Lallart M., Guyomar D., Jayet Y., Petit L., Lefeuvre E., Monnier T., Guy P. and Richard C., "Synchronized Switch Harvesting applied to Selfpowered Smart Systems : Piezoactive Microgenerators for Autonomous Wireless Receiver", *Sens. Actuators A: Phys.*, 147(1), pp. 263-272, 2008.

[4] Krikke J., "Sunrise for energy harvesting products", *IEEE Pervasive Comput.*, 4, pp. 4-35, 2005.

[5] Xu C.-N., Akiyama M., Nonaka K. and Watanabe T., "Electrical power generation characteristics of PZT piezoelectric ceramics", *IEEE Trans. Ultrason., Ferroelect., Freq. Contr.*, 45, pp. 1065-1070, 1998.

[6] Glynne-Jones P., Beeby S. P. and White N. M., "Towards a piezoelectric vibration-powered microgenerator", *Sci., Meas. Technol., IEE Proc.*, 148, pp. 6872, 2001.

[7] Guyomar D., Badel A., Lefeuvre E. and Richard C., "Towards energy harvesting using active materials and conversion improvement by nonlinear processing", *IEEE Trans. Ultrason., Ferroelect., Freq. Contr.*, 52, pp. 584-595, 2005.

[8] Lefeuvre E., Badel A., Richard C., Petit L. and Guyomar D., "A comparison between several vibration-powered piezoelectric generators for standalone systems", *Sens. Actuators A: Phys.*, 126, pp. 405-416, 2006.

[9] Anton S. R. and Sodano H. A., "A review of power harvesting using piezoelectric materials (2003-2006)"", *Smart Mater. Struct.*, 16 R1R21, 2007

[10] Lallart M. and Guyomar D., "An optimized self-powered switching circuit for non-linear energy harvesting with low voltage output", *Smart Mater. and Struct.*, 17, 035030, 2008.

[11] Lallart M., L. Garbuio L., Petit L., Richard C. and Guyomar D., "Double Synchronized Switch Harvesting (DSSH): A New Energy Harvesting Scheme For Efficient Energy Extraction", *IEEE Trans. Ultrason., Ferroelect., Freq. Contr.*, 55(10), pp. 2119-2131, 2008.

[12] Shearwood C. and Yates R. B., "Development of an electromagnetic microgenerator", *Electronics Letters*, 33, pp. 18831884, 1997.

[13] Saha C. R., ODonnell T., Loder H., Beeby S. and Tudor J., "Optimization of an Electromagnetic Energy Harvesting Device", *IEEE Trans. Magnetics*, 42(10), 2006, 3509-3511.

[14] Beeby S. P., Torah R. N., Tudor M. J., Glynne-Jones P., ODonnell T., Saha C. R. and Roy S., "A micro electromagnetic generator for vibration energy harvesting", *J. Micromech. Microeng.*, 17, 12571265, 2007.

[15] Wacharasindhu T. and Kwon J. W., "A micromachined energy harvester from a keyboard using combined electromagnetic and piezoelectric conversion", *J. Micromech. Microeng.*, 18, 104016, 2008.

[16] Lallart M., Magnet C., Richard C., Lefeuvre E., Petit L., Guyomar D. and Bouillault F., "New Synchronized Switch Damping methods using dual transformations", *Sensors and Actuators A: Physical*, 143, 302-314, 2008.

[17] Badel A., Lagache M., Guyomar D., Lefeuvre E. and Richard C., "Finite Element and Simple Lumped Modeling for Flexural Nonlinear Semi-passive Damping", *J. Intell. Mater. Syst. Struct.*, 18, pp. 727-742, 2007.

[18] Erturk A. and Inman D. J., "Issues in mathematical modeling of piezoelectric energy harvesters", *Smart Mater. Struct.*, 17, 065016, 2008.

Proceedings of the IMAC-XXVIII
February 1–4, 2010, Jacksonville, Florida USA
©2010 Society for Experimental Mechanics Inc.

Vibro-Impact Dynamics of a Piezoelectric Energy Harvester

K.H. Mak*, S. McWilliam, A.A. Popov, C.H.J. Fox
Materials, Mechanics & Structures Division, Faculty of Engineering,
ITRC Building, University Park, University of Nottingham, NG7 2RD, UK

Abstract

A common design of piezoelectric energy harvester consists of a piezoelectric monomorph cantilever which converts the ambient vibration into electrical energy. However, if the level of the ambient vibration is high, large bending stresses will develop in the monomorph, which can cause fatigue and mechanical failure. A stop is introduced into the harvester to reduce the bending stress by limiting the maximum amplitude of oscillation of the cantilever. The dynamics of such a system is complex and involves considerations of vibro-impact mechanics as well as electromechanical interactions. A theoretical model of a piezoelectric-vibro-impact system is demonstrated in this study. The theoretical model is able to predict the dynamical and electrical responses of an energy harvester. It also estimates the contact force between the cantilever and a stop. Typical simulation results are presented and the physical meaning of the results is explained. The simulation results also show that moving the position of a stop can significantly affect the electrical output from the monomorph.

Keywords: Energy harvesting, Vibration, Piezoelectricity, Impact dynamics

1. Introduction

The harvesting of vibration energy is a feasible solution to supplying electrical energy to small electronic devices, such as wireless sensors for monitoring. In practice, energy harvesters are subjected to periodic and random vibrations during operation. They may also be subjected to shock accelerations which can reduce service life. For a cantilever type of harvester, the maximum bending stress always acts at the clamped end. The use of a stop can restrict the displacement of a beam, thus limiting the bending stress, including the maximum stress at the root of the beam. However, there is a concern that limiting the displacement of a cantilever monomorph can lead to a reduction in generated electrical power from the energy harvester.

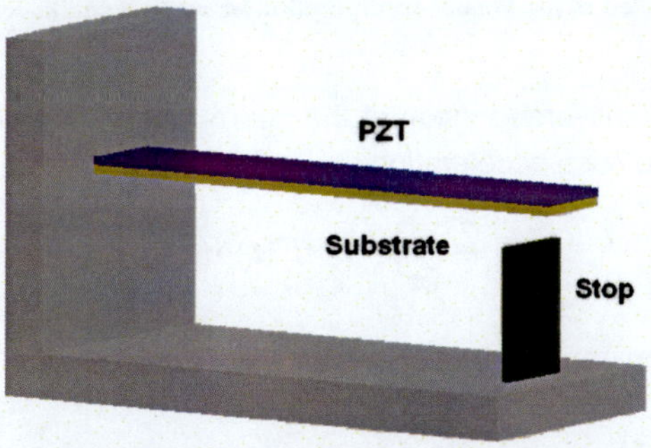

Figure 1 Illustration of the impact configuration between a cantilever beam and a stop

Mechanical impact is not a new area of research and a variety of studies have been reported for different kinds of impact systems [1]. Exact solutions are not normally possible and most studies use approximate solutions [2] – [5]. There are two basic approaches to modelling an impact system, like the one shown in Figure 1. One

* *Corresponding author*: Tel: +44 0115 8467682
Email address: epxkhm@nottingham.ac.uk (K.H. Mak)

T. Proulx (ed.), *Structural Dynamics and Renewable Energy, Volume 1*, Conference Proceedings of the Society for Experimental Mechanics Series 10, DOI 10.1007/978-1-4419-9716-6_25, © The Society for Experimental Mechanics, Inc. 2011

approach is to employ Newton's coefficient of restitution to predict the velocity after impact [6]. The benefit of this approach is to simplify the analysis by assuming a known relationship between the velocities before and after impact. This approach does not require the contact force to be determined [7]. The value of the coefficient of restitution can only be determined through experiments and it may vary if the contact duration varies and more high modes are excited [6]. The second approach is to estimate the contact force and use this to predict the dynamics after impact [2] – [5]. Higher mode transient waves will be induced in the beam by the impact force, excluding these will obscure some of the fine detail of the dynamics during contact.

A theoretical model for a piezoelectric monomorph cantilever with impact is derived in Section 2. Numerical simulation results will be presented and discussed in Section 3 and broad conclusions are provided in Section 4. The dynamical and electrical responses for the impact system are determined and the simulation results obtained offer an insight into the chattering impact and the contact force. This paper also suggests a simple way to minimise the reduction in electrical output from impact.

2. Impact model for a piezoelectric monomorph cantilever

The cantilever beam consists of a substrate layer on which a layer of piezoelectric material (e.g. PZT) is mounted. The structure is treated as an Euler-Bernoulli beam with a stop as shown in Figure 2. The stop can be modelled as a linear spring or a rod capable of longitudinal vibration. The displacement of a linear spring is always zero unless there is a contact force acting on it, so it is not possible for the spring to vibrate when the beam is out of contact with the stop. For this reason, it is more realistic to model the stop as a longitudinal rod, and this approach is used in the theoretical model.

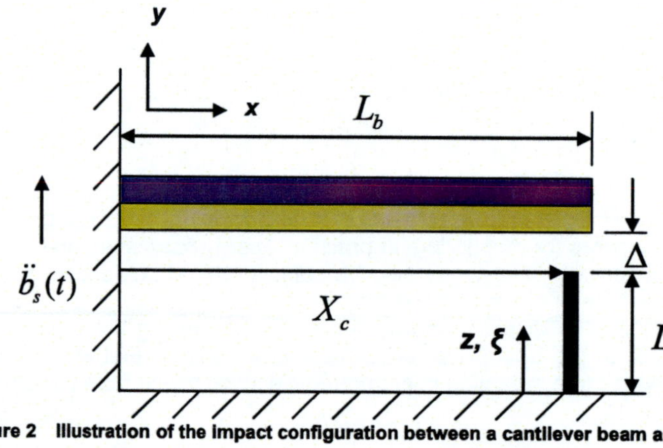

Figure 2 Illustration of the impact configuration between a cantilever beam and a rod

The governing equation for the transverse motion of the beam responding to the contact force, *F(t)*, between the beam and the stop, and to the base acceleration, $\ddot{b}_s(t)$, is:

$$E_b I_b \frac{\partial^4 y(x,t)}{\partial x^4} - C_b \frac{\partial y(x,t)}{\partial t} - \rho_b A_b \frac{\partial^2 y(x,t)}{\partial t^2} = F(t)\delta(x - X_c) - \rho_b A_b \frac{d^2 b_s(t)}{dt^2} \tag{1}$$

Similarly, the governing equation for the motion of the rod (stop) to the contact force and the external acceleration is:

$$E_r A_r \frac{\partial^2 z(\xi,t)}{\partial \xi^2} - C_r \frac{\partial z(\xi,t)}{\partial t} - \rho_r A_r \frac{\partial^2 z(\xi,t)}{\partial t^2} = -F(t)\delta(\xi - L_r) - \rho_r A_r \frac{d^2 b_s(t)}{dt^2} \tag{2}$$

In equations (1) and (2), $y(x,t)$ is the transverse displacement of the beam and $z(\xi,t)$ is the longitudinal displacement of the rod. The subscripts *b* and *r* denote the beam and the rod respectively and *E* and *ρ* are the Young's modulus and density. *L* and *A* are the length and uniform cross sectional area and I_b is the second moment of area about the neutral axis of the beam. *C* is the structural damping coefficient. *F(t)* is the contact force acting between the beam and the stop. X_c is the position of impact along the beam. $\delta(x - X_c)$ is a Dirac delta function and is used to specify the position of the contact force. Notice that the piezoelectric effect is not considered in (1), but it will be introduced into the equation of motion once the impact model is developed.

Theoretically, the solutions to equations (1) and (2) can be expressed as an infinite series of eigenfunctions and associated generalized coordinates:

$$y(x,t)=\sum_{i=1}^{\infty}\phi_{bi}(x)r_{bi}(t) \quad (3)$$

$$z(\xi,t)=\sum_{i=1}^{\infty}\phi_{ri}(\xi)r_{ri}(t) \quad (4)$$

where $\phi(x)$ and $r(t)$ are the mode shape functions and generalised coordinates respectively.

The basic solution procedure for the impact system defined by equations (1) and (2) is as follows. When the beam and the stop are out of contact, *F(t)* will be zero in both equations (1) and (2), which can therefore be solved separately. When the beam comes into contact with the stop, *F(t)*, is non-zero in (1) and (2) (see Fig.3) and the motions of the beam and the stop are then coupled. Equations (1) and (2) must be solved simultaneously to determine the size and the duration of the contact force. In this situation, the nature of the coupled equations means that they must be solved numerically using a time-stepping method.

2.1 Determining of contact force

To determine the unknown contact force, the motion of the beam has to be coupled with the motion of the stop because their displacements at the point of contact are identical when they are in contact. Before estimating the contact force, the time at which contact begins must be determined by using the following inequalities:

Out-of-contact: $y(X_c,t)>z(L_r,t)+\Delta \quad \Rightarrow \quad F(t)=0$ **(5)**

In-contact: $y(X_c,t)\le z(L_r,t)+\Delta \quad \Rightarrow \quad F(t)\neq 0$ **(6)**

These equations are used to check whether the beam and the stop are in contact. If contact is detected, the time at which contact takes place, t_e, must be determined. This can be done by equating the displacements of the beam and rod as equation (7).

$$y(X_c,t_e)\equiv z(L_r,t_e)+\Delta \quad (7)$$

This equation is solved in the time domain to calculate t_e. Once this has been achieved, the contact force can be estimated by coupling the displacements of the beam and stop at the contact position. The motion of the coupled system is then tracked until the contact force becomes negative, indicating that the beam has separated from the stop. The time at which separation takes place is determined by the same method as used to determine the time of contact.

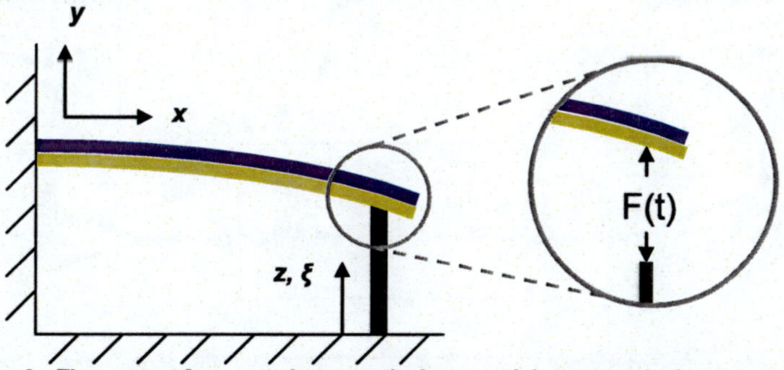

Figure 3 The contact force acts between the beam and the stop at the impact position

The displacements of the beam and the stop are separated into two parts as follows:

$$y(x,t)=y_s(x,t)+y_{imp}(x,t) \quad (8)$$

$$z(\xi,t)=z_s(\xi,t)+z_{imp}(\xi,t) \quad (9)$$

In both equations, the first term on the right hand side is the response to the base excitation and the second term is the response to the contact force. The displacements are such that $y=z+\Delta$ during contact and this is used to calculate the contact force, $F(t)$. These terms are expressed in the form of Duhamel's integral to facilitate estimation of the contact force [5].

2.2 Harvesting electrical energy

The monomorph is designed to work in the *d31* mode to generate electrical charge. The piezoelectric material is modelled as a current source $I(t)$ in parallel with a capacitor, C_P. To withdraw the generated electrical charge from the piezoelectric material, the energy harvester can be connected to a resistor, R, in series, as shown in Figure 4. Based on the electrical circuit in Figure 4, Kirchhoff's law is used to deduce the following governing equation.

$$C_p\dot{v}(t)+\frac{v(t)}{R}=-\sum_{i=1}^{\infty}\Theta_i\dot{r}_{bi}(t) \tag{10}$$

The electromechanical coupling coefficient, Θ, is a measure of the conversion of mechanical energy from the beam shape function into electrical energy. It is introduced to obtain with the voltage, $v(t)$, across the harvester, and it describes the piezoelectric effect in the theoretical model [8].

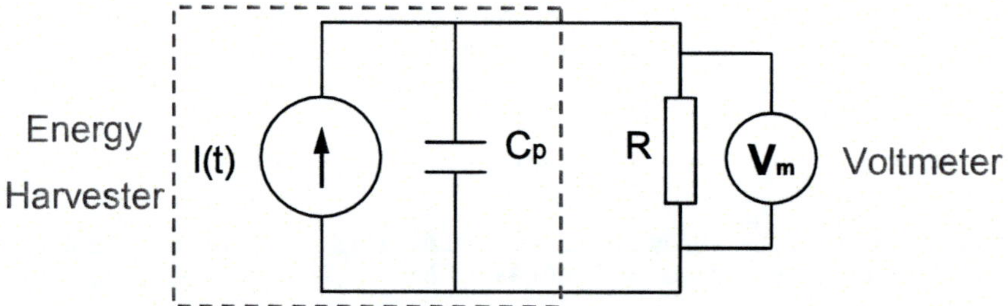

Figure 4 The energy harvester is connected in series to a resistor

The procedure and logic of the theoretical model are summarised in the flow chart provided in Figure 5. These procedures were used in a computer program that was developed.

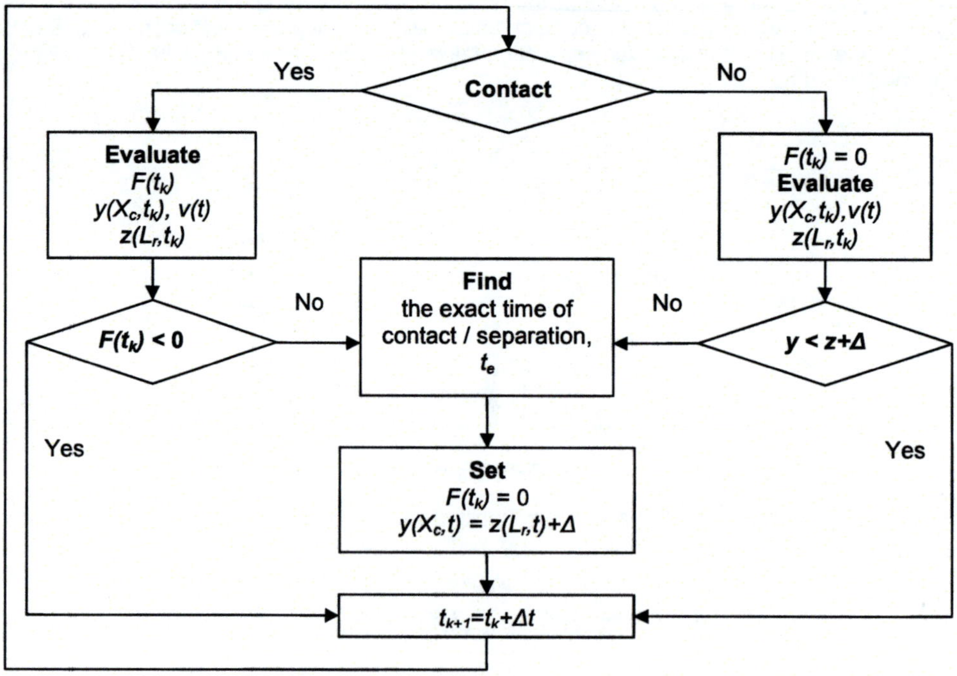

Figure 5 A flow chart for the theoretical model of a piezoelectric impact system

3. Numerical simulation results

Numerical simulation results, based on the model developed in Section 2, are presented here. A harmonic support motion is applied such that $\ddot{b}_s(t) = B_{max} \sin 2\pi ft$ where B_{max} is the maximum acceleration and f is the excitation frequency. A damping ratio of 0.01 is assumed for all beam and rod modes considered. The stop is located at the end of cantilever beam.

Table 1 shows the dimensions and mechanical properties of the energy harvester considered. The numbers of modes used for the beam and the rod were determined from an investigation of system convergence. Contact forces can excite high frequency modes significantly and these modes affect the predicted interaction between the beam and the stop. However, all the modes cannot be included in the simulation, so the minimum number of modes needed for practically useful predictions must be determined. A convergence study was also performed to calculate the required time step. There are restrictions on the size of time step for such an impact system [2], [3]. All of the key parameters used in the results are listed in Table 2.

	L (mm)	w (mm)	t (mm)	ρ (kg/m^3)	E (Gpa)	C_p (nF)
Substrate	50	10	1	7840	190	—
PZT	50	10	1.02	7800	66	8.59
Stop	8	10	1	7840	190	—

Table 1 The dimensions and mechanical properties of the beam and the rod used in the simulations

Number of mode		time step	Δ	F	B_{max}	X_c	R
Beam	Rod	(s)	(mm)	(Hz)	(m/s^2)	(mm)	(MΩ)
10	10	2x10^{-7}	0	100	100	L_b	1

Table 2 The key parameters of the impact model

An example from a numerical simulation is presented in Figure 6. It can be seen that the beam is excited by the support motion and it repeatedly impacts the stop. When impact occurs, the displacement of the beam at the impact position is limited by the stop and the contact force reacts against the beam. The calculated contact force is shown in Figure 6(b). The beam does not rest continuously on the stop during contact; instead a chattering impact occurs as shown in Figure 7. The duration of contact is relatively short compared to the first few oscillating periods of the beam, indicating that the higher frequency modes are excited by the contact force. These components can be seen in the electrical response (Figure 6(c)).

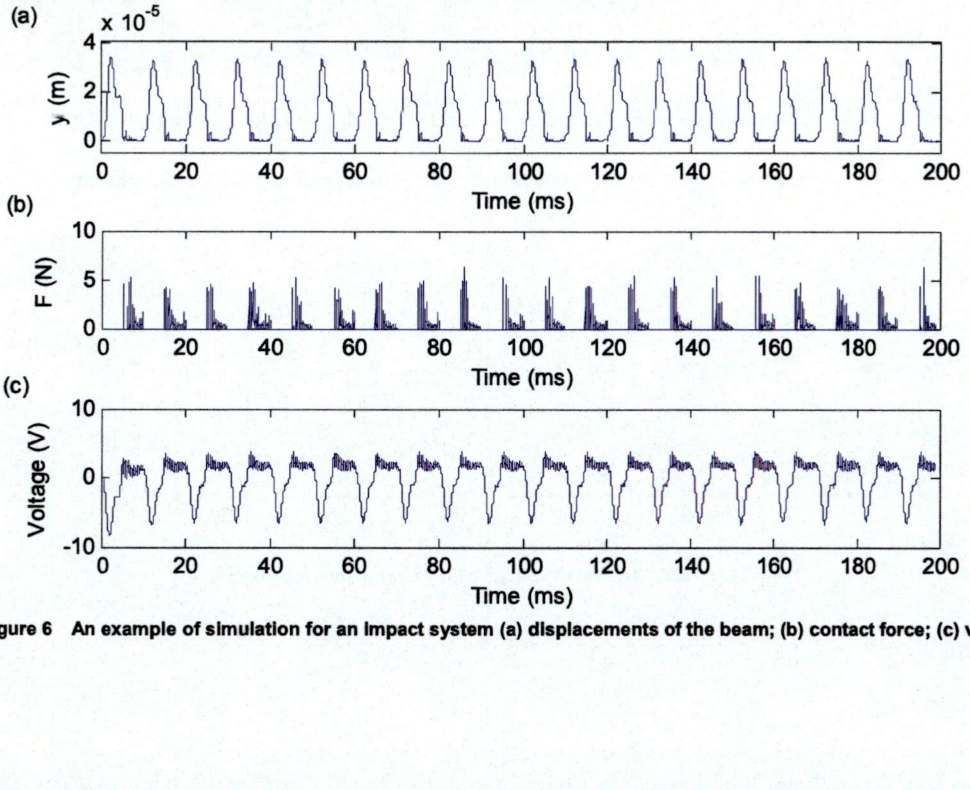

Figure 6 An example of simulation for an impact system (a) displacements of the beam; (b) contact force; (c) voltage

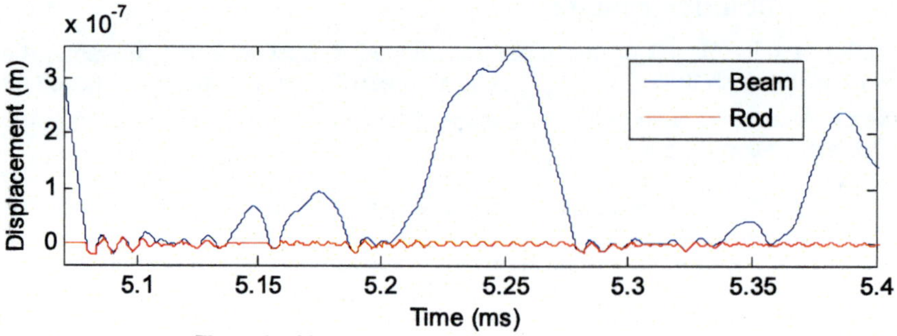

Figure 7 Chattering impact between the beam and the stop

The developed model can be used to assess the influence of the impact on the generated voltage. Figure 8 shows a comparison between the simulated voltages with and without (i.e. no stop) impact taking place. It is obvious that the voltage is reduced during the contact phase, but is maintained when the beam and the stop are out of contact. Clearly the overall generated electrical power would be reduced by the impact, which is undesirable. One way to alleviate this reduction in voltage might be to move the position of the stop, subject to consideration of the maximum induced stress.

If the stop is relocated so that it is not at the free end of the beam, the deflected shape will be different. Figure 9 compares the generated voltage with the stop located at the free end of the beam, at the mid-span, and without a stop. It shows that higher peak voltages are generated when the stop is located at the mid span of the beam, compared to the tip. The level of generated power from the harvester greatly relies on the deflected shape of the monomorph. Moving the position of the stop changes the deflected shape in the contact phase. Figure 10 shows the possible deflected shapes of the monomorph with the stop located at the end and the mid-span. The monomorph in Figure 10 (a) looks to have a clamped-pinned boundary condition and the bending stress across the entire monomorph is less than the one shown in Figure 10 (b). This is one of the reasons why the generated voltage is less when the stop is at the free end (i.e. $X_c = L_b$), but this is a subject for more detailed investigation.

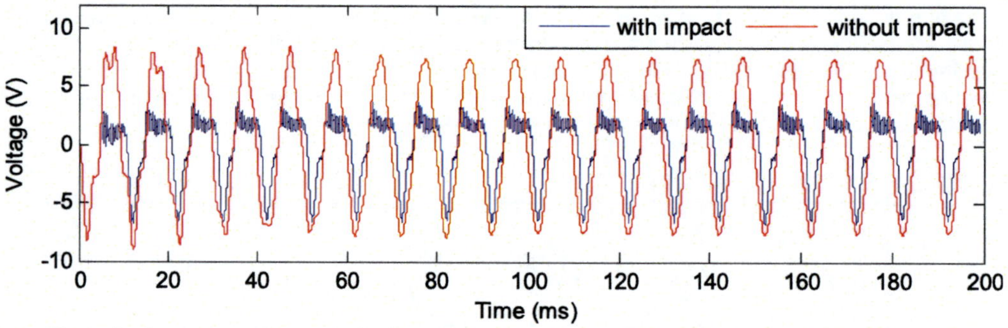

Figure 8 Comparison of generated voltages when impact is included and excluded in the simulation

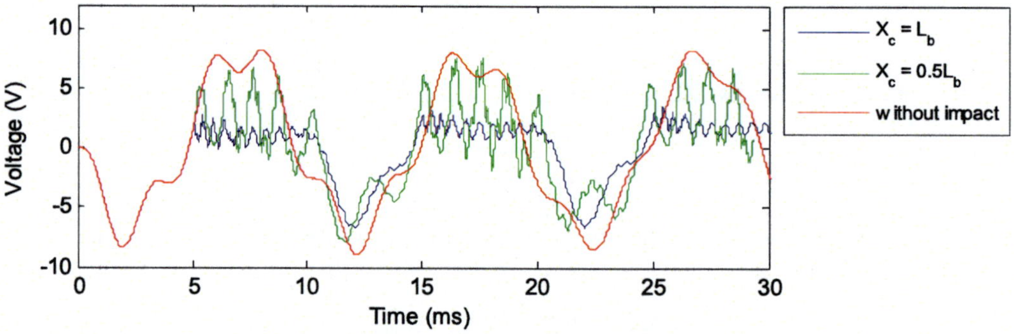

Figure 9 Generated voltage with different stop locations

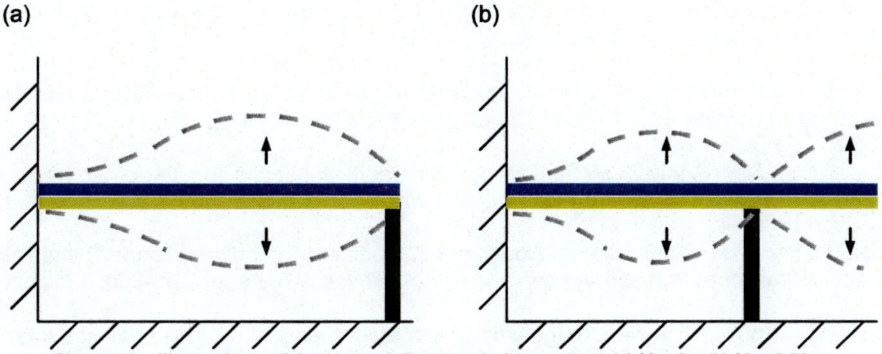

Figure 10 Illustrations of the beam deflection during contact (a) $X_c = L_b$; (b) $X_c = 0.5L_b$

4. Conclusions

A computer model of a piezoelectric vibro-impact system has been developed. The simulation is able to predict dynamical and electrical responses, as well as the impact interaction between a beam and a stop. The simulation gives insight into a beam-stop chattering impact. It is important to perform convergence conditions studies to ensure that the simulation results reflect the important responses in the system. It is also worth mentioning that one set of convergence conditions will not necessarily be valid for every combination of dimensions and parameters. It was found that moving the position of the stop can affect the electrical output from the energy harvester, and this is a matter for further investigation.

Further work is in progress to validate the simulation results against experimental measurements. This will include connection of the energy harvester to a charging circuit as shown in Figure 11, so that a comparison can be made for the energy stored in the storage capacitor, C_s, within a given period of time.

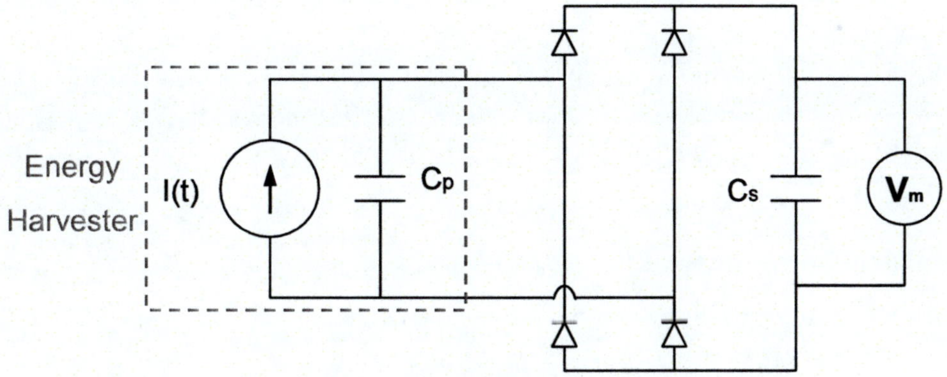

Figure 11 Energy harvester is connected AC-DC circuit to charge a capacitor

Acknowledgement

The authors gratefully acknowledge Atlantic Inertial Systems (AIS) for their financial and technical support for the work reported here.

Reference

[1] Babitsky V.I., Theory of Vibro-Impact Systems and Applications, 1998, Springer, Pages 245-255, 22 March 1980

[2] Lo C.C., A cantilever beam chattering against a stop, Journal of Sound and Vibration, Volume 69, Issue 2, Pages 245-255, 22 March 1980

[3] Fathi A., Popplewell N., Improved Approximations for a Beam Impacting a Stop, Journal of Sound and Vibration, Volume 170, Issue 3, Pages 365-375, 24 February 1994

[4] Tsai H. C., Wu M.K., Methods to compute dynamic response of a cantilever with a stop to limit motion, Computers & Structures, Volume 58, Issue 5, Pages 859-867, 3 March 1996

[5] Wang C., Kim J., New Analysis Method for a Thin Beam Impacting Against a Stop based on the Full Continuous Model, Journal of Sound and Vibration, Volume 191, Issue 5, Pages 809-823, 18 April 1996

[6] Wagg D.J., Bishop S. R., Application of Non-Smooth Modelling Techniques to the Dynamics of a Flexible Impacting Beam, Journal of Sound and Vibration, Volume 256, Issue 5, Pages 803-820, 3 October 2002

[7] Yin X.C., Qin Y., Zou H., Transient responses of repeated impact of a beam against a stop, International Journal of Solids and Structures, Volume 44, Issues 22-23, Pages 7323-7339, November 2007

[8] Sodano H.A., Park G. and Inman D.J., Estimation of Electric Charge Output for Piezoelectric Energy Harvesting, Strain (40), *Pages* 49–58, 2004

[9] Se J.A., Weui B.J., Wan S.Y., Improvement of impulse response spectrum and its application, Journal of Sound and Vibration, Volume 288, Issues 4-5, Pages 1223-1239, , 20 December 2005

Proceedings of the IMAC-XXVIII
February 1–4, 2010, Jacksonville, Florida USA

Experimental Vibration Analysis of the Zigzag Structure for Energy Harvesting

M.A. Karami[1], and D.J. Inman[2]

Abstract

The results of the previously developed analytical model for vibrations of a zigzag micro-structure is compared with the findings from Rayleigh's method and experiments. Cantilever type energy harvesters use a substrate to mount a piezoelectric transduction element on. The maximum power output of such device is when the substrate fundamental frequency is near to dominant frequency of ambient vibration insuring a resonance response maximizing the strain into piezoelectric material. The high natural frequencies of the existing designs of MEMS vibrational energy harvesters are due to their short length constraint and present a serious drawback in the development of MEMS scale energy harvesting devices. The zigzag design was proposed by the authors and was proved capable of reducing the natural frequency of the MEMS harvesters. The electromechanical vibrations of the zigzag structures have been analytically modeled. This paper verifies the findings from analytical model by remodeling the structure using Rayleigh's approach. It also compares the previous model with experimental results. The previous model neglected the mass of the links and that results in a noticeable underestimation of the natural frequencies. After modifying the analytical model and considering the link masses the model and the experiment match.

Keywords: Energy Harvesting, MEMS, Continuous Vibrations, Zigzag Structure, Piezoelectric

Nomenclature

I_p: mass axial moment of inertia of the beam per unit length
m_{tip}: mass of the tip mass
m_l: mass of the links
$Q(x,t)$: Shear force
$w_i(x,t) = W_i(x)T(t)$: Out of plane deformation of the ith beam
x_{end}: The x-coordinate of the free end
x^*: The x-coordinate of the connection of two beams
YI: Bending stiffness
$\beta_i(x)$: Twist angle of the ith beam
ρA: Mass per unit length

1. Introduction

The advances in low power micro electronics, micro-scale sensors and smart actuators have paved the way to independent remote sensing nodes. It is now almost possible to have sensor packs spread on bridges, to regularly do the structural health monitoring and to send the data wirelessly to the base. One component still needing improvement is the energy source needed to

[1] Graduate Research Assistant and ICTAS Doctoral Scholar, Department of Engineering Science and Mechanics, Virginia Tech, Blacksburg VA, 24061, karami@vt.edu
[2] George R. Goodson Professor, Center for Intelligent Material Systems and Structures, Department of Mechanical Engineering, Virginia Tech, Blacksburg VA, 24061, dinman@vt.edu

power the sensors and circuits. The traditional power sources, which are still widely used, are batteries. The necessity to change the batteries requires scheduled access to the devices limiting their placement and increasing cost. The need of a self sustained energy source has motivated many researchers in recent years [1] who have considered solar power, thermal gradients and ambient vibrations for generating electrical energy. The devices which convert ambient vibrations kinetic energy to electrical power are mostly either electrostatic, electromagnetic, electrostricitve or piezoelectric. We focus on conversion of vibrational energy to electricity using piezoelectric devices. A goal is to have the energy harvesting device as small as sensors and circuitry. Now that the MEMS sensors are being used it is desired to design a harvesting device in the micro scale.

The first design of a MEMS harvester was proposed by Lu et al. [2] where the thickness of the energy harvesting beam was one tenth of its length. That dimensions ratio seemed intuitive dealing with large scale structures, but it was too much for a micro cantilever and caused the beams fundamental natural frequency to be about 3 kHz. The more the deflection of the energy harvesting beam is, the more power results. Therefore, the beams should be vibrated at or near their natural frequencies. The frequencies of typical ambient vibrations are from 10 to 100 Hz. Having a natural frequency in orders of kHz simply means that the ambient vibration would not shake the structure at all. The improper choice of thickness to length ratio has resulted similar results in [3] and [4]. Zheng et al. [5] used two beams, one with a distant on top of the other to support the tip mass. The configuration of the beams would make the structure very stiff and light which translates to even higher natural frequency (10 kHz).

Fang et al. [6] were the first to try a low thickness to length ratio (1/100). That resulted in tremendous improvement in the natural frequency (600 Hz). The trend was followed by [7] and [8] and natural frequencies of about 460 Hz and 100 Hz were achieved. This is still not low enough in many situations. What seems to be needed is an improvement in the general design of the harvester. The cantilever beam fundamental frequency is dependent on its length and cannot be designed to a lower frequency in a MEMS device because of space limitation.

The zigzag design, illustrated in Fig. 1 was proposed by the authors [9] as a low frequency energy harvesting structure. The free vibrations of the structure was analytically modeled and the natural frequency and mode shapes were calculated. It was shown that utilizing the structure can increase the energy output of a MEMS energy harvester by 200 times [10].

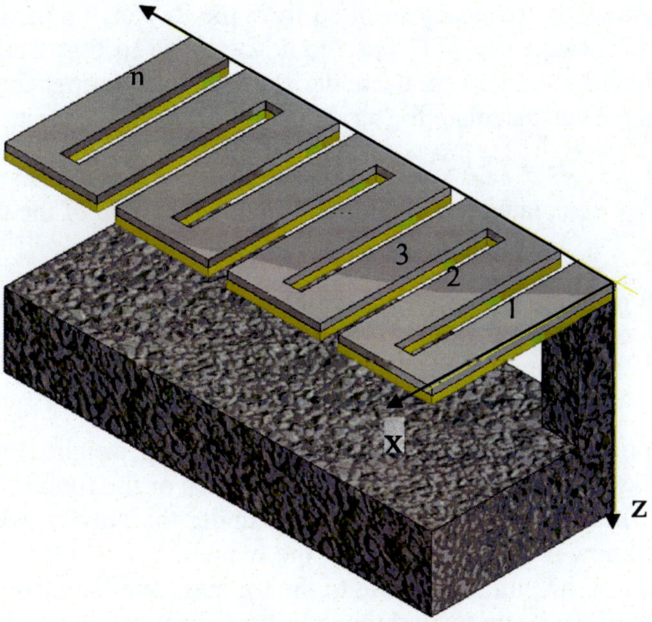

Fig. 1: The Zigzag Energy Harvesting Structure

In the following two approaches are taken to verify the analytical model derived. The first approach is using the Rayleigh's method to estimate the fundamental natural frequency of the structure. The frequency resulted from the Rayleigh's method should be higher but close to fundamental frequency of the analytical model. The second approach, which is more rigorous, is experimentally testing the natural frequency of a zigzag structure and observing how well the model results matches the experiments. Comparing the results of the previous model with experiments revealed some error in prediction of natural frequencies. The shortcoming was fixed by considering the link masses in the model. The modified model is then confirmed to predict the natural frequencies and mode shapes accurately.

2. Device Configuration

The main part of proposed piezoelectric energy harvesting device is a flat zigzag spring illustrated in fig. 2. The thin spring is fixed at one end and forms a cantilever structure. The plane that the zigzag structure lies in is called *the main plane* of the zigzag structure. The structure can deflect out of the main plane. The structure can be modeled as a few straight lateral beams, with rectangular cross sections, placed next to each other on the main plane. Each beam is connected to its neighbor beams at its ends. Each of the beams can bend out of the main plane and can twist. The portions of the structure which connects the elements are very small and can be modeled as rigid links. The torsion of each of the beams causes the next beam to move out of the main plane. The amount of relative motion is the torsion angle times the rigid arm length.

Each of the lateral beams is a uniform composite beam composed of a piezoelectric layer bonded to the substructure layer (this forms a Unimorph). When the beams are deflected some strain is generated in the piezoelectric layer which generates electrical energy.

3. Vibration analysis using Rayleigh's method

Rayleigh's method is commonly used to approximate the fundamental natural frequency of structures. We however use that method to check the natural frequency resulted from our analytical calculations. The frequency resulted from the Rayleigh's method is always greater that the exact fundamental frequency [11] but should be close to that frequency provided the trial function is close to the fundamental mode shape. Considering the bending and torsional deflection of beams we can calculate the strain energy in the structure as [12]:

$$\pi = \sum_{i=1}^{n} \frac{EI}{2} \int_0^L \left(\frac{\partial^2 w_i}{\partial x^2}\right)^2 dx + \sum_{i=1}^{n} \frac{GJ}{2} \int_0^L \left(\frac{\partial \beta_i}{\partial x}\right)^2 dx \tag{1}$$

The kinetic energy associated with the out of plane deflection of the beam is

$$T = \sum_{i=1}^{n} \int_{x=0}^{L} \rho A \left(\frac{\partial w_i}{\partial t}\right)^2 dx \tag{2}$$

Having chosen a trial function $\psi_i(x)$, by separating the variables as $w_i(x,t) = \psi_i(x)T(t)$ and assuming harmonic vibrations such that $\dot{T} = i\omega T$ we finally derive the natural frequency as

$$\omega^2 = \frac{\sum_{i=1}^{n} \int_0^L \left(EI(\psi_i'')^2 + GJ(\beta_i')^2\right) dx}{\sum_{i=1}^{n} \rho A \int_0^L \psi_i^2 dx + M_{tip}\psi_n(x_{end})^2} \tag{3}$$

The static deflection curve of the structure under its own weight is used as the trial function. This choice is well known to result in good approximates of the fundamental frequency of elastic structures [13]. To find the static deflection shape under the gravity, we start from the tip of the structure where we know that the shear force, the twist torque and the bending moment are zero. We add up the effect of the point force due to the tip mass and the distributed force resulted from the distributed mass of the beam to find the values of shear force, bending moment and the twist torque all along the last beam and its junction to the neighboring beam. Next, the model in [9] can be used to calculate the force and moments at the tip of the next neighboring member. The procedure described is repeated till forces and moments are calculated all along the members and at the junctions. Then we use the known force and moments to calculate the deflected shape under the gravity. This deflection curve is used as the trial function, $\psi_i(x)$ in Eq. (3) to find the approximate fundamental natural frequency.

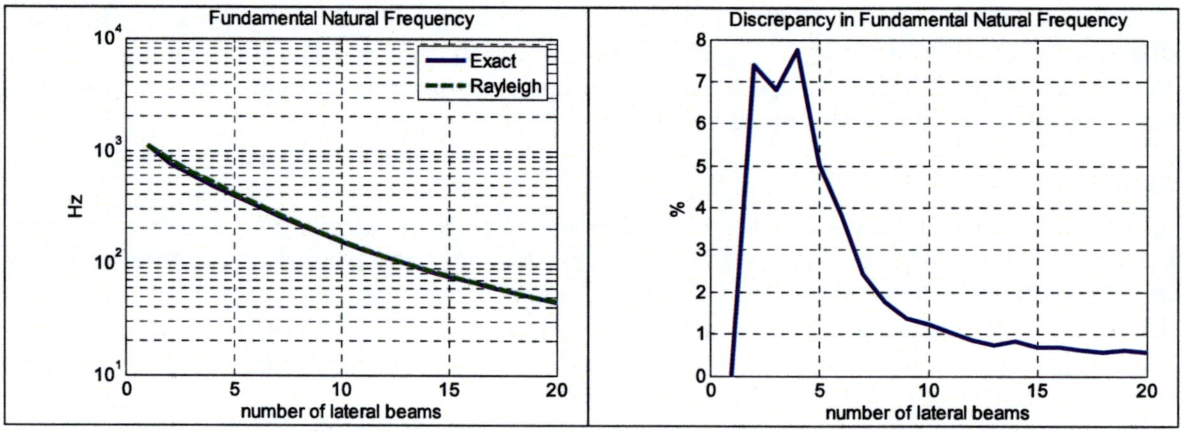

Fig. 2: The exact vs. the approximate fundamental frequencies

The fundamental natural frequency of a zigzag structure has been calculated using the exact and Rayleigh's method. The results are illustrated in Fig. 2. The estimation from Rayleigh's method is close to the exact fundamental frequency and as expected the approximate frequency is always higher than the exact value. The maximum discrepancy is about 8% which corresponds to the four-member structure.

4. Experimental Verification

Experimental measurements have been performed as a second test of the validity of the analytical model. To avoid fabrication complications a large size model of the zigzag structure has been tested. The structure is described in Table 1 and illustrated in Fig. 3.

Length of the beams, l (mm)	171.4
Width of each of the beams, b (mm)	19
Center to center lateral distance of two adjacent beams, d (mm)	22.2
Thickness of the substructure, h_s (mm)	4.76
Tip mass (including accelerometer) (gr)	7.8
Young's modulus of Aluminum 5083, Y_s (GPa)	73.1
Density of the Aluminum substructure, ρ_s (Kg m^{-3})	2770

Table 1: The specifications of the experimental structures

Fig. 3 :Experimental 3 and 5 member zigzag structures

To characterize the vibrations the accelerations of the base and the tip of structure are measured using two accelerometers. Considering the presented analytical solution, the transfer function between the base and the tip motion can be derived after perfuming a modal expansion:

$$w_i(x,t) = \phi_{i(m)}(x)\ \eta_{(m)}(t) \tag{4}$$

where $\phi_{i(m)}(x)$ is the mass normalized mode shape, i.e.

$$\sum_{i=1}^{n} \int_0^L \phi_{i(m)}^2(x) + m_{tip}\phi_{n(m)}^2(x_{end}) = 1 \tag{5}$$

The base vibrations, exciting the structure, are in and out of the main plane. Therefore the governing differential equations for the deflection of structure relative to its base are:

$$YI\frac{\partial^4 w_{i(rel)}}{\partial x^4} + \rho A\frac{\partial^2 w_{i(rel)}}{\partial t^2} = -[\rho A + m_{tip}\delta(x - x_{end})\delta(i-n)]a_b(t) \tag{6}$$

$$GJ\frac{\partial^2 \beta_i}{\partial x^2} - I_p\frac{\partial^2 \beta_i}{\partial t^2} = 0$$

Substituting Eq. (4) in Eq. (6), taking the Fourier transform, pre-multiplying by $\phi_{i(r)}(x)$ and integrating over all the structure results:

$$\frac{\eta_{rel}(\omega)}{a_b(\omega)} = -\sum_{r=1}^{\infty}\frac{\gamma_r}{\omega_r^2+2j\zeta_r\omega_r\omega-\omega^2} \tag{7}$$

where γ_r is defined as:

$$\gamma_r = \sum_{i=1}^{n}\int_0^L \phi_{i(m)}(x) + m_{tip}\phi_{n(m)}(x_{end}) \tag{8}$$

The relative tip acceleration is related to the base acceleration as

$$\frac{a_{rel}(\omega)}{a_b(\omega)} = \sum_{r=1}^{\infty}\frac{\gamma_r\phi_r(x_{end})\omega^2}{\omega_r^2+2j\zeta_r\omega_r\omega-\omega^2} \tag{9}$$

The transfer function from Eq. (9) has been compared against the experimental measurements in Fig. 4. Although the model accurately matches the experiments for a one member structure there is a meaningful discrepancy for multi member structures. The experimental natural frequencies are lower than their model predicted counterparts. This difference between the predicted and experimental natural frequencies is more dominant in higher modes. A lower natural frequency can have resulted from neglecting masses somewhere in the model or over estimating the stiffness of some parts. The major approximation in deriving the model has been assuming that the links are rigid and massless. We will improve the model by taking into account the mass of the members and see if this can reduce the discrepancy between the model and the experiment.

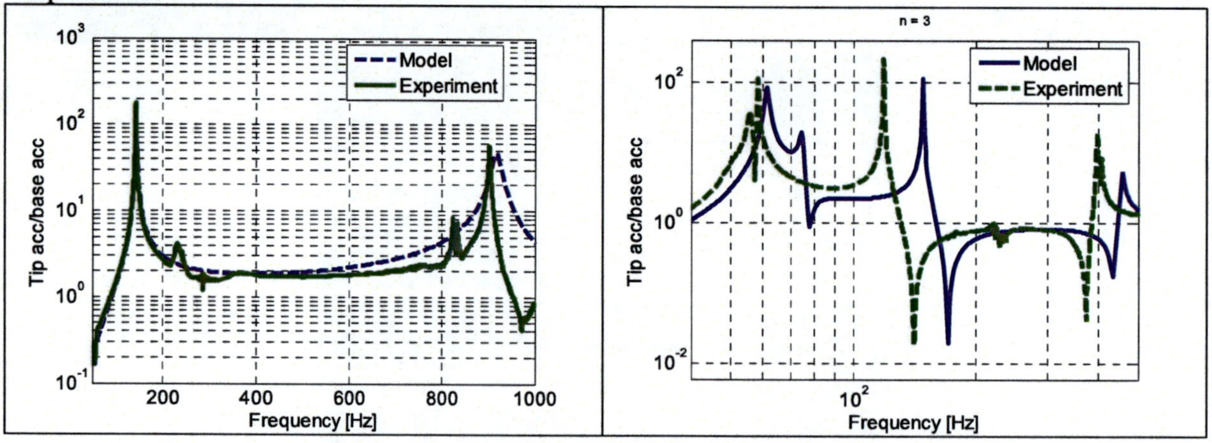

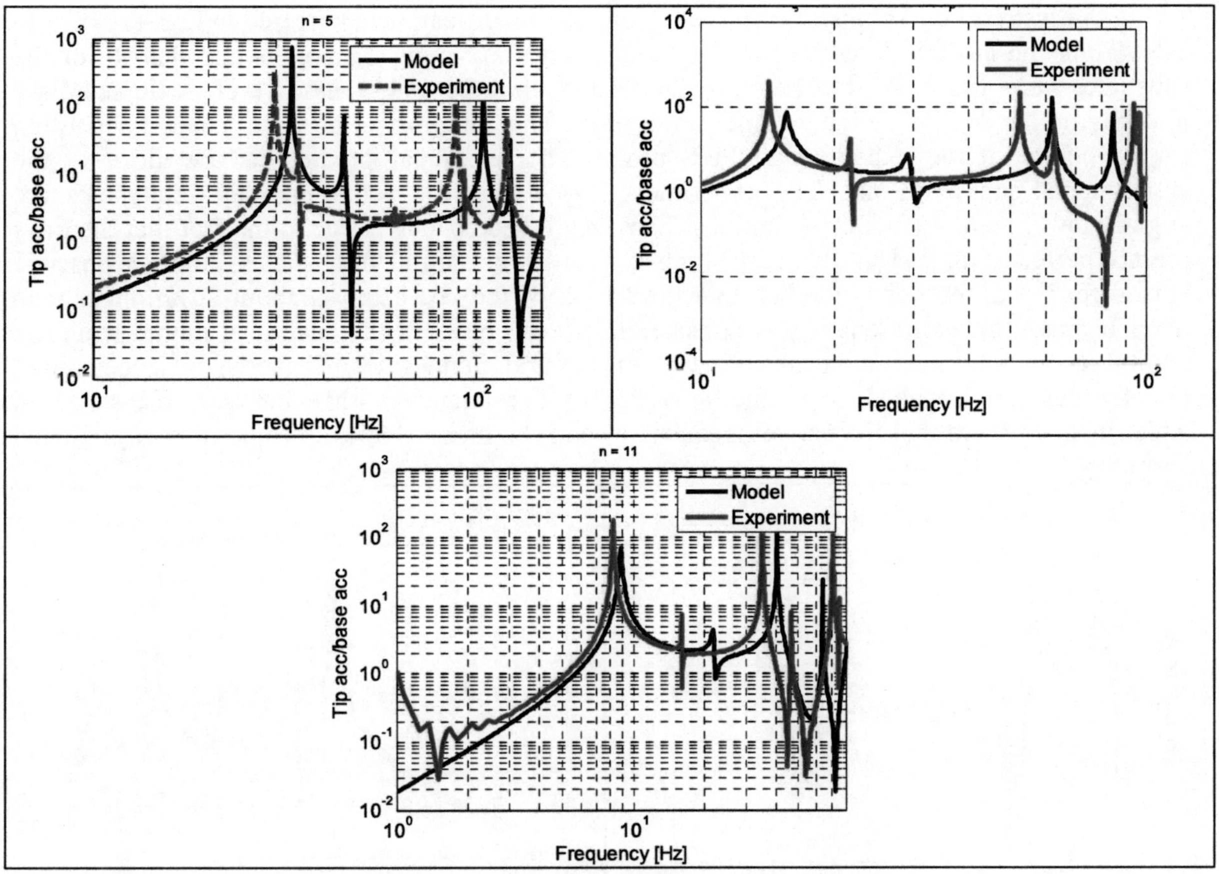

Fig. 4: Original model predictions vs. experiments

5. Consideration of the link masses in the model

To improve the model and consider the link masses in the calculations we start by modifying the derivations of natural frequencies and mode shapes. The tip mass can cause dynamic loading of the connections. Therefore its effects appear in the shear force equilibrium equations. Eq. 19 of [9] is modified to:

$$Q_{i-1}(x^*) = -Q_i(x^*) \pm m_l \ddot{w}_i(x^*, t)$$

where the plus sign in the right hand side corresponds to $x^* = 0$ and the minus is associated with $x^* = l$. This can be written as:

$$YIW_{i-1}^{(3)}(x^*) = -YIW_i^{(3)}(x^*) \mp m_l \omega^2 W_i(x^*) \tag{10}$$

This changes the matrix relations and consequently the characteristic equation. The new model would therefore result different natural frequencies and mode shapes from the previous model [9].

The consideration of link masses also changes the tip deflection transfer function. The first difference would be mass normalization equation (Eq. 5): the model shapes should now satisfy

$$\sum_{i=1}^{n} \int_0^L \phi_{i(m)}^2(x) + \sum_{i=1}^{n-1} m_l \phi_{i(m)}^2(x^*) + m_{tip} \phi_{n(m)}^2(x_{end}) = 1 \tag{11}$$

Also the definitions for γ_r(Eq. 8) should be modified to

$$\gamma_r = \sum_{i=1}^{n} \int_0^L \phi_{i(m)}(x) + \sum_{i=1}^{n-1} m_l \phi_{i(m)}(x^*) + m_{tip} \phi_{n(m)}(x_{end}) \tag{12}$$

Fig. 5 illustrates the results after implementing the modifications expressed in Eqs. (10-12). In calculating the analytical results for Fig. 5 the values of young's modulus and density of the substrate were found in the material data sheet and the link masses were calculated by multiplying the density by the volume of the links. The damping ratios were selected according to the height of the experimental FRF peaks. The analytical predictions are close to the experimental measurements. The experiment shows that the actual natural frequencies are slightly lower than the model predictions. That small discrepancy is due to the fact that the links are not perfectly rigid. This flexibility translates to lower natural frequencies. Another possible reason for the difference is the fact that the length of the last beam is about 10% longer than other beams. This extra length was considered as a tip mass in the analytical model which in reality results in slight over-estimations of the natural frequencies. That, the tip acceleration transfer function resulted from the analytical model matches that function measured in experiments, proves that the natural frequencies and the mode shapes have both been predicted correctly.

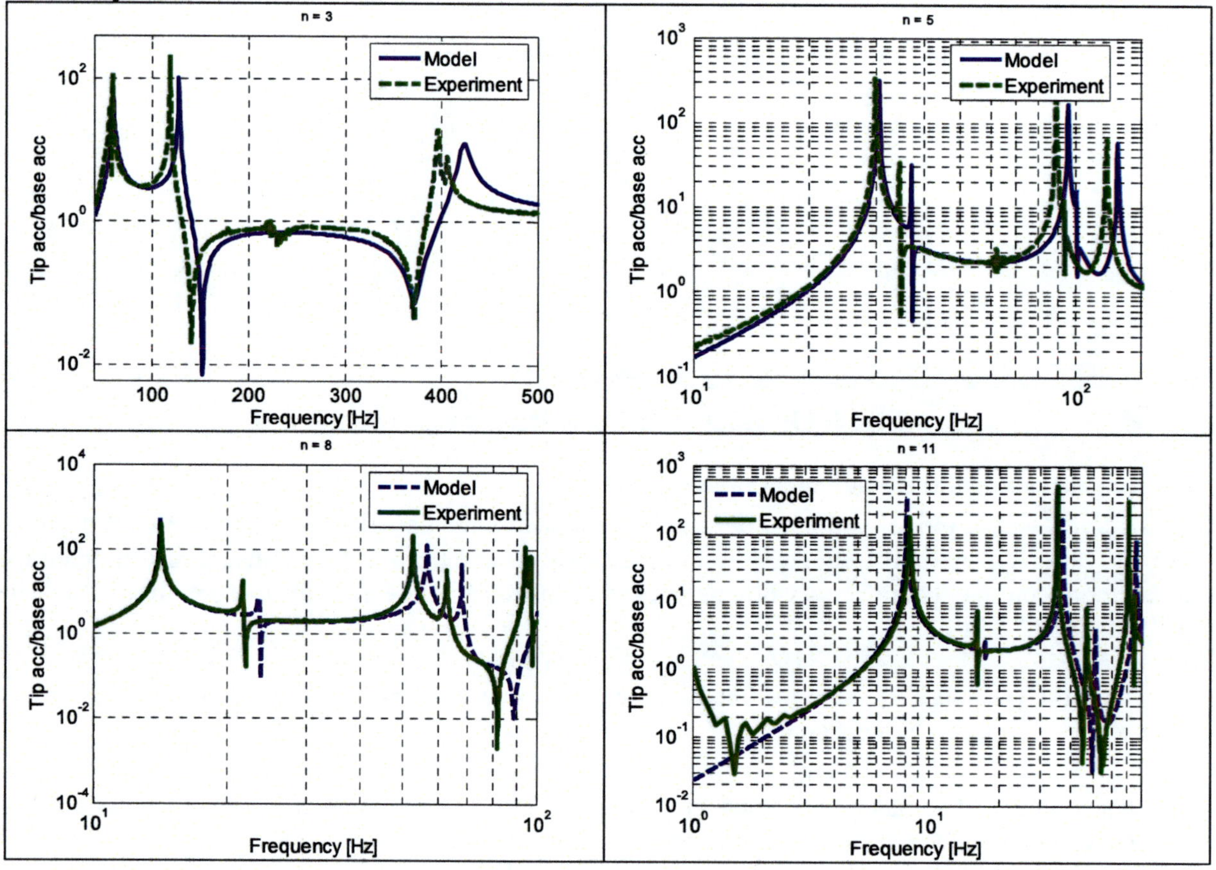

Fig. 5 : improved model vs. experimental measurements

6. Conclusion

We had proposed a new geometry to lower the frequency of a MEMS scale harvesting device without increasing its length, making the substrate device compatible with the frequency range of ambient vibrations. An analytical method had been developed to calculate the natural frequencies and mode shapes of the zigzag bi-layered structure. The predictions of the analytical model were validated by Rayleigh's method results and experimental tests. It was shown that neglecting the link masses can cause the natural frequencies to be over estimated. This error is more visible in the higher modes compared to the fundamental mode. Including the tip mass significantly reduced the discrepancy and made the model in good agreement with experiments.

7. Acknowledgement

This work was performed under the support of the US Department of Commerce, National Institute of Standards and Technology, Technology Innovation Program, Cooperative Agreement Number 70NANB9H9007".

The first author would like to thank Prof. Masoud Agah for his course on MEMS fabrication during which the idea of using the zigzag design occurred to the first author.

8. References

[1] SR Anton, and HA Sodano. A review of power harvesting using piezoelectric materials (2003-2006).(2007). *Smart Materials and Structures*, 16(3), 1.

[2] F Lu, HP Lee, and SP Lim. Modeling and analysis of micro piezoelectric power generators for micro-electromechanical-systems applications.(2004). *Smart Materials and Structures*, 13(1), 57-63.

[3] YB Jeon, R Sood, J Jeong, and SG Kim. MEMS power generator with transverse mode thin film PZT.(2005). *Sensors & Actuators: A. Physical*, 122(1), 16-22.

[4] I Kuehne, D Marinkovic, G Eckstein, and H Seidel. A new approach for MEMS power generation based on a piezoelectric diaphragm.(2008). *Sensors & Actuators: A. Physical*, 142(1), 292-297.

[5] Q Zheng, and Y Xu. Asymmetric air-spaced cantilevers for vibration energy harvesting.(2008). *Smart Materials and Structures*, 17, 055009.

[6] HB Fang, JQ Liu, ZY Xu, L Dong, L Wang, D Chen, BC Cai, and Y Liu. Fabrication and performance of MEMS-based piezoelectric power generator for vibration energy harvesting.(2006). *Microelectronics Journal*, 37(11), 1280-1284.

[7] D Shen, J Park, J Ajitsaria, S Choe, HC Wikle, and D Kim. The design, fabrication and evaluation of a MEMS PZT cantilever with an integrated Si proof mass for vibration energy harvesting.(2008). *Journal of Micromechanics and Microengineering*, 18(5), 55017.

[8] Jing-Quan Liu, Hua-Bin Fang, Zheng-Yi Xu, Xin-Hui Mao, Xiu-Cheng Shen, Di Chen, Hang Liao, and Bing-Chu Cai. A MEMS-based piezoelectric power generator array for vibration energy harvesting.(2008). *Microelectronics Journal*, 39(5), 802-806.

[9] M.A. Karami, and D. J. Inman. (2009). "Vibration Analysis of the Zigzag Micro-structure for Energy Harvesting." In: *16th SPIE Annual International Symposium on Smart Structures and Materials & Nondestructive Evaluation and Health Monitoring*, San Diego, CA.

[10] M. A. Karami, and D. J. Inman. (2009). "Electromechanical Modeling of the Low Frequency MEMS Energy Harvester." In: *IDETC*, ASME, San Diego, CA.

[11] L Meirovitch. Analytical methods in vibration.(1967). *Mecmillan Company London*.

[12] IH Shames. *Energy and finite element methods in structural mechanics* (1985). Hemisphere Pub.
[13] Singiresu S. Rao. *Vibration of Continuous Systems* (2007). John Wiley and Sons,.

Proceedings of the IMAC-XXVIII
February 1–4, 2010, Jacksonville, Florida USA

ENERGY HARVESTING TO POWER SENSING HARDWARE ONBOARD WIND TURBINE BLADE

Clinton P. Carlson[1], Alexander D. Schlichting[2], Scott Ouellette[3], Kevin Farinholt[4], Gyuhae Park[4]

[1]Department of Civil Engineering, University of Michigan
[2]Department of Mechanical and Aerospace Engineering, Cornell University
[3]Department of Structural Engineering, University of California – San Diego
[4]The Engineering Institute, Los Alamos National Laboratory

ABSTRACT

Wind turbines are becoming a larger source of renewable energy in the United States. However, most of the designs are geared toward the weather conditions seen in Europe. Also, in the United States, manufacturers have been increasing the length of the turbine blades, often made of composite materials, to maximize power output. As a result of the more severe loading conditions in the United States and the material level flaws in composite structures, blade failure has been a more common occurrence in the U.S. than in Europe. Therefore, it is imperative that a structural health monitoring system be incorporated into the design of the wind turbines in order to monitor flaws before they lead to a catastrophic failure. Due to the rotation of the turbine and issues related to lightning strikes, the best way to implement a structural health monitoring system would be to use a network of wireless sensor nodes. In order to provide power to these sensor nodes, piezoelectric, thermoelectric and photovoltaic energy harvesting techniques are examined on a cross section of a CX-100 wind turbine blade in order to determine the feasibility of powering individual nodes that would compose the sensor network.

1. INTRODUCTION

Global climate change has sparked renewed interest in domestic and renewable energy sources in the United States for both economic and environmental reasons. In 2008, the U.S. wind energy industry brought online over 8,500 megawatts (MW) of new wind power capacity, increasing the nation's cumulative total by 50% to over 23,000 MW – accounting for 1.5% of the total energy produced - and pushing the U.S. above Germany as the country with the largest amount of installed wind power capacity.[1] Currently, the U.S. has approximately 15,000 - 20,000 wind turbines in operation across 34 states. The largest wind farm in the U.S. is located in Taylor, Texas, where 421 wind turbines produce 735 MW of electric capacity. On average, wind farms cost $1 million per megawatt of installed capacity, and the annual maintenance cost for each wind turbine is approximately 1.5% - 2% of the original cost.[2, 3] With the U.S. looking to expand wind energy to account for 20% of the total energy output by 2030, the ability to transition from time-based maintenance to condition-based maintenance could potentially cut maintenance costs by 50%, resulting in a cost savings of approximately $2 - 3 billion dollars annually.

At present, turbine designs used in the U.S. follow European design criterion, which fail to meet the more severe loading conditions observed in the wind corridor of the Midwestern states. In addition, wind turbine blade lengths continue to grow (>50m) in an effort to capture more of the inbound wind energy. As such, unforeseen structural failures due to the complex loading along the length of the blade plague the industry. Also, in order to reduce the weight while still maintaining the necessary strength and stiffness characteristics, manufacturers use composite materials (e.g. fiberglass or carbon-fiber) to construct the blades (Figure 2). However, significant drawbacks exist from the manufacturing process as blades may possess material flaws such as voids in the epoxy, delamination, and surface wrinkles. Under sufficient loading these flaws grow and in some cases endanger the structural integrity of the blade and by extension the entire turbine as well.

T. Proulx (ed.), *Structural Dynamics and Renewable Energy, Volume 1*, Conference Proceedings of the Society for Experimental Mechanics Series 10, DOI 10.1007/978-1-4419-9716-6_27, © The Society for Experimental Mechanics, Inc. 2011

Table 1: Current and Power Draw for WID 3.0[5]

Mode	Current (mA)	Power (mW)
Measurement	26	72.8
Data Transmission	22	61.6
Sleep Mode	0.075	0.21

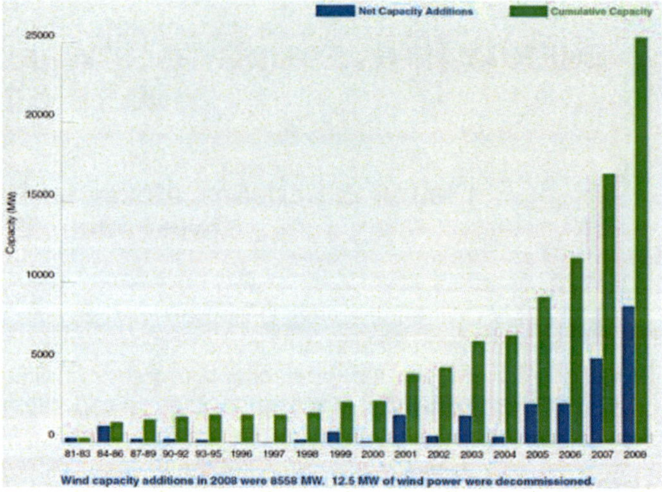

Figure 1: U.S. wind turbine growth[1]

The Structural Health Monitoring (SHM) process presents a possible solution to this issue. During this process, sensors embedded in the structure of the blade at critical locations actively monitor the structure for damage. Refer to the report from the Energy Harvesting for Structural Health Monitoring Sensor Networks workshop held at Los Alamos National Laboratory for a brief introduction to the SHM process.[4] Wind turbine blades present multiple implementation challenges due to the overall design of the wind turbine. An SHM sensor network with wires running along the length of the blades increases the turbine's vulnerability to lightning as it would create multiple conductive paths in addition to the existing lightning protection system already embedded in the blade. Therefore, any sensor network embedded in the blades must rely on wireless technology in order to extract data from the sensor nodes embedded within each blade. Also, the rotating hub of the wind turbine requires a decentralized active sensing and processing network with a hybrid design as detailed in Park et al. (2005). Since the sensor nodes themselves will be dispersed along the length of the blade, and wired power is impractical due to lightning issues, a long-term power management system must be implemented, with energy harvesting serving as one possible solution to this problem.

This research focuses on the overall goal of reducing maintenance costs by investigating various energy harvesting methods on the wind turbine blade to power the wireless impedance sensor (WID 3.0), developed at Los Alamos National Laboratory. This sensor node is designed specifically to operate with different power sources, ranging from batteries to energy harvesting sources to wireless energy transmission methods. Under normal operation the WID cycles through several different operating modes, including a measurement cycle, a data transmission cycle and a sleep cycle. The power requirements for each state are presented in Table 1. For energy harvesting sources the sensor node has an onboard power conditioning module that limits the power released to the system's onboard microcontroller to between 2.7V–3.5V, the stable operating range of the WID.[5]

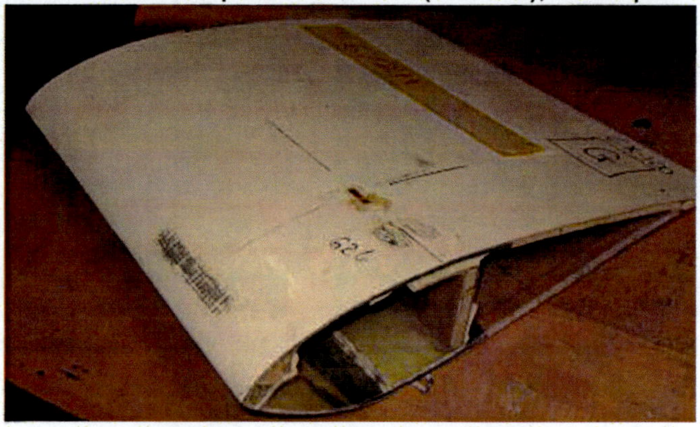

Figure 2: CX-100 Cross Section

1.1 Energy Harvesting Methods

Of the common energy harvesting methods, the three most common methods under investigation for their feasibility in powering the WID 3.0 sensor on a wind turbine blade are: ambient vibration harvesting with piezoelectric materials, thermal energy harvesting with thermoelectric materials, and solar energy harvesting with photovoltaic cells.

1.1.1 Vibration Energy Harvesting with Piezoelectric Materials

Piezoelectric materials belong to the family of ferroelectric materials whose molecular structure consists of electric dipoles. The piezoelectric effect states that when a mechanical strain is present in the material a potential difference is created across the dipoles. This behavior allows piezoelectric materials to be used as sensors or actuators as well as for energy harvesting purposes. For manufactured materials, the piezoelectric effect appears in the presence of a strong electric field while the material undergoes heating.

One material that exhibits a strong piezoelectric effect, or electromechanical coupling, is lead zirconate titanate (PZT). PZT sensors have been used in stack and patch configurations. In the stack configuration both the electrical field and the mechanical strain act in the same direction, the "3" direction. In the patch, or bender, configuration, the electrical field and the mechanical strain act in different directions, the "3" and the "1" direction. Therefore a stack configuration operates in the "33" mode and a bender operates in the "31" mode. Note that the poling for both configurations acts in the "3" direction, which by convention will always be the case. This naming convention mainly comes into play when referring to the electromechanical coupling coefficients for actuation, "d", and for sensing, "g". For example, an electromechanical coupling coefficient labeled "g_{31}" describes a piezoelectric sensor in the bending configuration.[8]

Each configuration has its own advantages and common applications. The stack configuration displays large electromechanical coupling coefficients but small deflections, limiting its use primarily to actuation applications which require large forces. The bender configuration amplifies the deflection of the material due to its geometry, making it ideal for energy harvesting from vibration or actuation applications which require larger deflections than the stack configuration can provide.

Testing on piezoelectric materials has shown that they can be utilized for a wide frequency range. Piezoelectric materials also have no moving parts, so they produce little noise and have low maintenance requirements. However, piezoelectric materials only undergo small displacements and, in turn, produce lower power outputs compared to other materials used in energy harvesting. Tests have shown that piezoelectric-based energy harvesters are capable of producing approximately 800 $\mu W/cm^3$ when mounted to machines that vibrate in the kHz range.[9]

1.1.2 Thermal Energy Harvesting using Thermoelectric Materials

Thermoelectric materials, such as bismuth telluride, are commonly found in cooling and temperature measurement applications. They behave according to the Seebeck effect, which relates the potential difference across the junction of two dissimilar metals, a p-type and an n-type, to the temperature difference between them by the Seebeck coefficient. When used for cooling applications, two ceramic faces house multiple p-type and n-type junctions connected electrically in series and thermally in parallel to form a thermoelectric module. Applying an electric current to the leads of the module causes a temperature difference to be formed across the faces, allowing the module to act as a heat pump. The opposite of this concept allows electricity to be produced from an existing temperature difference, producing a thermal energy harvester.

A significant advantage of thermoelectric materials is that they are a more mature technology, so the products on the market are more reliable and more consistent. Thermoelectric materials also do not have any moving parts. The disadvantages to thermoelectric materials include a high cost and low efficiency. The larger size of thermoelectric devices also prevents them from use in microscale applications. Thermoelectric-generators (TEG's) have been proven to produce 60 $\mu W/cm^2$ during experiments.[9]

1.1.3 Solar Energy Harvesting with Photovoltaic (PV) Cells

Photovoltaic cells consist of semiconductor materials such as crystalline silicon as well as single and polycrystalline thin films. Similar to thermoelectric modules, p-type and n-type materials are separated by a junction. When solar radiation energizes the molecules of the materials, a current is formed over the junction due to the differences between the p-type and n-type materials, more specifically the p-type materials have an abundance of "holes" where electrons could go and the n-type materials have an abundance of electrons.[10]

Two main system configurations have arisen during the technology's short lifespan, flat plate systems and concentrator systems. The flat plate system arranges the photovoltaic cells on a flat surface with a simple plastic or glass cover. Sometimes a tracking system angles the array so that the maximum amount of solar energy is collected at all times of the day. The concentrator system instead minimizes the amount of photovoltaic material needed by concentrating the solar radiation using plastic lenses and metal reflective housings. This system is more powerful than the flat plate system and operates at a much higher efficiency when it is exposed to concentrated light. However the drawbacks include, but are not limited to, cost, the risk of overheating and requiring a tracking system for optimal output throughout the day. Photovoltaic materials have produced 100 $\mu W/cm^2$ in common settings and a much larger 100 mW/cm^2 in direct sunlight during testing.[9]

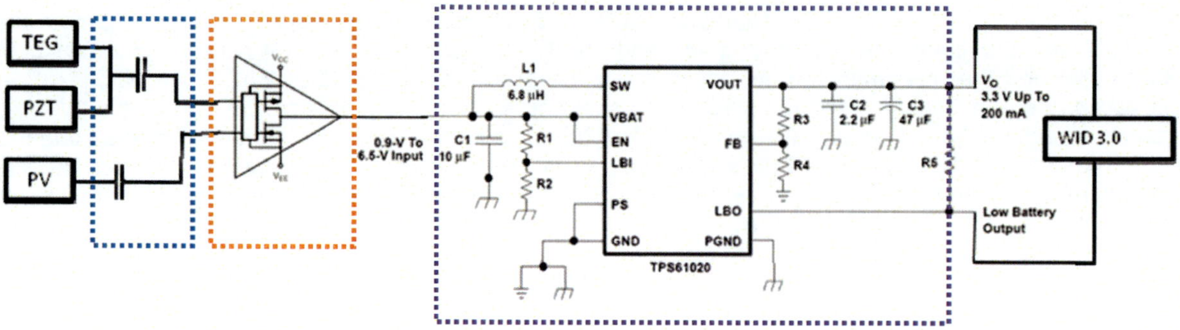

Figure 3: MISO Power Management Circuit

1.2 MISO Power Management Circuit

With multiple sources of energy available on a wind turbine blade, each of which produces unique electrical energy characteristics, combining multiple energy sources to power a single sensor node is necessary in order to offset operational and environmental variability. The power management circuit (Figure 3) is designed to sort a dynamic voltage range (0.9-6.5V) from the three proposed energy harvesting modules – TEG, PZT and PV cell – and provide DC voltage compliance (2.7-3.5V) to the WID 3.0 sensor node. The circuit is a three-stage multiple-input single-output (MISO) system.

- Stage 1 – highlighted in blue – stores the energy into two 1F super capacitors. The voltage output from TEG and PZT modules are relatively low compared to the PV module. In addition, the PZT module produces an AC voltage output, and therefore needs AC-DC rectification prior to charging the capacitor.
- Stage 2 – highlighted in orange – dynamically compares the voltage stored on each capacitor and passes the larger voltage. The comparator used has an operational voltage range of 0.85-6V.
    - *Note:* dynamic sorting means the highest reference voltage is continuously monitored due to the fact that energy harvesting modules are subject to operational and environmental variability. For example, a photovoltaic module might perform slightly better than a TEG module due to direct sunlight low thermal gradient; however, at night, the TEG might perform better due to residual heat and good insulation.
- Stage 3 – highlighted in violet – is a buck/boost converter that accepts voltages in the range of 0.9-6.5V. The converter can be customized for variable output voltages. The diagram shown in Figure 3 is designed for 3.3V and up to 200mA, which is compliant with the WID 3.0 power requirements.

2. EXPERIMENTAL SETUP

2.1 Piezoelectric Testing

In order to determine the power production of piezoelectric materials on the wind turbine blade, a previously studied L-shaped bracket is explored (Figure 4). This configuration possesses advantages over the standard cantilevered beam due to its two-to-one internal resonance. Instead of a theoretical ratio of 6.27 between the first and second resonance frequencies as with a cantilever without a lumped mass at the tip, the L-shaped bracket can obtain a ratio as low as 2. This presents an advantage when attempting to harvest energy from ambient vibrations because the frequency of vibration does not remain constant but travels across a small range.

Figure 4: Experimental setups of the cantilever, the cantilever with lumped mass and the L-shaped bracket.

Therefore tuning the L-shaped bracket so that the frequencies of the ambient vibrations lie between the first two modes will produce more power from the piezoelectric materials than with a cantilever energy harvester.[11]

Table 2: L-Shaped bracket parameters

	Length (cm)	Width (cm)	Thickness (mm)
Horizontal Beam	21.1	3.8	0.79
Vertical Beam	35.6	2.5	0.79
Lumped Mass 1	5.7	2.5	25.4
Lumped Mass 2	5.7	1.9	25.4
PZT Patch 1	3.8	3.2	0.27
PZT Patch 2	3.8	2.5	0.27

As Erturk et al.[11] stopped at an analytical model of the system, this work focuses on a finite element modal analysis of the base structure without the piezoelectric material as well as an experimental L-shaped bracket with piezoceramic patches on the beams. Stainless steel beams, polycarbonate lumped masses of 48.2 and 28.3 grams, and 5A4E PZT piezoceramic patches were used in the experimental setup (see Table 2 for parameters). The finite element modal analysis was performed on ABAQUS CAE with both beam and shell elements. The experimental testing consisted of modal analyses as well as capacitor charging tests on an electromagnetic shaker for a simple cantilever beam, a cantilever beam with a proof-mass at the tip and the L-shaped bracket.

2.2 Thermoelectric Testing

The capability of harvesting ambient energy from existing temperature gradients on the CX-100 wind turbine blade using thermoelectric materials was explored using multiple techniques. First, the available temperature gradient across the cross-section of the blade was experimentally measured. This information helped to identify which techniques for thermal gradient energy harvesting should be investigated. Tests were also conducted in order to examine the possibility of integrating a solar harvesting TEG into the blade based on the work of Sodano et. al.[9]

Figure 5: CX-100 Temperature Gradient Test Setup

Initial thermal gradient tests were conducted on the CX-100 blade cross-section by placing thermocouples on the top and underside of the blade (Figure 5). Two scenarios were tested simultaneously: 1) thermal gradient due to natural heating conditions – clear tape used to place thermocouple on top and bottom surface; and, 2) thermal gradient with black-body heating on the top side of the blade - black electrical tape served as the block-body surface. Temperature measurements for each thermocouple were taken every 10 minutes between 2-4 pm.

Three thermoelectric modules served as the benchmarks for the characterization study: the Tellurex G1-1.4-127-1.65 "127", and the Marlow Industries TG-12-8-01L and TG-12-4-01L. A hot plate produced a temperature gradient across the modules in order to determine the internal resistance and the respective Seebeck coefficient, which could then be used to approximate the expected power output of the module on the wind turbine blade itself. A resistor sweep test was used to verify the internal resistance of each thermoelectric module. Each module was placed on a hot plate set to ambient conditions measured the previous day during thermal gradient testing, and the output voltage was measured with a varying resistive load from 1.5 to 10 Ohms (Figure 6). The output power of each module was derived (Equation 1) and plotted against load resistance in order to maximize the output power density efficiency to the WID 3.0 sensor node.

$$P = \frac{V^2}{R} \quad (1)$$

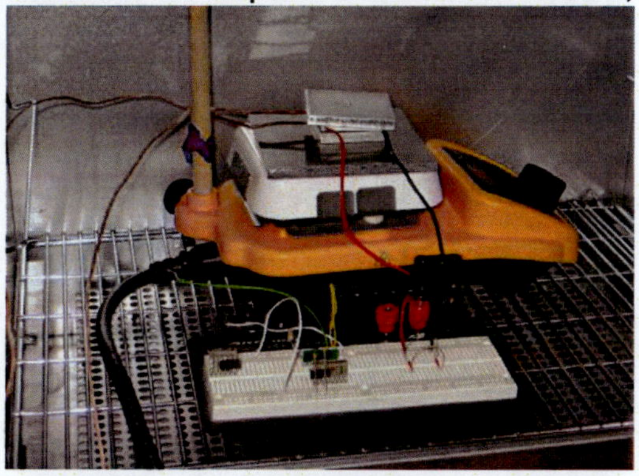

Figure 6: Thermoelectric Module Characterization Experimental Setup

The Seebeck coefficient was determined for each thermoelectric module by performing a voltage sweep detailed by Sodano et. al.[9] Increasing the temperature gradient across the thermoelectric

Table 3: Specifications for the PowerFilm photovoltaic cells.

Model	Size (l x w x h, in.)	Operating Voltage (V)	Operating Current (mA)	Weight (oz.)
S-MPT3.6-75	2.9 x 3.0 x 0.01	3.6	50	0.06
S-MPT3.6-150	2.9 x 5.9 x 0.01	3.6	100	0.10
S-MPT4.8-75	3.7 x 3.0 x 0.01	4.8	50	0.07
S-MPT4.8-150	3.7 x 5.9 x 0.01	4.8	100	0.10
S-MPT6-75	4.5 x 3.0 x 0.01	6.0	50	0.08

module causes the output voltage to increase linearly, and the resulting slope from plotting output voltage vs. temperature gradient represents an average Seebeck coefficient.

2.3 Photovoltaic Testing

The photovoltaic testing involved five different photovoltaic cells from the PowerFilm line and their properties can be seen in Table 3. The experimental setup consisted of the photovoltaic cells taped down to a single composite flat surface (Figure 7). The goal of the tests pertained to the effect of cell temperature, solar radiation level, composite substrate temperature, and solar incidence angle on the power output of the cells. The power output of the cells was determined by measuring the potential difference across the leads as well as the current output of the cells. Also, charging tests with various capacitors explored the ability of the cells to charge a capacitor that could be used to power the WID 3.0 sensor node.

In order to test the effects of the cell temperature and the composite substrate on power output, the experimental setup was placed outside on a sunny day. An infrared thermometer measured the temperature of each individual cell in the five photovoltaic arrays. A thermocouple attached to the composite substrate surface measured the temperature of the photovoltaic cell's base. These measurements occurred concurrently with the photovoltaic power output measurements.

The tests examining the effect of the solar incidence angle on the photovoltaic cell output utilized six different angles: 0°, 10°, 13°, 19°, 30° and 46° (Figure 8). Single point measurements of the voltage, current, and alignment angle of the experimental setup in the north-south direction eliminated the effect of the changing position of the sun during each of the tests. The effects of solar radiation on the power of the cells were tested by placing the cells in various lighting conditions. The experiment consisted of measuring the voltage and current output of the photovoltaic cells while placing the setup in direct sunlight on a flat surface and at a 43° solar incidence angle, a shaded area outside, a box with holes to allow sunlight through (Figure 9) and a lab with fluorescent lighting. The solar radiation could not be directly calculated in the different environments, but as the level of exposure to actual sunlight increases the solar radiation should as well.

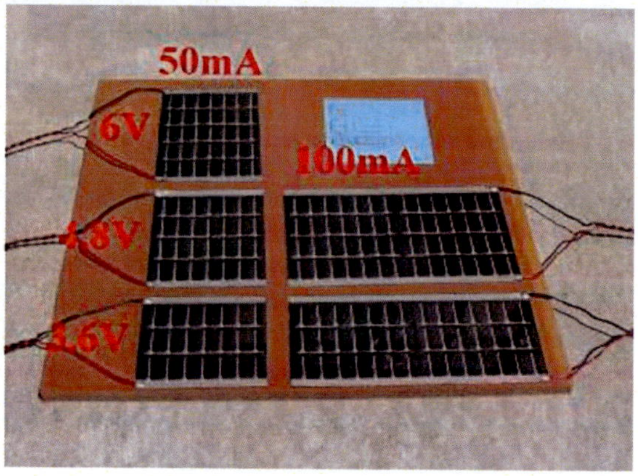

Figure 7: Photovoltaic cells experimental setup

Figure 8: Incidence angle tests

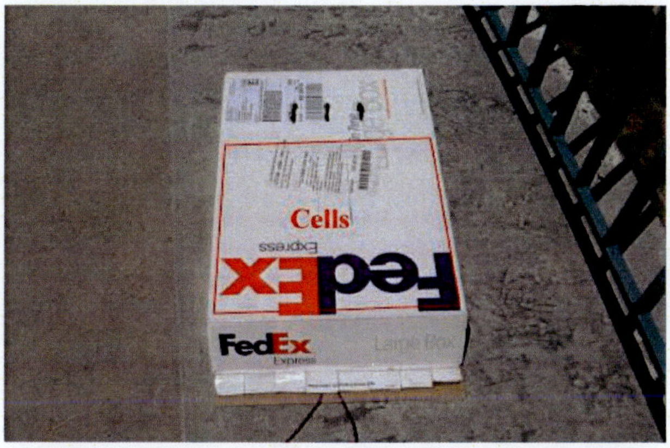

Figure 9: Low Radiation Experimental Test Setup

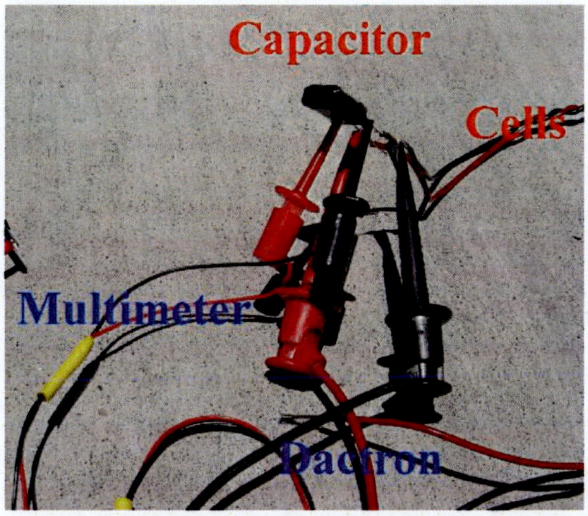

Figure 10: Wiring Setup for Capacitor Charging Tests

Figure 10 shows the wiring setup used for the capacitor charging tests. The capacitor was tied to the leads of one cell and grounded while a Dactron program recorded the voltage of the capacitor over a set time period. After the program initially started recording data, the ground was removed from the capacitor and the capacitor began charging. The energy (E) was calculated using

$$E = \frac{1}{2} C V^2 \tag{2}$$

where V is the measured voltage and C is the capacitance of the capacitor in the circuit. Normalizing the energy with respect to the area of the cells produced a truer representation of how well the cells performed. The slope of a linear fit of the energy density curve provided the power density of the cells. Three capacitors were initially tested, but a 0.1F 5V capacitor charged too fast and a 4.7F 2.5V capacitor did not charge enough, so they were removed from further testing. Therefore subsequent tests used only a 2.2F 2.5V capacitor.

In a follow-up test, a 1F 5V capacitor was charged using the same connections and procedure as the initial charging tests, but in two different conditions: solar incidence angles of 0° and 33°. The voltages, energy densities and power densities were calculated using the same methods as the 2.2F capacitor charging test. However, for this test, the average power density was calculated for each condition. The average power density for the 33° angle was subtracted from the average power density for the 0° angle to produce a difference plot to show the variation in the power density between the different conditions (Figure 26). The purpose of this test was to see what effects, if any, a less than optimal condition had on the charging of a capacitor.

2.4 Boost Converter Testing

The boost converter (Figure 11) tested was the Texas Instruments TPS61020; a low-power synchronous booster IC with input voltage range of 0.9V to 6.5V. The output voltage is adjustable by the voltage dividers on the output pin, and - for these tests - was 3.3V. The internal synchronous rectifier and the capacitors on the output voltage pin control the voltage ripple. The boost converter was soldered to a 72-pin Schmartboard with a 0.5mm pitch to match the surface-mount pin connections on the TPS61020. Wire leads were then soldered to the Schmartboard in order to connect the booster IC to a breadboard. The remaining components to the boost converter setup featured in Figure 11 were then connected to the breadboard.

Figure 11: Boost Converter Test Setup

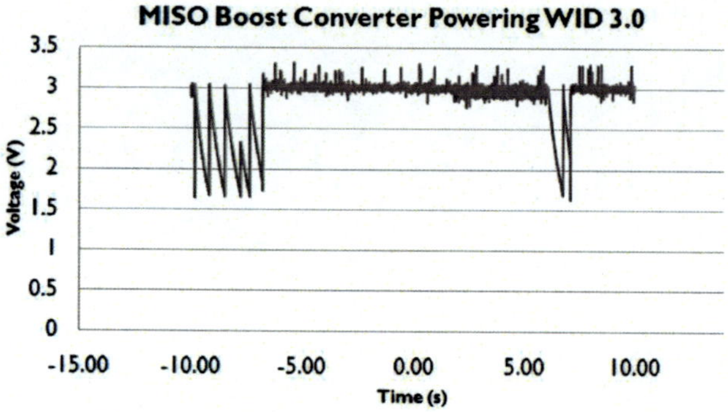

Figure 12: Output Voltage Signal while Powering WID 3.0

The output of the boost converter was connected to the WID 3.0, which was monitored by an oscilloscope, and the power was provided by an adjustable DC power supply. Preliminary testing showed the boost converter circuit was capable of continuously powering the WID 3.0 for measurement and transmission with only an input voltage of 1.3V and current of 50mA. The output voltage signal is a distinct "saw-tooth" pattern, which is a result of the charging and discharging of the capacitors on the output voltage pin. Once a load is applied to the circuit (i.e. WID 3.0 wakes-up and operates), the output voltage signal is more stable with a central output voltage of 3V (Figure 12).

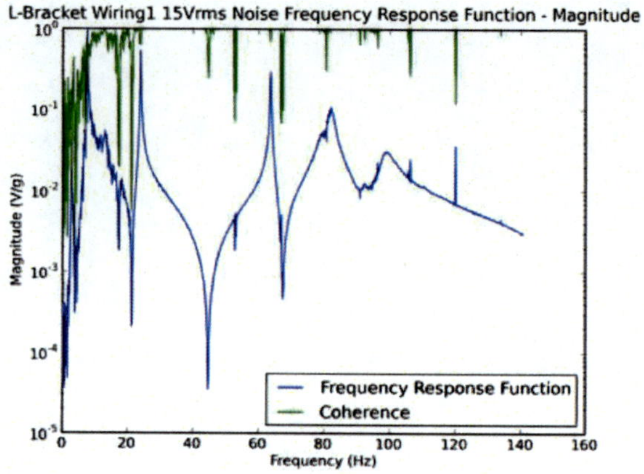

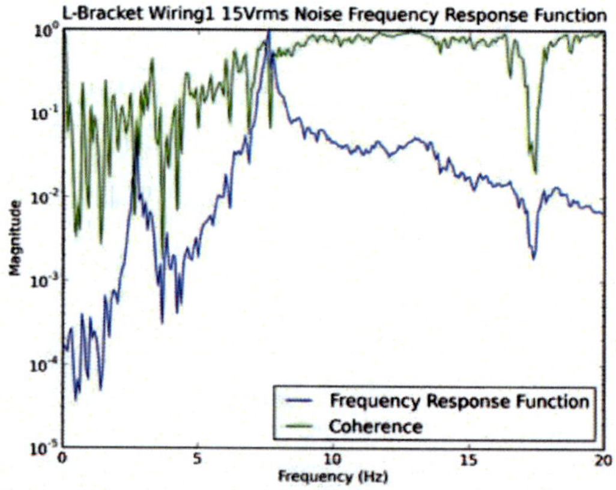

Figure 13: Normalized FRFs for L-Shaped Bracket

3. EXPERIMENTAL RESULTS

3.1 Piezoelectric Results

In order to optimize the voltage output of the various energy harvester configurations, the frequency of the base excitation should match the natural frequencies of the transfer function between output voltage and the base acceleration. The frequency response function of the L-shaped bracket (Figure 13) shows the first natural frequency of the beam around 2.7Hz and the second natural frequency around 7.5Hz. The poor coherence of the frequency response function below 20Hz corresponds with the significant amount of noise in the frequency response function itself, easily visible in the 0 – 160Hz plot. The experimental setup most likely generated this as the frequency response function is the linear average of 15 tests. The frequency response functions of all three of the harvester configurations illustrate the behavior of the natural frequencies as a tip mass is added to a cantilever and then when it is expanded into an L-shaped bracket:

From the normalized frequency response functions (Figure 14 & Table 4) of the three configurations in the frequency range of interest for the wind turbine blade, about 0-20Hz, it can be seen that the more mass added onto the end of the cantilever beam the lower the first natural frequency becomes, which is consistent with the theory. Also, the amount of damping in the system highly affects the response between the first and second natural frequencies of the L-shaped. Therefore a low amount of damping is necessary in the L-shaped bracket configuration.

Table 4: Frequency Response Function Results

Configuration	ω_1 (Hz)	ω_2 (Hz)	ω_2/ω_1 (Hz)
Cantilever	16	92	5.8
Cantilever w/Tip Mass	7.3	69	9.5
L-Shaped Bracket	2.7	7.5	2.8

Next, it was verified that all three harvester configurations would be able to reach the output voltage levels necessary to charge the capacitor for the WID 3.0 sensor node to operate. All three of the configurations obtained the necessary 2.7V output at relatively small excitation accelerations and most did or could obtain the 3.5V maximum (Table 5). Capacitor charging tests (Figure 15)

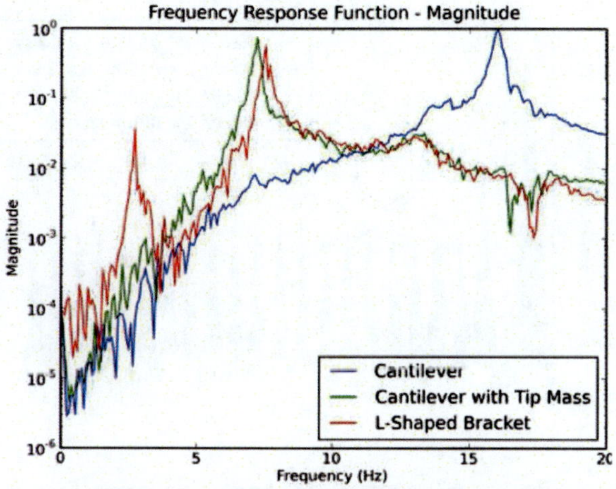

Figure 14: Normalized FRF Comparison

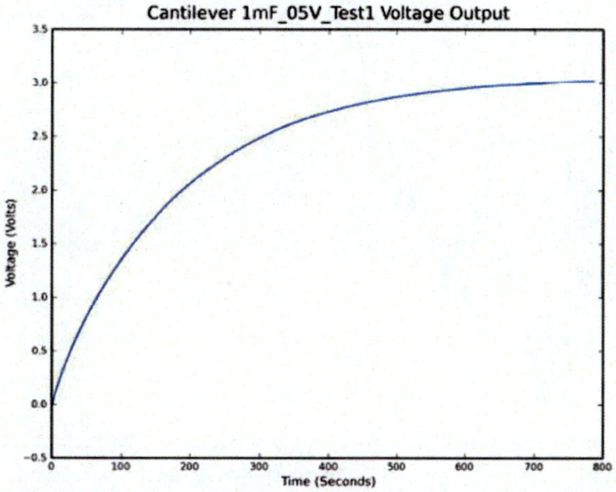

Figure 15: 1mF Cantilever Capacitor Charging Test

performed at both the first and second natural frequencies used a single frequency harmonic waveform excitation. A 1mF capacitor provided the most appropriate comparison between the three configurations based on the WID 3.0 power requirements as well as the performance of the configurations themselves. The tests lacking an rms base acceleration value used a lower measurement rate in order to increase the length of the test and therefore the resolution was not high enough to produce an accurate measurement. Also, the time values in Table 6 denoted with an asterisk indicate tests where the voltage across the capacitor peaked below the goal of 3.5 volts.

Table 5: Straight and Rectified Output Voltage Tests

Configuration	Excitation Voltage (V)	Freq. (Hz)	Voltage Cond.	$A_{B,RMS}$ (g)	V_{max} (V)
Cant.	0.25	16.0	Straight	0.004	4.5
Cant.	0.25	16.0	Rectified	0.004	6.7
Cant.	1.00	92.0	Straight	0.087	4.1
Cant.	1.00	92.0	Rectified	0.086	7.4
w/Tip Mass	0.75	7.3	Straight	0.006	3.8
w/Tip Mass	0.75	7.3	Rectified	0.006	5.1
w/Tip Mass	0.50	69.0	Straight	0.012	3.5
w/Tip Mass	0.50	69.0	Rectified	0.013	6.2
L-Bracket	0.50	2.7	Straight	0.003	3.3
L-Bracket	0.50	2.7	Rectified	0.002	3.3
L-Bracket	0.25	7.5	Straight	0.003	3.2
L-Bracket	0.25	7.5	Rectified	0.003	3.1

Table 6: 1mF Capacitor Charging Tests

Configuration	Freq. (Hz)	Excitation Voltage (V)	RMS Base Accel. (g)	Max Output Voltage (V)	Time (s)
Cant.	16.0	0.5	0.007	3.5	350
Cant.	16.0	1.0	0.016	3.5	160
Cant.	92.0	0.5	N/A	3.0	800*
Cant.	92.0	0.75	N/A	3.5	300
w/Tip Mass	7.3	1.5	0.012	3.2	1300*
w/Tip Mass	69.0	0.5	N/A	3.5	300
w/Tip Mass	69.0	1.0	N/A	3.5	90
L-Bracket	2.7	7.0	0.021	1.0	1000
L-Bracket	7.5	7.0	0.063	1.9	150*

From Table 6, there is a consistent trend where as the frequency of excitation decreases, the ability of the harvester to charge the capacitor lowers as well. Also, the amount of base excitation required to charge the capacitor increases as the overall mass of the system increases, especially with the transition to the L-shaped bracket harvester configuration.

Optimizing the design of the L-shaped structure to significantly decrease the mass would dramatically increase the output of the piezoceramic patches.

Another issue of concern comes from the difference in the mode shapes between the first and second natural frequencies. As mentioned in Erturk et

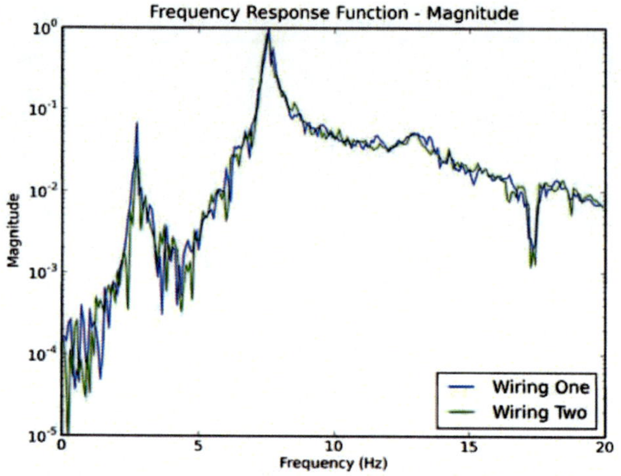

Figure 16: L-Shaped Bracket Wiring Comparison

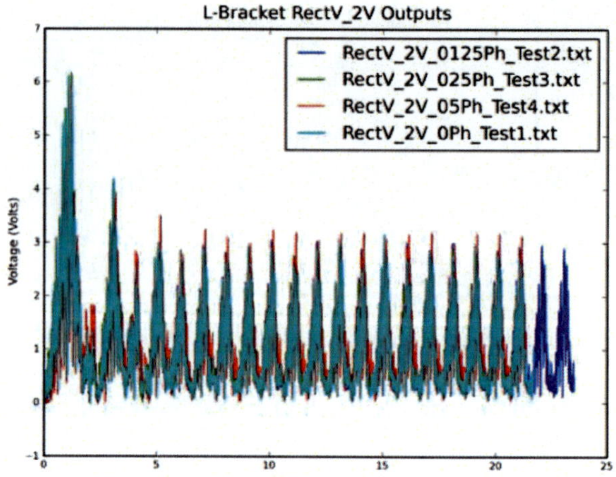

Figure 17: Dual Frequency Excitation Phase Variation Tests.

al[11], one wiring configuration cancellation occurs between the voltage outputs of the two piezoelectric patches in the first frequency, and the other wiring configuration cancellation occurs in the second frequency. Comparing the frequency response functions between the two different wiring configurations shows only small changes in magnitude of the system response for either frequency (Figure 16).

Also, tests using a dual-frequency excitation function targeting both the first and the second natural frequencies performed on the L-shaped bracket harvester explored the possibility of amplifying the voltage output of the energy harvester by coupling the first two modes. The results showed no change when the phase shift between the two frequencies varied between 0 and 90 degrees (Figure 17).

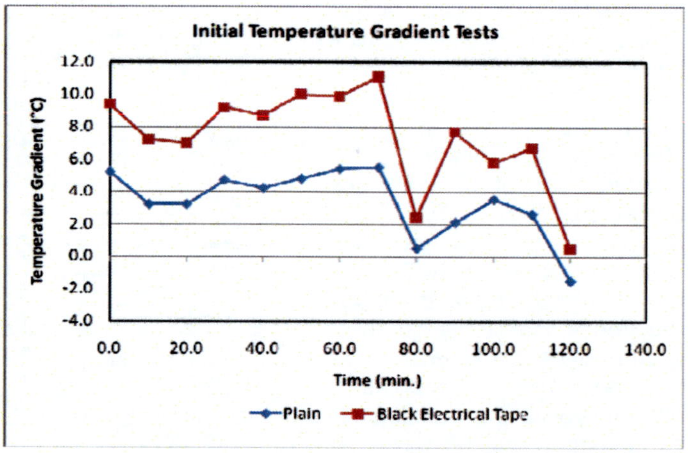

Figure 18: Initial Temperature Gradient Tests

3.2 Thermoelectric Results

The results show that a temperature gradient of approximately 10°C is available across the blade with the black electrical tape and 5°C with natural heating conditions (Figure 18). As shown in the plot below, the thermal gradient on the turbine blade cross-section is highly subject to environmental variability, which is a cause for concern when considering the use of thermoelectric modules for powering sensing hardware. Fortunately, there are methods to improve the thermal gradient in addition to mitigating the sensitivity to environmental conditions.

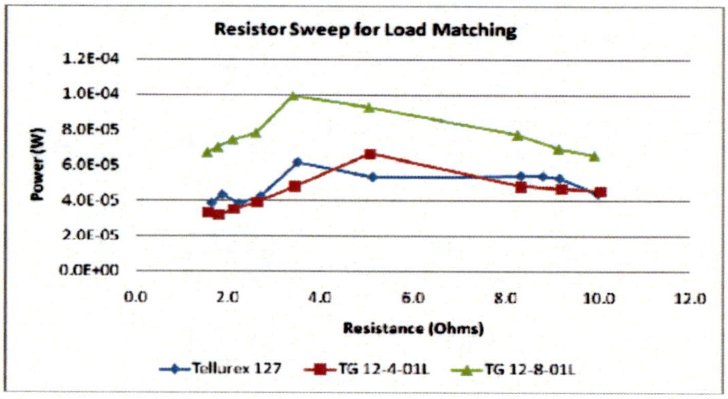

Figure 19: Thermoelectric Module Resistor Sweep Tests

A resistor sweep test was performed to verify the internal resistance of each thermoelectric module in order to match the electrical load of WID 3.0, maximizing the power density efficiency. Each module was placed on a hot plate set to ambient conditions measured the previous day during thermal gradient testing, and the output voltage was measured with a varying resistive load from 1.5 to 10 Ohms. The output power of each

Table 7: Seebeck Coefficients for each of the thermoelectric modules used in this study.

Thermoelectric Module	Seebeck Coefficient (mV/°C)
Tellurex 127	5.38
TG 12-4-01L	7.78
TG 12-8-01L	7.93

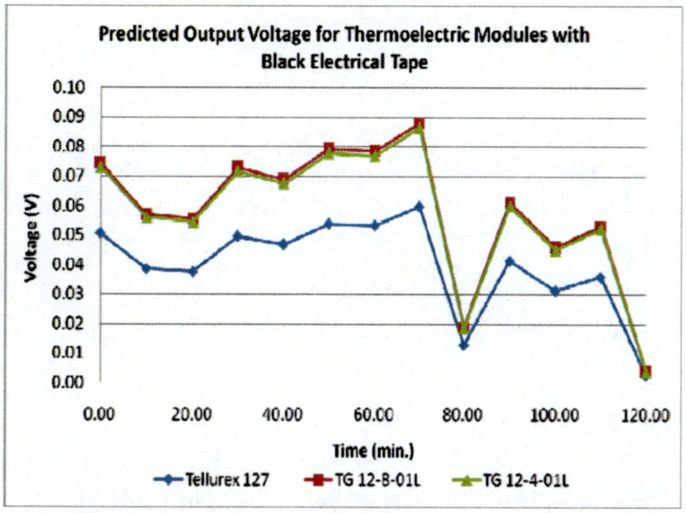

Figure 20: Predicted output voltage for each module assuming a black-body heat source.

module was plotted against load resistance, and the matched load is the peak output power value (Figure 19). As shown in the figure below, the Tellurex 127 and Marlow TG 12-8-01L had an internal resistance of 3.5 Ohms and the Marlow TG 12-4-01L had an internal resistance of 5.1 Ohms. The Marlow TG 12-8-01L resistance sweep lacks a defined peak, which either illustrates the need for more data points or poor contact between the thermoelectric module and the heat source.

The measured Seebeck coefficients for each module are shown in the Table 7. The results show the Marlow modules have a better potential for output voltage relative to the Tellurex module. These measurements were made with the same temperature gradient range and heat source temperature settings. Using these values for the Seebeck coefficient and the thermal gradient measurements, Figure 20 shows the predicted output voltage for each thermoelectric module with the black-body heat source. As shown in the figure, the average output voltage for the three thermoelectric modules was approximately 0.6 volts DC.

3.3 Photovoltaic Results

As the temperatures of the PV cells and the mounting surface increased, a marginal increase in the output power from the cells resulted and can reasonably be attributed to the rising of the sun in the sky and decreasing solar incidence angle. There were large temperature variations on each individual cell and this was seen for all cells. The temperature of the surface matched the temperature of the cells according to the thermocouple reading. However, when the surface temperature was measured using the IR thermometer, it was measured as 10°F less than what the thermocouple was reading.

As the solar incidence angle increased, the current and power of all the cells decreased. Overall the voltage did not drop noticeably from 0° to 46°, only decreasing by 2%. Drops in the current and power were not seen until after the 13° angle measurement. From the 13° angle measurement, the percentage drops in the current and power for all of the cells followed a linear pattern as can be seen in Figures 21 and 22. The overall percentage drop in current is about 25% and the overall percentage drop in power is about 27%. Increasing the solar incidence angle to a 45° angle has little effect on the voltage of our photovoltaic cells, but significantly decreases current and power. The condition with the solar incidence angle of 43° had similar results to the 46° angle

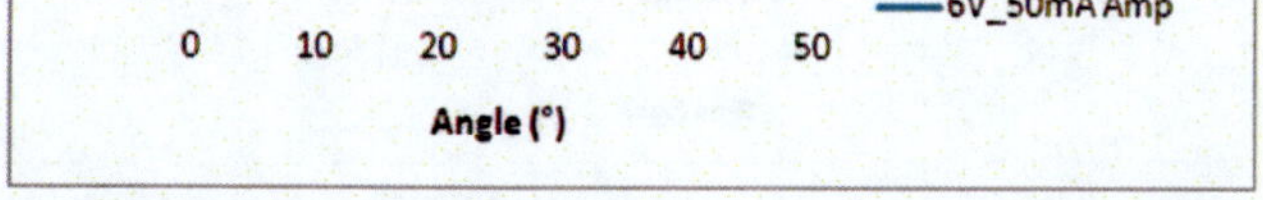

Figure 21: Percentage drop in current from initial value

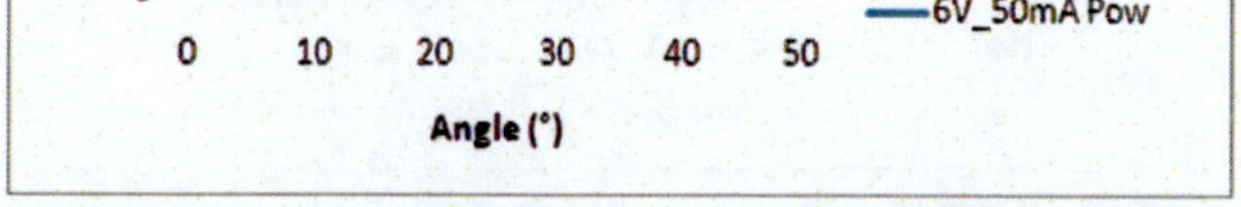

Figure 22: Percentage drop in power from initial value

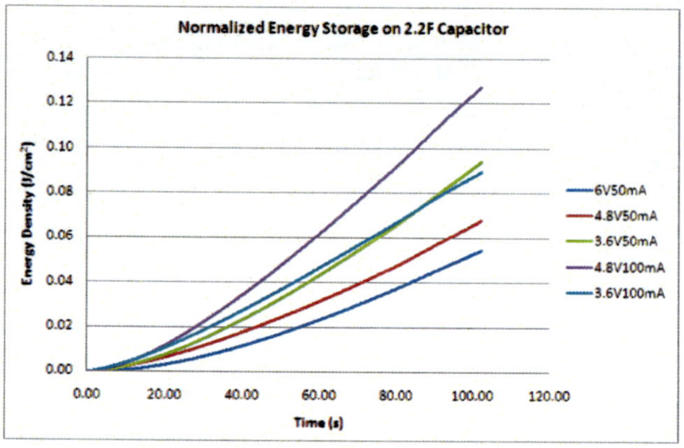

Figure 23: Energy density stored to 2.2F capacitor. Power density is represented by the slope.

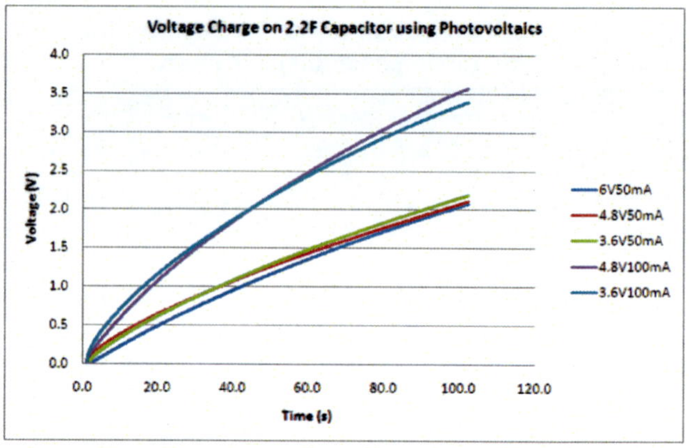

Figure 24: Voltage charged to 2.2F capacitor for different cells

discussed previously. The power of the cells placed in the sun at a solar incidence angle of 0°was the highest of the four environments. The power values were still in the hundreds of mW range for the 43° angle, but were about 32% less than the 0° angle. Although dependent on the voltage and current capacity of the cell, all of the power values were in the hundreds of mW range.

When the cells were placed in the lab with the fluorescent lights, little to no power was recorded. The exact solar radiation in the lab was not known, but it is probably the lowest of the five environments. A box with holes to let in light was the other low solar radiation condition. Similar values for power were seen in this condition although there was current present in every cell. The measurements taken when the cells were in the shade showed power in the range of tens of mW. In the shade, the voltage dropped by 40% and the current dropped to about 2 mA for the 50 mA cells and 5 mA for the 100 mA cells. Higher solar radiation conditions produced more power for the same cells, but there is still power available in conditions with lower solar radiation including shade. However, some amount of sunlight must be present to produce a usable level of power.

The power density delivered by the 100 mA cells to the 2.2F 2.5V capacitor was higher than the power density delivered by the 50 mA cells; the power density was calculated from the slope of the fitted line of the energy density curve (Figure 23). The largest power density seen was 1.67 mW/cm^2 in the 4.8V 100mA and the lowest power density seen was 0.75 mW/cm^2 in the 6V 50mA. The power densities for the 3.6V cells were approximately the same for both current ratings. Over the span of 100 seconds, the capacitor was charged to 3.5 V with the 100 mA cells and 2 V with the 50 mA cells (Figure 24). The 4.8V 50mA cell produced the largest power density, but the power densities for all of the cells were in a very close range.

For the charging tests performed with solar incidence angles of 0° and 33°, the average difference in the power densities oscillated around zero after 40 seconds. Plots of the energy densities for both conditions (Figures 25 and 26) and the plot of the average power density difference are shown(Figure 27). The largest power density

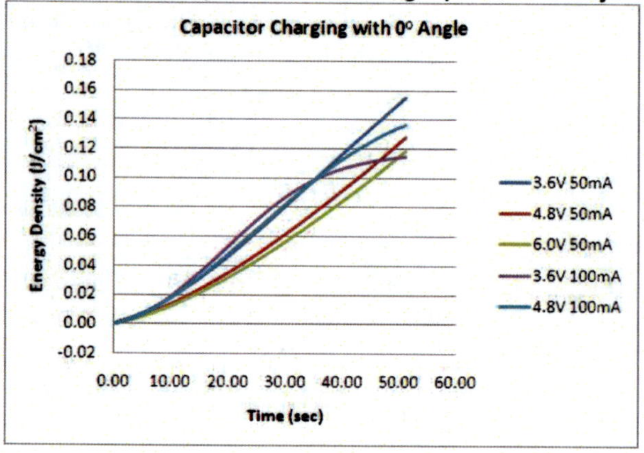

Figure 25: Energy density with 0˚ incidence angle.

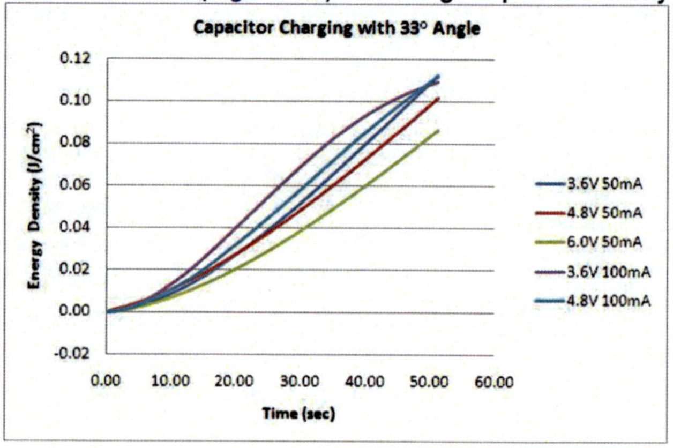

Figure 26: Energy density with 33˚ incidence angle.

seen for the 33° angle condition was 2.8 mW/cm^2 in the 3.6V 100mA cell. The largest power density seen for the 0° angle was 3.5 mW/cm^2 in the 3.6V 100mA cell, which is quite a large difference from the 33° angle. However, when looking at the average power density difference plot, a 0.7 mW/cm^2 difference is the maximum value and after 40 seconds the average difference is approximately zero. Therefore, the solar cells can charge a capacitor at roughly the same rate in the range of -30° to 30° solar incidence angle.

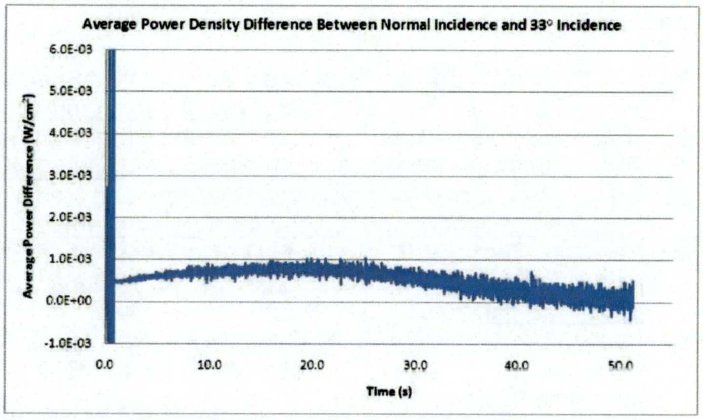

Figure 27: Average difference in power density for 0° and 33° Incidence angles.

4. SUMMARY AND CONCLUSIONS

The goal of this investigation has been to examine the feasibility of several different energy harvesting approaches to power embedded sensing hardware that could be used to interrogate the operational state and health of wind turbine blades. The three harvesting devices considered in this study include piezoelectric elements, thermoelectric modules, and photovoltaic cells. Of these three harvesters, the photovoltaic system was able to scavenge the most energy, as expected, due to the high energy density (per unit area) associated with the photovoltaic cells. The piezoelectric system used in this study was based upon an L-shaped bracket design that capitalizes on the nonlinear interaction between each leg of the bracket to reduce the spacing between the 1st and 2nd resonant modes to a theoretical limit of 2. Such spacing is advantageous in this application as some experimental turbine blades have demonstrated a harmonic response in the 1st and 2nd transverse modes, generating a kinetic environment that could be capitalized on by the L-shaped energy harvester. Lab tests indicated that a spacing of 2.78 could be obtained for the test structure used in this study. Further tests indicated that the hardware was capable of generating enough energy to power the wireless sensor node; however the time required was considerably longer than that of the photovoltaic cells. The final harvesting approach was the use of thermoelectric modules to capitalize on the thermal gradients that have been observed across the thickness of the wind turbine blades. Tests indicated that ~ 10°C could be expected across the blade thickness; however this was not sufficient to produce the power needed to operate one of the WID 3.0 sensor nodes. To address this issue a multi-input, single-output power conditioning circuit was developed to control how the power that is generated by each harvester is conditioned prior to being released to the sensor node. The circuit also has an integrated boost converter that allows the system to provide a 3.3V output given an input as low as 1.2V. Bench tests have shown that the system operates as intended, however further tests will be conducted to actually integrate the full suite of energy harvesters into one power supply system.

The result of this study is that a multi-source energy harvesting system has been demonstrated to be a feasible solution to the energy needs of wireless sensor nodes embedded within wind turbine blades. This is particularly true for photovoltaic and piezoelectric devices which can each generate the energy necessary to operate nodes such as the WID 3.0 sensor node. While each system has the potential to power the sensor node independently, a MISO power conditioning circuit will allow multiple harvesters to be used to power a single sensor node, providing a more robust power solution that would allow the integration of several transducers, such as thermoelectrics, to provide supplemental energy that could augment performance when direct solar or kinetic energy is not readily available. The proof of concept has been demonstrated in this initial study, and further development and testing of the MISO power conditioning circuit should result in a multi-source energy harvesting solution that will be integrated within test blades that are currently being proposed for field tests on a 19.8m wind turbine. The proposed test blades are currently scheduled to be flown in late 2011 at the U.S. Department of Agriculture – Sandia National Laboratory Wind Energy Technology Test Site in Bushland, Texas.

REFERENCES

1. American Wind Energy Association, Annual Wind Energy Report Year Ending 2008, Annual Statistics on U.S. Wind Energy, January 2009. http://www.awea.org/publications/reports/AWEA-Annual-Wind-Report-2009.pdf
2. U.S. Department of Energy, Landowners' Frequently Asked Questions about Wind Development, August 2003. http://www.windpoweringamerica.gov/pdfs/wpa/34600_landowners_faq.pdf
3. Danish Wind Industry Association, Operation and Maintenance Costs for Wind Turbines, May 2003. http://www.windpower.org/EN/tour/econ/oandm.htm
4. Park, G., Farrar, C.R., Todd, M.D., Hodgkiss, W. & Rosing, T., "Energy Harvesting for Structural Health Monitoring Sensor Netorwks," Los Alamos National Labs Workshop, www.lanl.gov/projects/ei/pdf_files/LA-14314-MS.pdf, 2007.
5. Farinholt, K.M., Taylor, S.G., Overly, T.G., Park, G. & Farrar, C.R., "Recent Advances in Impedance-based Wireless Sensor Nodes," Proceedings of the ASME Conference on Smart Materials, Adaptive Structures and Intelligent Systems, 2008.
8. Sodano, H.A., Inman, D.J. & Park, G., "A Review of Power Harvesting from Vibration using Piezoelectric Materials," *The Shock and Vibration Digest*, **36**(3) pp. 197-205.
9. J.A. Paradiso and T. Starner, 2005, "Energy Scavenging for Mobile and Wireless Electronics," *IEEE Pervasive Computing*, **4**, pp.18–27.
10. U.S. Department of Energy, The Photoelectric Effect, Solar Energy Technologies Program, July 2009. http://www1.eere.energy.gov/solar/photoelectric_effect.html
11. A. Erturk, J. M. Renno, and D. J. Inman, "Piezoelectric Energy Harvesting from an L-Shaped Beam-Mass Structure," in *Proceedings of SPIE Conference on Active and Passive Smart Structures and Integrated Systems*, (San Diego, CA), 2008.
12. H. A. Sodano et al, "Recharging Batteries using Energy Harvested from Thermal Gradients," in *Journal of Intelligent Material Systems and Structures*, Vol. 18, Jan. 2007.

Proceedings of the IMAC-XXVIII
February 1–4, 2010, Jacksonville, Florida USA

Limit Cycle Oscillations of a Nonlinear Piezo-magneto-elastic Structure for Broadband Vibration Energy Harvesting

A. Erturk[1,*], J. Hoffmann[2], and D.J. Inman[2]
Center for Intelligent Material Systems and Structures
[1] Department of Engineering Science and Mechanics
[2] Department of Mechanical Engineering
Virginia Polytechnic Institute and State University, Blacksburg, VA 24061
[*] Corresponding author: erturk@vt.edu

ABSTRACT

Vibration-based energy harvesting has been investigated by several researchers over the last decade. Typically, devices employing piezoelectric, electromagnetic, electrostatic and magnetostrictive transductions have been designed in order to convert ambient vibrations into electricity under resonance excitation. Regardless of the transduction mechanism, a primary issue in resonant energy harvesters is that the best performance of the device is limited to resonance excitation. That is, the power output is drastically reduced if the excitation frequency slightly deviates from the resonance frequency of the generator. In order to overcome this issue of the conventional cantilever configuration, a nonlinear piezo-magneto-elastic energy harvester is introduced in this paper. First the electromechanical equations describing the nonlinear system are given along with theoretical simulations to demonstrate the existence of large-amplitude limit cycle oscillations (LCO) at different frequencies. In agreement with the theory, the experiments show that the transient chaotic vibrations of the generator can turn into large-amplitude LCO on a high-energy orbit, which can also be realized by applying a disturbance to the structure oscillating on a low-energy orbit around one of its foci. It is shown experimentally that the open-circuit voltage output of the piezo-magneto-elastic configuration can be three times that of the conventional piezo-elastic cantilever configuration without magnetic buckling. Therefore the device proposed here generates an order of magnitude larger power output over a range of frequencies. The magneto-elastic structure is discussed here for piezoelectric energy harvesting and it can also be utilized to design enhanced energy harvesters using electromagnetic, electrostatic and magnetostrictive transductions as well as their hybrid configurations.

1. INTRODUCTION

The idea of vibration-to-electricity conversion to generate low-level electrical power has received great attention over the last decade [1-5]. The main motivation in research related to vibration-based energy harvesting is to convert the vibration energy available in the environment of small electronic components (such as wireless sensor nodes in remote locations) into electrical energy so that the requirement of battery replacement (and therefore the relevant maintenance requirement) is minimized.

The basic transduction mechanisms that can be used for vibration-to-electricity conversion are piezoelectric [6], electromagnetic [7], electrostatic [8] and magnetostrictive [9] transductions. Regardless of the transduction mechanism, a primary issue in vibration-based energy harvesting is that the best performance of a generator is usually limited to excitation at its fundamental resonance frequency. If the applied ambient vibration deviates slightly from the resonance condition then the power output is drastically reduced. A major issue in vibration energy harvesting is therefore to enable broadband energy

T. Proulx (ed.), *Structural Dynamics and Renewable Energy, Volume 1*, Conference Proceedings of the Society for Experimental Mechanics Series 10, DOI 10.1007/978-1-4419-9716-6_28, © The Society for Experimental Mechanics, Inc. 2011

harvesters. Researchers have recently focused on the concept of broadband energy harvesting to solve this issue with different approaches [10,11].

This paper introduces a broadband piezoelectric energy harvester that exhibits large-amplitude oscillations over a frequency range. First the lumped-parameter electromechanical equations of the piezo-magneto-elastic energy harvester are presented. Theoretical simulations are then given for demonstrating the dramatically large voltage generation from the large-amplitude LCO on high-energy orbits of the piezo-magneto-elastic configuration. Comparisons are provided against voltage generation from the LCO of the conventional piezo-elastic configuration and it is shown that the piezo-magneto-elastic configuration can generate much larger voltage. Experimental verifications are presented and it is shown that an order of magnitude larger power output can be obtained with the piezo-magneto-elastic structure over a range of frequencies. The outstanding broadband power generation performance of the magneto-elastic configuration is discussed here for piezoelectric energy harvesting and it can easily be extended to electromagnetic, electrostatic and magnetostrictive energy harvesting techniques as well as to their hybrid combinations with similar devices.

2. THE PIEZO-MAGNETO-ELASTIC POWER GENERATOR

The magneto-elastic structure shown in Fig. 1a was first investigated by Moon and Holmes [12] as a mechanical structure that exhibits *strange attractor* motions. The device consists of a ferromagnetic cantilevered beam with two permanent magnets located symmetrically near the free end and it is subjected to harmonic base excitation. The bifurcations of the static problem are described by a *butterfly catastrophe* with a sixth order magneto-elastic potential. Depending on the magnet spacing, the ferromagnetic beam may have five (with three stable), three (with two stable) or one (stable) equilibrium positions. For the case with three equilibrium positions, the governing lumped-parameter equation of motion has the form of the Duffing equation:

$$\ddot{x} + 2\zeta\dot{x} - \frac{1}{2}x\left(1 - x^2\right) = f\cos\Omega t \qquad (1)$$

where x is the dimensionless tip displacement of the beam in the transverse direction, ζ is the mechanical damping ratio, Ω is the dimensionless excitation frequency, f is the dimensionless excitation force due to base acceleration ($f \propto \Omega^2 X_0$ where X_0 is the dimensionless base displacement amplitude) and an over-dot represents differentiation with respect to dimensionless time. The three equilibrium positions obtained from Eq. (1) are $(x,\dot{x}) = (0,0)$ (a saddle) and $(x,\dot{x}) = (\pm 1, 0)$ (two centers). Detailed nonlinear analysis of the magneto-elastic structure shown in Fig. 1a can be found in the papers by Moon and Holmes [12,13].

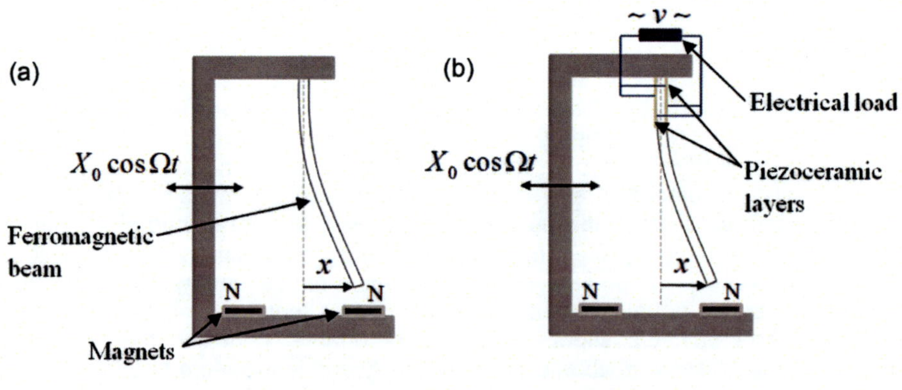

Figure 1. Schematics of the (a) magneto-elastic structure investigated by Moon and Holmes [12] and the (b) piezo-magneto-elastic power generator proposed here

In order to use this device as a piezoelectric power generator, we attach two piezoceramic layers to the root of the cantilever and obtain a bimorph generator as depicted in Fig. 1b. The piezoceramic layers are connected to an electrical load (a resistor for simplicity) and the voltage output of the generator across the load due to seismic excitation is the primary interest in energy harvesting. Introducing piezoelectric coupling [6] into Eq. (1) and applying the Kirchhoff laws to the circuit with a resistive load (Fig. 1b) leads to the following electromechanical equations:

$$\ddot{x}+2\zeta\dot{x}-\frac{1}{2}x\left(1-x^2\right)-\chi v=f\cos\Omega t \tag{2}$$

$$\dot{v}+\lambda v+\kappa\dot{x}=0 \tag{3}$$

where v is the dimensionless voltage across the load resistance, χ is the dimensionless piezoelectric coupling term in the mechanical equation, κ is the dimensionless piezoelectric coupling term in the electrical circuit equation and λ is the reciprocal of the dimensionless time constant ($\lambda \propto 1/R_l C_p$ where R_l is the load resistance and C_p is the equivalent capacitance of the piezoceramic layers) [6]. Note that the possible nonlinearity coming from piezoelectric coupling is ignored in Eqs. (2) and (3), assuming the standard form of the linear piezoelectric constitutive relations [14].

The state-space form of Eqs. (2) and (3) can be expressed as

$$\begin{Bmatrix}\dot{u}_1\\ \dot{u}_2\\ \dot{u}_3\end{Bmatrix}=\begin{Bmatrix}u_2\\ -2\zeta u_2+\frac{1}{2}u_1\left(1-u_1^{\,2}\right)+\chi u_3+f\cos\Omega t\\ -\lambda u_3-\kappa u_2\end{Bmatrix} \tag{4}$$

where the state variables are $u_1=x$, $u_2=\dot{x}$ and $u_3=v$. The electromechanically coupled equations given by Eq. (4) can be used in an ordinary differential equation solver for numerical simulations.

The time-domain voltage simulations in Figs. 2 and 3 are obtained for $\Omega=0.8$, $\zeta=0.01$, $\chi=0.05$, $\kappa=0.5$ and $\lambda=0.05$ (close to open circuit conditions). In the first case (Fig. 2a), the forcing term is $f=0.083$ and the motion starts with an initial deflection at one of the stable equilibrium positions ($x(0)=1$ with zero initial velocity and voltage: $\dot{x}(0)=v(0)=0$). The resulting vibratory motion is on a chaotic strange attractor (yielding the chaotic voltage history shown in Fig. 2a) and the Poincaré map of this strange attractor motion is shown in Fig. 2b on its phase portrait.

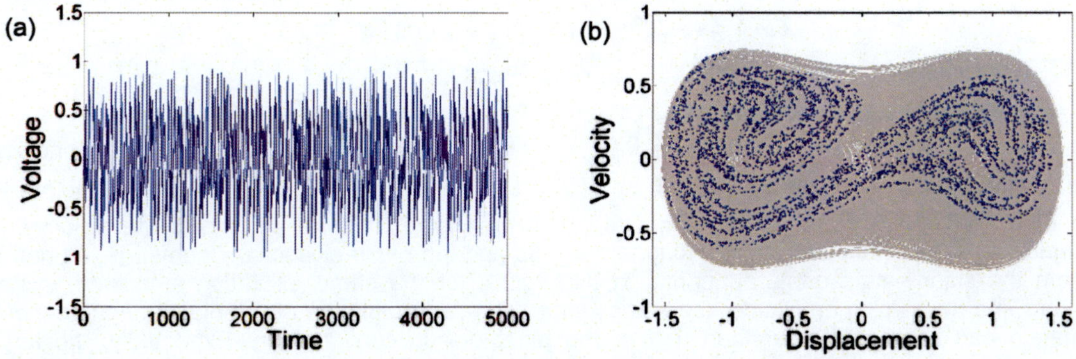

Figure 2. (a) Theoretical voltage history exhibiting the strange attractor motion for and (b) its Poincaré map ($x(0)=1$, $\dot{x}(0)=0$, $v(0)=0$, $f=0.083$, $\Omega=0.8$)

If the excitation amplitude is increased by keeping the same initial conditions, the transient chaotic behavior is followed by a large-amplitude LCO on a high-energy orbit with improved voltage response (Fig. 3a). More importantly, Fig. 3b shows that this type of large-amplitude voltage response can be obtained with the original excitation amplitude (of Fig. 2a) using different initial conditions (simply by imposing an initial velocity condition so that $x(0) = \dot{x}(0) = 1$, $v(0) = 0$).

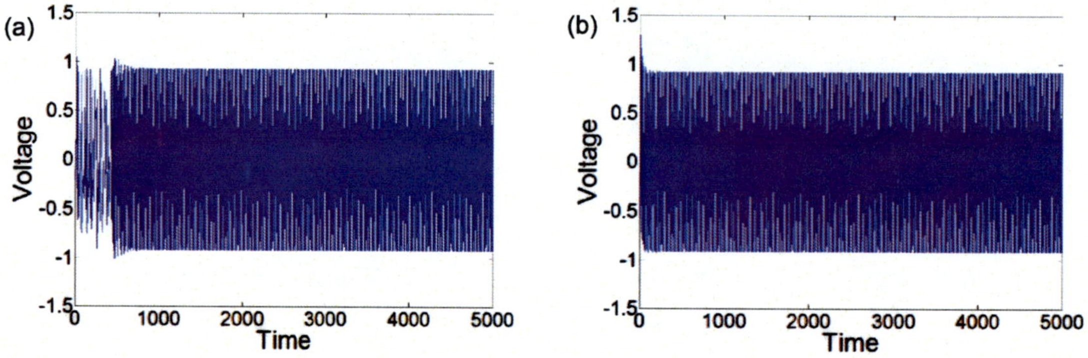

Figure 3. Theoretical voltage histories: (a) Large-amplitude periodic response due to the excitation amplitude ($x(0) = 1$, $\dot{x}(0) = 0$, $v(0) = 0$, $f = 0.115$, $\Omega = 0.8$); (b) Large-amplitude periodic response due to the initial conditions for a lower excitation amplitude ($x(0) = 1$, $\dot{x}(0) = 1$, $v(0) = 0$, $f = 0.083$, $\Omega = 0.8$)

Having observed the large-amplitude electromechanical response on high-energy orbits of the piezo-magneto-elastic energy harvester configuration described by Eqs. (2) and (3), a simple comparison can be made against the conventional piezo-elastic configuration (which is the *linear* cantilever configuration without the magnets causing bi-stability). The lumped-parameter equations of the linear piezo-elastic configuration are

$$\ddot{x} + 2\zeta\dot{x} + x - \chi v = f\cos\Omega t \tag{5}$$

$$\dot{v} + \lambda v + \kappa\dot{x} = 0 \tag{6}$$

which can be given in the state-space form as

$$\begin{Bmatrix} \dot{u}_1 \\ \dot{u}_2 \\ \dot{u}_3 \end{Bmatrix} = \begin{Bmatrix} u_2 \\ -2\zeta u_2 - u_1 + \chi u_3 + f\cos\Omega t \\ -\lambda u_3 - \kappa u_2 \end{Bmatrix} \tag{7}$$

For the same numerical input ($\Omega = 0.8$, $\zeta = 0.01$, $\chi = 0.05$, $\kappa = 0.5$ and $\lambda = 0.05$), initial conditions and the forcing amplitude of Fig. 3b ($x(0) = 1$, $\dot{x}(0) = 1$, $v(0) = 0$, $f = 0.083$), one can simulate the voltage response of the piezo-elastic configuration using Eq. (7). Figure 4a shows the velocity vs. displacement phase portrait of the piezo-magneto-elastic and the piezo-elastic configurations. As can be seen from the steady-state orbits appearing in this figure, for the same excitation amplitude, system parameters and the forcing amplitude, the steady-state vibration amplitude of the piezo-magneto-elastic configuration can be much larger than that of the piezo-elastic configuration. Expectedly, the large-amplitude LCO on the high-energy orbit is also observed in the velocity vs. voltage phase portrait* shown in Fig. 4b.

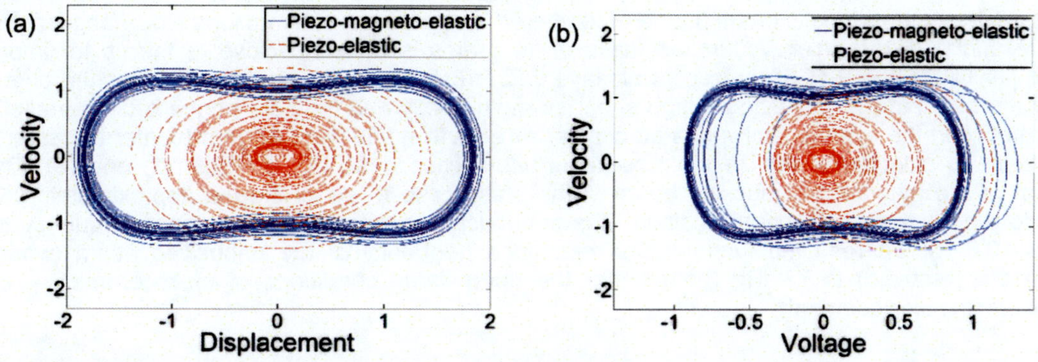

Figure 4. Comparison of the (a) velocity vs. displacement and the (b) velocity vs. voltage phase portraits of the piezo-magneto-elastic and piezo-elastic configurations ($x(0)=1$, $\dot{x}(0)=1$, $v(0)=0$, $f=0.083$, $\Omega=0.8$)

The superiority of the piezo-magneto-elastic configuration over the piezo-elastic configuration can be shown by plotting these trajectories at several other frequencies except for the resonance ($\Omega \cong 1$) case of the linear problem (for which the piezo-elastic configuration generates more voltage). However, at several other frequencies (e.g. $\Omega=0.6$, $\Omega=0.7$, $\Omega=0.9$), a substantially amplified response similar to the case of $\Omega=0.8$ given by Fig. 4 can be obtained. The velocity vs. open-circuit voltage phase portraits of the piezo-magneto-elastic and the piezo-elastic configurations are plotted for $\Omega=0.7$ and $\Omega=0.9$ in Figs. 5a and 5b, respectively. Therefore the theoretical simulations of the lumped-parameter electromechanical model show promising results to utilize the piezo-magneto-elastic configuration as a broadband electric generator. Experimental verifications are given in the following sections.

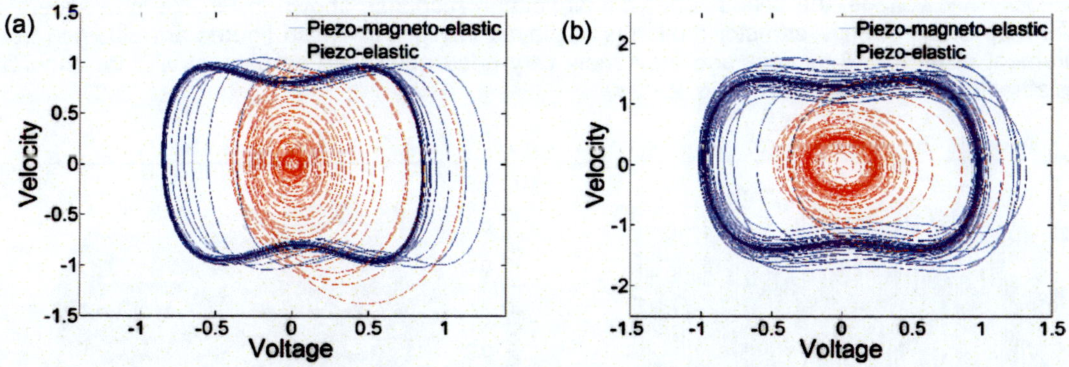

Figure 5. Comparison of the velocity vs. voltage phase portraits of the piezo-magneto-elastic and piezo-elastic configurations for (a) $\Omega=0.7$ and (b) $\Omega=0.9$ ($x(0)=1$, $\dot{x}(0)=1$, $v(0)=0$, $f=0.083$)

* For the system parameters used in these simulations, the phase between the voltage and the velocity is approximately 90 degrees because the system is close to open-circuit conditions. Therefore, in open-circuit conditions, it is reasonable to plot the velocity vs. voltage output as the electromechanical phase portrait (as an alternative to the conventional velocity vs. displacement phase portrait).

3. EXPERIMENTAL SETUP AND PERFORMANCE RESULTS

The piezo-magneto-elastic generator and the setup used in the first set of experiments are shown in Fig. 6. Harmonic base excitation is provided by a seismic shaker, acceleration at the base of the cantilever is measured by a small accelerometer and the velocity response of the cantilever is recorded by a laser vibrometer. The ferromagnetic beam (made of tempered blue steel) is 145 mm long (overhang length), 26 mm wide and 0.26 mm thick. A lumped mass of 14 grams is attached close to the tip for improved dynamic flexibility. Two PZT-5A piezoceramic layers (QP16N, Midé Corporation) are attached onto both faces of the beam at the root using a high shear strength epoxy and they are connected in parallel. The spacing between the symmetrically located circular rare earth magnets is 50 mm (center to center) and this distance is selected to realize the three equilibrium case described by Eqs. (2) and (3). The tip deflection of the magnetically buckled beam in the static case to either side is approximately 15 mm relative to the unstable equilibrium position. The post-buckled fundamental resonance frequency of the beam is 10.6 Hz whereas the fundamental resonance frequency of the unbuckled beam (when the magnets are removed) is 7.4 Hz (both under the open-circuit conditions of piezoceramics – *i.e.* at constant electric displacement).

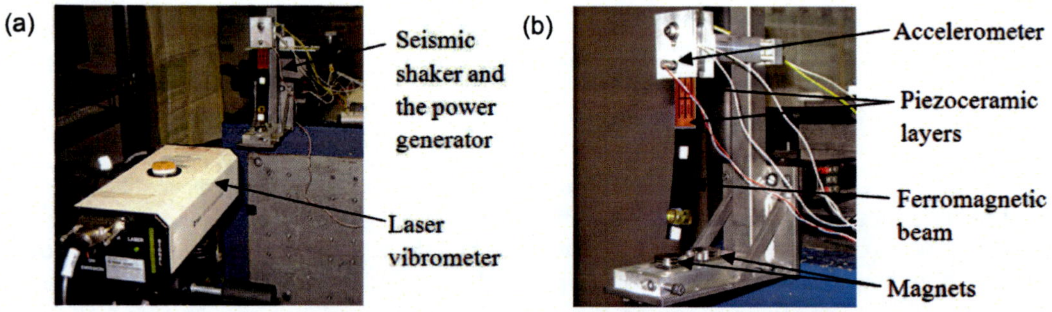

Figure 6. (a) A view of the experimental setup and (b) the piezo-magneto-elastic power generator

For a harmonic excitation amplitude of 0.5*g* (where *g* is the gravitational acceleration: *g* = 9.81 m/s^2) at 8 Hz with an initial deflection at one of the stable equilibrium positions (15 mm to the shaker side), zero initial velocity and voltage, the chaotic open-circuit voltage response shown in Fig. 7a is obtained. The Poincaré map of the strange attractor motion is displayed in Fig. 7b. These figures are obtained from a measurement taken for 15 minutes and they show very good qualitative agreement with the theoretical strange attractor simulation given by Fig. 2.

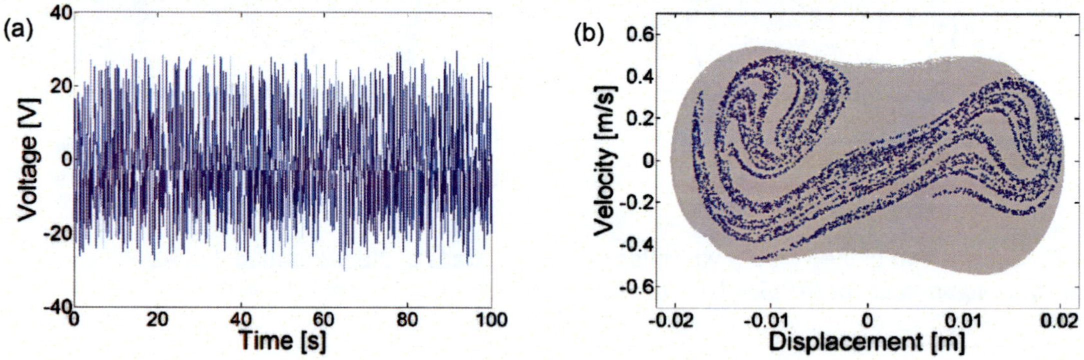

Figure 7. (a) Experimental voltage history exhibiting the strange attractor motion for and (b) its Poincaré map (excitation: 0.5*g* at 8 Hz)

If the excitation amplitude is increased to 0.8*g* (at the same frequency), the structure goes from transient chaos to a large-amplitude periodic motion with a strong improvement in the voltage response as shown in Fig. 8a. A similar improvement is obtained in Fig. 8b where the excitation amplitude is kept as the original one (0.5*g*) and a disturbance (hand impulse) is applied at $t = 11$ s (as a simple alternative to creating a velocity initial condition). Such a disturbance can be realized in practice by applying an impulse type voltage input through one of the piezoceramic layers for once. The experimental evidence given with Fig. 8 is in agreement with the theoretical discussion given with Fig. 3. Noticing the large-amplitude steady-state voltage response obtained at an *off-resonance* frequency in Fig. 8, the broadband performance of the device is investigated next and comparisons against the piezo-elastic configuration are given.

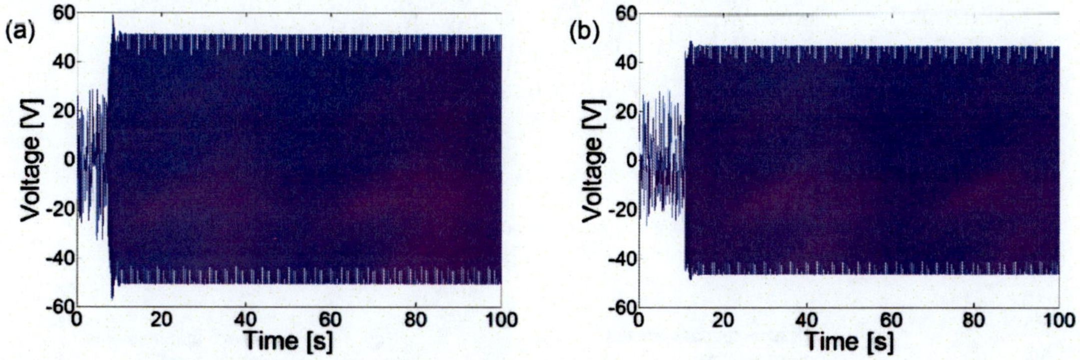

Figure 8. Experimental voltage histories: (a) Large-amplitude periodic response due to the excitation amplitude (excitation: 0.8*g* at 8 Hz); (b) Large-amplitude periodic response due to a disturbance $t = 11$ s (excitation: 0.5*g* at 8 Hz)

4. PERFORMANCE COMPARISON AGAINST THE PIEZO-ELASTIC CONFIGURATION

Before the comparisons of the piezo-magneto-elastic and piezo-elastic† configurations are given over a frequency range, reconsider the voltage history of Fig. 8b in two parts. The time history until the instant of the disturbance is chaotic, which would yield a strange attractor motion similar to what was shown in Fig. 7 if no disturbance was applied. After the disturbance is applied at $t = 11$ s, the LCO on a high-energy orbit is obtained as the steady-state response. In order to understand the advantage of the second region in the response history of Fig. 8b, the open-circuit voltage histories of the piezo-magneto-elastic and piezo-elastic configurations are compared for the same harmonic input (0.5*g* at 8 Hz). Figure 9a shows the acceleration input to the piezo-magneto-elastic and piezo-elastic configurations at an arbitrary instant of time. The voltage input to the seismic shaker is identical for both configurations, yielding very similar acceleration amplitudes (according to the signal output of the accelerometer) for a fair comparison. Figure 9b displays the comparison of the piezo-magneto-elastic and piezo-elastic configurations where the former exhibits chaotic response (5s-6s time interval in Fig. 8b) and the latter has already reached its harmonic steady-state response amplitude at the input frequency. As a rough comparison, from Fig. 9b, it is not possible to claim that the chaotic response of the nonlinear configuration has any advantage over the harmonic response of the linear configuration as their amplitudes look very similar (a more accurate comparison can be made through the RMS – root mean square – amplitudes). Besides, one would definitely prefer a periodic signal to a chaotic signal when it comes to processing the harvested energy using an efficient energy harvesting circuit [15-17]. Figure 9c shows the voltage histories of these

† The piezo-elastic configuration is simply the conventional configuration obtained by removing the magnets of the piezo-magneto-elastic configuration in the experiments.

configurations some time after the disturbance is applied to the piezo-magneto-elastic configuration (80s-81s time interval in Fig. 8b). Obviously if the same disturbance is applied to the piezo-elastic configuration, the trajectory (on the phase portrait) returns to the same low-amplitude limit cycle after some transients. Therefore the response amplitude of the piezo-elastic configuration is identical in Figs. 9b and 9c. Although the chaotic response of the piezo-magneto-elastic structure has no considerable advantage according to Fig. 9b, the large-amplitude LCO of this structure can give 2-3 times larger voltage output according to Fig. 9c.

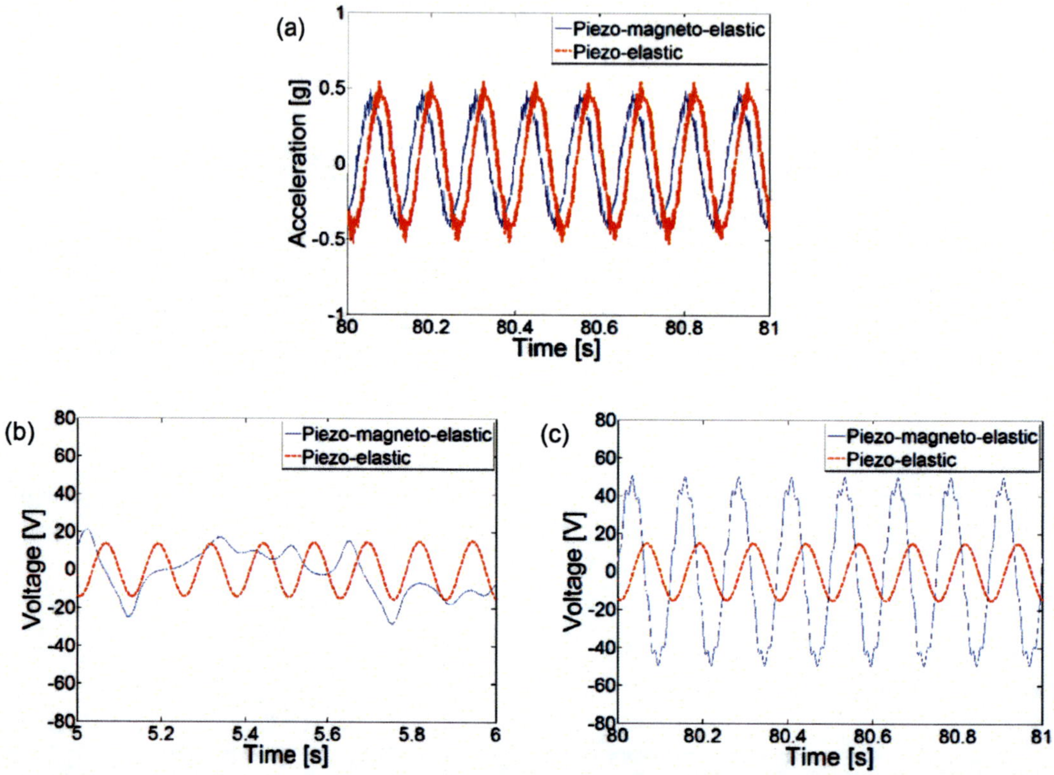

Figure 9. Comparison of the input and output time histories of the piezo-magneto-elastic and piezo-elastic configurations: (a) Input acceleration histories; (b) Voltage outputs in the chaotic response region of the piezo-magneto-elastic configuration; (c) Voltage outputs in the large-amplitude periodic response region of the piezo-magneto-elastic configuration (excitation: 0.5*g* at 8 Hz)

Figure 10 compares the velocity vs. voltage phase portraits of the piezo-magneto-elastic and piezo-elastic configurations, showing the advantage of the large-amplitude orbit clearly. This figure is therefore analogous to the theoretical demonstration given by Fig. 4b (additional harmonics are present in the experimental data of the continuous structure). One should then investigate different frequencies to see if similar high-energy orbits can be reached at other frequencies as well (as in the theoretical case of Fig. 5).

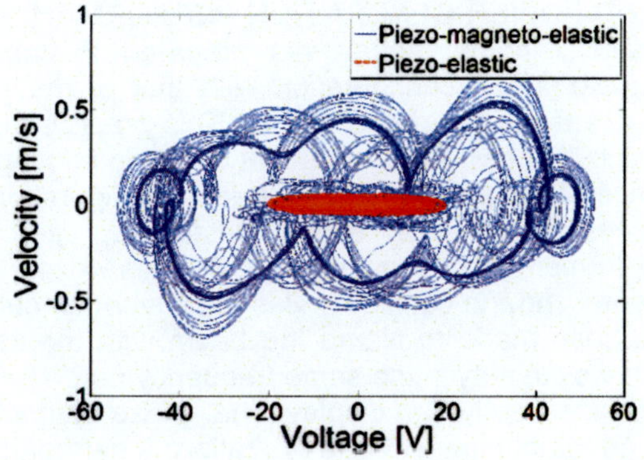

Figure 10. Comparison of the velocity vs. open-circuit voltage graphs (phase portraits) for the piezo-magneto-elastic and piezo-elastic configurations (excitation: 0.5*g* at 8 Hz)

For a harmonic base excitation amplitude of 0.5*g* (yielding an RMS acceleration of 0.35*g*), experiments are conducted at 4.5 Hz, 5 Hz, 5.5 Hz, 6 Hz, 6.5 Hz, 7 Hz, 7.5 Hz and 8 Hz. At each frequency, a large-amplitude periodic response is obtained the same way as in Fig. 8b with a disturbance around $t = 11$ s. Then the magnets are removed for comparison of the device performance with that of the conventional piezo-elastic configuration and the base excitation tests are repeated for the same frequencies with approximately the same input acceleration. The open-circuit RMS voltage outputs of the piezo-magneto-elastic and piezo-elastic configurations at each frequency are obtained considering the steady-state response in the 80s-100s time interval. The RMS values of the input base acceleration are also extracted for the same time interval.

Figure 11a shows that the excitation amplitudes of both configurations are indeed very similar (with an average RMS value of approximately 0.35*g*). The broadband voltage generation performance of the piezo-magneto-elastic generator is shown in Fig. 11b. The resonant piezo-elastic device gives larger voltage output only when the excitation frequency is at or very close to its resonance frequency (7.4 Hz) whereas the voltage output of the piezo-magneto-elastic device can be 3 times that of the piezo-elastic device at several other frequencies below its post-buckled resonance frequency (10.6 Hz). It should be noted that power output is proportional to the square of the voltage. Hence an order of magnitude larger power output over a frequency range can be expected with this device.

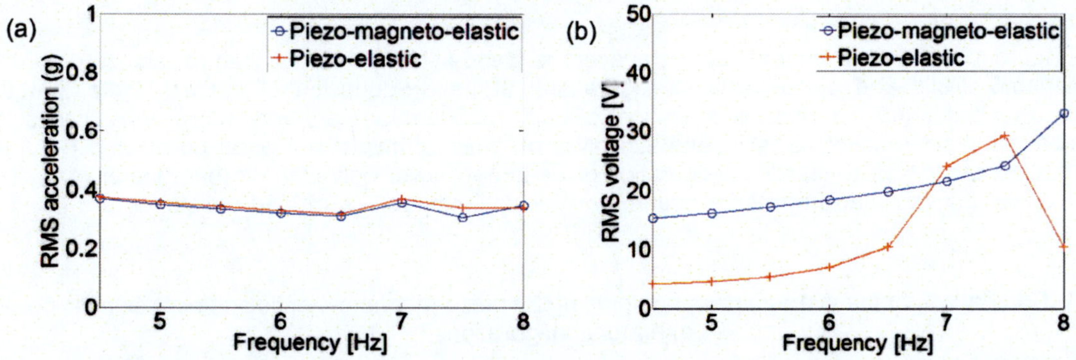

Figure 11. (a) RMS acceleration input at different frequencies (average value: 0.35*g*); (b) open-circuit RMS voltage output over a frequency range showing the broadband advantage of the piezo-magneto-elastic generator

5. OUTSTANDING BROADBAND POWER GENERATION PERFORMANCE

After the first set of experiments, another setup was prepared to compare the power generation performance of the piezo-magneto-elastic configuration with that of the piezo-elastic configuration to verify the order of magnitude increase in the power output. This practice also shows whether or not the presence of a resistive load (which is known to create shunt damping effect [6] due to Joule heating in the resistor) reduces the performance of the piezo-magneto-elastic configuration by modifying the amplitude of the large-amplitude LCO discussed here. The experimental setup used for this purpose is shown in Fig. 12a and it is similar to the former setup shown in Fig. 6. These experiments have been conducted two months after the previous ones and the cantilever was unclamped in between and the magnets were removed. Therefore, effort has been made to clamp the beam with the same overhang length and to relocate the magnets in a similar way to stay in the same frequency range.

Figures 12b and 12c, respectively, display the piezo-magneto-elastic and piezo-elastic configurations tested for power generation under base excitation. A harmonic base excitation amplitude of 0.5*g* (yielding an RMS value of approximately 0.35*g*) is applied at frequencies of 5 Hz, 6 Hz, 7 Hz and 8 Hz. From the former discussion related to the open-circuit voltage output given with Fig. 11b, it is expected to obtain an order of magnitude larger power with the piezo-magneto-elastic device at three of these frequencies (5 Hz, 6 Hz and 8 Hz). However, it is anticipated to obtain larger power from the piezo-elastic configuration around its resonance and 7 Hz is the frequency that is close to the resonance of the linear system (as can be noted from Fig. 11b).

(a) (b) (c)

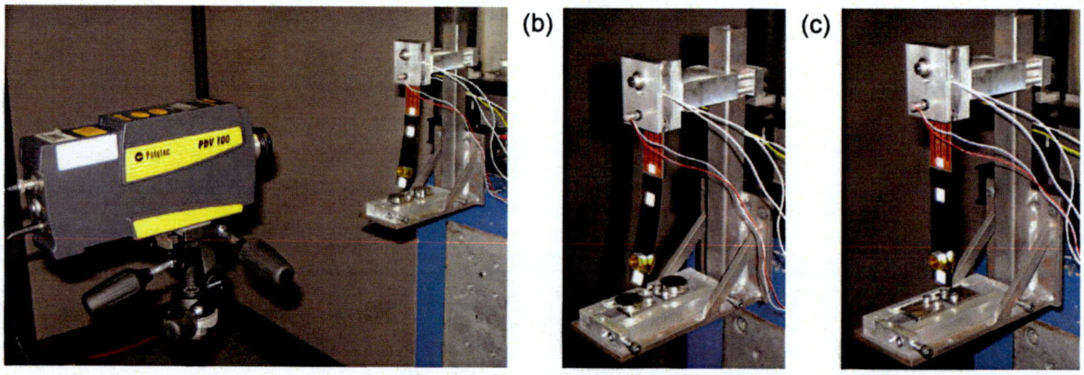

Figure 12. (a) Experimental setup used for investigating the power generation performance of the piezo-magneto-elastic structure; (b) Piezo-magneto-elastic configuration; (c) Piezo-elastic configuration

Figure 13 shows the comparison of the average steady-state power vs. load resistance graphs of piezo-magneto-elastic and piezo-elastic configurations at the frequencies of interest. Note that the excitation amplitudes (i.e. the base acceleration) of both configurations are very similar in all cases. As anticipated, the piezo-magneto-elastic generator gives an order of magnitude larger power at 5 Hz, 6 Hz and 8 Hz whereas the piezo-elastic configuration gives larger power only at 7 Hz (by a factor of 2). The average power outputs read from these graphs for the optimum values of load resistance are listed in Table 1.

Table 1. Comparison of the average power outputs of the piezo-magneto-elastic and piezo-elastic power generator configurations

Excitation frequency [Hz]	5	6	7	8
Piezo-magneto-elastic configuration [mW]	1.57	2.33	3.54	8.45
Piezo-elastic configuration [mW]	0.10	0.31	8.23	0.46

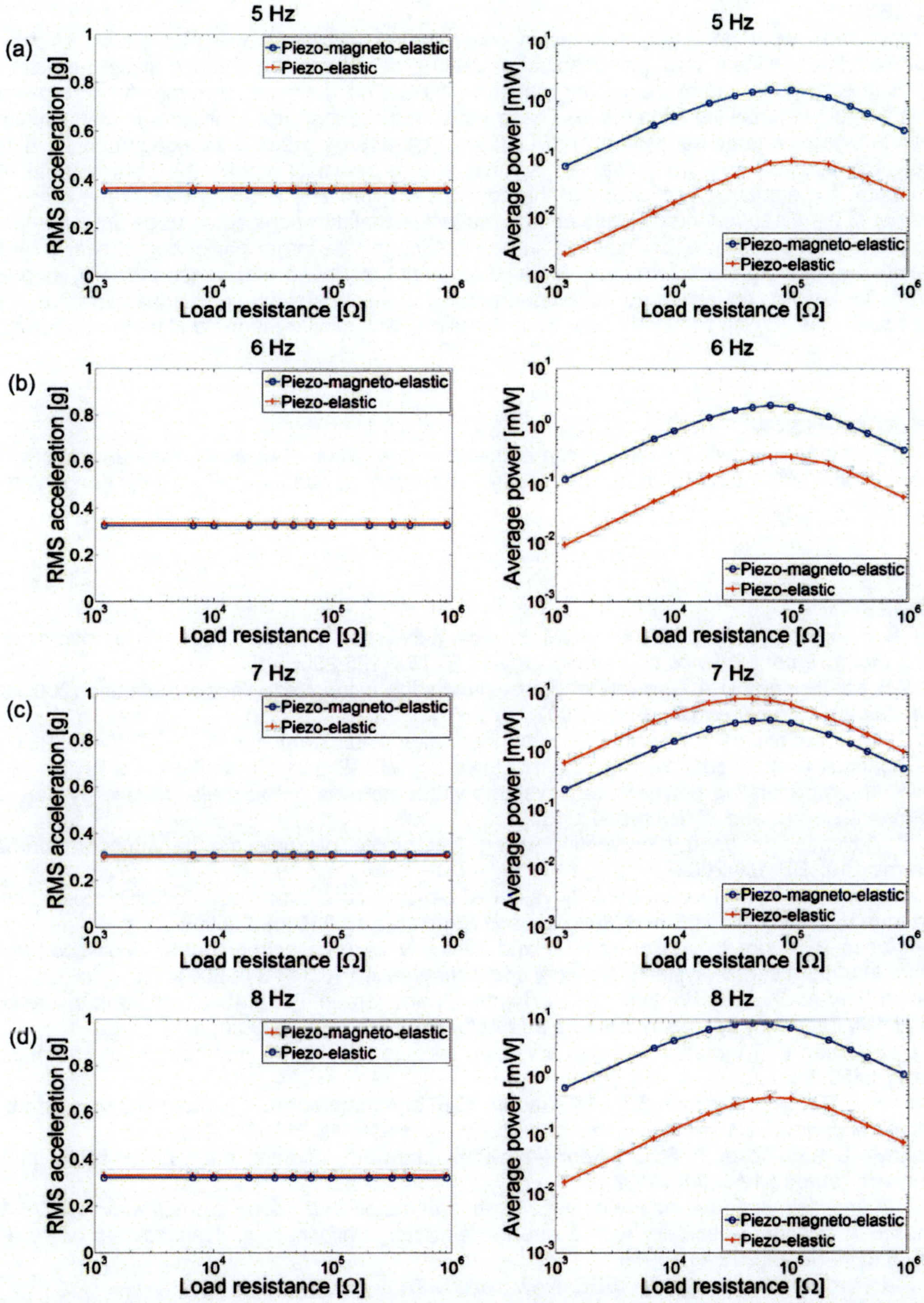

Figure 13. Comparison of the acceleration input and power output of the piezo-magneto-elastic and piezo-elastic configurations at steady state for a range of excitation frequencies: (a) 5 Hz; (b) 6 Hz; (c) 7 Hz; (d) 8 Hz

6. CONCLUSIONS

A novel broadband vibration energy harvester configuration utilizing a magneto-elastic structure is presented. First the nonlinear electromechanical equations describing the coupled system dynamics of the piezo-magneto-elastic energy harvester are given based on the lumped-parameter approximation considering the fundamental vibration mode. These equations are then used for time-domain simulations and the advantage of energy harvesting from LCO on high-energy orbits is demonstrated for different frequencies and comparisons are presented against energy harvesting from the conventional piezo-elastic structure. Experimental verification of the concept is given and it is shown that the open-circuit voltage output of the piezo-magneto-elastic energy harvester configuration can be three times larger than that of the piezo-elastic configuration, yielding an order of magnitude larger power output over a range of frequencies. The magneto-elastic structure is discussed for broadband piezoelectric energy harvesting here and it can easily be extended to electromagnetic, electrostatic and magnetostrictive energy harvesting techniques as well as to their hybrid combinations with similar magneto-elastic configurations.

ACKNOWLEDGEMENTS

The authors gratefully acknowledge the support of the Air Force Office of Scientific Research MURI under grant number F 9550-06-1-0326 "Energy Harvesting and Storage Systems for Future Air Force Vehicles" monitored by Dr. B. L. Lee.

REFERENCES

[1] Beeby S P, Tudor M J and White N M Energy harvesting vibration sources for microsystems applications *Measurement Science and Technology* **13**:R175-R195 2006

[2] Anton S R and Sodano H A A review of power harvesting using piezoelectric materials (2003-2006) *Smart Materials and Structures* **16**:R1-R21 2007

[3] Arnold D Review of microscale magnetic power generation *IEEE Trans. on Magn.* **43**:3940–3951 2007

[4] Cook-Chennault K A, Thambi N, Sastry A M Powering MEMS portable devices – a review of non-regenerative and regenerative power supply systems with emphasis on piezoelectric energy harvesting systems *Smart Materials and Structures* **17**:043001:1-33 2008

[5] Priya S Advances in energy harvesting using low profile piezoelectric transducers *Journal of Electroceramics* **19**:167–184 2007

[6] Erturk A and Inman D J An experimentally validated bimorph cantilever model for piezoelectric energy harvesting from base excitations *Smart Materials and Structures* **18** 025009 2009

[7] Glynne-Jones P, Tudor M J, Beeby S P and White N M An electromagnetic, vibration-powered generator for intelligent sensor systems *Sensors and Actuators A* **110**:344-349 2004

[8] Mitcheson P, Miao P, Start B, Yeatman E, Holmes A and Green T MEMS electrostatic micro-power generator for low frequency operation *Sensors and Actuators A* **115**:523-529 2004

[9] Wang L and Yuan F G Vibration energy harvesting by magnetostrictive material *Smart Materials and Structures* **17** 045009 2008

[10] Soliman M S M, Abdel-Rahman E M, El-Saadany E F and Mansour R R A wideband vibration-based energy harvester *Journal of Micromechanics and Microengineering* **18** 115021 2008

[11] Marinkovic B and Koser H Smart Sand—a wide bandwidth vibration energy harvesting platform *Applied Physics Letters* **94** 103505 2009

[12] Moon F C and Holmes P J A magnetoelastic strange attractor *J. of Sound and Vibration* **65** 275 1979

[13] Holmes P A nonlinear oscillator with a strange attractor, *Philosophical Transactions of the Royal Society of London, Ser. A* **292** 419 1979

[14] *IEEE Standard on Piezoelectricity*, IEEE, New York 1987

[15] Ottman G K, Hofmann H F, Bhatt A C and Lesieutre G A Adaptive piezoelectric energy harvesting circuit for wireless remote power supply *IEEE Transactions on Power Electronics* **17**:669-676 2002

[16] Guan M J and Liao W H On the efficiencies of piezoelectric energy harvesting circuits towards storage device voltages *Smart Materials and Structures* **16**:498-505 2007

[17] Shu Y C, Lien I C and Wu W J 2007 An improved analysis of the SSHI interface in piezoelectric energy harvesting *Smart Materials and Structures* **16** 2253–2264

Proceedings of the IMAC-XXVIII
February 1–4, 2010, Jacksonville, Florida USA
©2010 Society for Experimental Mechanics Inc.

Applicability Limits of Operational Modal Analysis to Operational Wind Turbines

D. Tcherniak[+], S. Chauhan[+], M.H. Hansen[*]

[+] Bruel & Kjaer Sound and Vibration Measurement A/S
Skodsborgvej 307, DK-2850, Naerum, Denmark

[*] Wind Energy Division, Risø DTU National Laboratory for Sustainable Energy,
Frederiksborgvej 399, DK-4000, Roskilde, Denmark

Email: dtcherniak@bksv.com, schauhan@bksv.com, mhha@risoe.dtu.dk

Nomenclature

f	Frequency
ω	Circular frequency
t	Time
$\{F(\omega)\}$	Vector of excitation force spectra (input)
$\{X(\omega)\}$	Vector of response spectra (output)
$[G_{FF}(\omega)]$, $[G_{XX}(\omega)]$	Cross-spectra matrices of forces and responses respectively
$\underline{S_u}(f)$	Auto- or cross-spectra of incident wind fluctuations
U	Mean wind speed
σ_u^2	Standard deviation of wind speed fluctuations
r_K	Radius of point K on a blade
$\kappa(\tau)$	Auto- or cross-correlation functions
$\gamma_{AB}^2(\omega)$	Coherence of aerodynamic forces (or wind speed fluctuations) at points A and B

Abstract

Operational Modal Analysis (OMA) is one of the branches of experimental modal analysis which allows extracting modal parameters based on measuring only the responses of a structure under ambient or operational excitation which is not needed to be measured. This makes OMA extremely attractive to modal analysis of big structures such as wind turbines where providing measured excitation force is an extremely difficult task.

One of the main OMA assumption concerning the excitation is that it is distributed randomly both temporally and spatially. Obviously, closer the real excitation is to the assumed one, better modal parameter estimation one can expect. Traditionally, wind excitation is considered as a perfect excitation obeying the OMA assumptions. However, the present study shows that the aeroelastic phenomena due to rotor rotation dramatically changes the character of aerodynamic excitation and sets limitations on the applicability of OMA to operational wind turbines.

The main purpose of the study is to warn the experimentalists about these limitations and discuss possible ways of dealing with them.

Introduction

The ability of Operational Modal Analysis to extract modal parameters of a structure under operation and without measuring excitation forces makes the technique extremely attractive to the application on wind turbines. First of all, a wind turbine is a huge structure where providing measured excitation is a challenging task. Secondly, since

OMA can be applied to a structure under operation, it promises to provide experimental values for *operational* natural frequencies and *operational* aerodynamic damping which differ from the ones at standstill and are otherwise difficult to obtain. At the same time a wind turbine is naturally subjected to aerodynamic forces due to wind turbulence, which have a broadband spectrum and properly excite wind turbine modes.

It's not incorrect to state that wind turbines have played a key role in developing OMA to a level where it currently stands. Though utilization of output only measurements for system identification goes back to 1970s, it wasn't till early 1990s that the science of OMA developed to an extent where it started getting applied to huge civil structures. This was result of research activities carried out in early 1990s at Wind Energy Research Organization, Sandia National Labs, USA; where by, while trying to understand the dynamics of wind turbines, James et al. proposed the NExT framework [1], that laid down the foundations of Operational Modal Analysis as it is understood now. As a part of this research, NExT was applied to vertical and horizontal axis wind turbines [1, 2], with varying degree of success. It was following this research that immense potential of output-only system identification techniques was realized, and subsequently applied to variety of civil structures including buildings, bridges, stadiums etc.

Having played such an important role in shaping OMA techniques, it's surprising to note that there wasn't much follow up to this pioneering work in understanding wind turbine dynamics by utilizing OMA techniques. It turns out that application of OMA to wind turbines is not a straight forward task due to a number of reasons, including presence of considerable aeroelastic effects, presence of rotational components in the excitation forces and time-varying nature of the wind turbine structure. These effects present considerable challenge for application of OMA techniques to wind turbines as they push the limits of these techniques by stretching the very basic assumptions that form the core of OMA.

A closer look onto OMA application to operational wind turbines reveals a number of inherent problems. First of all, the assumption of *structure time invariance* is violated. This assumption is distinctive to any kind of experimental modal analysis methods, and states that the object under test must not change during the test duration (or at least these changes should not be significant). In case of operational wind turbine this is not true: a wind turbine structure consists of substructures which move with respect to each other while the wind turbine operates: the nacelle rotates about the tower (which is characterized by the yaw angle); the rotor rotates about its axis; the blades' pitch may change depending on wind turbine type and operating conditions. Different methods can be applied to deal with time varying nature of the structure: For yaw and pitch, for example, periods of time when these angles do not change or change insignificantly can be selected for the analysis; a simple coordinate transformation can be performed in order to account for yaw angle; averaged characteristic value of pitch angle can be used while accepting that the obtained modal parameters are "smeared" due to blades pitching during the test duration. It is more difficult to deal with rotor rotation. Including rotor rotation into the equations of motion of entire wind turbine causes the mass, stiffness and gyroscopic matrices to be dependent on time. Formulating and solving the corresponding eigenvalue problem yields to time-dependent eigenvalues and eigenvectors, which do not have a meaning as modal frequencies, damping and mode shapes in traditional sense. Fortunately, by applying Coleman transformation [4-6], one can get rid of time dependencies in the equation of motion and obtain meaningful modal parameters. The combination of Coleman transformation and Operational Modal Analysis is demonstrated in [7].

Another inherent problem of application of OMA to operational wind turbines is the violation of OMA assumption concerning the operational excitation forces. OMA sets three quite specific requirements to the excitation: the forces should have broadband frequency spectra; they have to be distributed over entire structure, and they have to be uncorrelated. Seemingly, the excitation due to wind turbulence is ideal for OMA. Indeed, the forces due to wind turbulence have almost uniform broadband spectra; they are uncorrelated and obviously excite entire structure. However, the effect of rotor rotation dramatically changes the nature of aerodynamic forces. First of all, the shape of the spectra transforms from being flat to a curve with prominent peaks at rotation frequency and its harmonics; the peaks' shape is not sharp but characterized by thick tails [8, 9]. Secondly, the forces appear to be quite correlated around the rotation frequency and its harmonics. The abovementioned reasons make the aerodynamic excitation much less suitable for OMA. The presented paper addresses this phenomenon.

The paper is built as follows: Section 1 briefly reminds the reader the main ideas and assumptions OMA is based on. Section 2 uses the analytical model of wind excitation based on von Karman spectrum, and derives the spectra of wind excitation forces and coherence between them; Section 3 demonstrates the same using the results of numerical simulation of wind/blades interaction of wind turbine under operation. In Section 4, the results are discussed and some conclusions are drawn.

1. Operational Modal Analysis: theory and main assumptions

OMA is a system identification technique based only on measured output responses. It does not require input force information but makes certain assumptions about the nature of the input excitation forces. Like any other technique, adherence to basic assumptions is the key to successful application of OMA techniques. Importance of these assumptions can be gauged by the fact that modal parameters obtained using OMA are obviously affected depending on how closely the actual conditions resemble the one supported by these basic assumptions. These assumptions are listed below:

1. Power spectra of the input forces are assumed to be broadband and smooth;
2. The input forces are assumed to be uncorrelated;
3. The forces are distributed over entire structure

In other words, *the excitation is assumed to be randomly distributed both temporally and spatially.*

Expression (1) gives the mathematical relationship between the vector of measured responses, $\{X(\omega)\}$ and vector of input forces, $\{F(\omega)\}$ in terms of the frequency response function (FRF) matrix $[H(\omega)]$ [3]:

$$\{X(\omega)\} = [H(\omega)]\{F(\omega)\}; \tag{1}$$

$$\{X(\omega)\}^H = \{F(\omega)\}^H [H(\omega)]^H . \tag{2}$$

Now multiplying (1) and (2)

$$\{X(\omega)\}\{X(\omega)\}^H = [H(\omega)]\{F(\omega)\}\{F(\omega)\}^H [H(\omega)]^H , \tag{3}$$

or with averaging,

$$[G_{XX}(\omega)] = [H(\omega)][G_{FF}(\omega)][H(\omega)]^H , \tag{4}$$

where $[G_{XX}(\omega)]$ is the matrix of output power spectra and $[G_{FF}(\omega)]$ is the input force power spectra matrix.

From the first two assumptions, it follows that

$$[G_{FF}(\omega)] \propto [I], \tag{5}$$

therefore the output power spectra $[G_{XX}(\omega)]$ is proportional to the product $[H(\omega)][H(\omega)]^H$ and the order of output power spectrum is twice that of the frequency response functions. Thus $[G_{XX}(\omega)]$ can be expressed in terms of frequency response functions as

$$[G_{XX}(\omega)] \propto [H(\omega)][I][H(\omega)]^H . \tag{6}$$

Partial fraction form of G_{XX} for particular locations p and q is given as

$$G_{pq}(\omega) = \sum_{k=1}^{N} \frac{R_{pqk}}{j\omega - \lambda_k} + \frac{R^*_{pqk}}{j\omega - \lambda^*_k} + \frac{S_{pqk}}{j\omega - (-\lambda_k)} + \frac{S^*_{pqk}}{j\omega - (-\lambda^*_k)} , \tag{7}$$

where R_{pqk} and S_{pqk} are k^{th} mathematical residue terms and are not to be confused with residue terms obtained using FRF based model as these term do not contain modal scaling information (since input force is not measured). It is important to note that this expression shows that power spectra *contains all information needed to define the modal model* of the system (except for modal scaling factor), *provided the loading assumptions are true.*

2. Frequency analysis of aerodynamic forces

In wind turbine engineering [8], it is typical to separate the loading due to the steady wind and loading due to wind speed fluctuation (i.e. turbulence). The first one is often called *deterministic* load component while the second is *stochastic.* The first component is characterized by the mean wind speed, which is considered to be time-constant in the time scale over about 10 minutes. In this study, we consider it not possessing any broadband frequency content. The second component *has* broadband frequency content, and therefore has a key importance for Operational Modal Analysis due to its ability to provoke modal behavior of the wind turbine structure. The presented paper concerns only the stochastic loading.

The source of the stochastic loading is turbulence, a fluctuation in wind speed on a relatively fast time-scale typically less than about 10 minutes [8]. In other words, we assume the fluctuations of wind speed have a zero mean when averaging over 10 minutes. There are two main sources of turbulence: flow disturbance due to topographic features and thermal effects causing air masses to move vertically. These are obviously complex processes; therefore the description of turbulence is typically developed using its statistical properties.

Turbulence frequency spectrum is an important property of turbulence. There are several turbulence models used in wind turbine engineering; in this study we will refer to two of them; the first one is described by Kaimal spectrum (used in Section 3), the second one – by von Karman spectrum. In this section we use the formulation suggested by von Karman [10]:

$$\frac{f S_u(f)}{\sigma_u^2} = \frac{4L_u^x}{\overline{U}\left(1+70.8\left(\frac{f L_u^x}{\overline{U}}\right)^2\right)^{5/6}}, \tag{8}$$

where f is frequency, $S_u(f)$ is the autospectral density of the turbulence (its longitudinal component which is denoted by symbol u), σ_u is the standard deviation of wind speed variations about mean wind speed in longitudinal direction $\overline{U}$. Standard deviation σ_u and mean speed $\overline{U}$ are linked via *turbulence intensity* $I_u = \sigma_u / \overline{U}$. L_u^x is a turbulence *length scale*. Different national standards provide empiric expressions defining the values for the turbulence intensity and length scale; these values mainly depend on the elevation above the ground and the surface roughness. The blue line on Figure 1 shows a typical spectrum calculated according to (1) with L_u^x = 64.4 m (corresponding to the elevation greater then 30 m above the ground) and $\overline{U}$ = 8 m/s^2.

As it can be seen from the plot, most of the energy is distributed on the very lower frequencies, much below typical rotation frequencies (here set equal to 0.202Hz and denoted by “1p” on the frequency axis). On the higher frequencies the turbulent energy tend to dissipate as a heat, and the spectral density asymptotically approaches the limit $f^{-5/3}$ Hz.

As it was mentioned in Section 1, Operational Modal Analysis sets a number of requirements on structure excitation. In the case of wind turbine, the main excitation is due to wind-structure interaction; therefore it is important to understand the relation between the turbulence and the resulting aerodynamic forces. As it is stated in [8], it is usual to assume a *linear relation* between the *fluctuation* of the wind speed u incident on the aerofoil and the resulting loading *fluctuations* $\Delta L \propto u$. This assumption is correct for low wind speeds and becomes inaccurate for pitch regulated turbines as wind speed approaches cut-out value and breaks down for stall-regulated machines when wind speed can cause stall. In this section we will assume low wind speeds and therefore the linear relation between the wind speed fluctuation and the resulting loading, so the results will be given for wind speed fluctuation u, implying the forces behave similarly.

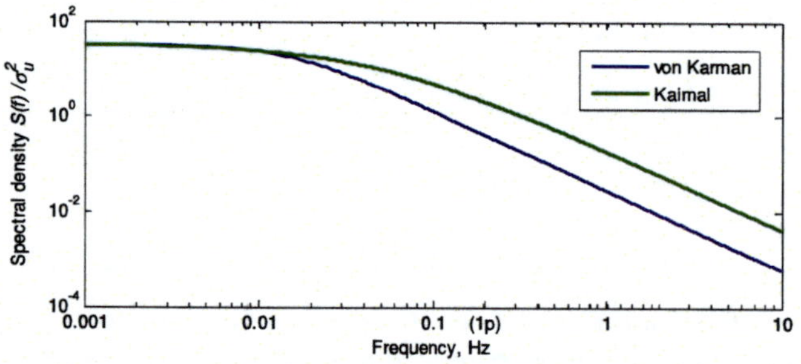

Figure 1: Normalized spectra: according to von Karman (blue) and Kaimal (green)

Following [8] and assuming that the fluid is incompressible, the turbulence is homogeneous and isotropic and using Taylor's hypothesis of “frozen turbulence”, it is possible to obtain an analytical expressions for cross-correlation function of wind speed computed for points A and B on the blades (Figure 2).

In brief, the derivation is based on the use of Wiener-Khinchin theorem on power spectrum (8) which yields to analytical expression for the autocorrelation function at any fixed point in space; and then computing the cross-correlation function between points A_t and $B_{t+\tau}$ (Figure 2). Using the hypothesis of “frozen turbulence”, instantaneous wind fluctuation at $B_{t+\tau}$ is assumed to be equal to the wind fluctuation at point B_t'. The derived expression for correlation along vector s at $t=0$ approximates the cross-correlation function between points A_t and $B_{t+\tau}$: $\kappa_u(\vec{s},0) = \kappa_u^o(r_1, r_2, \tau)$.

$$\kappa_u^o(r_1, r_2, \tau) = $$
$$= \frac{2\sigma_u^2}{\Gamma(\frac{1}{3})}\left(\frac{s/2}{1.34L_u^x}\right)^{1/3}\left(K_{1/3}\left(\frac{s}{1.34L_u^x}\right) - \frac{s/2}{1.34L_u^x}\left(\frac{r_1^2 + r_2^2 - 2r_1 r_2\cos(\Omega\tau + k\cdot 2\pi/3)}{s^2}\right)K_{2/3}\left(\frac{s}{1.34L_u^x}\right)\right) \tag{9}$$

where interpretations for s, r_1, r_2 can be found on Figure 2a,b; K_ν is Bessel function of second kind; and $k = 0$ if points A and B are on the same blade, $k = -1$ if blade with point A is moving before the blade with B and $k = 1$ for the opposite situation. As it can be seen from the triangle on Figure 2c,

$$s^2 = (\bar{U}\tau)^2 + r_1^2 + r_2^2 - 2r_1 r_2\cos(\Omega\tau + k\cdot 2\pi/3). \tag{10}$$

Figure 3a shows autocorrelation function computed for the fixed point on the blade (i.e. points A and B coincide, $r_1 = r_2$ and $k = 0$) for different radii. The peaks at multiples of the period of rotor revolution T are due to high correlation of the incident wind speed with itself when a blade passes the same region of the rotor swept area.

Spectra of aerodynamic forces

Application of the Wiener-Khinchin theorem to expression (9) yields the power spectrum density (PSD) of the incident wind speed fluctuations; there is no implicit expression for the resulting integral, instead it can be approximated and estimated numerically. The PSDs of the corresponding autocorrelation functions are shown on Figure 3b.

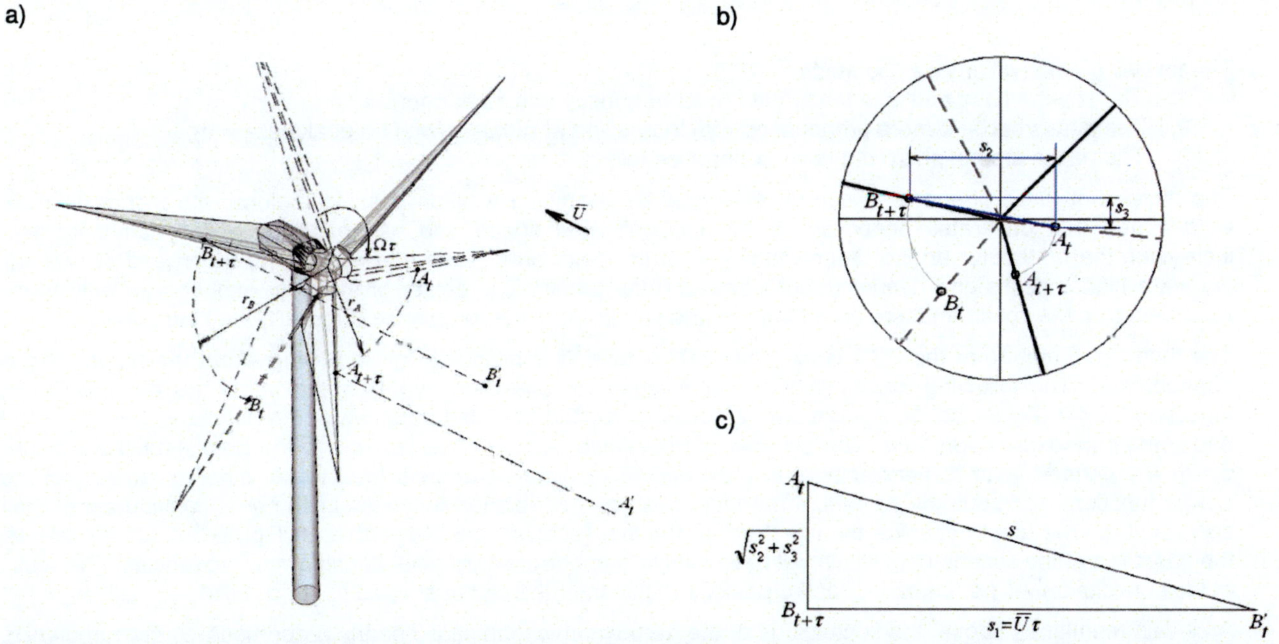

Figure 2. Points A and B are located at radii r_1 and r_2 respectively. A_t, B_t are the positions of the corresponding points at time t. a) Geometry of the wind turbine; b) View along the wind speed; c) Position triangle used for cross-spectrum calculation.

a)

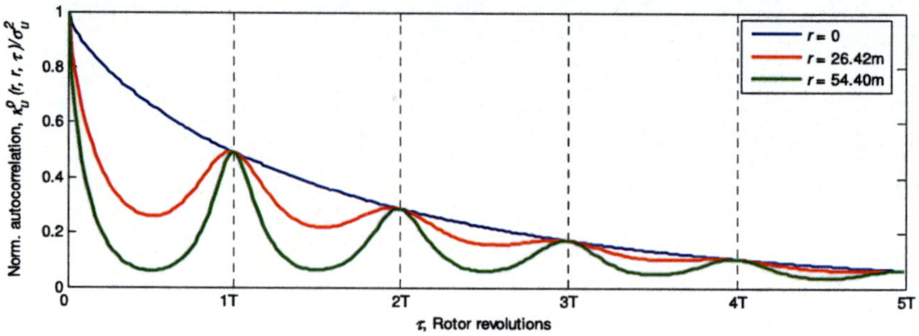

b)

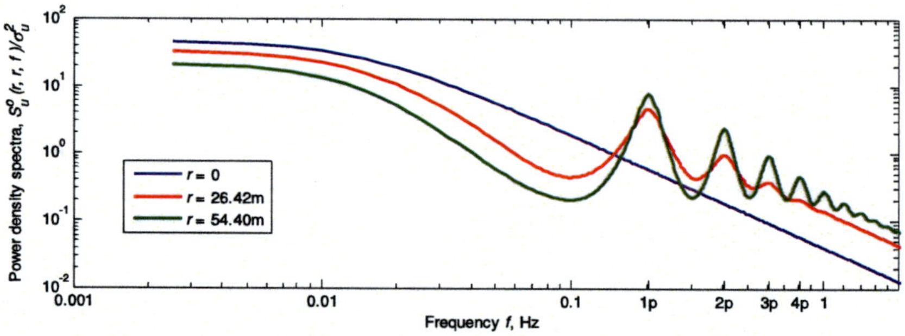

Figure 3. a) Normalized autocorrelation functions computed for the *fixed* point on the blade ($r_2 = r_1$ and $k = 0$), on the different radii (0, 26.42 and 54.40 m); b) Normalized PSD calculated for the autocorrelation functions shown above.

The following observations can be made:

1. The spectra have peaks at the fundamental frequency and its harmonics;
2. The peaks become more pronounced with increasing of radius r (with no peaks at $r = 0$);
3. The peaks are not sharp but have rather thick tails.

The physics of these phenomena can be explained by imagining a blade passing regions with higher or lower wind speed that generates periodicity of the incident wind speed and hence of the aerodynamic loading; increasing the ratio between the tangential speed of the point on a blade and the mean wind speed $\Omega r / \overline{U}$ will generate higher peaks and deeper troughs between the peaks. The center of the rotor experiences only mean wind therefore the spectrum there coincides with the spectrum of the turbulence (Figure 1, the blue curve).

The third listed feature of the PSD is the peak's thick tails. It is difficult to get a feeling about the nature of this phenomenon. The following explanation can be suggested: one can roughly approximate the autocorrelation function at $r \neq 0$ (Figure 3a) by a product of a decaying function like the blue curve (the autocorrelation at $r = 0$) and some periodic function. The latter consists of the infinite number of terms (due to the Bessel functions in (9)) which are periodic (due to periodicity of s, (10)) and have different periods (due to the different powers of the cosine function). In frequency domain, a product of two autocorrelation functions becomes a convolution of their spectra [11]. One has to imagine the spectrum of the first function (the blue curve on Figure 3b) convolved with the spectrum of the second one, which will have infinite number of sharp peaks at different frequencies. The result of the convolution will be "smearing" of the peaks into the smooth spectra shaped as "thick tails".

As it was mentioned above, the fluctuations of the aerodynamic loading is linearly proportional to the turbulence, meaning that the PSD of the aerodynamic forces will have the same shape as curves shown on Figure 3b. Thus, we can see that the excitation spectra are not flat in the frequency range of interest, and this therefore violates the first OMA assumption.

Correlation between aerodynamic forces

The second important OMA assumption is that the forces acting at different points of a structure must be uncorrelated (see Section 1). In this subsection we will estimate the correlation of the excitation forces by calculating the coherence between the wind speed fluctuations at different points on the same and different blades.

Coherence γ_{AB}^2 between incident wind speed fluctuation at points A and B on the rotating blades is

$$\gamma_{AB}^2(f) = \frac{\left|S_u^{AB}(f)\right|^2}{S_u^{AA}(f)S_u^{BB}(f)}, \tag{11}$$

where S_u^{AA} and S_u^{BB} are autospectra of wind turbulence at points A and B respectively and S_u^{AB} is cross-spectra between these points. The auto- and cross-spectra can be calculated from auto- and cross-correlation functions (9) applying Wiener-Khinchin theorem. Figure 4 shows the auto- and cross-correlation functions calculated at different points on different blades, corresponding auto- and cross-spectra, and the coherence.

The correlation functions are presented on Figure 4a. Cross-correlation functions computed for the points on different blades pass ahead the autocorrelation function: indeed, if for example, blade #1 experiences a gust at some segment of the rotor swept area, it will take it a full revolution (and time T) to experience the same gust again. However, for blade #2 which follows blade #1, it will take only $T/3$ to experience the same gust, and for blade #3 it will happen after $2T/3$. This explains the corresponding time lags of the black and magenta curves on Figure 4a.

The auto- and cross-spectra are obtained numerically for the corresponding auto- and cross-correlation functions; their magnitudes are shown on Figure 4b[1]. The cross-spectra for the cases when blade A follows blade B and B follows A are the same.

Coherence functions between the wind turbulence at points A and B if the points are located on different blades, are presented on Figure 4c. Similar to the spectra, the coherences have peaks at rotation frequency and its harmonics; the peaks are more pronounced for low harmonics and gradually decay at higher frequencies. Further, the peaks are not sharp but have rather thick tails; this makes the troughs between them quite narrow.

Thus, here one can claim the violation of the second OMA assumptions concerning the correlation between the excitation forces. As one can see, the assumption is strongly violated in the frequency range between 0.5p - 4.5p, leaving only narrow deeps between the peaks where the assumption is valid.

3. Analysis of the synthesized time data

As it was mentioned in Section 2, the presented analytical approach is only suitable for weak wind where the linear relation between aerodynamic force fluctuations and wind turbulence is valid. For higher wind speeds the analysis is usually performed in time domain by numerical simulation of wind-blade interaction.

Similar to [12], the simulations were performed using the nonlinear aeroelastic multi-body code HAWC2 [13]. A turbulence box with resolution 32 x 32 x 16384 was generated based on Kaimal turbulence spectrum with mean wind speed 18 m/s and turbulence intensity 5%. Figure 1 (green line) shows the corresponding spectrum calculated for this speed and L_{1u} = 150 m. The turbulence box corresponded to 12000 s of simulation. Aerodynamic forces at radii 26.42, 39.95 and 54.40 m in X- and Y- directions were calculated as functions of time; the analysis was done on the generated time histories, for X (along the wind) direction only.

---

[1] Note that the numerical approach suggested in [8] is suitable here since it assumes the symmetry of the correlation functions about $t = 0$. This is correct for auto-correlation functions and cross-correlations between the points on the same blade. The cross-correlation functions between the points on different blades (i.e. for $k \neq 0$) are not symmetric, see Figure 4a.

a)

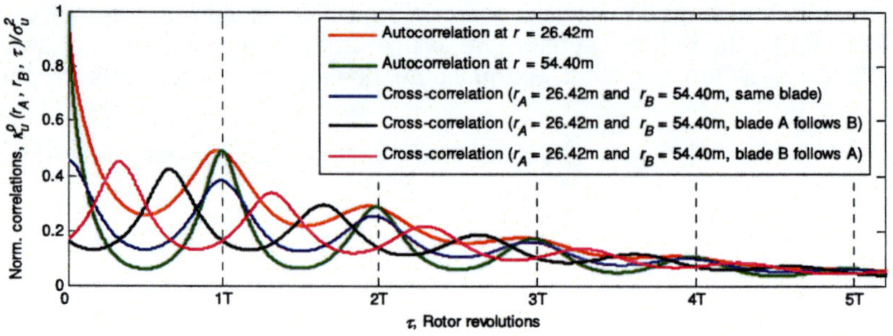

b)

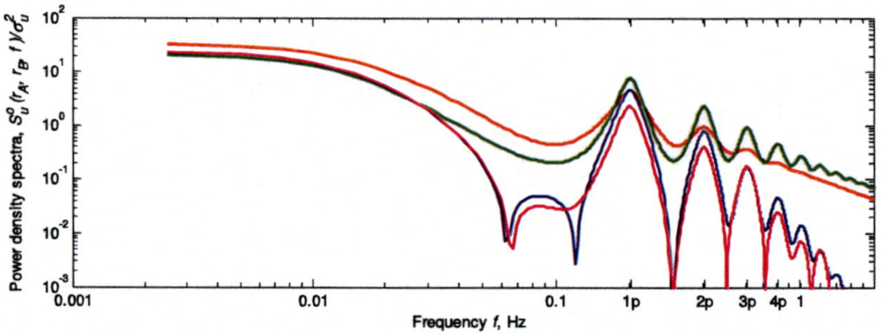

c)

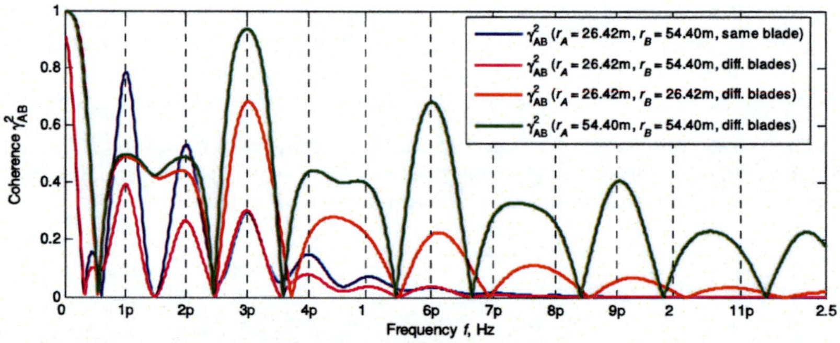

Figure 4. a) Auto- and cross-correlation functions; b) auto- and cross-spectra; c) coherence.

Figure 5 shows the autospectra of the aerodynamic forces acting in X-direction, and Figure 6 shows the coherence between aerodynamic forces simulated at different points on the blades.

Obviously, the results obtained analytically and numerically are not directly comparable. First of all, for analytical case the spectra and coherence of *wind fluctuations* are shown where for the numerical case the resulting *aerodynamic forces* are presented. Secondly, the wind speed is different (modest 8 m/s for analytical case vs. strong 18 m/s wind for the numerical case). Thirdly, two different wind turbulence spectra are used: von Karman for the analytical case and Kaimal for the numerical simulations. Despite this, qualitatively, the shape of the spectra and coherence functions look very similar. In both case they are characterized by peaks at rotation frequency and its harmonics, with their heights decreasing with frequency increase. At low frequencies (1p-4p) the peaks clearly exhibit the "thick tails".

The shape of the coherence functions computed for the two cases is more different: in the numerical case the sharp peaks present even on high harmonics. Otherwise, there are similarities, too. For example, the coherence between the points on the same radius on the different blades (Figure 6c,d) is generally higher compare to the one between the points at different radii (Figure 6a,b).

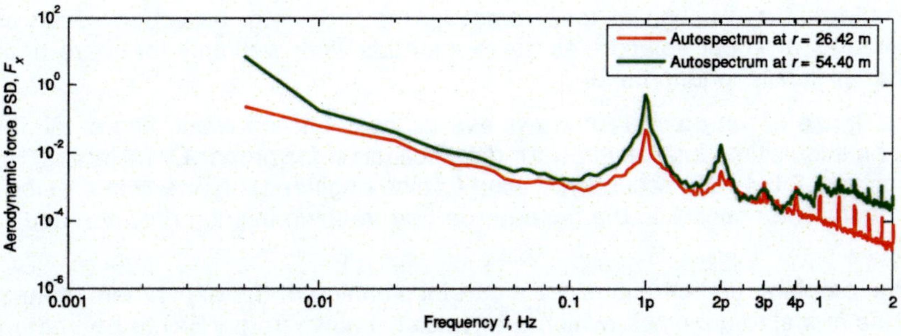

Figure 5. Power spectral density of the simulated aerodynamic forces in X direction.

4. Discussion and Conclusion

In the study it is demonstrated both analytically and numerically, for weak and strong wind excitation that for *operational* wind turbine:

1) The spectra of aerodynamic forces are not flat but are characterized by peaks at rotational frequency and few lower harmonics. The peaks are not sharp but have thick tails.
2) The forces acting at different points on the blades are highly correlated on the rotational frequency and its lower harmonics.

This means that two important OMA assumptions are not satisfied in a number of frequency regions around the rotational frequency and its lower harmonics. Thus one must not expect OMA to provide correct results in these frequency regions. Some important modes of modern wind turbines (e.g. drivetrain mode, first tower modes, first flapwise modes) are located in these regions for the nominal rotational speeds [14], so one has to be careful trying to identify them using OMA.

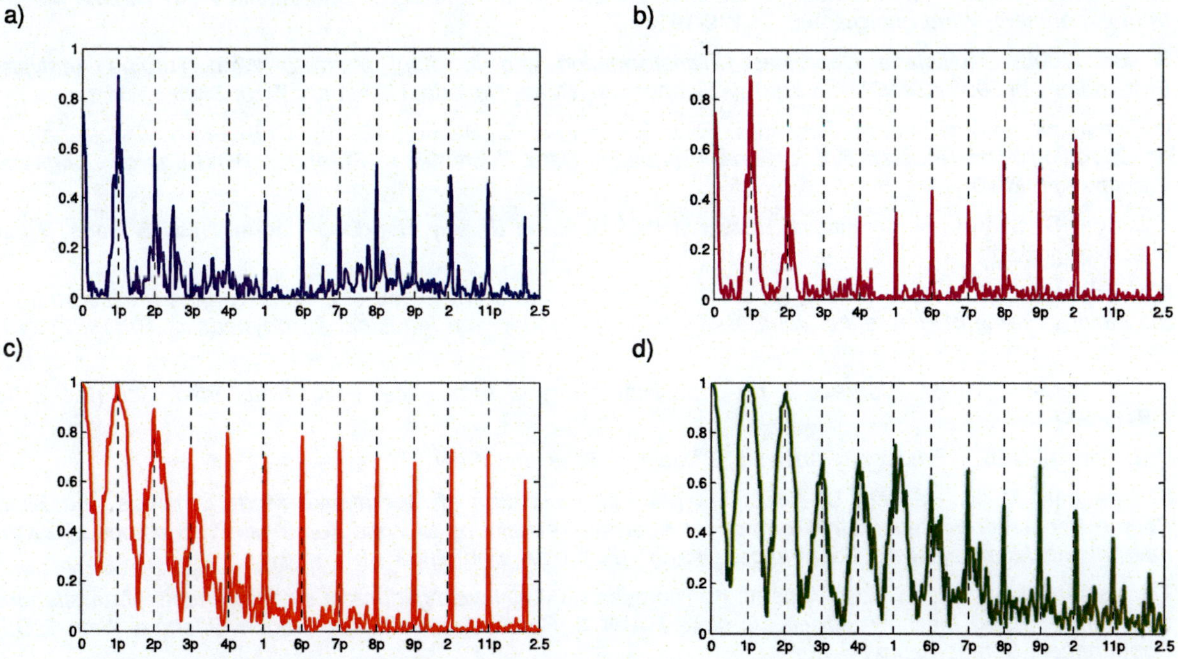

Figure 6. Coherence of the simulated aerodynamic forces. Same colour scheme as on Figure 4c.

It is important to note that application of *tone removing* methods (e.g. based on synchronous averaging, [15]) cannot be considered as a proper solution, as these methods work well only for sharp peaks but will not help in this case due to the "thick tails" phenomenon.

From a first glance, a use of run-up and run-down events looks attractive but, first of all, these events are rather short compare to the acquisition time required for data collection for proper OMA (about 10 minutes if the lowest frequency of interest is 0.2-0.4 Hz). Secondly, a wind turbine engineer is typically interested in the dependency of modal parameter on the rotor speed; in the case of run-up/run-down events, only averaged modal characteristics can be obtained.

In order to solve the problem, one can consider a careful planning of the experiment, constructing the test matrix in a way to avoid the modal frequencies (which are typically known from FEA) to be in the vicinity of rotor speed and its lowest harmonics. This means that only few modes can be estimated with higher degree of confidentiality for a given rotor speed, while another rotor speed will be suitable for another set of modes.

Amongst recently suggested methods, operational modal analysis based on transmissibility functions [16] appears very appealing. One of the main advantages of this method is its insensitivity to colored excitation spectra. However, so far this method is still under development and is not ready for industrial applications.

Literature

[1] G.H. James, T.G. Carne, J.P. Lauffer (1995), *The Natural Excitation Technique (NExT) for Modal Parameter Extraction from Operating Structures, Modal analysis: The International Journal of Analytical and Experimental Modal Analysis 10, 260-277.*

[2] G.H. James (1994), *Extraction of Modal Parameters from an Operating HAWT using the Natural Excitation Technique (NExT), Proceedings of the 13th ASME Wind Energy Symposium, New Orleans, LA.*

[3] J. Bendat, A. Piersol (1986), *Random Data: Analysis and Measurement Procedures*, 2nd edition, Wiley, New York.

[4] M.H. Hansen (2003), *Improved Modal Dynamics of Wind Turbines to Avoid Stall-induced Vibrations, Wind Energ.* 2003; **6**: 179-195

[5] M.H. Hansen (2006), *Two Methods for Estimating Aeroelastic Damping of Operational Wind Turbine Modes from Experiment, Wind Energ.* 2006; **9**: 179-191

[6] G. Bir (2008), *Multiblade Coordinate Transformation and its Application to Wind Turbine Analysis, Proceedings of 2008 ASME Wind Energy Symposium, Reno, Nevada, USA, Jan. 7-10, 2008*

[7] D. Tcherniak, S. Chauhan, M. Rosseti, I. Font, J. Basurko, O. Salgado (2010), *Output-only Modal Analysis on Operating Wind Turbines: Application to Simulated Data*, Submitted to *European Wind Energy Congress, Warsaw Apr. 2010*

[8] T.Burton, D. Sharpe, N. Jenkins, E.Bossanyi (2001), *Wind Energy Handbook*, John Wiley & Sons, West Sussex, England

[9] L. Kristensen, S. Frandsen (1982), *Model for Power Spectra of the Blade of a Wind Turbine Measured from the Moving Frame of Reference, Journal of Wind Engineering and Industrial Aerodynamics, 10 (1982) 249-262*

[10] T. von Karman (1948), *Progress in the statistical theory of turbulence, Proc. Natl. Acad. Sci. (U.S.), 34 (1948) 530-539*

[11] R.B. Randall (1987), *Frequency Analysis*, 3rd edition, Bruel and Kjaer

[12] S. Chauhan, M. H. Hansen, D. Tcherniak (2009), *Application of Operational Modal Analysis and Blind Source Separation /Independent Component Analysis Techniques to Wind Turbines, Proceedings of XXVII International Modal Analysis Conference, Orlando (FL), USA, Feb. 2009*

[13] T.J. Larsen, H.A. Madsen, A.M. Hansen, K. Thomsen (2005), *Investigations of stability effects of an offshore wind turbine using the new aeroelastic code HAWC2, Proceedings of Copenhagen Offshore Wind 2005, Copenhagen, Denmark, p.p.25–28.*

[14] M.H. Hansen (2007), *Aeroelastic Instability Problems for Wind Turbines, Wind Energ.* 2007; **10**: 551-577

[15] B. Peeters, B. Cornelis, K. Janssens, H. Van der Auweraer (2007), *Removing disturbing harmonics in operational modal analysis, Proceedings of International Operational Modal Analysis Conference, Copenhagen, Denmark, 2007*

[16] C. Devriendt, T. De Troyer, G. De Sitter, P. Guillaume (2008), *Automated operational modal analysis using transmissibility functions, Proceedings of International Seminar on Modal Analysis, Leuven, Belgium Sep. 2008*

Proceedings of the IMAC-XXVIII
February 1–4, 2010, Jacksonville, Florida USA

Excitation Methods for a 60 kW Vertical Axis Wind Turbine

D. Todd Griffith
Randy L. Mayes
Patrick S. Hunter

Sandia National Laboratories*
Albuquerque, NM 87185-0557

A simple modal test to determine the first tower bending mode of a 60 kW (82 feet tall) vertical axis wind turbine was performed. The minimal response instrumentation included accelerometers mounted only at easily accessible locations part way up the tower and strain gages near the tower base. The turbine was excited in the parked condition with step relaxation, random human excitation, and wind excitation. The resulting modal parameters from the various excitation methods are compared.

Introduction

Resonant behavior is one of the drivers in the design of wind turbine structures. Wind turbines are the largest rotating structures in the world, and resonant conditions can exist near the operating speed and at its multiples. Finite element models, which include the rotational effects, can be used to evaluate resonance; however, especially for new designs field modal tests are necessary for validation. In this work, a field modal test was performed to assess a potential resonant condition identified from finite element analysis.

The paper is organized as follows. First, the turbine is described. Then, the pre-test analysis performed to identify the modes of the rotating turbine is discussed. The test design and execution are then discussed including a description of the instrumentation and excitation methods. Finally, the data analysis and test outcomes for the parked case is described. Wind excitation was also used while the turbine was rotating and producing power; however, analysis of this data is not reported in this paper.

Description of Test Article and Objective

The test article is a vertical axis wind turbine (VAWT) located in Clines Corners, New Mexico. The turbine is a prototype being developed by VAWTPower Management, Inc [1] and is termed the VP60. A photo of the field turbine is shown in Figure 1.

The VP60 prototype is rated at 60 kW, and is 82 feet tall. The lower housing of the turbine is fixed (non-rotating), is 31 feet tall, and has a diameter of 66 inches. A ladder is welded onto the outside of the housing for climbing access. The turbine has three blades which are connected to the upper rotating shaft (torque tube) of the turbine. The shaft is supported by bearings at two locations: (1) the lower bearing is inside the lower housing at approximately 10 feet from ground level, and (2) the upper bearing is at the top of the housing. The turbine generator and gearbox are located at the lower end of the torque tube inside the housing.

---

* Sandia is a multiprogram laboratory operated by Sandia Corporation, a Lockheed Martin Company for the United States Department of Energy's National Nuclear Security Administration under contract DE-AC04-94AL85000.

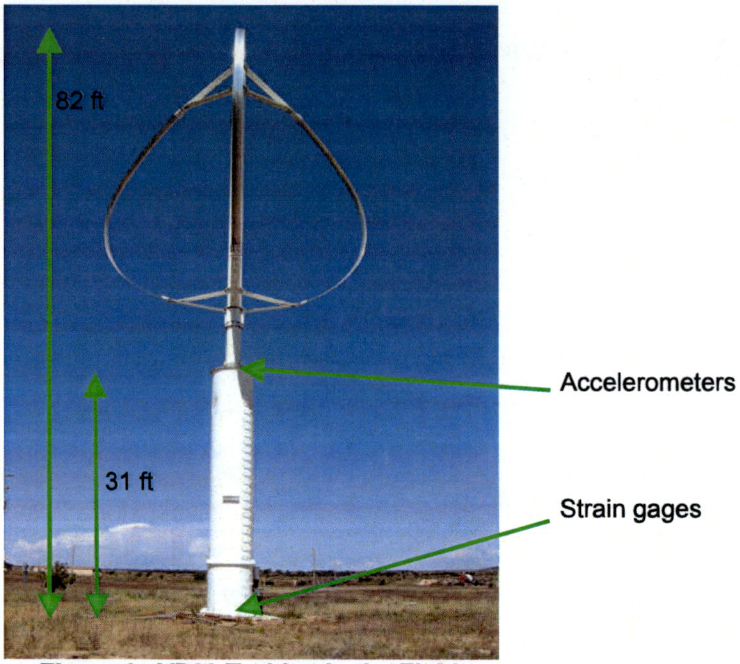

Figure 1. VP60 Turbine in the Field

Pre-test Finite Element Analysis

A pre-test finite element analysis was conducted to assess the potential for resonant conditions for this turbine design. This requires not only consideration of the parked condition for the modal analysis, but also modal analysis including the effects of rotation. Several decades ago, a code was developed at Sandia Labs to perform this analysis for vertical axis wind turbines [2]. First, a finite element model of the turbine is developed in Nastran. Typically, a turbine is modeled using beam-type elements and concentrated masses. Then, the rotational effects of the rotor are included through the use of DMIG (Direct Matrix Input at a Grid) cards. These rotational effects include tension stiffening, centrifugal softening, and coriolis terms. The inclusion of these effects results in modifications to the damping and stiffness matrices of the non-rotating model of the structure as noted in Reference 2. A complex eigenvalue analysis is then used to compute the modes of the turbine. For a range of operating speeds of interest, the rotational effects for a particular constant operating speed are computed and included in the system matrices with the modes computed using the complex eigen solver. The result is a plot of natural frequency versus operating speed, which is typically referred to as a Campbell diagram or simply as a fan plot.

In the initial investigation, a baseline model was developed based on design drawings for the turbine. The geometry for the baseline finite element model is shown in Figure 2. The turbine housing, shafts, and blades were modeled using beam elements. Concentrated masses were added at flange locations and other connection points. All six degrees of freedom at the base of the lower housing were fixed. Additional constraints were applied at the lower and upper bearing locations of the rotor torque tube, and at the locations in which the blades are attached to the torque tube at their two ends. It should be noted that the drive train dynamics were not modeled in detail, thus all degrees of freedom were constrained at the lower bearing.

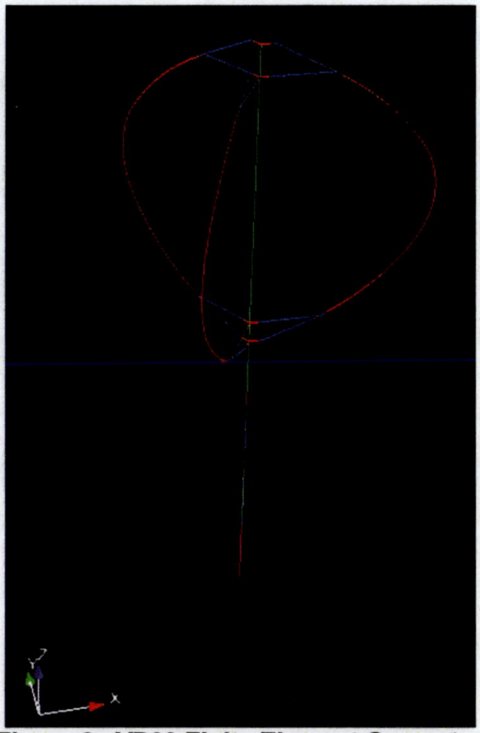

Figure 2. VP60 Finite Element Geometry

Subsequent to this analysis, efforts were made to refine the finite element model. Visual observations of the turbine when brought to an abrupt stop indicated that the frequency of first tower bending mode for the parked condition was about 0.9 Hz, which was about 10% lower than predicted in the baseline model. It was decided that an initial model refinement should consider the compliance at the two bearing locations; therefore, lateral direction springs (x and y directions) were included in the model and were valued such that the first tower mode frequency matched the visual observation. Then, a new fan plot was generated to determine the location of potential resonance conditions with respect to operating speed. A plot of the mode shape for the first tower bending mode is provided in Figure 3.

In order to evaluate the finite element model, a simple and quick test was designed to measure the first tower bending mode. The remainder of this paper describes the design and execution of the tests as well as the data analysis.

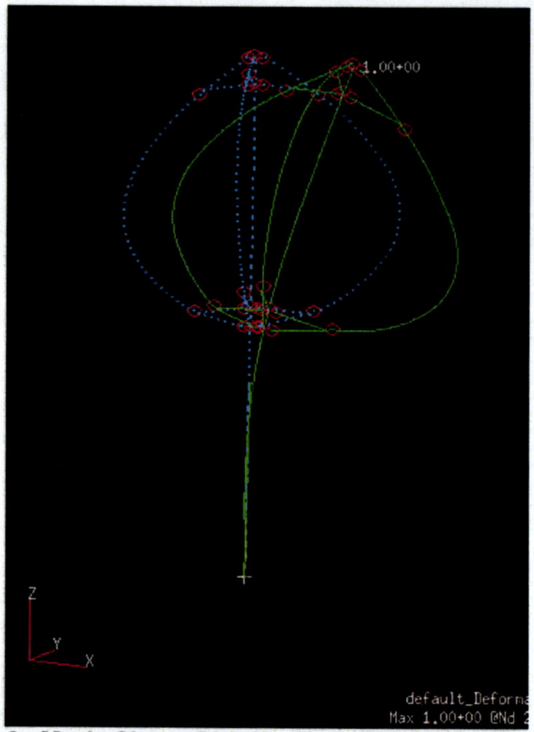

Figure 3. Mode Shape Plot for First Tower Bending Mode

Test Design and Test Execution

The primary objective of the field modal test was determination of the frequency of the first tower bending mode. Clearly from Figure 3, the largest motion for this mode occurs at the top of the tower. Ideally, one would place accelerometers in this location. However, this would require a man-lift capable of reaching 82 feet and a number of accompanying safety procedures to be evaluated. Therefore, we opted to simplify the instrumentation effort by placing sensors only at easily accessible locations. Note in Figure 1 that a ladder is located on the exterior of the lower housing. This provides access to the top of the lower housing as well as the torque tube as it exits the housing above the upper bearing. High sensitivity accelerometers (500 mV/g) were placed above the upper bearing on the upper torque tube of the rotor and below the upper bearing on the top of the housing in order to investigate motion in bearing along with the first tower mode. Although the largest acceleration occurs at the top of the tower, the hope was that high sensitivity sensors would provide sufficient acceleration measurement only part of the way up the tower. On the other hand, the maximum strain is present at the base of the structure. The hope was that high output piezoelectric strain sensors could be used to measure the dynamic strain response at the turbine base. Of course, there are actually two tower modes to be considered which occur in the lateral X and Y directions, or fore-aft and side-side. Acceleration and strain sensors were mounted at 90 degrees about the circumference in order to observe both modes. The blades were positioned for the tests so that they aligned with the analysis coordinate system.

One challenge in modal testing of a wind turbine structure using measured force (artificial) excitation is the presence of non-quantified force input from the wind. A considerable amount of research has been pursued for operational modal analysis (OMA) methods to overcome this issue. Additionally, many times it is difficult to provide a sufficiently large artificial input to a very large, stiff structure whereas natural excitation such as the wind is usually sufficient to excite the structure. Reference 3 is one of the early works in OMA and was applied to a wind turbine structure. Much additional work has been done on OMA in recent years -- a review of stochastic identification for OMA has reported in Reference 4. In this work, several different methods of excitation were examined including measured force (artificial) excitation and natural wind excitation for the parked rotor condition. In the remainder of this section the test approaches are detailed and in the following section results are presented for our choices of acceleration and strain measurements.

First we examine artificial excitation. We considered 1) impact testing with a large modal hammer, 2) step relaxation, and 3) human random excitation. Photos of the instrumentation and force measurement sensor are shown in Figure 4. In Figure 4(a) strain sensors mounted at the turbine base are shown. In Figure 4(b), one of the team members is climbing up the tower to install accelerometers. In Figure 4(c), the force measurement sensor can be seen attached at the top of the ladder, which was used to record the input provided during step relaxation and random human excitation. The opposite end of the cable shown in Figure 4(c) was fixed at the ground.

Figure 4. Photos of Instrumentation and Force Sensor
a) Strain gauges at the base (top left), b) Mounting of accelerometers (top right), and c) Force Sensor (bottom, covered with foam)

As mentioned before, wind input results in non-quantified force input. Therefore, tests with artificial excitation were performed in the early to late morning as this is the time of day in which the winds were most calm. Typically, the winds at this time were too low to begin operation of the machine. All of these tests were performed for a parked rotor. In Figure 5, photos of the artificial excitation are shown. In Figure 5(a), a photo of an impact excitation is provided. Here, the input was provided at the highest point possible on the tower that could be easily reached. A photo taken during random human excitation is provided in Figure 5(b). During this test, the turbine was excited by randomly pulling down on the cable during the measurement. The step relaxation input, shown in Figure 5(c), was performed by attaching weights to the cable and cutting a string.

Figure 5. Photos of Artificial Excitation
a) Impact Hammer Excitation (top left), b) Random Human Excitation (top right), and c) Step Relaxation (bottom)

Additionally, an emergency stop of the rotor was executed. For this case, the rotor was accelerated to a fraction of the full speed and the brakes were used to bring the turbine to rest in a matter of seconds. The resulting motion response for the parked rotor was recorded; however, analysis of this data is not reported in this paper.

For the wind excited modal tests, no forced excitation was provided. Here a strong wind capable of exciting the turbine for the parked configuration was desired. A plot of the wind speed at the test site for one approximate 24 hour period during the test is shown in Figure 6. This data was recorded from 2:10 PM until 1:30 PM the following day.

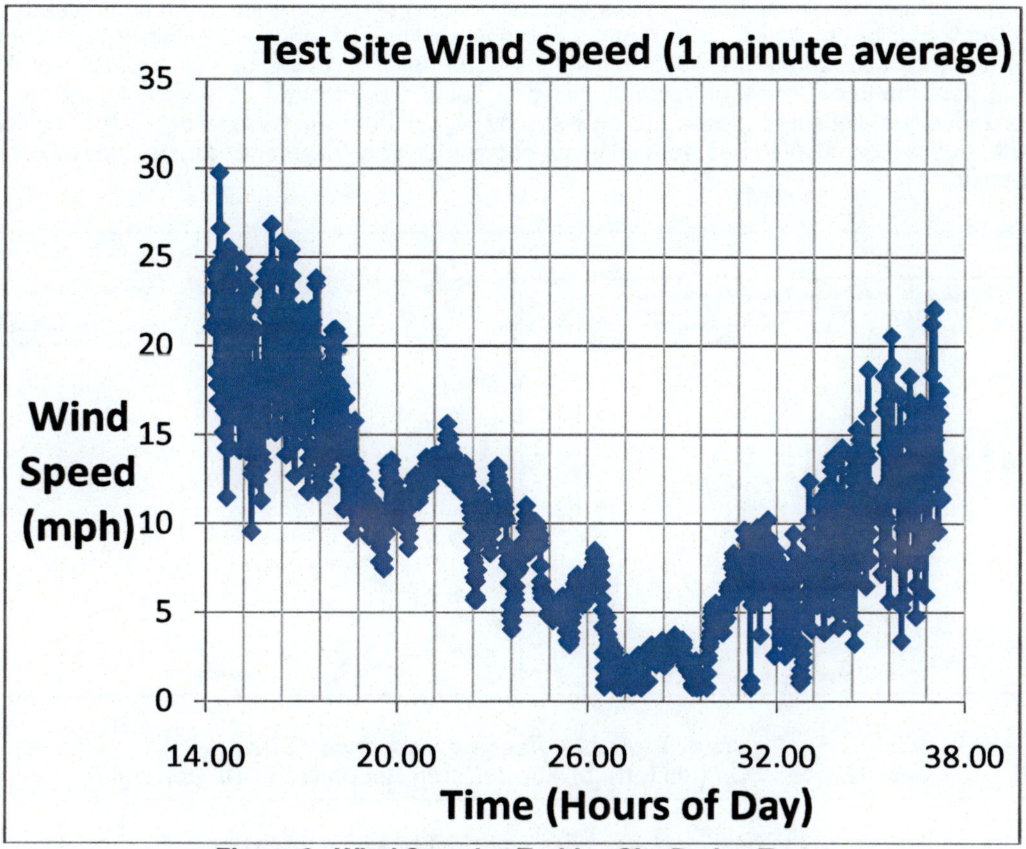

Figure 6. Wind Speed at Turbine Site During Test

Data Analysis and Test Outcomes

The primary objective of these tests was measurement of the frequency of the tower bending modes. A more simple, but less rigorous approach to identify the frequency is peak picking from the auto-spectra measurements. This can be done for each excitation method including artificial excitation, natural excitation, and the emergency stop. A more rigorous approach is fitting of the data to estimate the complete set of modal parameters.

For the step relaxation and random human excitations FRFs (frequency response functions) were measured. The I-DEAS Direct Parameter algorithm was used to compute the modal parameters [5]. For the wind excited case, a special version of the SMAC (Synthesis Modes and Correlate) algorithm developed at Sandia was used to estimate the modal parameters from the cross-spectra of the outputs [6]. The results from using I-DEAS Direct Parameter to fit FRFs and SMAC to fit the cross-spectra are listed in Table 1. Only the accelerometer data was considered in this analysis. The predominant direction of the wind for these tests was in the Y-direction. Forced excitation was provided in the X-direction.

Table 1. Modal Parameter Estimates for Parked Rotor

Excitation	Frequency (Hz)	Damping (%)
Step Relaxation (X direction)	0.81 Hz	0.26%
Random Human (X direction)	0.81 Hz	1.5%
Wind Excitation (Y direction)	0.82 Hz	1.3%

In Figure 6(a), a comparison of the PSDs (power spectral density) of the force inputs from the step relaxation and random human excitation is plotted. The random human excitation is about 2.5 decades higher than the step relaxation at the frequency of the first tower mode. In Figure 6(b), a comparison of the PSDs of the response (acceleration) from the three excitation methods listed in Table 1 are plotted. Surprisingly, the random human excitation provided the highest response but only slightly higher than wind excitation. Wind excited data was much smoother because 100 averages were taken as opposed to only 3 averages for step relaxation and random human excitation.

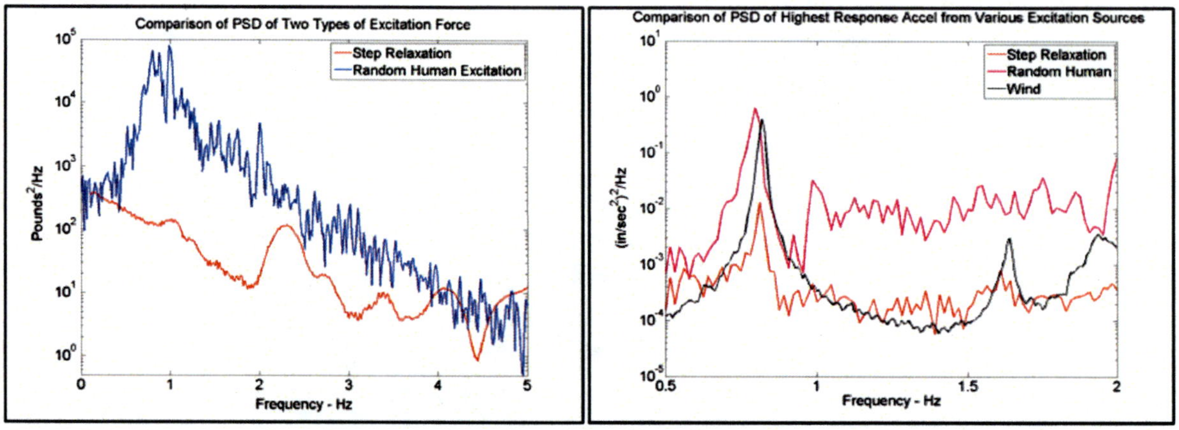

Figure 6. Plots of Force and Response PSDs
a) Force PSDs (top left), b) Acceleration Response PSDs (top right)

Generally, the wind provided the best input as it was sufficient to excite the modes of the turbine. Also, it was easier to record a large number of averages with wind excitation therefore providing less noisy measurements. Random human excitation provided similar modal parameter estimates. Step relaxation provided a good estimate of the natural frequency; however, the damping estimate was low and not in good agreement with the other excitation techniques. Examination of the mode shapes showed a poor estimate of the mode shape for the step relaxation input in the combined X and Y directions. On the other hand, the mode shape from the random human excitation followed the direction of the input. The mode shape for the wind excited data was mostly in the Y direction, the direction of the wind, for the accelerometers mounted on the upper torque tube and in the combined X and Y directions for the co-located accelerometers mounted on the housing.

A plot of the data and synthesis of the CMIF (complex mode indicator function) for the wind excited case is provided in Figure 7. Again, this fit was performed using the SMAC code. The CMIF is applied to the cross-spectra data in this application, as opposed to standard frequency response functions. The CMIF collapses all the cross-spectra into a single curve which is more easily compared to the analytical results synthesized from the extracted modal parameters. With 100 averages, the auto and cross-spectra were much smoother than FRFs from only 3 averages.

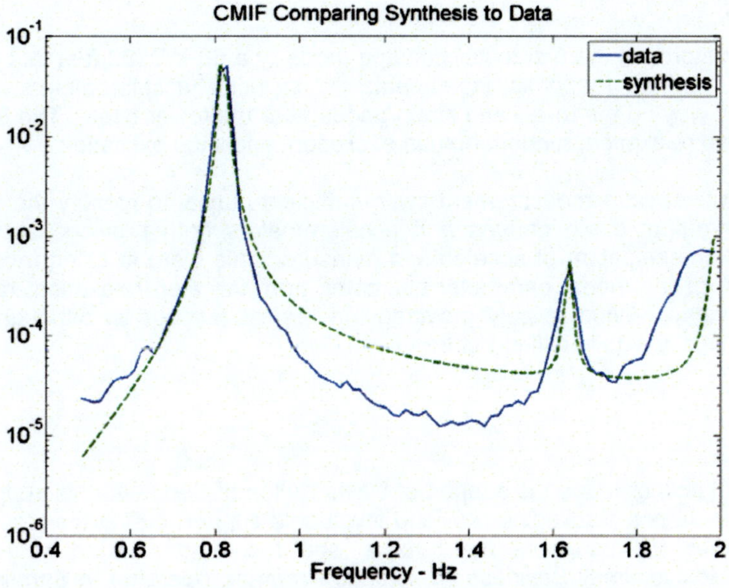

Figure 7. Plots of Data and Synthesis for Wind Excited Response

From the data analysis, there were two bending modes almost at the same frequency and in all cases only one root for which a mode shape could be estimated was found. None of the analyses result in high quality data reconstruction and it is not clear whether this is because the modes are complex when they were fit real or because only one of the two roots was extracted. Again, this is based on analysis of accelerometer data only.

The strain sensors also gave strong signals as compared to the accelerometers. A comparison of strain and acceleration PSDs for wind excitation is plotted in Figure 8. The responses are plotted with non-comparable units. However, the amplitudes of the peaks at the frequency of the first tower mode are very similar with respect to their threshold values.

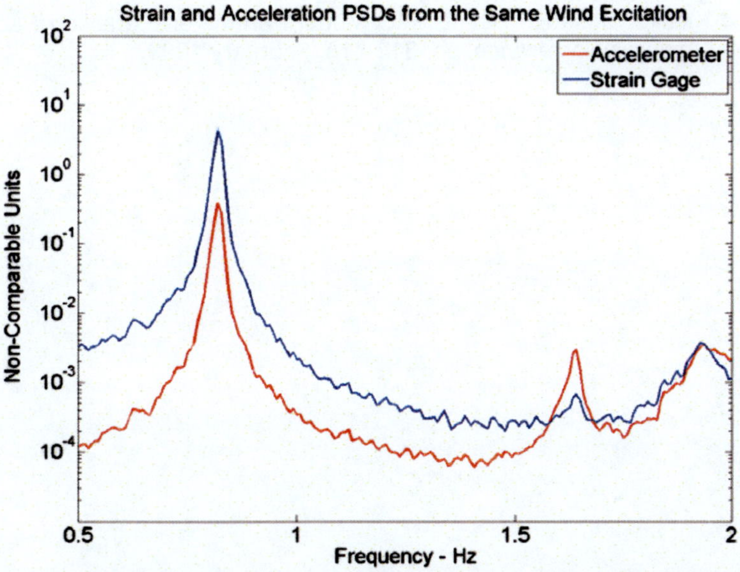

Figure 8. Comparison of Strain and Acceleration PSDs

Conclusions

A simple modal test to determine the first tower bending mode of a 60 kW (82 feet tall) vertical axis wind turbine was performed. The minimal response instrumentation included accelerometers mounted only at easily accessible locations part way up the tower and strain gages near the tower base. The turbine was excited in the parked condition with step relaxation, random human excitation, and wind excitation.

All methods provided auto-spectra measurements with sufficient quality to identify the peak at the frequency of the tower mode. High-output strain gauges and accelerometers both provided strong signals. Frequency response functions and cross-spectra of acceleration measurements were fit using modal parameter estimation routines. A comparison of the modal parameter estimates from the step relaxation, random human, and wind excited cases was provided. Wind reliably provided the largest responses, although artificial excitation was successful even though this is a quite stiff structure.

Acknowledgments

The authors would like to acknowledge the support of David Kelton of Sandia for his support in preparation for the test. Jerry Berglund and James Mackenzie provided invaluable support during the testing days in operating the turbine, performing climbs to instrument the turbine, and recording of wind speed data. We gratefully acknowledge the New Mexico Small Business Assistance Program, operated at Sandia Labs, for providing the funding for this work.

References

[1] Company website: http://www.vawtpower.blogspot.com/.
[2] T.G. Carne, D.W. Lobitz, A.R. Nord, and R.A. Watson, "Finite Element Analysis and Modal Testing of a Rotating Wind Turbine", Sandia National Laboratories, SAND82-0345.
[3] James, G.H., Carne, T.G., and Lauffer, J.P., "The Natural Excitation Technique (NExT) for Modal Parameter Extraction from Operation Structures," *Modal Analysis: the International Journal of Analytical and Experimental Modal Analysis*, v 10, n4, Oct 1995, p 260-277, Society for Experimental Mechanics.
[4] B. Peeters and G. De Roeck, "Stochastic Identification for Operational Modal Analysis: A Review," Journal of Dynamic Systems, Measurement, and Control, ASME, 123, 2001, pp 659-667.
[5] I-DEAS Test Software, by Siemens PLM Software.
[6] Mayes, Randall L. and Klenke, Scott E., "The SMAC Modal Parameter Extraction Package," *Proceedings of the 17th International Modal Analysis Conference*, pp. 812-818, February 1999.

Proceedings of the IMAC-XXVIII
February 1–4, 2010, Jacksonville, Florida USA

Comparison of System Identification Techniques for Predicting Dynamic Properties of Large Scale Wind Turbines by Using the Simulated Time Response

Fanzhong Meng, Muammer Ozbek, Daniel J. Rixen, Michel J.L. van Tooren

Faculty of Aerospace Engineering — Faculty of Mechanical Engineering
Delft University of Technology
Kluyverweg 1, 2629HS Delft, The Netherlands — Mekelweg 2, 2628CD Delft, The Netherlands

ABSTRACT

Accurate prediction of the dynamics of large wind turbines in state-space model is very important in analyzing the aero-elastic stability problem which is known as one of the design issues associating with increase in size of wind turbines. In this work, two different system identification techniques known as Least Square Complex Exponential (LSCE) method and Sub-space System Identification (SSI) will be investigated in terms of their efficiencies in predicting the dynamic characteristics of a wind turbine blade by using the simulated responses of a reference wind turbine. The results obtained through two different methods are then compared in order to discuss their performance and sensitivity to the simulation data and identification parameters. It shows that these two methods are able to identify the frequencies and damping ratios of the aero-elastic modes for large wind turbine blade when the time domain data set contains enough number of cycles.

NOMENCLATURE

X, Y, U	state vector, output vector, input vector
A, B, C, D	system matrix, input matrix, output matrix, direct feedback matrix
U_p, U_f	input "past" block Hankel matrix, input "future" block Hankel matrix
Y_p, Y_f	output "past" block Hankel matrix, output "future" block Hankel matrix
$\mathcal{O}_i$	matrix representation of the oblique projection
Γ_i	observability matrix
L,Q	LQ decomposition matrix
Λ, Ψ	eigenvalue matrix, eigenvector matrix
v_k, w_k	measurement noise, system noise
$(\cdot)^{\dagger}, (\cdot)^{+}$	Moore-Penrose pseudo-inverse, shift one block row down between "past" and "future"
Δt	sampling time interval
R_{ij}	correlation vector
u, k	response vector, discrete sample step
ϕ_{ri}, ψ_{ri}	mode shape vectors
A_{rj}, C_{rij}	modal constant
s_r, β	Laplace variable, polynomial coefficient
N	modal order
p, q	number of response, reference signals
ℓ	number of linear LSCE equations per R_{ij}
m, ω, ζ	modal mass, frequency, damping

T. Proulx (ed.), *Structural Dynamics and Renewable Energy, Volume 1*, Conference Proceedings of the Society for Experimental Mechanics Series 10, DOI 10.1007/978-1-4419-9716-6_31, © The Society for Experimental Mechanics, Inc. 2011

1 INTRODUCTION

During the past years, wind turbines considerably increased in size. This increase is beneficial in terms of reducing the manufacturing costs as well as the ground surface occupied by the wind farms. Compared to the sizes currently available, further increases are still desirable but not easily achievable. In fact, there are technical issues to be considered such as the stability problem, which can put the operation of the wind turbines at serious risks. Thus, improving the understanding of stability of large wind turbines can provide useful information for their design. A significant increase in power generation efficiency can easily be reached by a better aeroelastic design. Therefore, an accurate representation of the dynamics of large wind turbines in state-space model is very important to analyze the aero-elastic stability phenomena. However, finding the modes of the coupled fluid-structure model is not straightforward. One strategy is to perform transient simulation with the model and analyze the transient response to identify the dominant modes. The same identification techniques are also used on measured experimental responses to verify a construction or monitor it over time.In this work, two different system identification techniques known as Least Square Complex Exponential (LSCE) method and Sub-space System Identification (SSI) will be investigated in terms of their efficiencies in predicting the dynamic characteristics of a wind turbine blade.

Over the past several decades, there have been basically two methods used in representing the dynamics of wind turbines in a linear way to directly get the frequencies and the damping ratios of the nonlinear dynamic system. As the first approach, the assumed mode shape method[1] is very popular in the wind energy industry because it has a relatively low computational cost and it provides a compact system matrix for the control design. It is a direct way which means that the equations of motion of a dynamic system are linearized directly. However, this will cause the loss of the nonlinear effects directly from the beginning. So this type of approach is not sufficient for the large diameter rotors with flexible blades. Secondly, instead of linearizing the equations of motion of a dynamic system directly from the beginning, system identification techniques can be applied after finishing the nonlinear simulation to get a linear model which can approximately represent the original nonlinear dynamic system. Bauchau and Nikiskov[2] proposed a computationally efficient approach in 1998. It allows to evaluate a limited number of eigenvalues of the state matrix with the largest modulus using Arnoldi's iterative algorithm. This method can correctly assess the stability boundary, but spurious eigenvalues associated with algebraic constraints may arise which will cause a wrong evaluation of stability limits.

Two different system identification methods are investigated in this paper. The basic theory of these two methods is shown. And then, these two methods are applied on the response calculated by using a 5 degrees-of-freedom analytical model with known frequencies and damping ratios. In order to investigate the efficiency of the utilized algorithms, the system properties identified by the two methods are then compared with the known real values. In addition, a nonlinear aero-elastic code based on flexible multi-body dynamics called MBDyn-AeroDyn[9] is used to generate the time domain responses of the blades at different operational conditions. After that, both of these two system identification techniques are applied to obtain a linear model around an equilibrium state from which the frequencies, damping ratios of the blade are extracted. The identified linear model can be used to analyze the aero-elastic stability phenomena of future large wind turbines.

2 SUBSPACE SYSTEM IDENTIFICATION

In this section, a sub-space system identification algorithm for processing the time domain simulation data obtained by the above mentioned aero-elastic simulation tool is presented. This system identification model is based on the work of Peter van Overschee[3]. Figure 1 shows the general approach of subspace system identification and the classical approach. As can be seen from the figure, subspace identification method, firstly, estimates state sequences directly from the given time domain input-output data through an orthogonal or oblique projection of the row spaces of certain block Hankel matrices of data into the row spaces of other block Hankel matrices, followed by singular value decomposition(SVD) to determine the order of the system and the state sequence. Then the state space model is achieved through the solution of a least squares problem. While the classical approach obtains the system matrices first and then estimates the states.

2.1 State-Space Model

The wind turbine dynamic system can be described by the discrete state space equation shown in Equation 1.

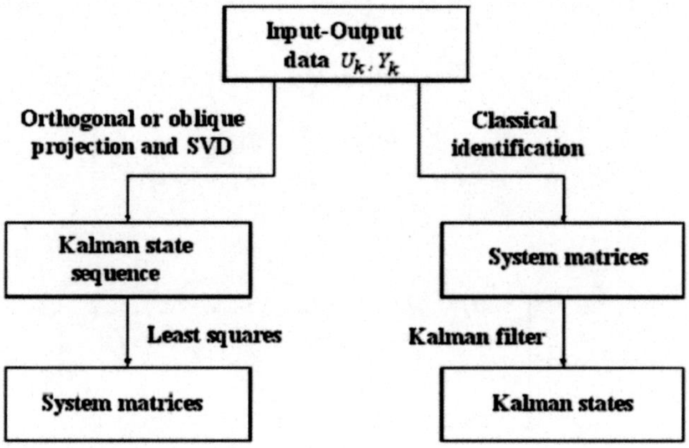

Figure 1: Difference between subspace identification and classical identification.

$$\begin{cases} X_{k+1} = AX_k + BU_k + w_k \\ Y_k = CX_k + DU_k + v_k \end{cases} \quad (1)$$

In the above equation, the vectors $U_k \in \mathbb{R}^m$ and $Y_k \in \mathbb{R}^l$ are the observations at time instant k of respectively m input channels and l output channels.of the system. The vector $X_k \in \mathbb{R}^n$ is called the state vector of the system at time k and contains n states. $v_k \in \mathbb{R}^l$ and $w_k \in \mathbb{R}^n$ are unobservable vector signals called measurement noise and system noise. The matrix $A \in \mathbb{R}^{n \times n}$ is called the (dynamic) system matrix. It describes the dynamics of the system as characterized by its eigenvalues. $B \in \mathbb{R}^{n \times m}$ is the input matrix, which represents the linear transformation by which the deterministic input signals influence the next state. $C \in \mathbb{R}^{l \times n}$ is the output matrix, which describes how the internal states are transferred to the output y_k. Matrix $D \in \mathbb{R}^{l \times m}$ is called the direct feedback term. In this paper, we assume that, for a noiseless model, the v_k and w_k are equal to zero. A Simulink style representation of the state space system equation can be found in Figure 2.

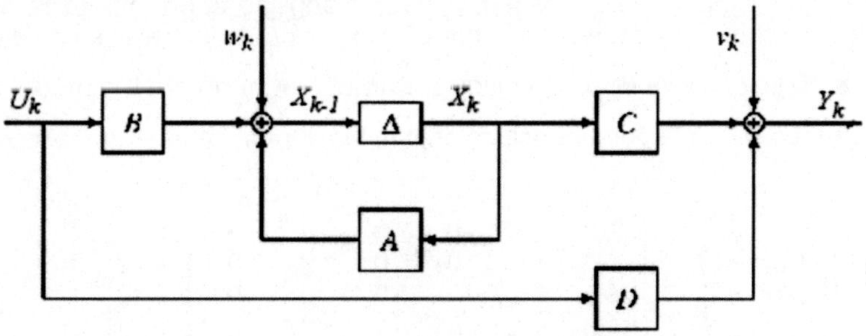

Figure 2: State-space system chart.

Subspace system identification begins by forming the block Hankel matrices with input and output data. They play an important role in the subspace identification algorithm. These matrices can be easily constructed from the given input and output data. Here the input "past" and "future" block Hankel matrices are defined as:

$$U_{0|i-1} = U_p = \begin{bmatrix} u_0 & u_1 & u_2 & \cdots & u_{j-1} \\ u_1 & u_2 & u_3 & \cdots & u_j \\ \vdots & \vdots & \vdots & \cdots & \vdots \\ u_{i-1} & u_i & u_{i+1} & \cdots & u_{i+j-2} \end{bmatrix} \tag{2}$$

$$U_{i|2i-1} = U_f = \begin{bmatrix} u_i & u_{i+1} & u_{i+2} & \cdots & u_{i+j-1} \\ u_{i+1} & u_{i+2} & u_{i+3} & \cdots & u_{i+j} \\ \vdots & \vdots & \vdots & \cdots & \vdots \\ u_{2i-1} & u_{2i} & u_{2i+1} & \cdots & u_{2i+j-2} \end{bmatrix} \tag{3}$$

where, i is the number of block rows which should at least be larger than the maximum order of the system to be identified. Since each block row contains m rows of inputs, the matrix $U_{0|i-1}$ consists of mi rows. j is equal to $j = s - 2i + 1$, which implies that all s data samples are used. The matrices $U_p^+ = U_{0|i}$ and $U_f^- = U_{i+1|2i-1}$ are defined similarly by shifting the border between "past" and "future" one block row down. In a similar way, the output "past" and "future" block Hankel matrix are defined as $Y_{0|i-1} = Y_p$ and $Y_{i|2i-1} = Y_f$, and also the matrices $Y_p^+ = Y_{0|i}$ and $Y_f^- = Y_{i+1|2i-1}$. Following the same notation used above, the block Hankel matrices consisting of inputs and outputs are defined as W_p and W_p^+, which denote the "past" inputs and outputs joint block Hankel matrix and the shifted one.

$$W_{0|i-1} = W_p = \begin{bmatrix} U_{0|i-1} \\ Y_{0|i-1} \end{bmatrix} \tag{4}$$

$$W_{0|i} = W_p^+ = \begin{bmatrix} U_{0|i} \\ Y_{0|i} \end{bmatrix} \tag{5}$$

2.2 Calculation of a State Sequences and System Matrices

The state sequences can be obtained from input-output data without any knowledge of the system matrices in before with two steps. First, the future output row space is projected along the future input row space into the joint row space of past input and past output. This projection is called oblique projection which can be written in matrix form: $Y_f/_{U_f}\begin{pmatrix} U_p \\ Y_p \end{pmatrix}$. Second, a singular value decomposition is carried out to obtain the model order, the observability matrix and the state sequences. The LQ decomposition of joint block past input and past output matrix is computed as follows

$$\begin{bmatrix} U_{0|i-1} \\ U_{i|i} \\ U_{i+1|2i-1} \\ Y_{0|i-1} \\ Y_{i|i} \\ Y_{i+1|2i-1} \end{bmatrix} = \begin{bmatrix} L_{11} & 0 & 0 & 0 & 0 & 0 \\ L_{21} & L_{22} & 0 & 0 & 0 & 0 \\ L_{31} & L_{32} & L_{33} & 0 & 0 & 0 \\ L_{41} & L_{42} & L_{43} & L_{44} & 0 & 0 \\ L_{51} & L_{52} & L_{53} & L_{54} & L_{55} & 0 \\ L_{61} & L_{62} & L_{63} & L_{64} & L_{65} & L_{66} \end{bmatrix} \begin{bmatrix} Q_1^T \\ Q_2^T \\ Q_3^T \\ Q_4^T \\ Q_5^T \\ Q_6^T \end{bmatrix} \tag{6}$$

Then, the matrix representation of the oblique projection can be obtained as

$$\mathcal{O}_i = Y_f/_{U_f}\begin{pmatrix} U_p \\ Y_p \end{pmatrix} = L_{U_p} L_{11} Q_1^T + L_{Y_p} \begin{pmatrix} L_{41} & L_{42} & L_{43} & L_{44} \end{pmatrix} \begin{pmatrix} Q_1^T \\ Q_2^T \\ Q_3^T \\ Q_4^T \end{pmatrix} \tag{7}$$

where,

$$\left[\begin{array}{ccc} L_{U_p} & L_{U_f} & L_{Y_p} \end{array}\right] \left[\begin{array}{c|cc|c} L_{11} & 0 & 0 & 0 \\ \hline L_{21} & L_{22} & 0 & 0 \\ L_{31} & L_{32} & L_{33} & 0 \\ \hline L_{41} & L_{42} & L_{43} & L_{44} \end{array}\right] = \left[\begin{array}{c|cc|c} L_{51} & L_{52} & L_{53} & L_{54} \\ L_{61} & L_{62} & L_{63} & L_{64} \end{array}\right] \tag{8}$$

from which L_{U_p} and L_{Y_p} can be calculated. Similarly, the oblique projection $Y_f^- /_{U_f^-} \begin{pmatrix} U_p^+ \\ Y_p^+ \end{pmatrix}$, denoted by $\mathcal{O}_{i-1}$ is equal to

$$\mathcal{O}_{i-1} = L_{U_p^+} \begin{bmatrix} L_{11} & 0 \\ L_{21} & L_{22} \end{bmatrix} \begin{bmatrix} Q_1^T \\ Q_2^T \end{bmatrix} + L_{Y_p^+} \begin{bmatrix} L_{41} & L_{42} & L_{43} & L_{44} & 0 \\ L_{51} & L_{52} & L_{53} & L_{54} & L_{55} \end{bmatrix} \begin{bmatrix} Q_1^T \\ Q_2^T \\ Q_3^T \\ Q_4^T \\ Q_5^T \end{bmatrix} \tag{9}$$

The matrix $\mathcal{O}_i$ is also equal to the product of extended observability matrix and the state sequences, obtained from the non-steady state Kalman filters[4].

$$\mathcal{O}_i = \Gamma_i X_i \tag{10}$$

A singular value decomposition is applied on matrix $\mathcal{O}_i$ which is represented in the follow

$$\mathcal{O}_i = \begin{bmatrix} U_1 & U_2 \end{bmatrix} \begin{bmatrix} S_1 & 0 \\ 0 & 0 \end{bmatrix} \begin{bmatrix} V_1^T \\ V_2^T \end{bmatrix} \tag{11}$$

The order of the system is equal to the number of singular values in Equation(11) different from zero. The extended observability matrix Γ_i can be taken to be $\Gamma_i = U_1 S_1^{\frac{1}{2}}$ without losing any generality. Substituting Γ_i to Equation(10), the state sequence X_i can be obtained by

$$X_i = \Gamma_i^{\dagger} \mathcal{O}_i \tag{12}$$

where $\Gamma_i^{\dagger}$ is called the Moore-Penrose pseudo-inverse of the matrix Γ_i. Similarly, the "shifted" state sequence X_{i+1} can be obtained as

$$X_{i+1} = \Gamma_{i-1}^{\dagger} \mathcal{O}_{i-1} \tag{13}$$

As soon as the order of the system and the state sequences X_i and X_{i+1} are known. the state space matrices A, B, C, D can be solved from the following equation:

$$\begin{bmatrix} X_{i+1} \\ Y_{i|i} \end{bmatrix} = \begin{bmatrix} A & B \\ C & D \end{bmatrix} \begin{bmatrix} X_i \\ U_{i|i} \end{bmatrix} \tag{14}$$

by using a linear least squares method. Here $Y_{i|i}$ and $U_{i|i}$ are block Hankel matrices with only one block row of input channel and output channel.

The identified A, B, C and D matrices can then be transformed to get $\Lambda, \Psi^{-1}B$ and $C\Psi$. The diagonal matrix Λ contains the information of modal damping rates and damped natural frequencies. The matrix $C\Psi$ defines the mode shape at the calculated points. Therefore, all the modal parameters of this wind turbine dynamic system can be identified by the three matrices. The desired modal damping rates and damped natural frequencies are simply the real and imaginary parts of the eigenvalues. A MATLAB code has been built to realize the above mentioned algorithm.

3 LEAST SQUARE COMPLEX EXPONENTIAL METHOD

3.1 NEXT(Natural Excitation Technique)

If a system is assumed to be excited by a stationary white noise the correlation function $R_{ij}(t)$ between the response signals u_i and u_j at a time interval of t can be considered as similar to the response of the structure at i due to an impulse on j. This relation can be expressed by the equation [5]

$$\begin{aligned} R_{ij}(t) &= \lim_{T\to\infty}\frac{1}{T}\int_{-T/2}^{T/2} u_i(\tau)u_j(\tau-t)d\tau \\ &= \sum_{r=1}^{N}\frac{\phi_{ri}A_{rj}}{m^r\omega_d^r}e^{(-\zeta^r\omega_n^r t)}\sin(\omega_d^r t+\theta^r) \end{aligned} \tag{15}$$

where ϕ_{ri} is the ith component of the rth eigenmode of the system, A_{rj} is a constant associated to the jth response signal taken as reference, m^r is the rth modal mass, ζ^r and ω_n^r are respectively the rth modal damping ratio and non-damped eigenfrequency, $\omega_d^r = \omega_n^r\sqrt{1-\zeta^{r^2}}$ and θ^r is the phase angle associated with the rth modal response. Hence the correlation between signals is a superposition of decaying oscillations having damping and frequencies equal to the damping and frequencies of the structural mode [5, 6].

3.2 Complex Exponential Method

The relation mentioned above enables modal parameter identification techniques like the Least Square Complex Exponential method (LSCE) to be used to extract the modal parameters from the correlation functions between measured responses to the white noise input. In terms of the complex modes of the structure, the correlation function (15) can be written as

$$R_{ij}(k\Delta t) = \sum_{r=1}^{N}\psi_{ri}e^{s_r k\Delta t}C_{rj} + \sum_{r=1}^{N}\psi_{ri}^{*}e^{s_r^{*}k\Delta t}C_{rj}^{*} \tag{16}$$

where $s_r = \omega_r\zeta_r + i\omega_r\sqrt{(1-\zeta_r^2)}$ and where C_{rj} is a constant associated with r^{th} mode for the jth response signal, which is the reference signal. Δt is the sampling time-step and the superscript * denotes the complex conjugate. It should be noted that in conventional modal analysis, these constant multipliers are modal participation factors. Numbering all complex modes and eigenvalues in sequence, Equation (16) can be written as

$$R_{ij}(k\Delta t) = \sum_{r=1}^{2N} C'_{rij}e^{s_r k\Delta t} \tag{17}$$

As s_r appears in complex conjugate forms in this last form, there exists a polynomial of order $2N$ (known as Prony's equation) of which $e^{s_r\Delta t}$ are roots:

$$\beta_0 + \beta_1 V_r^1 + \beta_2 V_r^2 + ... + \beta_{2N-1}V_r^{2N-1} + V_r^{2N} = 0 \tag{18}$$

where $V_r = e^{s_r\Delta t}$ and where $\beta_{2N} = 1$. $\{\beta\}$ are the coefficients matrix of the polynomial. To determine the values of β_i, let us multiply the impulse response (17) for sample k by the coefficients β_k and sum up these values for $k = 0, \ldots 2N$ [7, 8, 6]:

$$\begin{aligned} \sum_{k=0}^{2N}\beta_k R_{ij}(k\Delta t) &= \sum_{k=0}^{2N}\left(\beta_k\sum_{r=1}^{2N}C'_{rij}V_r^k\right) \\ &= \sum_{r=1}^{2N}\left(C'_{rij}\sum_{k=0}^{2N}\beta_k V_r^k\right) = 0 \end{aligned} \tag{19}$$

Hence the coefficients $\{\beta\}$ satisfy a linear equation which coefficients are the impulse response (or correlation functions) at $2N+1$ successive time samples. In order to determine those coefficients, relation (19) is written at least for $2N$ times, starting at successive time samples. In other words, many Prony's equations can be written to build up a linear system that determines the coefficients $\{\beta\}$:

$$\beta_0 R_{ij}^n + \beta_1 R_{ij}^{n+1} + ... + \beta_{2N-1}R_{ij}^{n+2N-1} = -R_{ij}^{n+2N}$$

where we note $R_{ij}^{k} = R_{ij}(k\Delta t)$ and $n = 1...\ell$. Therefore, a linear system of ℓ equations based on the Hankel matrix to determine the coefficients can be built. $\{\beta\}$:

$$[R]_{ij} \ \{\beta\} = \ -\{R'\}_{ij} \tag{20}$$

The procedure explained above uses a single correlation function. Since $\{\beta\}$ are global quantities related to modal parameters, the same parameters can be derived from different correlation functions between any two response signals and from the auto-correlation functions. Hence, if we consider p response stations and adopt q of them as reference locations we can group the associated equations (20) for all qp correlation functions in one system yields

$$\begin{bmatrix} [R]_{11} \\ [R]_{12} \\ \vdots \\ [R]_{qp} \end{bmatrix} \{\beta\} = - \begin{Bmatrix} \{R'\}_{11} \\ \{R'\}_{12} \\ \vdots \\ \{R'\}_{qp} \end{Bmatrix} \tag{21}$$

A solution in a least square sense can be found for $\{\beta\}$ from (21) using pseudo-inverse techniques, provided $\ell qp \geq 2N$.

The LSCE method as explained above is different from standard LSCE since we process correlation functions between any two response locations whereas the standard LSCE considers only one reference. Note however that, although correlation functions with respect to several references are considered, the procedure described here is different from the so-called Polyreference LSCE [7] where the response at one location is assumed to result from simultaneous excitations at multiple points. This special LSCE version described by Equation (21) allows us to explicitly introduce harmonic components in the identification procedure [6].

4 VERIFICATION AND APPLICATION

In this section, firstly, both of these two system identification methods are applied on a simple five degrees-of-freedom mass-spring-damper dynamic system called the reference dynamic system with known frequencies and damping ratio values. In order to investigate the efficiency of the utilized algorithms, the system properties identified by the two methods are then compared with the known real values. The purpose is to find the model sensitivity to the identification parameters used and to the length of the time-series. Secondly, an application on a 5MW reference wind turbine will be shown. It had been known that the number of cycles contained in the time-series data set will affect the accuracy of identified frequencies and the damping ratios. When the time-series data set is too short, the identified system will be inaccurate.

4.1 Model Verification

The system which is being used for the verification is the mass-spring-damper system shown in Figure 3. It is a five degrees-of-freedom dynamic system. By tuning the mass, the stiffness of the spring and the damping value of the system properly, the frequencies and damping ratio values can be adjusted to the required values. Table 1 lists the frequencies and damping ratio values of this dynamic system used in this identification. The system used in the analysis can be seen in Figure 3. The input excitation force is a white noise signal with the frequency bandwidth of $[0, 32]$Hz. The total simulation time is 1000 seconds. In order to eliminate the transient response first 300 seconds period was ignored and the remaining 700 seconds part was used to investigate the effect of the length of the simulation time on the identification results. This data set has been cut into seven pieces with different durations. Figure 4 and 5 show the damping ratios of the 1st mode and the 5th mode. It can be seen that the identified damping ratio is very unstable when the duration of data set is less than 200 seconds for both of these two system identification methods. When the duration of data set is longer than 200 seconds, the identified damping ratio is getting stable and converges to the real value.

In fact, the real reason that affects the accuracy of results is expected to be the number of cycles coming from the lowest mode. The results indicate that the minimum number of cycles has to be larger than 200. In Figure 6, only the results from the SSI method is presented because LSCE method failed to identify this mode because the damping ratio of this mode is very high. In order to determine when the LSCE method would be able to detect the highly damped mode the length of the measurement duration was then progressively increased. This frequency and the

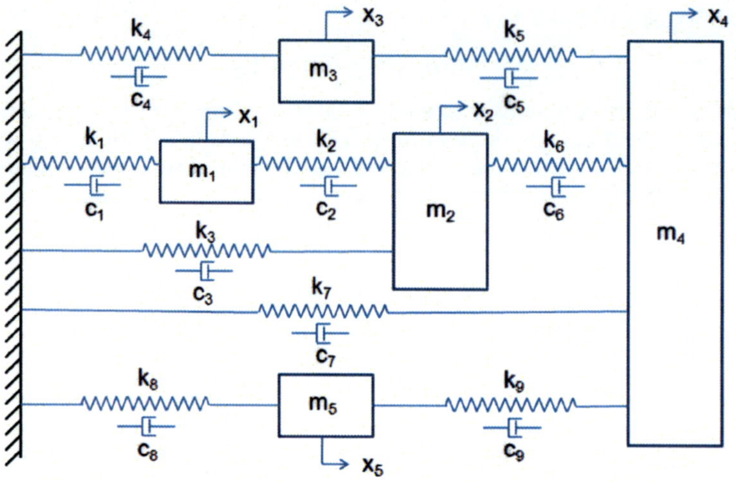

Figure 3: Reference mass-spring-damper dynamic system

	F-Exact(Hz)	D-Exact	F-SSI(Hz)	D-SSI	F-LSCE(Hz)	D-LSCE
mode 1	0.752	0.010	0.752	0.008	0.750	0.009
mode 2	0.920	0.300	0.875	0.298	-	-
mode 3	1.286	0.014	1.282	0.015	1.282	0.014
mode 4	1.650	0.009	1.649	0.010	1.650	0.010
mode 5	1.884	0.017	1.886	0.016	1.884	0.016

TABLE 1: Identified and exact frequencies and damping ratio of the reference system using 700 seconds data

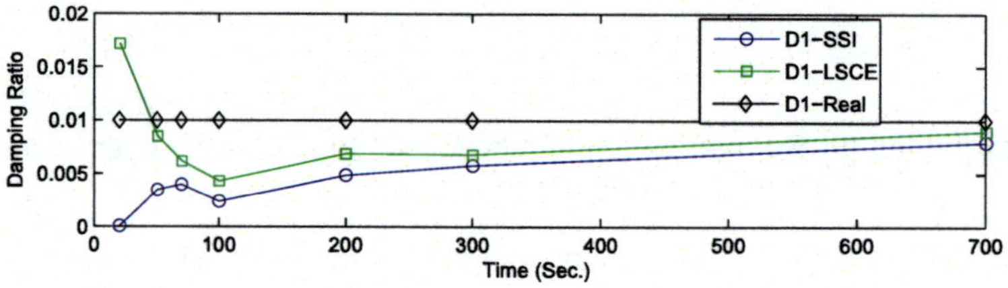

Figure 4: Comparison of exact damping ratio with identified damping for mode 1

corresponding damping ratio could be identified by LSCE only when the length of the data series were increased up to 5000 seconds. Table 1 shows that both system identification methods perform well on finding the frequencies and damping ratio with 700 seconds data set.

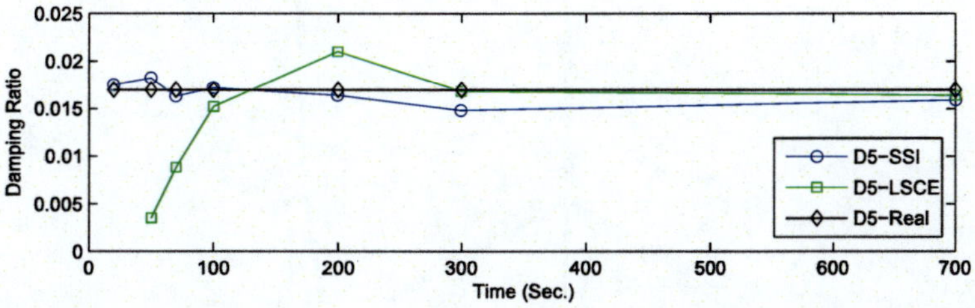

Figure 5: Comparison of exact damping ratio with identified damping for mode 5

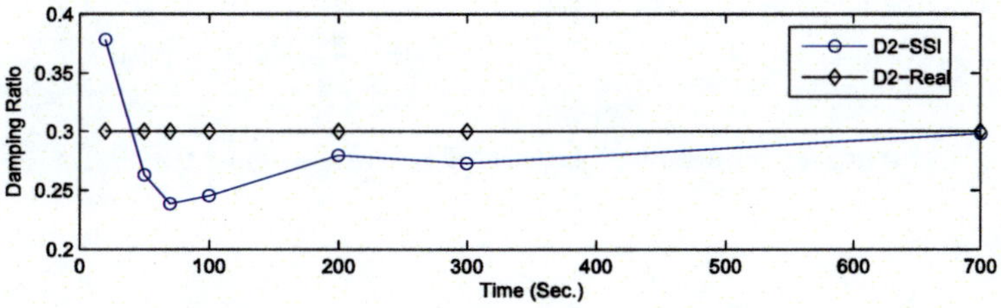

Figure 6: Comparison of exact damping ratio with identified damping for mode 2

4.2 Application

The two system identification methods presented in the previous sections were applied on a 5MW reference wind turbine. A flexible multibody dynamics based aero-elastic code for wind turbine simulation called MBDyn-AeroDyn[9] is used to get the blade deformations in time domain. In this code, the tower is modeled by five beam elements, each containing three nodes. The rotor hub is described by a rigid body element and can rotate around a horizontal revolution hinge located at the free end of the main shaft, which is rigidly connected on the nacelle element having 5 degrees tilt angle. Three blades are mounted on the hub with a 1.5 meters offset from the hub center and with 2.5 degrees pre-cone angle(toward the upwind direction). They are clamped on the hub while the pitch angle can be adjusted. The blades are 61.5 meters long and have a chord length variation in spanwise direction. They are modeled by five three-node beams. The overall assembling result can be seen in Figure 7.

In the simulation, an uniform wind field of 12m/s is used. The input excitation is a white noise pitch input signal. The amplitude is 0.5 degree and the frequency bandwidth is from 0Hz to 10Hz. The total simulation time is 660 seconds. the first 60 seconds period was just ignored and the remaining 600 seconds part was used for the identification. Figure 8 and Figure 9 shows the comparison of original tip deflection calculated by the aero-elastic code with the identified one in flapwise and edgewise direction. It can be seen that the identified signal matches quite well with the original signal. In Table 2 the five lowest damped eigen-frequencies and damping ratios calculated by use of these two system identification methods are listed.

mode	F-SSI(Hz)	D-SSI	F-LSCE(Hz)	D-LSCE
1st flapwise	0.72	0.640	-	-
1st edgewise	1.07	0.026	1.07	0.020
2nd flapwise	2.05	0.035	2.05	0.026
3rd flapwise	4.37	0.047	4.47	0.011
2nd edgewise	-	-	5.43	0.013

TABLE 2: Five lowest damped eigenfrequencies in Hz and damping ratio for a 5MW wind turbine

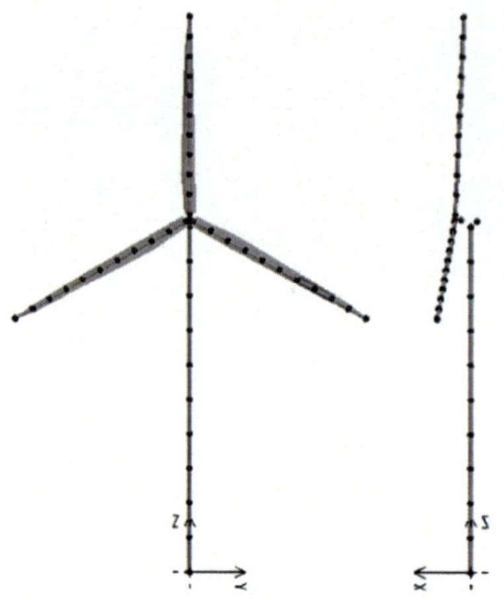

Figure 7: The overall layout of reference wind turbine.

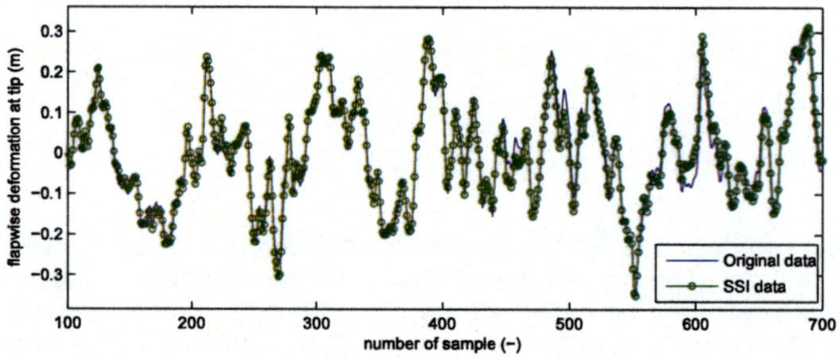

Figure 8: Comparison of original response with identified response in flapwise

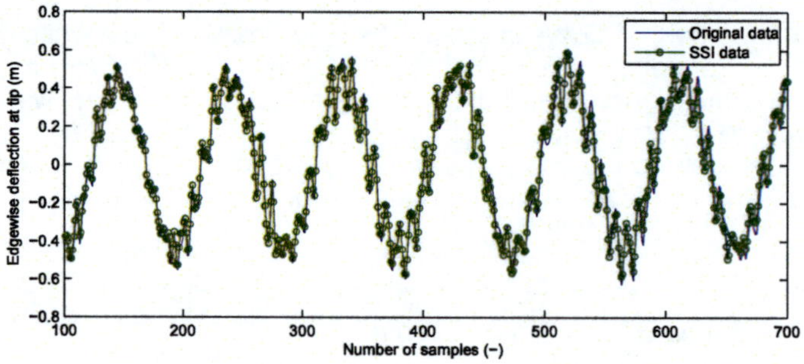

Figure 9: Comparison of original response with identified response in edgewise

5 CONCLUSION

The results demonstrate that sub-space system identification method and least square complex exponential identification method are capable of identifying the frequencies and damping ratios of the aero-elastic mode for a large wind turbine blade. However, LSCE is not very efficient in identifying the highly damped modes. The frequencies having high damping ratios could not be identified either for the analytical model or the wind turbine simulations. These frequencies could be identified by LSCE only when the length of the data series were increased up to 5000 seconds which are approximately 10 times longer than the data series used in this work. This problem is not expected to be related to the LSCE algorithm but the basic assumptions of NEXT technique. For very highly damped systems correlation functions are not able to represent the impulse functions accurately unless the the data set is very long. The required number of cycles of the lowest frequency can be as high as 1000 while for a lightly damped system 200 cycles can provide quite satisfactory results.

Even for the lightly damped modes, the identified results will be very uncertain and incorrect when the time domain data set does not contain enough number of cycles. At least 200 cycles have to be used in both of these two system identification methods in order to achieve a good result. The number of rows of input and output block Hankel matrix is another important parameter for sub-space system identification method because it affects the number of cycles which has been involved in each column of input and output block Hankel matrix. This will eventually affect the identification quality. At least a half cycle long data set has to be used to create the input and output block Hankel matrix. It is also found that the aero-elastic modes are always coupled strongly in flapwise and edgewise directions.

ACKNOWLEDGMENTS

We acknowledge the UPWIND project which is an EU 6th framework programme of Europe Union(Contract No.:019945) for the financial support.

REFERENCES

[1] **Lindenburg, C.**, *BLADMODE Program for Rotor Blade Mode Analysis*, Ecn c–02-050, 2003.

[2] **Bauchau, O. A.**, *Computational schemes for flexible, nonlinear multi-body systems*, Multibody System Dynamics, Vol. 2, pp. 169–225, 1998.

[3] **van Overschee, P.** and **Moor, B. D.**, Subspace Identification for Linear Systems: Theory - Implementation - Applications, New York: Springe, 1st edition, 1996.

[4] **Cock, K. D.** and **Moor, B. D.**, *Subspace identification methods*, Control systems robotics and automation, Vol. 1, No. 3, pp. 933–979, 2003.

[5] **James, G. H.**, **Carne, T. G.** and **Lauffer, J. P.**, *The Natural Excitation Technique (NExT) for Modal Parameter Extraction from Operating Structures*, Journal of Analytical and Experimental Modal Analysis, Vol. 10, No. 4, pp. 260–277, October 1995.

[6] **Mohanty, P.** and **Rixen, D. J.**, *Operational modal analysis in the presence of harmonic excitation*, Journal of Sound and Vibration, Vol. 270, pp. 93–109, 2004.

[7] **Maia, N.** and **Silva, J.** (editors), Theoretical and Experimental Modal Analysis, Research Studies Press Ltd., Somerser, England, 1997, isbn 0 86380 208 7.

[8] **Hermans, L.** and **der Auweraer, H. V.**, *Modal Testing and Analysis of Structures Under Operational Conditions: Industrial Applications*, Mechanical System and Signal Processing, Vol. 13, No. 2, pp. 193–216, October 1999.

[9] **Meng, F.**, **Masarati, P.** and **van Tooren, M.**, *Free/Open Source Multibody and Aerodynamic Software for Aeroelastic Analysis of Wind Turbines, 28th ASME Wind Energy Symposium, (47th Aerospace Science Meeting and Exhibit)*, Orlando,Florida, U.S.A., Jan. 2009.

Proceedings of the IMAC-XXVIII
February 1–4, 2010, Jacksonville, Florida USA

Identification of the Dynamics of Large Wind Turbines by Using Photogrammetry

Muammer Ozbek, Fanzhong Meng†, Daniel J. Rixen, Michel J.L. van Tooren†

Department of Mechanical, Materials and Maritime Engineering
Delft University of Technology, Mekelweg 2, 2628CD Delft, the Netherlands
†Faculty of Aerospace Engineering
Delft University of Technology, Kluyverweg 1, 2629HS Delft, the Netherlands

ABSTRACT

In this work, the initial results of the infield tests performed to investigate the feasibility of using photogrammetry in monitoring the dynamics of large scale wind turbines in operation are presented. Within the scope of the work, the response of a wind turbine, with a rotor diameter of eighty meters, was captured by using four CCD cameras simultaneously while the turbine was in operation.

The captured response was then analyzed by using two different system identification techniques based on Least Square Complex Exponential (LSCE) method and Sub-space System Identification (SSI) while the dynamic characteristics (the frequencies, damping ratios and mode shapes) of the turbine were derived. Possible effects of very high modal damping ratios and relatively short measurement periods on the identified results were also considered.

1 INTRODUCTION

Photogrammetry is a measurement technique where 3D coordinates or displacements of an object can be obtained by using the 2D images of the same object taken from different locations and orientations. Although each picture provides 2D information only, very accurate 3D information related to the coordinates and/or displacements of the object can be obtained by simultaneous processing of these images as displayed in Figure1 [1].

Although close range photogrammetry is widely used in measuring the coordinates and displacements of the objects, its use is limited to small measurement volumes and the measurements are usually performed in laboratories or other controllable environments. Within this small measurement volume it can be efficiently used with different purposes such as vibration analysis, identification of dynamic properties of objects or 3D modeling [2]. The size of the object to be tracked is restricted due to the problems such as insufficient illumination, difficulties in calibration, or low resolution of the cameras. If properly adapted on large scale structures, this new measurement technique can provide very useful information in monitoring the dynamic behavior of wind turbines and understanding the interaction among different parts of these systems [3].

It is state of the art to use accelerometers and/or strain gauges placed inside the blade or tower in dynamic tests performed on wind turbines. However, these measurement systems are sensitive to lightning and electromagnetic fields. Besides, some extra installations inside the blades such as placement of cables for power supply and data transfer are required for this application. The signals from rotating sensors on the blades are transferred to a stationary computer via slip rings or by radio transmission. For large commercial turbines the required installations and preparations (sensor calibration) might be very expensive and time consuming [4].

T. Proulx (ed.), *Structural Dynamics and Renewable Energy, Volume 1*, Conference Proceedings of the Society for Experimental Mechanics Series 10, DOI 10.1007/978-1-4419-9716-6_32, © The Society for Experimental Mechanics, Inc. 2011

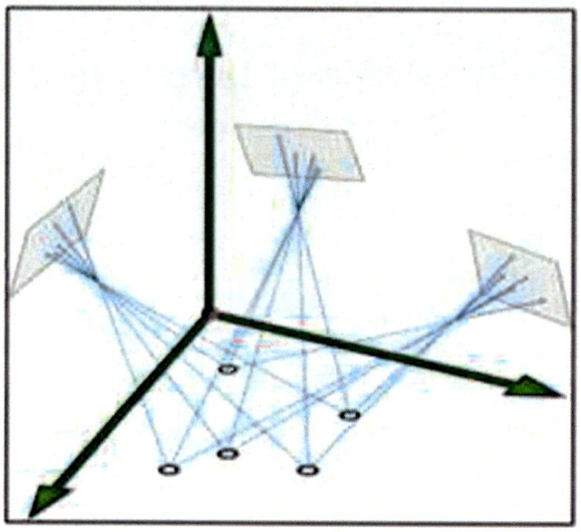

Figure 1: Simultaneous process of 2D images taken from different locations

Since the deflections under the action of wind loading can be considered as the sum of slowly changing static part and dynamically changing fluctuating part, identification of low frequency vibrations play a crucial role in predicting the response of structures under wind loading [5]. Several researches reported that in wind response measurements, accelerometers should be used together with other systems such as GPS which are able to detect these low frequency motions [6].

Fiber optic strain gauges are proposed to be a promising alternative to accelerometers and conventional strain gauges since optical sensors are not prone to electro-magnetic fields or lightning. However, it is reported that some additional feasibility tests are still needed to ensure the effective and cost efficient use of this measurement system. The factors affecting the performance of the fiber optic sensors such as sensitivity to temperature variations and related error compensation methods should also be further investigated [7].

Photogrammetric measurement techniques are expected to be a common solution to all these problems, considering the rapid progress in optical sensor technology and computer hardware, it can be said that these techniques will certainly replace most of the conventional measurement techniques that are currently in use.

2 OPERATIONAL MODAL ANALYSIS

This article does not aim at reviewing the theoretical background of the utilized algorithms but evaluating whether or not these methods can be used to monitor the dynamic behavior of wind turbines with sufficient accuracy. The two important assumptions of the conventional system identification procedures, namely the time invariant system and steady state white noise excitation are not always easy to satisfy due to the nature of wind loads, the complicated interaction among different parts of the turbine (nacelle, rotor, tower and generator) and strong aerodynamic coupling between the blades and the wind. Besides, the low frequency vibrations dominating the overall response and very high aeroelastic damping ratios make the proper application of the identification methods to these specific structures very difficult.

Two different identification methods, namely the Least Square Complex Exponential (LSCE) method and Sub-space System Identification (SSI)are utilized to estimate the dynamic properties (eigenfrequencies, damping ratios and mode shapes) of the system.

LSCE is a time domain identification method mainly using the impulse response functions to estimate the modal parameters [8]. NEXT (Natural Excitation Technique) enables LSCE and other similar time domain methods to be used to analyze the ambient vibration data by demonstrating the fact that correlation functions between signals is a superposition of decaying oscillations having damping and frequencies equal to the damping and frequencies of the structural mode and therefore can substitute the impulse response functions if the structure is excited by steady state white noise [9].

LSCE has been recently extended to extract system parameters in the presence of harmonic excitations even with frequencies close to the eigenfrequencies of the system [10]. This recent improvement makes it a more useful technique in identifying the response of various structures such as wind turbines, cars and ships where harmonic excitations are also present in addition to random loads. This method was also utilized to analyze the vibration measurements taken on the same turbine (at standstill) by using a Doppler laser vibrometer in a previous test [3]. Results indicate that it can be successfully used to obtain the dynamic properties of the system.

SSI is another time domain identification method utilizing state space representation. It estimates the time-invariant matrices of a linear dynamic system that best fits the processed data. Once the system matrix has been estimated, traditional eigenvalue analysis can be used to compute the natural frequencies, damping and mode shapes for the state space modes. The identification algorithm is based on the work of Peter van Overschee [11]. Several forms and algorithms of this method have been in use for testing the wind turbines in operation since the early 90s [12],[9],[13]. Hansen et al. recently applied this method to the strain gauge data measured on a 2.75 Mw wind turbine and could extract the modal parameters related to the first tower and first edgewise whirling modes [14].

3 MEASUREMENT SETUP

The measurements were performed on a pitch controlled, variable speed Nordex N80 wind turbine with a rated power of 2.5 Mw. The turbine has a rotor diameter and tower height of 80 meters and can be considered as one of the largest wind turbines that are commercially available at the time the tests were conducted. The measurements were performed by GOM mbH [15], at the ECN (Energy Research Center of the Netherlands) wind turbine test site located in Wieringermeer, the Netherlands.

Four CCD cameras were used simultaneously to monitor the dynamic behavior of the turbine in operation. The whole turbine structure was aimed to be captured in all the pictures taken which resulted in a very large area (120 m height - 80 m width) to be viewed by each camera continuously during the entire measurement period. The distance between the camera-flash light systems and the turbine was 220 meters. In order to ensure a sufficient depth accuracy the cameras were placed at a distance of 120 meters from each other which resulted in an average intersection angle of 30 degrees between the axes of cameras.

Photogrammetry usually requires some markers to be placed on the object to be tracked, the camera systems just follow the motion of these markers from different orientations and construct the 3D deformation vectors, these markers are made up of retro-reflective materials (1000 times more reflective than the background material) to increase the reflectivity of the target and to provide a better visibility. In our testing, a total of 55 markers (11 markers on each blade and 22 markers on the tower) were placed on the turbine. Since they are in the form of very thin stickers (round shaped - 400 mm diameter), they are not expected to affect the structural or aerodynamic properties of the blade. The final distribution of the markers on the turbine can be seen in Figure2. Although the picture shown in Figure2 was captured by a normal camera using its flash light only, the markers can be seen easily.

4 DATA PROCESSING AND ANALYSIS

Since the turbine was not kept at a fixed yawing angle during the measurements, the relative angle between the normal of the rotation plane and the 4 camera axes changed during the measurements. In order to prevent possible sources of error, rotation plane and the corresponding axes attached to this plane were continuously updated throughout the measurements. All the deformations shown below were defined with respect to this continuously updated rotation plane and coordinate axes.

Since a marker selected to be tracked experiences very large displacements during a complete cycle of the blade, a rigid body correction was applied to the actual deformation vectors measured and the displacement component coming from rigid body motion of the rotor was subtracted from the measured real displacement of each marker. Therefore, although the results displayed below include rotational effects they can be evaluated as if they were

Figure 2: The layout of the markers on the turbine

measured on a standstill turbine vibrating under the action of wind loads. In this way, a better visualization and an easier interpretation of the recorded data were obtained.

The tip displacements measured in flapwise direction for the 3 blades during measurement 1 can be seen in Figure3.

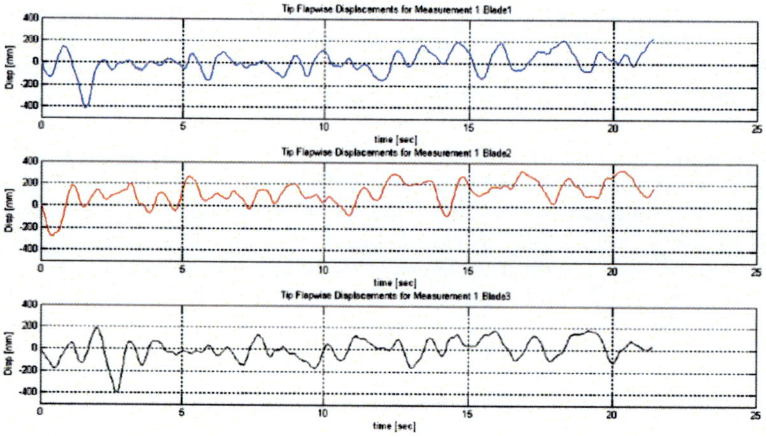

Figure 3: Tip flapwise displacements recorded during measurement 1

Figure4 and Figure5 are typical examples of PSD (Power Spectral Density) graphs of edgewise and flapwise vibrations, respectively. It should be noted that the frequencies displayed in the PSD plots below are normalized with respect to the rotational frequency. In order to protect the manufacturer's interests, the real frequencies are not given explicitly in this article. Figure4 and Figure5 are presented to provide a 3D frequency distribution which also includes the information related to the measurement location. X axis represents the normalized frequencies identified from the analyzed data; Y axis shows the marker from which the displayed data series were measured and Z axis represents the computed PSD amplitude. 1P 2P components and first edgewise mode can be identified from Figure4.

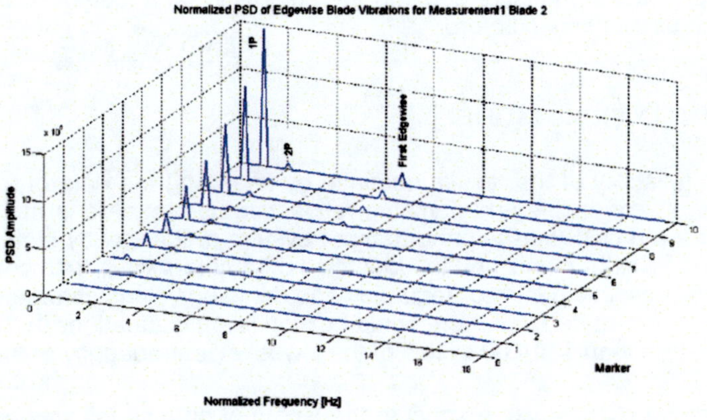

Figure 4: Normalized PSD of edgewise blade vibration for measurement 1 Blade 2

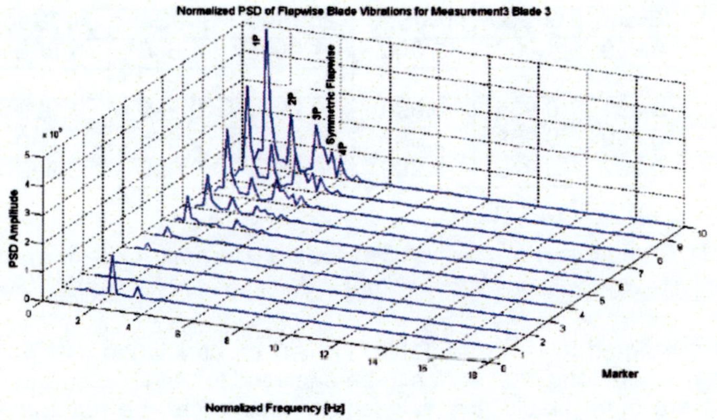

Figure 5: Normalized PSD of flapwise blade vibration for Measurement 3 Blade 3

As can be seen in Figure5, flapwise vibration data enables more frequencies to be identified. The integer multiples of rotation frequency up to 4P can be detected from the corresponding PSD graph. Besides these P components, first symmetric flapwise and some other important eigenfrequencies can also be detected from the flapwise vibrations. However, since the response is mainly dominated by the P components, the identification of these system frequencies usually requires the investigation of different time series.

It should be noted that the duration of the measurements is only 21 seconds which is much shorter than the common measurement periods used in conventional OMA applications. Generally, the measurements are expected to include at least 300-400 cycles of the lowest frequency of the system. Considering the fact that for a wind turbine similar to the one used in this project, the lowest frequency may vary between 0.2 and 0.3 Hz, the measurements should be taken for at least 1500-2000 seconds. However, the rapidly changing wind speed and/or direction and blade pitch angle make it almost impossible to have a time invariant system and steady state excitation which are the most important assumptions of operational modal analysis procedures.

Since the capacity of the hardware used was limited, no measurements longer than 21 seconds could be recorded at the time the tests were conducted. It is almost certain that the required hardware capacity will be reached very soon. On the other hand, these hardware memory limitations are not related to the quality or the accuracy of the measurements. In other words, the deformation of the blade can be measured very precisely at different locations on each blade but the use of these measurements in the current system identification procedures requires time series that are long enough as mentioned before.

5 RESULTS OF THE ANALYTICAL MODEL

In order to investigate the reliability of the results obtained by analyzing the data of relatively short durations, an analytical model with known frequencies and damping ratios was used by Meng et.al [16]. Within the scope of this work, the response of the 5 DOF mass-spring-damper system to the steady state white noise excitation was analyzed by both LSCE and SSI algorithms and the identified parameters were then compared with the known real values. The authors showed how the identified frequency and damping values change with respect to the length of the data series analyzed and concluded that there could be significant fluctuations in the identified damping ratio if the data analyzed contains less than 150-200 cycles of the lowest eigenfrequency of the system.

The change of extracted mode shapes with respect to the length of data series was also investigated. The mode shapes calculated for a sufficiently long data set were considered as reference shapes. For each measurement duration the calculated mode shapes were compared with the reference modes and MAC values showing the similarity between two modes were computed. Table1 shows the MAC values obtained for different measurement durations.

F-Exact(Hz)	D-Exact	25Sec	50Sec	100Sec	200Sec	300Sec	700Sec	2500Sec	5000Sec
0.752	0.010	0.999	1.000	1.000	1.000	1.000	1.000	1.000	1.000
0.920	0.300	-	-	-	-	-	-	0.839	1.000
1.286	0.014	0.991	0.995	0.999	0.999	0.999	1.000	1.000	1.000
1.650	0.009	0.962	0.988	0.987	0.990	0.995	1.000	1.000	1.000
1.884	0.017	0.882	0.979	0.989	0.995	0.996	0.999	1.000	1.000

TABLE 1: Comparison of the MAC (Model Assurance Criterion) values [16]

Depending on the results presented in [16] and Table1, it can be concluded that damping values are generally difficult to estimate and they require long data series to be analyzed to obtain accurate results. However, compared to damping, mode shapes are more stable and satisfactory results can be reached even for short data series. This conclusion makes it possible to use mode shapes as reliable system parameters in further analyses such as health monitoring and damage identification even if the measurement is not long enough to get accurate damping estimations.

6 IDENTIFIED SYSTEM PARAMETERS

The 3D vibration data measured by using photogrammetry while the turbine is rotating was analyzed by both SSI and LSCE. It should be noted that in order to protect the interests of the manufacturer the frequencies mentioned in this section are normalized with respect to the previously known value of the tower frequency, resulting in a dimensionless frequency. Although tower mode cannot be identified from the results presented here, this normalization makes it possible to compare the results presented in Table2 with the previous results calculated by using laser vibrometer measurements performed on the same turbine [3]. It can be concluded that all the frequencies summarized in Table2 are consistent with those reported in [3]. The increase varying between 3 % and 10 % can be attributed to tension stiffening of the blade during rotation. The first flapwise mode at the normalized frequency of 3.09 could not be identified by LSCE because of the very high damping ratio. However, it can be concluded that more frequencies can be identified with this method. The extracted mode shapes are shown through Figure6 to 9

The non-smooth shape of the mode shapes can be attributed to the relatively low accuracy of the correlation functions. It should be noted that the correlation functions used in this analysis were generated by using only

Blade Mode	F-LSCE	F-SSI
1st flapwise	-	3.09
1st edgewise	5.42	5.39
2nd flapwise	8.24	-
3rd flapwise	15.78	-
2nd edgewise	18.81	-

TABLE 2: The normalized frequencies identified by LSCE and SSI by using photogrammetric measurements

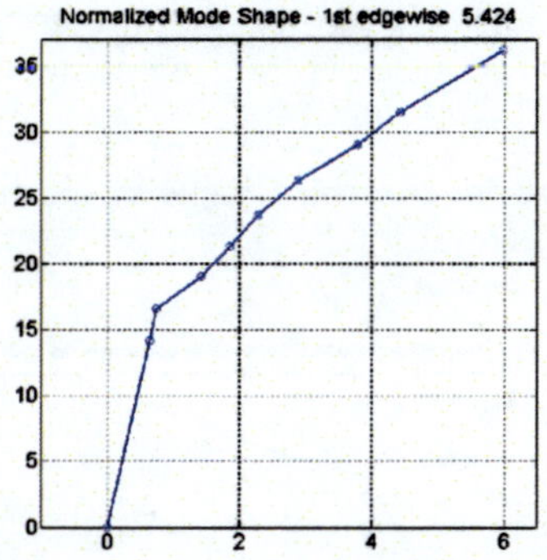

Figure 6: The normalized mode shape for the first edgewise mode(displacements in edgewise direction)

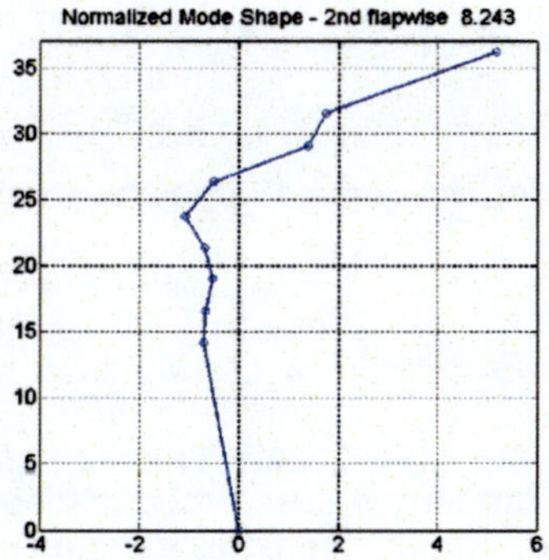

Figure 7: The normalized mode shape for the second flapwise mode(displacements in flapwise direction)

600 data points (measured for 21 seconds at 28 Hz sampling rate) which is much less than those used in similar calculations. Although it is still possible to recognize different mode shapes, depending on the purpose of the analysis, smoother shapes might be needed. However, it is always possible to improve the quality of the correlation functions and mode shapes just by increasing the measurement duration.

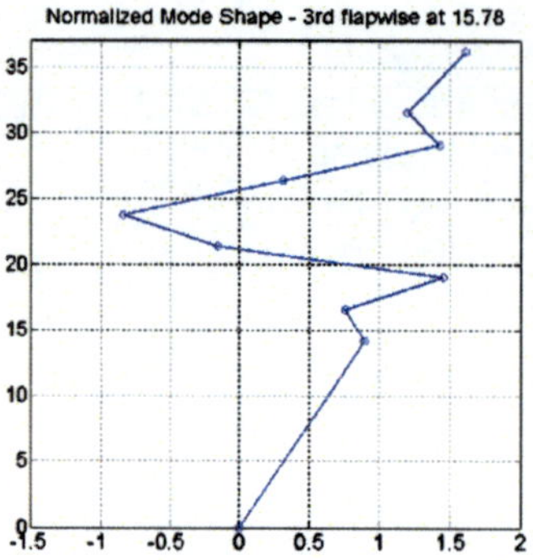

Figure 8: The normalized mode shape for the third flapwise mode (displacements in flapwise direction)

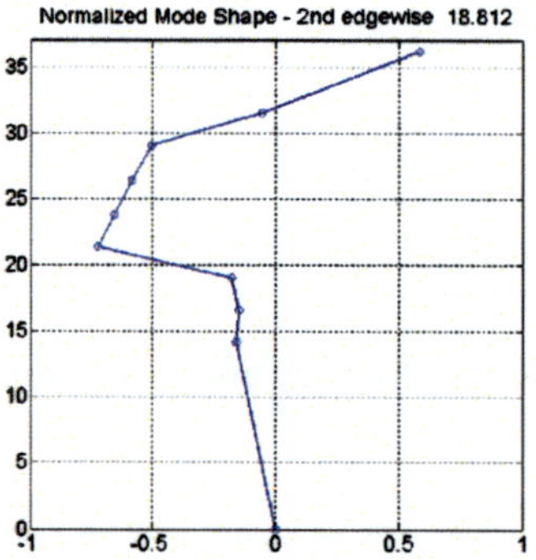

Figure 9: The normalized mode shape for the second edgewise mode (displacements in edgewise direction)

7 CONCLUSION

The results demonstrate that sub-space system identification method and least square complex exponential identification method are capable of identifying the frequencies and damping ratios of the aero-elastic mode for a large wind turbine. LSCE is not very efficient in identifying the highly damped modes. However, it can successfully identify the modes with relatively low damping even if their contribution to the overall motion is weak. SSI algorithm is able to detect the highly damped modes however, the frequencies that have a minor contribution in the response cannot be identified.

Depending on the results presented in Meng et al. [16], highly damped frequencies could be identified by LSCE only when the length of the data series were increased up to 5000 seconds. This problem is not expected to be related to the LSCE algorithm but to the basic assumptions of NEXT technique. For very highly damped systems correlation functions are not able to represent the impulse functions accurately unless the the data set is very long.

It is usually difficult to identify damping accurately and requires the analysis of long data series. However, when compared to damping, mode shapes are more stable and satisfactory results can be reached even for short data series. This conclusion makes it possible to use mode shapes as reliable system parameters in further analyses such as health monitoring and damage identification even if the measurement is not long enough to get accurate damping estimations.

Photogrammetric measurement technique provides very useful information in measuring the dynamic behavior of large wind turbines with sufficient accuracy. The initial results are quite promising. By the help of the rapid progress in hardware and corresponding sensor technology, photogrammetry is expected to replace most of the conventional measurement methods that are currently in use and become one of the most important measurement techniques utilized in monitoring wind turbine dynamics.

ACKNOWLEDGMENTS

The research is partly funded by We@Sea research program, financed by Dutch Ministry of Economical Affairs. We also acknowledge GOM mbH for performing the photogrammetric measurements [15].

REFERENCES

[1] **Mikhail, E. M.**, **Bethel, J. S.** and **McGlone, J. C.**, Introduction to Modern Photogrammetry, Wiley, Newyork, 2001.

[2] **Chang, C.** and **Ji, Y.**, *Flexible videogrammetric technique for three-dimensional structural vibration measurement*, ASCE Journal of Engineering Mechanics, Vol. 133, No. 6, 2007.

[3] **Ozbek, M.**, **Rixen, D. J.** and **Verbruggen, T. W.**, *Remote monitoring of wind turbine dynamics by laser interferometry: Phase1, 28thIMAC, International Modal Analysis Conference Proceedings*, Orlando,Florida, U.S.A., Feb. 2009.

[4] **Corten, G.** and **Sabel, J.**, *Optical motion analysis of wind turbines*, Technical Report, SV Research Group, Delft University of Technology, 1995.

[5] **Tamura, Y.**, **Matsui, M.**, **Pagnini, L.**, **Ishibashi, R.** and **Yoshida, A.**, *Measurement of wind-induced response of buildings using RTK-GPS*, J. of Wind Eng Ind Aerodyn, Vol. 90, pp. 1783–93, 2002.

[6] **Nakamura, S.**, *GPS measurement of wind-induced suspension bridge girder displacements*, Journal of Structural Engineering ASCE, Vol. 126, No. 12.

[7] **Rademakers, L.**, **Verbruggen, T.**, **van der Werff, P.**, **Korterink, H.**, **Richon, D.**, **Rey, P.** and **Lancon, F.**, *Fiber optic blade monitoring, European Wind Energy Conference*, London, November 2004.

[8] **Maia, N.** and **Silva, J.** (editors), Theoretical and Experimental Modal Analysis, Research Studies Press Ltd., Somerser, England, 1997, isbn 0 86380 208 7.

[9] **James, G. H.**, **Carne, T. G.** and **Lauffer, J. P.**, *The Natural Excitation Technique (NExT) for Modal Parameter Extraction from Operating Structures*, Journal of Analytical and Experimental Modal Analysis, Vol. 10, No. 4, pp. 260–277, October 1995.

[10] **Mohanty, P.** and **Rixen, D. J.**, *Operational modal analysis in the presence of harmonic excitation*, Journal of Sound and Vibration, Vol. 270, pp. 93–109, 2004.

[11] **van Overschee, P.** and **Moor, B. D.**, Subspace Identification for Linear Systems: Theory - Implementation - Applications, New York: Springe, 1st edition, 1996.

[12] **James, G. H.**, **Carne, T. G.** and **Lauffer, J. P.**, *The Natural Excitation Technique (NExT) for modal parameter extraction from operating wind turbines*, Technical Report SAND92-1666, Sandia National Laboratories, 1993.

[13] **James, G. H.**, **Carne, T. G.** and **Veers, P.**, *Damping measurements using operational data*, ASME Journal of Solar Energy Engineering, Vol. 118, pp. 190–193, 1996.

[14] **Hansen, M. H.**, **Thomsen, K.** and **Fuglsang, P.**, *Two methods for estimating aeroelastic damping of operational wind turbine modes from experiments*, Wind Energy, Vol. 9, pp. 179–191, 2006.

[15] ***GOM Optical Measuring Techniques***, *www.gom.com*.

[16] **Meng, F.**, **Ozbek, M.**, **Rixen, D. J.** and **van Tooren, M. J. L.**, *Comparison of System Identification Techniques for Predicting Dynamic Properties of Large Scale Wind Turbines by Using the Simulated Time Response, 28thIMAC, International Modal Analysis Conference Proceedings*, Jacksonville, Florida, U.S.A., Feb. 2010.

Proceedings of the IMAC-XXVIII
February 1–4, 2010, Jacksonville, Florida USA

Output-Only Modal Analysis of Linear Time Periodic Systems with Application to Wind Turbine Simulation Data

Matthew S. Allen,
Assistant Professor
msallen@engr.wisc.edu

Michael W. Sracic
Graduate Research Assistant, Ph.D. Candidate
mwsracic@wisc.edu
Department of Engineering Physics
University of Wisconsin-Madison
535 Engineering Research Building
1500 Engineering Drive
Madison, WI 53706

Shashank Chauhan
Brüel & Kjær
Sound & Vibration Measurement A/S
Skodsborgvej 307
DK-2850 Nærum, Denmark
schauhan@bksv.com

&

Morten Hartvig Hansen
Wind Energy Department
Risø National Laboratory
P.O. Box 49, DK-4000 Roskilde
Denmark

Abstract:

Many important systems, such as turbomachinery, helicopters and wind turbines, must be modeled with linear time-periodic equations of motion to correctly predict resonance phenomena. Time periodic effects in wind turbines might arise due to stratification in the velocity of the wind with height, changes in the aerodynamics of the blades as they pass the tower and/or blade-to-blade manufacturing variations. These effects may cause parametric resonance or other unexpected resonances, so it is important to properly characterize them so that these machines can be designed that achieve the high reliability, safety, and long lifetimes demanded to provide economical power. This work presents an output-only system identification methodology that can be used to identify models for linear, periodically time-varying systems. The methodology is demonstrated for the simple Mathieu oscillator and then used to interrogate simulated measurements from a rotating wind turbine, revealing that at least one of the turbine's modes seems to be time-periodic. The measurements were simulated using a state-of-the-art wind turbine model called HAWC2 that includes both structural dynamic and aerodynamic effects. Simulated experiments such as this may be useful both to validate dynamic models for a turbine or to obtain a set of time-periodic equations of motion from a numerical model; these are not readily available by other means due to the way that the aeroelastic effects are treated in the simulation code.

T. Proulx (ed.), *Structural Dynamics and Renewable Energy, Volume 1*, Conference Proceedings of the Society for Experimental Mechanics Series 10, DOI 10.1007/978-1-4419-9716-6_33, © The Society for Experimental Mechanics, Inc. 2011

1. Introduction

Linear time-periodic (LTP) differential equations are important for quite a wide range of systems including helicopters [1-7], turbomachinery [8-10] wind turbines [11-13] and virtually any system that can be linearized about a periodic trajectory (see, e.g. [14-17]). Allen & Sracic have even proposed that a general nonlinear system might be more easily identified if excited such that an LTP model is appropriate [18]. Other researchers have sought to exploit LTP effects to detect defects, such as a cracked shaft, in a rotating system [8-10].

This work explores LTP identification for wind turbines. Time-periodic models are necessary to model wind turbine dynamics if the turbine has only two blades, or if the blades of a 3+ blade turbine are not perfectly identical, since the turbine tower is always anisotropic. In practice there are always differences between the blades but these are often not considered in design, analysis or even when performing tests. Time-periodicity can also arise due to aeroelastic effects such as interaction between the flow around the blade and tower or variation in the wind speed with altitude. Most of these effects are difficult to model and there is considerable uncertainty in the parameters that should be used, so experimental methods are needed both to validate existing models and to probe experimental measurements to see whether time-periodic effects should be included in future models.

A few system identification strategies have been proposed for linear time-periodic systems. Many of them are extensions of system realization methods (e.g. the Eigensystem Realization Algorithm [19].) For example, Liu presented a discrete-time approach of this form in [20], Verhagen presented a subspace method [21], and others have investigated this as well [22, 23]. Peters and Fuehne independently developed a related approach that they call Generalized Floquet Theory [2, 7, 24, 25], which seeks to derive the Floquet exponents of the time-periodic system from measured time series. Other researchers have explored optimization-based approaches [26, 27]. While those works are valuable, frequency domain system identification approaches are preferred in many instances because one can more easily make sense of complicated measurements and averaging can be performed to reduce long, cumbersome time series to a compact representation of the system's response. Wereley and Hall extended the concept of the frequency response function (FRF) to LTP systems [28-30], paving the way for frequency domain identification of LTP systems. A number of other researchers subsequently presented many similar concepts [31], one group taking a complex frequency response approach [32]. However, to date only a few works have sought to implement these concepts experimentally and sometimes the results have been disappointing [3, 5, 33]. One exception is continuous-scan laser Doppler vibrometry (CSLDV), which can also be thought of as a special case of LTP system identification. A number of effective methods have been developed for CSLDV and they have been used with real measurements demonstrating very good results in many cases [34-42]. On the other hand, many of those methods are specialized to CSLDV and not immediately applicable to time-periodic systems in general.

This work uses the harmonic transfer function concept developed by Wereley [28] to create an output-only system identification method for linear time-periodic systems. The proposed method can be thought of as an extension of Operational Modal Analysis (OMA) [43-45] or Output-only modal analysis [46-48] to time-periodic systems. One of the first OMA algorithms in the structural dynamics community was Natural Excitation Technique (NExT) [49] by James & Carne, and in the intervening years there have been hundreds of papers on OMA and there is even now a conference devoted to the topic. As with OMA for time-invariant systems, the theory presented here can be used to extract the time-periodic mode vectors of a system from an (augmented) auto-spectrum of the response if the input obeys certain assumptions.

The rest of this paper is organized as follows. Section 2 derives the proposed output-only identification technique with particular emphasis on how to interpret the response spectra of an LTP system. In Section 3 the proposed methodology is demonstrated by identifying a model for a simple time-periodic system (the Mathieu oscillator) from simulated measurements. The method is then applied to simulated measurements from a wind turbine in Section 4, and Section 5 summarizes the conclusions.

2. Theoretical Development

The state space equations of motion of a linear time-periodic (LTP) system can be written as follows,

$$\dot{x} = A(t)x + B(t)u \quad , \quad y = C(t)x + D(t)u \tag{1}$$

where $A(t+T_A) = A(t)$ and the other matrices are periodic as well with the same period. The fundamental frequency of the time-periodic system is denoted $\omega_A = 2\pi/T_A$. The state transition matrix gives the free response of such a system at time t via the relationship,

$$x(t) = \Phi(t,t_0)x(t_0)\,. \tag{2}$$

One can also write the forced response of the system in terms of the STM as follows.

$$y(t) = C(t)\Phi(t,t_0)x(t_0) + \int_{t_0}^{t} C(t)\Phi(t,\tau)B(\tau)u(\tau)d\tau \tag{3}$$

Floquet theory reveals that, in the absence of degenerate roots, the state transition matrix of an LTP system can be represented as a modal sum [9, 10],

$$\Phi(t,t_0) = \sum_{r=1}^{N} \psi_r(t)L_r(t_0)^T \exp\left(\lambda_r(t-t_0)\right) \tag{4}$$

where λ_r is the rth Floquet exponent of the state transition matrix, ψ_r is the rth time-periodic mode vector of the STM and L_r is the rth column of $\begin{bmatrix}\psi_1(t) & \psi_2(t) & \cdots\end{bmatrix}^{-T}$. The Floquet exponents of an LTP system are analogous to the eigenvalues of a Linear Time Invariant (LTI) system, which can be written in terms of the damping ratio ζ_r and natural frequency ω_r as $\lambda_r = -\zeta_r\omega_r + i\omega_r\sqrt{1-\zeta_r^{\ 2}}$ for an underdamped mode. (Note that the eigenvectors of $A(t)$ have not been mentioned here; the modal parameters of the STM are more useful for determining the stability of the system and understanding its response, and they are more convenient to identify experimentally, so we shall focus on them. For an LTI system the STM and $A(t)$ share the same eigenvectors. Most system identification routines for LTI systems also seek to find the modal parameters of the STM, for example those based on a discrete-time representation, which is a special case of eq. (3) for a constant time step (see, e.g. [45, 50].)

Since the eigenvectors of the STM are periodic, it is convenient to expand them in a Fourier series. In practice the state vector is usually not measured but some vector of outputs ($y(t)$ in eq. (1)), so we shall concern ourselves only with the observed mode vectors, $C(t)\psi_r(t)$. As is often the case for LTI systems, $C(t)$ may simply be a matrix of ones and zeros indicating which of the states were measured, or one may have more measurement points than states and then $C(t)$ has more rows than columns. In any event the expansion is,

$$C(t)\psi_r(t) = \sum_{n=-\infty}^{\infty} \bar{C}_{r,n}e^{jn\omega_A t} \quad , \quad L_r(t)^T B(t) = \sum_{n=-\infty}^{\infty} \bar{B}_{r,n}e^{jn\omega_A t} \tag{5}$$

where $\bar{C}_{r,n}$ is the nth vector of Fourier coefficients of the rth observed mode vector, $C(t)\psi_r(t)$, and similarly for $L_r(t)^T B(t)$.

It is well known that the response of a LTI system to a complex exponential input, $u(t)$, results in a complex exponential output, $y(t)$, although with different magnitude and phase. In contrast, a linear time-periodic system may respond to a single sinusoid at an infinite number of frequencies, which caused some to believe that the concept of a transfer function couldn't be extended to LTP systems [31]. Wereley addressed this difficulty using an exponentially modulated periodic (EMP) signal space, in which u and y are described as follows.

$$u(t) = \sum_{n=-\infty}^{\infty} u_n e^{(j\omega+jn\omega_A)t} \quad , \quad y(t) = \sum_{n=-\infty}^{\infty} y_n e^{(j\omega+jn\omega_A)t} \tag{6}$$

These signals are useful because an EMP input signal causes an EMP output, although with different magnitude and phase at each frequency in the comb, so a concept that is completely analogous to the LTI transfer function can be derived. This is done by defining vectors of the harmonics, y_n, of output, $\mathbf{y}(\omega) = \begin{bmatrix}\cdots & y_{-1}^{\ T} & y_0^{\ T} & y_1^{\ T} & \cdots\end{bmatrix}^T$, and similarly for the input, then writing a transfer function relating input and output EMP signals, which is dubbed the Harmonic Transfer Function (HTF) [28].

$$\mathbf{y}(\omega) = \mathbf{G}(\omega)\mathbf{u}(\omega) \tag{7}$$

One can also write $\mathbf{y}$ as,

$$\mathbf{y}(\omega) = \begin{bmatrix} \cdots & Y(\omega - 2\omega_A)^T & Y(\omega - \omega_A)^T & Y(\omega)^T & Y(\omega + \omega_A)^T & Y(\omega + 2\omega_A)^T & \cdots \end{bmatrix}^T \tag{8}$$

where $Y(\omega)$ is the Fourier transform of $y(t)$. Hence, one can construct the EMP signals $\mathbf{u}$ and $\mathbf{y}$ from measurements of the input and response $u(t)$ and $y(t)$. Theoretically, one must consider an infinite number of harmonics to characterize an LTP system, yet one would expect that most systems can be well approximated with a finite, perhaps even small number.

For LTI systems, the FRF matrix has a well-known form with peaks near the natural frequencies of each mode, and with the FRF matrix approximating the shape of the corresponding eigenvector near that frequency. A similar relationship must be derived for LTP systems. This was done by substituting eq. (4) into eq. (3), dropping transient terms (assuming steady state) and using a Fourier series expansion to simplify the expression. Wereley's thesis shows an example of this type of calculation. After much algebra, one obtains the following expression for $\mathbf{G}$ in terms of the modal parameters of the STM.

$$\begin{aligned} \mathbf{G}(\omega) &= \sum_{r=1}^{N} \sum_{l=-\infty}^{\infty} \frac{\overline{\mathbf{C}}_{r,l} \overline{\mathbf{B}}_{r,l}}{i\omega - (\lambda_r - il\omega_A)} + \mathbf{D} \\ \overline{\mathbf{C}}_{r,l} &= \begin{bmatrix} \cdots & \overline{C}_{r,-1-l}{}^T & \overline{C}_{r,-l}{}^T & \overline{C}_{r,1-l}{}^T & \cdots \end{bmatrix}^T \\ \overline{\mathbf{B}}_{r,l} &= \begin{bmatrix} \cdots & \overline{B}_{r,l+1} & \overline{B}_{r,l} & \overline{B}_{r,l-1} & \cdots \end{bmatrix} \end{aligned} \tag{9}$$

where the nth term in the vector $\overline{\mathbf{C}}_{r,l}$, is $\overline{C}_{r,n-l}$, the $(n\text{-}l)$th Fourier coefficient of $C(t)\psi_r(t)$, $\overline{B}_{r,l-m}$ is the $(l\text{-}m)$th Fourier coefficient of $L_r(t)^T B(t)$ and the(n,m)th element of $\mathbf{D}$ is the $(n\text{-}m)$th Fourier coefficient of $D(t)$. This expression has exactly the same mathematical form as the expression for the FRF matrix in terms of the modal parameters for a linear time-invariant system, so the same algorithms can be applied to identify the parameters of the LTP system and the same intuition that one uses to interpret FRFs can also be used to interpret HTFs. There are, however, a few differences that must be noted:

- An LTP system of order N can potentially have an infinite number of peaks in its HTF, depending on how many terms in $\overline{\mathbf{C}}_{r,l}$ and $\overline{\mathbf{B}}_{r,l}$ are nonzero. Each peak will occur near the imaginary part of the Floquet exponent λ_r plus some integer multiple of the fundamental frequency ω_A. If the mode shapes of the system are constant in time, then $\overline{\mathbf{C}}_{r,l}$ and $\overline{\mathbf{B}}_{r,l}$ contain only one nonzero term and eq. (9) reduces to the familiar relationship for an LTI system.
- Whereas the mode shapes of an LTI system describe the spatial pattern of deformation of a mode, the observed mode shapes of an LTP system, $\overline{\mathbf{C}}_{r,l}$, also reveal how that shape changes with time. A single-output LTP system still has a mode shape that is a vector, with each element in that vector being a Fourier coefficient describing how the mode shape at that one output point changes with time.

In the applications that are of interest in this work the input is not known, so we desire to determine the modal parameters of the system from the output signal only. To do this, we write the power spectrum of the output in terms of the HTF matrix and the input, resulting in the following,

$$S_{yy}(\omega) = \mathbf{G}(\omega) S_{uu}(\omega) \mathbf{G}(\omega)^H \tag{10}$$

where the power spectra can be found by applying the usual estimation algorithm to the EMP signals,

$$S_{yy}(\omega) = \frac{1}{M} \sum_{k=1}^{M} \mathbf{y}_k(\omega) \mathbf{y}_k(\omega)^H \tag{11}$$

with $()^H$ denoting the Hermitian or complex-conjugate transpose. Replacing $\boldsymbol{G}$ with its modal representation in eq. (10), one obtains the following for the case where $D(t) = 0$,

$$\left[S_{yy}(\omega)\right]=\sum_{r=1}^{N}\sum_{l=-\infty}^{\infty}\sum_{s=1}^{N}\sum_{k=-\infty}^{\infty}\frac{\overline{\mathbf{C}}_{r,l}\mathbf{W}(\omega)_{r,s,l,k}\overline{\mathbf{C}}_{s,k}^{\ H}}{\left[i\omega-(\lambda_r-il\omega_A)\right]\left[i\omega-(\lambda_s-ik\omega_A)\right]^H}$$
$$\mathbf{W}(\omega)_{r,s,l,k}=\overline{\mathbf{B}}_{r,l}S_{uu}(\omega)\overline{\mathbf{B}}_{s,k}^{\ H} \tag{12}$$

where $\mathbf{W}(\omega)$ is a function of the input spectrum and the input characteristics of the system. This shows that the autospectrum of the output can be approximated by a sum of modal contributions if $\mathbf{W}(\omega)$ is reasonably flat. (For LTI systems this requires that the input spectrum $S_{uu}(\omega)$ be flat. Here the requirement is similar, yet perhaps more stringent.) The dominant terms in the summation above are those for which $j\omega-(\lambda_r-jl\omega_A)$ and $j\omega-(\lambda_s-jk\omega_A)$ are both minimum at the same frequency. If the sidebands for mode r do not overlap with those for mode s, then this occurs when $r = s$ and $l = k$ and the expression becomes,

$$\left[S_{yy}(\omega)\right]=\sum_{r=1}^{N}\sum_{l=-\infty}^{\infty}\frac{\overline{\mathbf{C}}_{r,l}\mathbf{W}(\omega)_{r,l}\overline{\mathbf{C}}_{r,l}^{\ H}}{\left[i\omega-(\lambda_r-il\omega_A)\right]\left[i\omega-(\lambda_r-il\omega_A)\right]^H} \tag{13}$$

so the shape of the autospectrum is proportional to the dominant mode shape and a particular element of the autospectrum plotted versus frequency has the shape of an LTI mode squared. The extension of these ideas to nonzero $D(t)$ is straightforward. As with operational modal analysis for LTI systems, there are additional scale factors that must be determined to find the response of the LTP system to an input or an initial condition, but the modal parameters are still useful for evaluating the response of the system qualitatively.

The proposed output-only system identification routine can be summarized as follows:

1. Record the response, $y(t)$, of an LTP system to a broadband input.
2. Form a vector, $\mathbf{y}$, of frequency shifted copies of the output as shown in eq. (8) by using the fact that $Y(\omega-n\omega_A)=\mathrm{FFT}[y(t)*\exp(-in\omega_A t)]$.
3. Compute the autospectrum of $\mathbf{y}$ by averaging multiple blocks of the time history as shown in eq. (11).
4. Use an LTI output only system identification routine to identify the modal parameters from the autospectrum, using eqs. (12) and (13) to interpret the results and reconstruct the modal parameters of the STM. One of the simplest applicable identification methods is the peak picking method.
5. Reconstruct $C(t)\psi_r(t)$, using the identified Fourier coefficients, $\overline{\mathbf{C}}_{r,l}$, if desired.

3. Simulated Application: Mathieu Oscillator

A single degree-of-freedom Mathieu Oscillator was used to evaluate the proposed identification methodology on a very simple time-periodic system. Figure 1 shows a schematic of the Mathieu Oscillator.

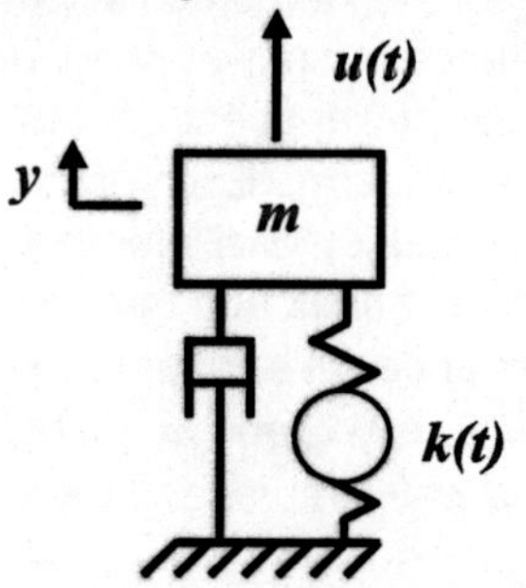

Figure 1: Mathieu Oscillator with a time varying spring stiffness.

The spring stiffness for the system varies periodically with frequency ω_A according to eq. (14).

$$k(t)=k_0+k_1\cos(\omega_A t) \tag{14}$$

After normalizing by dividing out the oscillator mass, the system has the following equation of motion

$$\ddot{y} + 2\zeta\omega_0\dot{y} + \left(\omega_0^2 + \omega_1^2 \cos(\omega_A t)\right)y = \frac{u(t)}{m} \tag{15}$$

with $2\zeta\omega_0 = c/m$, $\omega_0^2 = k_0/m$, $\omega_1^2 = k_1/m$, and input $u(t)$.

The Mathieu oscillator was used to simulate an output-only response. The parameters used for the oscillator were m=1, k_0=1, k_1=0.4, and ω_A=0.8 rad/s. A damping ratio of 0.02 was desired so the damping coefficient was set to c=0.4. The system was subjected to wide-band random white noise input, which was generated by passing a vector of randomly generated numbers through an 8th-order low-pass Butterworth filter with a cutoff frequency of 10 rad/s. Figure 2 shows a plot of the input signal.

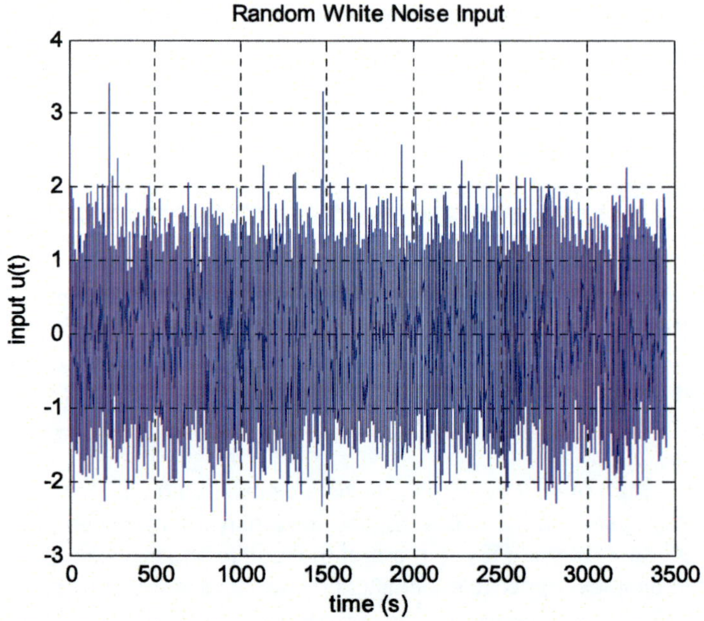

Figure 2: Random white noise input signal.

The response of the Mathieu oscillator was then determined by integrating the system's state space equations from zero initial conditions with MATLAB's adaptive Runge-Kutta (ode45) routine over the duration of the white noise input signal. The input was then discarded and the output of the integration treated as a measurement. The output measurement was then exponentially modulated with harmonics n=-2,-1,0,1, and 2 to form y in eq. (8), the EMP output signal. The output power spectrum was then calculated according to eq. (11). Figure 3 shows a plot of all of the terms in the power spectrum matrix along the primary row (e.g. with m=0 and n = -2…2). The legend gives the n-value for each term. Each of the spectra show strong peaks near ω=1 rad/s, which would be the natural frequency of the Mathieu oscillator in the case where k_1=0 (e.g. a time-invariant system). The n=0 spectrum contains the largest peak and the remaining become progressively smaller for n=-1,1,-2, and then 2. Two other strong groups of peaks appear in all of the spectra at ω=0.2 rad/s, where the peaks have the following order of dominance, n=[1,0,-1,2,-2] and ω=1.8 rad/s with ordering n=[-1,-2,0,1,2]. The remaining groupings of peaks occur at ω=2.6 rad/s and ω=0.6 rad/s, and there is evidence of a peak in the n=-2 curve near ω=3.4 rad/s.

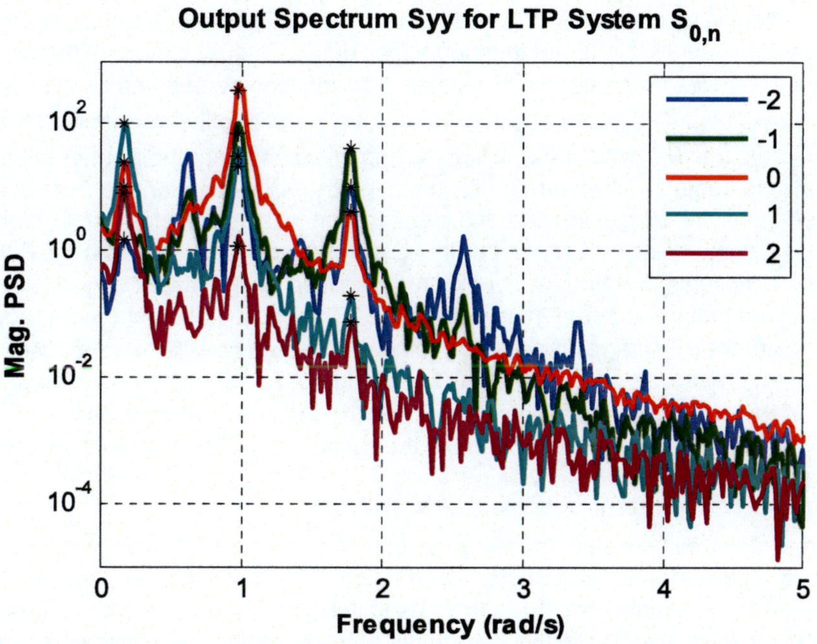

Figure 3: Harmonic PSD of the output for the Mathieu oscillator response computed using eqs. (8) and (11). Five curves are shown corresponding to the m,nth element of S_{yy} for m = 0 and n = -2…2.

The peak picking method was used to find the shape of the power spectrum matrix at the peaks at ω=0.2, 1, and 1.8 rad/s. The values of the spectra from each of those peaks were normalized by dividing by the largest value so that the shape could be compared, and the result is shown in Table (1). The analytical Fourier Coefficients for the Mathieu oscillator were also found by integrating the LTP equations of motion over the fundamental period, solving for $\psi(t)$ and then expanding it in a Fourier series to obtain the vector $\bar{\mathbf{C}}_{r,l}$ in eq. (9). Those Fourier coefficients were then normalized by the maximum and are also shown. (Additional detail regarding how $\bar{\mathbf{C}}_{r,l}$ was derived is provided in Appendix A.)

n	*Fourier Coefficients of the Mathieu Oscillator Mode Shape:*			
	Analytical	**0.2 rad/s**	**1 rad/s**	**1.8 rad/s**
-3	0.0135	0.0047 + 0.014i	-	-
-2	0.067	0.076 + 0.024i	0.064 + 0.015i	-
-1	-0.220	-0.26 + 0.0009i	-0.21 - 0.0021i	-0.24 + 0.0011i
0	1	1	1	1
1	0.092	0.097 + 0.0065i	0.095 + 0.0043i	0.10 + 0.0068i
2	0.0032	-	0.0036 + 0.0006i	0.0044 + 0.0009i
3	0.00004	-	-	-0.0011 - 0.0014i

Table 1 Identified mode shape coefficients at peaks occurring at ω=0.2, 1.0 and 1.8 rad/s for spectra corresponding to n=-3,-2,-1,0,1, 2 and 3. The analytical Fourier coefficients of the Floquet mode shape of the Mathieu oscillator are also provided.

Equations (12) and (13) revealed that each mode of an LTP system is manifest at multiple frequencies in the output PSD (i.e. for various *l*). If the linear term in the mode shape is the largest, then one can surmise that the peak occurring in the primary (0,0) element of the PSD matrix will be at the true natural frequency. (Mathematically, it does not matter which peak is the primary, so this assumption can always be made and one will still obtain a valid model.) Following this line of thinking, the dominant peak in Figure (3) occurs at ω=1.0

rad/s, so the Floquet exponent is taken to be λ=1.0i, and the corresponding natural frequency is ω_0 =1.0. The remaining peaks in the spectra can be attributed to modulations of this Floquet exponent by the system's fundamental frequency ω_A=0.8 rad/s. For example, the peaks 0.2 rad/s can be seen as ω_0-1*ω_A while the peaks at 1.8 rad/s can be seen as ω_0+1*ω_A. The peaks at 0.6 rad/s are due to the second harmonic of the conjugate of the Floquet exponent at -1.0 rad/s (i.e. 0.6 rad/s = -ω_0+2*ω_A). Using this reasoning, the Fourier coefficients can be arranged in rows as shown in Table 1. There are multiple estimates for the Fourier coefficients for n = -1...1, and all of the estimates agree quite well suggesting that the identification was successful. Other coefficients were only estimated from a few of the frequencies (columns), but those are small so it appears that the Fourier series expansion is converging. Comparing the Fourier coefficients identified by the output only technique with the analytical ones, one observes that the agreement is quite good. One could now average the Fourier coefficients that were obtained, reconstruct $C(t)\psi(t)$ and diagnose troublesome frequencies in the response. All of the above was surmised from the output response and without any knowledge of the system except the fact that it was time-periodic with period ω_A. If the scale factors on the mode shape $C(t)\psi(t)$ could be estimated by some means, then one could also reconstruct the time-varying state matrix of the system using the procedure in [10].

4. Simulated Application: Wind Turbine modeled with HAWC2

The proposed methodology was also applied to simulated measurements from a representative 5MW wind turbine (3-blade, 126 meter diameter, 100 meter tower height) [51] modeled using the HAWC2 simulation tool [52-54]. This same turbine and simulation code were used in a previous paper where Subspace Identification and Blind Source Separation were used to identify the modes of the turbine when it was in a parked condition [55]. That work identified the lower modes of the turbine, which had the following frequencies: 0.273 Hz (lateral tower bending), 0.275 Hz (longitudinal tower bending), 0.564 Hz (drive train torsion), 0.604 Hz (yaw), 0.635 Hz (tilt), 0.698 Hz (symmetric flapwise bending), 0.951 Hz (edgewise vertical bending), 0.975 Hz (edgewise horizontal bending), 1.526 Hz, 1.655 Hz, 1.74 Hz. Representative mode shapes are also shown in [55]. Here the rotating turbine will be studied. The HAWC2 model includes a structural model based on Timoshenko beam elements for the bending of the blades and tower, aerodynamic excitation of the wind turbine from turbulence in the flow, dynamic inflow, dynamic stall, skew inflow, shear effects on the induction and effects from large deflections. The wind speed for this model is 18 m/s at hub height with a logarithmic wind shear. Because of the wind shear, the flow around each blade changes as the blade rotates, so one would expect that the system might exhibit time-periodic dynamics due to the periodically changing aeroelasticity of the blades.

Measurements were simulated from five points along each blade and ten points along the tower, in all three directions at each point for a total of 75 measurement locations. The rotational speed of the turbine and its instantaneous angle were also "measured", the former averaging 0.20157 Hz (12.1 rev. per minute) with a standard deviation of 0.000312 Hz. The traditional power spectrum of the response was found using eight sensors as references: the lateral and longitudinal sensors at the tip of the tower and the edgewise and flapwise sensors at the tip of each blade, so the resulting PSD matrix was 75 by 8. Figure 4 shows the average of all of these power spectra, as well as the average of only those elements of the PSD matrix corresponding to the blades and the tower individually.

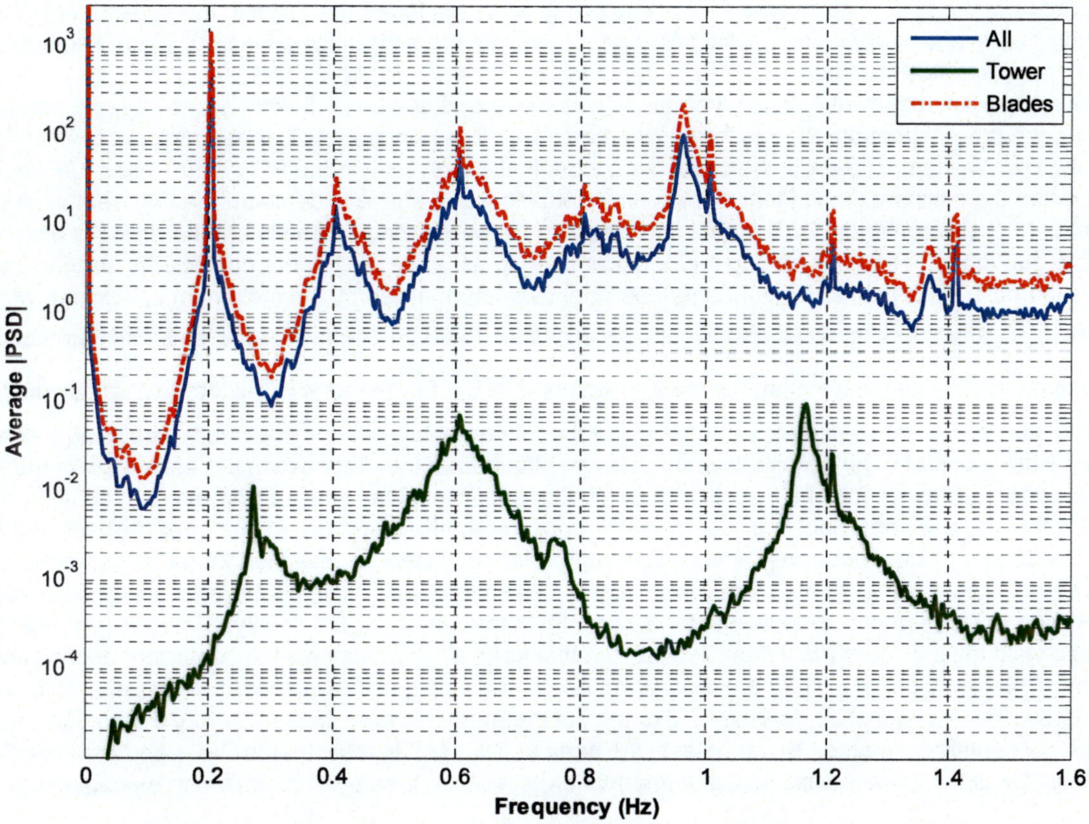

Figure 4: Average of Traditional Power Spectra of rotating wind turbine over all measurement points on the tower and over all of the blades. The average rotational speed of the turbine is 0.20157 Hz.

The effect of the rotation of the blades is readily apparent in the PSD, especially if one compare Figure 4 with the power spectra shown in [55] for the parked condition. There is a strong harmonic at the rotation frequency 0.2 Hz, and other significant ones at integer multiples of the rotation frequency. There are also broad, triangular-shaped peaks in the blade spectra at 0.2, 0.4 and 0.6 Hz, as well as a weaker one at 0.8 Hz. In the tower spectra there is only one such peak at 0.6 Hz, which is the frequency with which blades pass the tower. These peaks dominate the spectra at those frequencies, so none of the modes of the turbine are visible except for the peaks near 0.27 and 1.15 Hz in the tower response and the one near 0.95 Hz in the blade response. Comparing these frequencies with those from the parked condition, it seems reasonable to ascribe the peak near 0.27 Hz to the tower bending modes and the peak near 0.95 Hz to the edgewise blade modes. The parked turbine did not have any modes between 0.9 and 1.5 Hz, but the peak in the tower spectrum at 1.15 Hz can be understood to be a forward whirling mode due to the two edgewise blade modes at 0.95 Hz (0.95 Hz + ω_A). One would also expect to see the other of the two edgewise modes turn into a backward whirling mode as the blade rotates and appear at 0.95 Hz - ω_A. There is a very weak peak near 0.75 Hz so this is probably the case.

There is one coherent peak in the PSD that is not explained by the linear time invariant system, the peak occurring at 1.35 Hz in the blade responses. In the blade's reference frame the edgewise modes should not shift in frequency due to the rotation of the propeller, and the fact that this peak is spaced from the 0.95 Hz modes by an integer multiple of the rotation frequency suggests that it might be due to time-periodic effects. This warrants further investigation using the output-only identification presented in Section 2. Before doing so the following filtering was performed to attempt to minimize the noise in the signal due to the harmonics of the rotation frequency. First, the simulated measurement was resampled synchronous with the best-fit rotation frequency, which was found to be 0.20167 Hz. Then the FFT was found and all of the harmonics of the rotation frequency were deleted from the FFT. The inverse FFT of this signal was then found and used in the subsequent processing. This is potentially a non-causal filtering process, but the harmonics were troublesome so such a drastic approach seemed justified.

The edgewise response of blade 3 at the tip was expanded using eq. (8) with $n = -2\ldots2$ (so $\mathbf{y}$ was a 5 by 1 vector), and the power spectrum S_{yy} was found using eq. (11). The primary row in that matrix is shown in Figure 5 from 0 to 1.6 Hz. The largest peak in the spectrum occurs in the (0,0) term in the PSD matrix near 0.95 Hz, suggesting that the constant term in that mode's time-periodic mode shape is dominant. A coherent peak also occurs in the (0,2) element of the PSD at that same frequency, suggesting that a second harmonic is also important to that mode. The peak picking method was used to estimate the Fourier coefficients for each mode at the 0.95, 1.15 and 1.35 Hz peaks. Each Fourier coefficient was normalized by the maximum coefficient, which was ascribed to $\overline{C}_{r,0}$, and the normalized coefficients are shown in Table 2. The five coefficients estimated at each frequency are offset in the table so that the rows correspond to $\overline{C}_{r,n}$ for a certain n. All three columns consistently give the real part of the n=2 coefficient as about 5% of the fundamental, with the imaginary part being a consistent proportion of the real part. The second column (1.15 Hz) differs from the other two in the estimation of the n=1 coefficient, but visual inspection of the data from which that coefficient was derived (the red (0,0) curve in Figure 5) suggests that the true coefficient is buried in the response of the fundamental, so it should probably not be trusted. The 1st and 3rd columns consistently give that Fourier coefficient as less than 2% of the fundamental.

The coefficients for $n < 0$ are more problematic. The spectra that these coefficients are derived from are very noisy, due to the harmonic response of the turbine that was discussed previously. For example, the (0,-2) curve would yield the n=-2 coefficient near 0.95 Hz, but that peak is very noisy, so the coefficient doesn't seem trustworthy. The peaks below 0.95 Hz are incoherent due to the noise, so they cannot be used either. It seems that an improved method must be devised for filtering out the harmonic contributions before the coefficients for n<0 can be confidently estimated. Nevertheless, the peak at 0.95 Hz suggests that the n=-1 and n=-2 coefficients are less than 2% of the fundamental, so if that information is reliable, then perhaps they can be neglected.

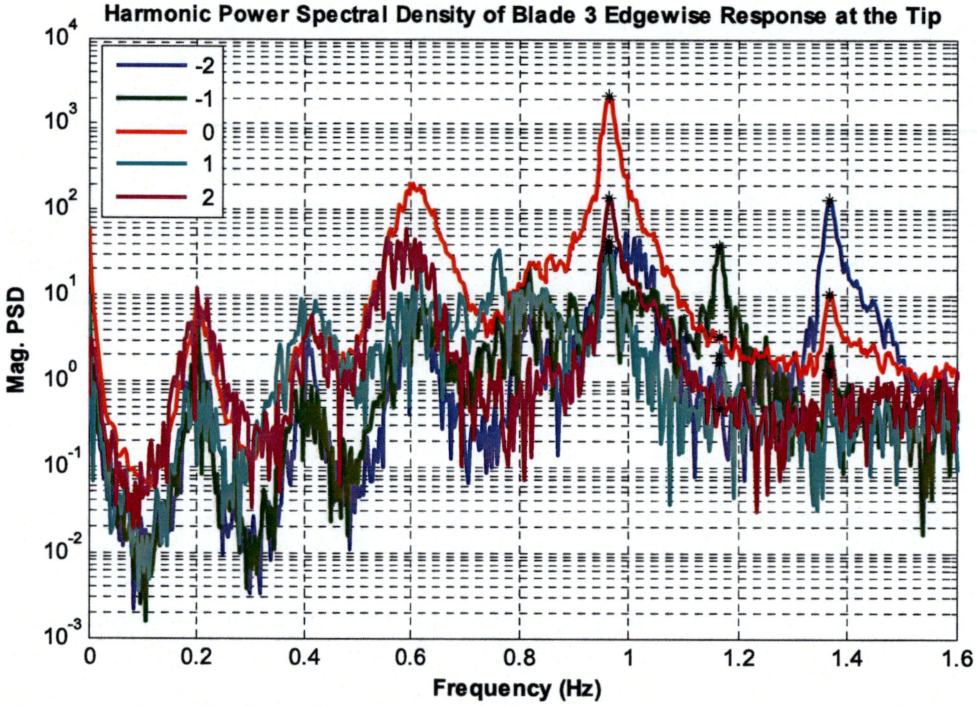

Figure 5: Harmonic PSD of the edgewise response of the tip of blade 3, computed using eqs. (8) and (11). Five curves are shown corresponding to the m,nth elements of S_{yy} for m = 0 and n = -2…2.

n	*Fourier Coefficients of B3-Tip EW Mode Shapes:* $\bar{C}_{r,n}$		
	0.95 Hz	**1.15 Hz**	**1.35 Hz**
-2	-0.0013 + 0.020i	-	-
-1	-0.00052 + 0.015i	0.011 + 0.043i	-
0	1	1	1
1	0.017 + 0.0041i	0.090 + 0.021i	0.015 + 0.0014i
2	-0.053 - 0.033i	-0.043 - 0.035i	-0.066 - 0.041i
3	-	-0.0028 - 0.013i	0.0068 + 0.0056i
4	-	-	0.0039 + 0.0095i

Table 2: Normalized Fourier Coefficients for the mode shape at the tip of blade 3 in the edgewise direction. Each column gives an estimate of the same Fourier coefficients.

4.1. Discussion

In [55] it was noted that the turbulence box used in the wind turbine model had a finite length and was repeated several times during the course of the simulation, so for the parked turbine sharp harmonic content was observed at 1.354 Hz and its multiples. It is possible that the peak in the blade response spectra at 1.35 in Figure 4 might also be due to this effect, but there is considerable evidence to the contrary. First, the turbine is rotating at $\omega_A = 0.20157$ Hz, and 1.354 Hz is not evenly divisible by 0.20157 Hz, so each time the turbulence box repeats the blades are in a different orientation causing them to see a non-periodic pattern of turbulence. Second, the measurements from the rotating turbine do not show peaks at 2.708 Hz, 4.062 Hz or any other integer multiples of 1.354 Hz, further confirming that the turbine's response is not periodic due to the repeating turbulence box. Third, the shape of the peak seen in Figure 4 at 1.35 Hz is not at all like the peaks due to the repeating turbulence box in the parked measurements [55]. Those peaks were sharp, nearly single-line peaks and the spectra around them had a strange character as well. Finally, when the LTP identification method is applied to the turbine responses, the spectra of the exponentially modulated response in Figure 5 show numerous peaks at integer multiples of 0.20157 Hz. This means that the turbine's response at distinct frequencies that are separated by 0.20157 Hz is coherent, evidence that the system is time periodic. If the response at $(\omega \pm \omega_A)$ were not coherent then the spectra would be weak and have the appearance of noise at these frequencies. For example, notice the peak in the (0,-2) element of the PSD matrix (blue). At 0.95 Hz that element of the PSD matrix is the product of the response at 0.95 Hz and that at $0.95-2*0.20157 = 0.55$ Hz. Figure 4 shows that the response at 0.55 Hz is large due to bleeding of the harmonics of the rotational frequency, yet this peak does not particularly stand out in the (0,-2) element of the PSD, presumably because the response at 0.55 Hz is not systematically related to that at 0.95 Hz.

5. Conclusions

A new output-only system identification routine was proposed that uses only the output spectra to estimate the modal parameters of a linear time-periodic system up to a scale factor. The proposed method makes used of the theory developed by Wereley regarding Harmonic Transfer Functions, and an analytical expression for the output spectrum in terms of the modal parameters was developed and used to interpret the spectra. The method was validated using simulated measurements from a Mathieu Oscillator, illustrating the simplicity of the approach and the similarity with familiar output-only identification approaches for linear time-invariant systems. The methodology was also applied to simulated measurements from a large, detailed model of a rotating wind turbine. The results suggest that the first edgewise modes of the turbine were measurably time-periodic. The dominant term in the Fourier expansion of the mode shape for the mode in question was the second term, which seems reasonable considering that the boundary condition imparted by the tower changes its effective stiffness twice per revolution of the rotor (in the blade's reference frame). Further results were not yet available at the time this article was submitted to the conference, but future works will explore these issues in more detail and seek to better understand why this time-periodicity occurs and how it might affect the stability and fatigue life of the wind turbine.

References

[1] J. Dugundji and J. H. Wendell, "SOME ANALYSIS METHODS FOR ROTATING SYSTEMS WITH PERIODIC COEFFICIENTS," *AIAA journal,* vol. 21, pp. 890-897, 1983.

[2] C. Fuehne, "Application of Generalized Floquet Theory to Ground Resonance Data," in *56th Annual National Forum of the American Helicopter Society* Virginia Beach, VA, 2000.

[3] S. Hwang, "Frequency Domain System Identification of Helicopter Rotor Dynamics Incorporating Models with Time Periodic Coefficients," in *Department of Aerospace Engineering*. vol. Doctor of Philosophy: University of Maryland - College Park, 1997.

[4] D. A. Peters, "Fast Floquet theory and trim for multi-bladed rotorcraft," *Journal of the American Helicopter Society,* vol. 39, pp. 82-89, 1994.

[5] A. Siddiqui, "Identification of the Harmonic Transfer Functions of a Helicopter Rotor," in *Department of Aeronautics and Astronautics*. vol. M.S.: Massachusetts Institute of Technology, 1999, p. 88.

[6] V. Verdult, M. Lovera, and M. Verhaegen, "Identification of linear parameter-varying state-space models with application to helicopter rotor dynamics," *International Journal of Control,* vol. 77, pp. 1149 - 1159, 2004.

[7] C. P. Fuehne, "Application of Generalized Floquet Theory to Numerical and Experimental Data," in *Mechanical Engineering*. vol. M.S. St. Louis, Missouri: Washington University in St. Louis,, 2000.

[8] I. Bucher and O. Shomer, "Detecting asymmetry in rotating structures a combined acutation and signal processing approach," in *23rd International Modal Analysis Conference (IMAC XXIII)* Orlando, Florida, 2005.

[9] M. Allen and J. H. Ginsberg, "Floquet Modal Analysis to Detect Cracks in a Rotating Shaft on Anisotropic Supports," in *24th International Modal Analysis Conference (IMAC XXIV)* St. Louis, MO, 2006.

[10] M. S. Allen, "Frequency-Domain Identification of Linear Time-Periodic Systems using LTI Techniques," *Journal of Computational and Nonlinear Dynamics* vol. 4, 24 Aug. 2009 2009.

[11] G. Bir and K. Stol, "Operating Modes of a Teetered-Rotor Wind Turbine," in *17th International Modal Analysis Conference (IMAC XVII)* Kissimmee, Florida, 1999.

[12] D. J. Malcolm, "Modal response of 3-bladed wind turbines," *Journal of Solar Energy Engineering, Transactions of the ASME,* vol. 124, pp. 372-377, 2002.

[13] K. Stol, M. Balas, and G. Bir, "Floquet modal analysis of a teetered-rotor wind turbine," *Journal of Solar Energy Engineering-Transactions of the ASME,* vol. 124, pp. 364-371, Nov 2002.

[14] J. B. Dingwell and J. P. Cusumano, "Nonlinear time series analysis of normal and pathological human walking," *Chaos,* vol. 10, pp. 848-63, 2000.

[15] J. B. Dingwell and K. Hyun Gu, "Differences between local and orbital dynamic stability during human walking," *Transactions of the ASME. Journal of Biomechanical Engineering,* vol. 129, pp. 586-93, 2007.

[16] Y. Hurmuzlu and C. Basdogan, "On the measurement of dynamic stability of human locomotion," *Journal of Biomechanical Engineering, Transactions of the ASME,* vol. 116, pp. 30-36, 1994.

[17] Y. Hurmuzlu, C. Basdogan, and D. Stoianovici, "Kinematics and dynamic stability of the locomotion of post-polio patients," *Transactions of the ASME. Journal of Biomechanical Engineering,* vol. 118, pp. 405-11, 1996.

[18] M. S. Allen and M. W. Sracic, "System Identification of Dynamic Systems with Cubic Nonlinearities Using Linear Time-Periodic Approximations," in *ASME 2009 International Design Engineering Technical Conferences & Computers and Information in Engineering Conference IDETC/CIE 2009* San Diego, California, USA, 2009.

[19] S. Maia, J. M. M. Silva, J. He, N. A. Lieven, R. M. Lin, G. W. Skingle, W. M. To, and A. P. V. Urgueira, *Theoretical and Experimental Modal Analysis*. Taunto, Somerset, England: Research Studies Press Ltd., 1997.

[20] K. Liu, "Identification of linear time-varying systems," *Journal of Sound and Vibration,* vol. 206, pp. 487-505, 1997.

[21] M. Verhaegen and Y. Xiaode, "A class of subspace model identification algorithms to identify periodically and arbitrarily time-varying systems," *Automatica,* vol. 31, pp. 201-16, 1995.

[22] F. Felici, J. W. van Wingerden, and M. Verhaegen, "Subspace identification of MIMO LPV systems using a periodic scheduling sequence," *Automatica,* vol. 43, pp. 1684-97, 2007.

[23] J. J. Hench, "A technique for the identification of linear periodic state-space models," *International Journal of Control,* vol. 62, pp. 289-301, 1995.

[24] D. A. Peters and X. Wang, "Generalized Floquet Theory for Analysis of Numerical or Experimental Rotor Response Data," in *24th European Rotorcraft Forum* Marseilles, France, 1998.

[25] O. A. Bauchau and J. Wang, "Efficient and robust approaches to the stability analysis of large multibody systems," *Journal of Computational and Nonlinear Dynamics,* vol. 3, 2008.

[26] P. O. Arambel and G. Tadmor, "Robust H infinity identification of linear periodic discrete-time systems," *International Journal of Robust and Nonlinear Control,* vol. 4, pp. 595-612, 1994.

[27] V. Deshmukh, "Spectral Collocation-Based Optimization in Parameter Estimation of Nonlinear Time-Varying Dynamical Systems " in *ASME 2007 International Design Engineering Technical Conference*. vol. DETC2007-35262 Las Vegas, Nevada, 2007.

[28] N. M. Wereley, "Analysis and Control of Linear Periodically Time Varying Systems," in *Department of Aeronautics and Astronautics*. vol. PhD Cambridge: Massachusetts Institute of Technology, 1991.

[29] N. M. Wereley and S. R. Hall, "Linear time periodic systems: transfer functions, poles, transmission zeroes and directional properties," in *Proceedings of the 1991 American Control Conference* Boston, MA, USA: American Autom. Control Council, 1991, pp. 1179-84.

[30] N. M. Wereley and S. R. Hall, "Frequency response of linear time periodic systems," Honolulu, HI, USA, 1990, pp. 3650-3655.

[31] H. Irretier, "MATHEMATICAL FOUNDATIONS OF EXPERIMENTAL MODAL ANALYSIS IN ROTOR DYNAMICS," *Mechanical Systems and Signal Processing,* vol. 13, pp. 183-191, 1999.

[32] C.-W. Lee, D.-J. Han, J.-H. Suh, and S.-W. Hong, "Modal analysis of periodically time-varying linear rotor systems," *Journal of Sound and Vibration,* vol. 303, pp. 553-574, 2007.

[33] S. J. Shin, C. E. S. Cesnik, and S. R. Hall, "System identification technique for active helicopter rotors," *Journal of Intelligent Material Systems and Structures,* vol. 16, pp. 1025-38, 2005.

[34] M. S. Allen and M. W. Sracic, "A Method for Generating Pseudo Single-Point FRFs from Continuous Scan Laser Vibrometer Measurements," in *26th International Modal Analysis Conference (IMAC XXVI)* Orlando, Florida, 2008.

[35] M. S. Allen and M. W. Sracic, "Mass Normalized Mode Shapes Using Impact Excitation and Continuous-Scan Laser Doppler Vibrometry," in *8th International Conference on Vibration Measurements by Laser Techniques* Ancona, Italy, 2008, pp. 7098-3.

[36] M. S. Allen and M. W. Sracic, "A New Method for Processing Impact Excited Continuous-Scan Laser Doppler Vibrometer Measurements " *Mechanical Systems and Signal Processing,* vol. submitted Jan. 2009, 2009.

[37] M. Martarelli, "Exploiting the Laser Scanning Facility for Vibration Measurements," in *Imperial College of Science, Technology & Medicine*. vol. Ph.D. London: Imperial College, 2001.

[38] R. Ribichini, D. Di Maio, A. B. Stanbridge, and D. J. Ewins, "Impact Testing With a Continuously-Scanning LDV," in *26th International Modal Analysis Conference (IMAC XXVI)* Orlando, Florida, 2008.

[39] P. Sriram, S. Hanagud, and J. I. Craig, "Mode shape measurement using a scanning laser doppler vibrometer," *International Journal of Analytical and Experimental Modal Analysis,* vol. 7, pp. 169-178, 1992.

[40] A. B. Stanbridge and D. J. Ewins, "Modal testing using a scanning laser Doppler vibrometer," *Mechanical Systems and Signal Processing,* vol. 13, pp. 255-70, 1999.

[41] A. B. Stanbridge, M. Martarelli, and D. J. Ewins, "Measuring area vibration mode shapes with a continuous-scan LDV," *Measurement,* vol. 35, pp. 181-9, 2004.

[42] S. Vanlanduit, P. Guillaume, and J. Schoukens, "Broadband vibration measurements using a continuously scanning laser vibrometer," *Measurement Science & Technology,* vol. 13, pp. 1574-82, 2002.

[43] L. Hermans and H. Van Der Auweraer, "Modal Testing and Analysis of Structures Under Operational Conditions: Industrial Applications," *Mechanical Systems and Signal Processing,* vol. 13, pp. 193-216, 1999.

[44] E. Parloo, B. Cauberghe, F. Benedettini, R. Alaggio, and P. Guillaume, "Sensitivity-Based Operational Mode Shape Normalization: Application to A Bridge," in *22nd International Modal Analysis Conference (IMAC XXII)* Dearborn, Michigan, 2004.

[45] B. Peeters and G. D. Roeck, "Stochastic System Identification for Operational Modal Analysis: A Review," *Journal of Dynamic Systems, Measurement, and Control,* vol. 123, December 2001.

[46] A. Fasana, L. Garibaldi, E. Giorcelli, and D. Sabia, "Z24 Bridge Dynamic Data Analysis by Time Domain Methods," in *International Modal Analysis Conference (IMAC XIX)* Kissimmee, Florida, 2001.

[47] A. Guyader and L. Mevel, "Covariance driven subspace methods : input/output vs output-only," in *International Modal Analysis Conference (IMAC XXII)* Dearborn, Michigan, 2004.

[48] M. Richardson and B. Schwarz, "Modal Parameter Estimation from Operating Data," in *Sound and Vibration*, 2003, pp. 28-36.

[49] G. H. James and T. G. Carne, "Damping measurements on operating wind turbines using the natural excitation technique (NExT)," in *11th ASME Wind Energy Symposium presented at the Energy Sources Technology Conference and Exhibition*. vol. 12 Houston, TX, USA: Publ by ASME, New York, NY, USA, 1992, pp. 75-81.
[50] P. Van Overschee and B. De Moor, *Subspace Identification for Linear Systems: Theory-Implementation-Applications*. Boston: Kluwer Academic Publishers, 1996.
[51] J. Jonkman, "NREL 5 MW Baseline Wind Turbine," Technical report, NREL/NWTC, 1617 Cole Boulevard; Golden, CO 80401-3393, USA 2005.
[52] M. H. Hansen, "Aeroelastic stability analysis of wind turbines using an eigenvalue approach," *Wind Energy,* vol. 7, pp. 133-143, 2004.
[53] M. H. Hansen, "Improved modal dynamics of wind turbines to avoid stall-induced vibrations," *Wind Energy,* vol. 6, pp. 179-195, 2003.
[54] M. O. L. Hansen, J. N. Sørensen, S. Voutsinas, N. Sørensen, and H. A. Madsen, "State of the art in wind turbine aerodynamics and aeroelasticity," *Progress in Aerospace Sciences,* vol. 42, pp. 285-330, 2006.
[55] S. Chauhan, M. H. Hansen, and D. Tcherniak, "Application of Operational Modal Analysis and Blind Source Separation / Independent Component Analysis Techniques to Wind Turbines," in *27th International Modal Analysis Conference (IMAC XXVII)* Orlando, Florida: SPIE, 2009.

Appendix A

5.1. Analytical Fourier Coefficients of the Floquet Mode Shapes for the Mathieu Oscillator

The Mathieu oscillator system can be easily represented in state space according to eq. (1), with the following system matrix, $A(t)$,

$$A(t)=\begin{bmatrix} 0 & 1 \\ -\omega_0^2-\omega_1^2\cos(\omega_A t) & -2\zeta\omega_0 \end{bmatrix}. \tag{16}$$

Recalling the time derivative relationship of the STM to the $A(t)$ matrix,

$$\frac{\partial}{\partial t}(\Phi(t,t_0))=A(t)\Phi(t,t_0) \tag{17}$$

one can obtain the analytical STM by using initial conditions of [1; 0] and [0; 1] and integrating the equations of motion over one fundamental period. The STM at the end of the fundamental period is also known as the Floquet transition matrix, and its eigenvalues are the Floquet multipliers of the system, λ_{STM}, which are related to the Floquet exponents λ_{FI} by $\lambda_{STM}=\exp(\lambda_{FI}*T_A)$. After extracting the Floquet exponent, λ_{FI} the periodic Floquet mode vectors can be calculated by rearranging eq. (4) to get eq. (18),

$$\psi_{FI}(t)=\Phi(t,t_0)\psi_{STM}(t_0)\exp(-\lambda_{FI}(t-t_0)) \tag{18}$$

where $\psi_{STM}(t_0)$ is the eigenvector of the STM at $t=T_A$. Then, because the Floquet mode vectors are periodic, they can be expanded with a Fourier Series as given in eq. (5).

Proceedings of the IMAC-XXVIII
February 1–4, 2010, Jacksonville, Florida USA

Validation of Wind Turbine Dynamics

Michele Rossetti, Closed-loop Control Leader, Ecotècnia Energìas Renovables
Roc Boronat 78, 08860 Barcelona, Spain, michele.rossetti@power.alstom.com

Nomenclature:

f = Modal frequency
K = Stiffness
m = Modal Mass

Abstract

Wind turbine dynamics is of crucial importance for the design of safe and reliable wind turbines. Imprecise modelling of the dynamics may lead to incorrect design of the controller and raises the probability of resonance to occur between subsystems. This paper resumes the main mode of vibrations of a wind turbine and describes how experimental data is used at Alstom Wind to increase the accuracy of the model. Results show that is possible to accurately tune the model with a few simple measurements. Conclusions are then drawn on acquisition of measurement data, identificability of turbine modes and validity of the numerical model, and the potentials of system identification techniques are addressed.

1 Introduction

The design of a wind turbine speed controller is intimately related to the wind turbine dynamic response. The controller design is based on a numerical model of the wind turbine, but usually the real turbine dynamics differ slightly form the model, and this may cause overall performance not to prove as expected. For this reason Ecotècnia Energìas Renovables continuously performs field tests which bring new and valuable information on the turbine dynamics, such as the coupling which may occur between modes, the real influence of excitations and the potential sources of an unexpected or underestimated modal frequency. The measurements are treated with tools such as waterfall plots, which basically are contour plots of many different power spectral densities at different operating points. These tools have been widely used in industry for vibration analysis of rotating machinery, and most often they are applied to the rotational speed signal to detect possible resonances [1]. This paper demonstrates that their application to other signals (e.g. nacelle accelerations and loads) can deliver very rich information on the turbine response throughout its full operating range, and these plots can be used successfully for both model tuning and controller optimisation. The work presented is an extension of previous work by the same author [2].

T. Proulx (ed.), *Structural Dynamics and Renewable Energy, Volume 1*, Conference Proceedings of the Society for Experimental Mechanics Series 10, DOI 10.1007/978-1-4419-9716-6_34, © The Society for Experimental Mechanics, Inc. 2011

2 Wind turbine modes and excitations

In a variable speed and variable pitch wind turbine modal natural frequencies as well as excitations vary with the wind speed. Figure 1 shows how the modal frequencies vary with wind speed, and figures 2 and 3 show the mode shapes for the main modes of vibration.

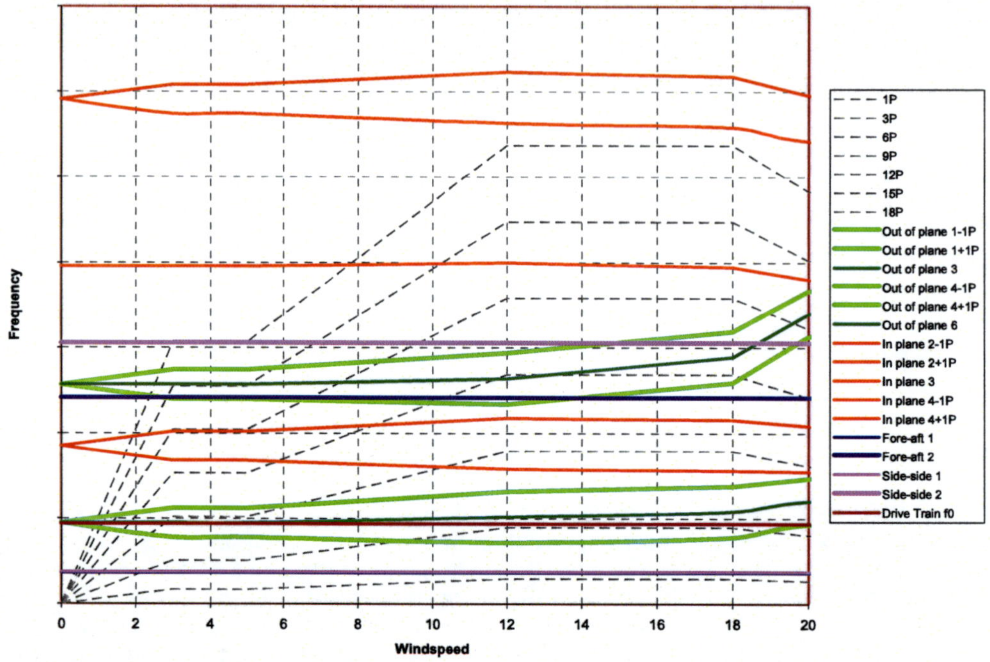

Figure 1: Main natural frequencies and excitations of a typical wind turbine.

The design of a wind turbine requires precise numerical models that correctly represent how the modes vary with the operating point. The accuracy of these models can be verified and validated against field tests, and data from experiments can be used to improve simulations. The main aim for validation is that the dynamics defined in the numerical model used for simulations resemble the dynamics of the real turbine as accurately as possible. This is achieved by verifying and tuning the dynamic characteristics of the numerical model using experimental data, a process here named model tuning.

1st Tower fore-aft mode   2nd Tower fore-aft mode   1st Tower side-to-side mode   2nd Tower side-to-side mode

Figure 2: Tower dominated modes of vibration of a wind turbine.

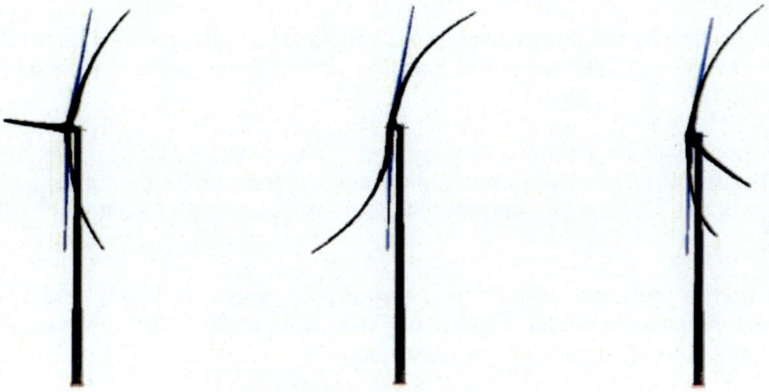

1st rotor flapwise (out of plane) modes

1st rotor edgewise (in-plane) modes

Figure 3: Rotor dominated modes of vibration of a wind turbine.

3 Model tuning

3.1 General

The aim of model tuning is that simulations match as closely as possible experimental data. This is the only way to ensure that a precise model is used as a base for the controller design. Matching simulations from a numerical model with real data from experiments implies an understanding of the real system, where the natural frequencies lay, how the modes couple with one another, and what sort of damping is present.

Classical experimental modal analysis techniques can provide this information at the expense that sometimes a large external excitation is needed, with the risk of inducing the machine in a non-standard condition and causing a potential malfunctioning that may be costly to repair.

More recently operational modal analysis techniques have been applied successfully to wind turbines to identify the system. These techniques have the advantage that the data is collected while the machine is operating in its standard regimes, so the risk of components failure is minimised. These tools, although they have the advantage that can predict the damping, are relatively demanding and require a certain degree of user interaction to identify the modes, plus they have some deficiency in analysing a mode when its frequency lays exactly on an excitation frequency. A practical application of both methods has been presented by Hansen [3].

3.2 A practical approach

The alternative procedure chosen by the author uses more simple tools, requires less user interaction to identify a modal frequency, and can still be applied to the real machine while it's operating in standard regimes without the need for external excitations.

For tuning the numerical model initially a contour plot is used . This is a 3-d plot of power spectral density (PSD) versus wind speed and frequency, which gives an overall picture of the turbine dynamics throughout its operating range. The big advantage of this plot is that it shows how both frequencies and excitations (1p and its harmonics) vary with the wind speed.

As an example the data from a particular wind turbine is analysed, showing how a contour plot of PSD can be used for model tuning and dynamic validation. The usual step is to analyse first a waterfall plot and select the operating point at which model tuning is going to be performed.

Figure 2 gives an example of such plot from measured blade root bending moment.

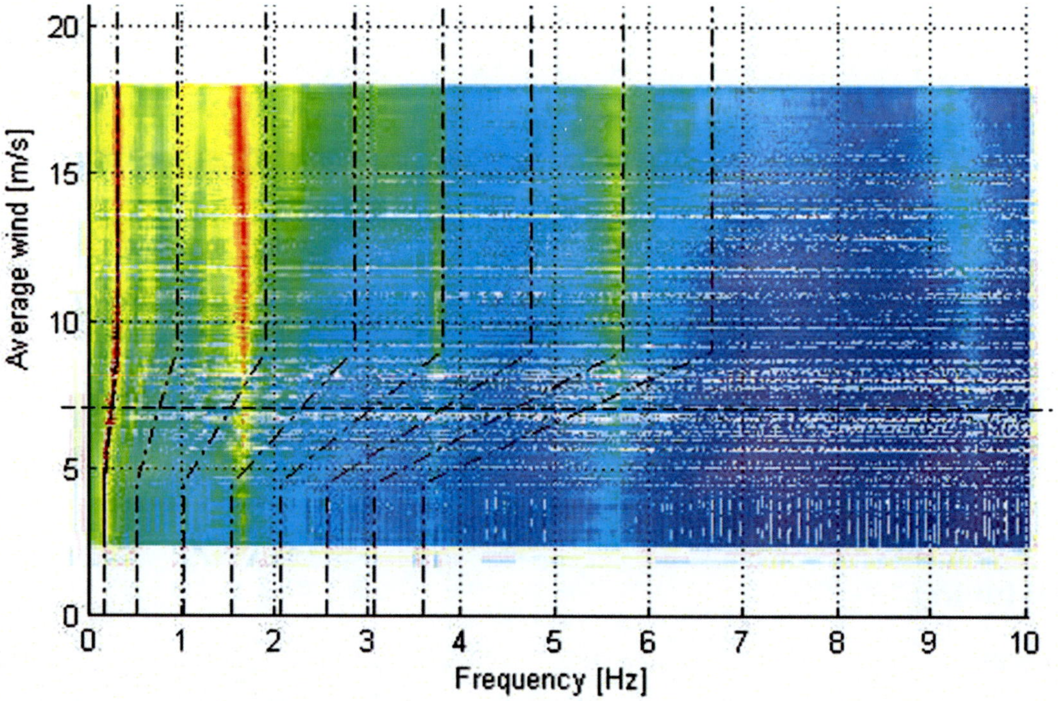

Figure 4: PSD Contour plot of edgewise root bending moment. Colours indicate PSD amplitude, the dashed vertical lines represent the excitations (1p and its harmonics). Note how is clearly visible the change in frequency with increasing wind speed. The operating point chosen for model tuning is shown as dashed horizontal line.

At the operating point chosen a comparison is made between simulation and experiments: in this case the 2-D PSD of blade root bending moment from measurements is compared against a the PSD of the same signal form a simulation using an un-tuned model (Figure 3).

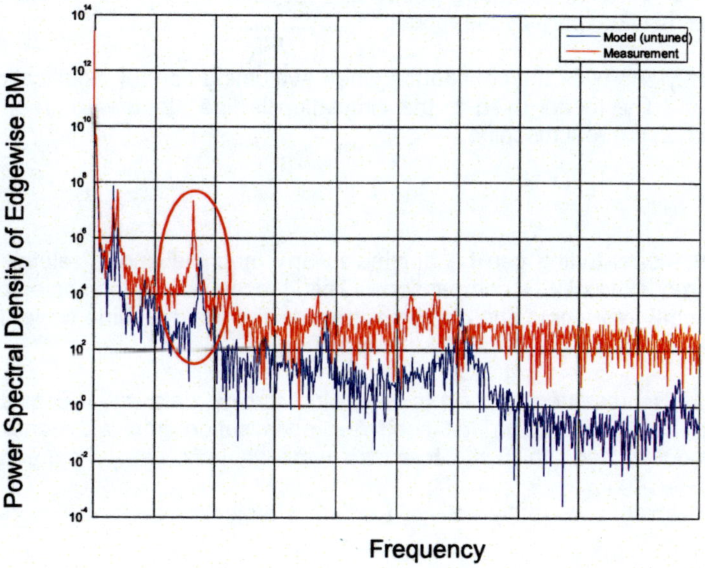

Figure 5: Comparison of edgewise spectrum from measurements (red) and simulation using an untuned model (blue).

Based on this comparison the edgewise stiffness in the model is modified to achieve a matching spectrum. This is simply done by adjusting the edgewise stiffness K in the model to match the experimental frequency by assuming that

$$f \propto \sqrt{\frac{K}{m}}$$

Figure 4 shows the result of tuned model which now matches the measurement.

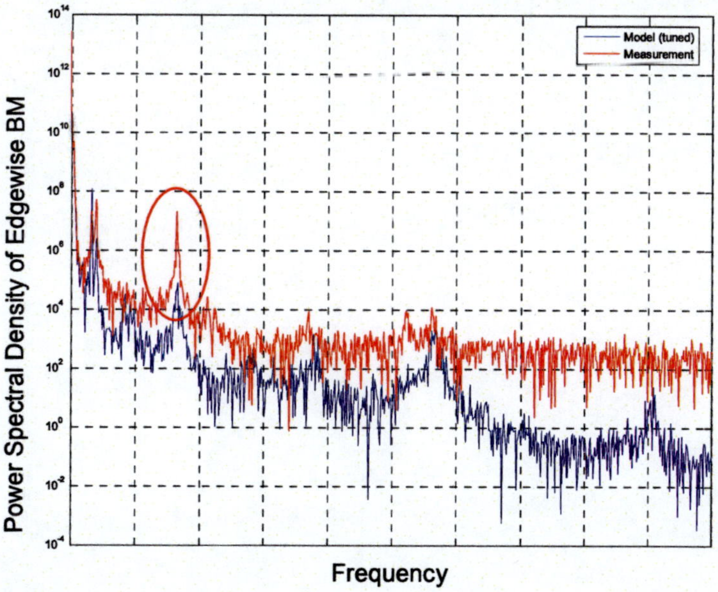

Figure 6: Comparison of edgewise spectrum from measurements (red) and simulation with tuned model (blue) Now the 1st edgewise frequency coincides with measurements.

The process is repeated for different operating points and for different signals individually, and assuming that coupling is between modes is taken care by the model.

This procedure considerably improves the confidence when evaluating control modifications via simulations prior to implementation on filed. The model used in the simulation is fine tuned with real data and hence similar behaviour will be expected on the real machine.

4 Conclusions

The simple procedure for model tuning can be applied to any machine and is relatively easy to evaluate the accuracy of the tuned model via PSD comparisons. This procedure uses simple tools, requires less user interaction to identify a modal frequency, and can be applied to the real machine while it's operating in standard regimes without compromising its performance or safe operation.

In general the analysis tools presented applied to real wind turbine data allow consistent assessment of the dynamic performance, provide information for controller optimisation and aid considerably the process of validating the numerical models used, which is a core task in the whole process of speed controller design.

Ecotècnia Energìas Renovables is currently active in different projects developing more advanced methods for system identification.

5 References

[1] Swanson, Powell and Weissman, "A practical review of rotating machinery critical speeds and modes", Sound and Vibration, 2005, http://www.sandv.com/downloads/0505swan.pdf.
[2] Rossetti, Guadayol, Santos and Simon, "Dynamic Tuning at Ecotècnia", poster presented at the European Wind Energy Conference, Milan, 2007.
[3] Hansen, Fuglsang, Thomsen and Knudsen ,"Two Methods for Estimating Aeroelastic Damping of Operational Wind Turbine Modes from Experiments", European Wind Energy Conference, London, 2004.